π CHAPMAN & HALL/CRC
Monographs and Surveys in
Pure and Applied Mathematics   125

UNBOUNDED FUNCTIONALS

IN THE

CALCULUS OF VARIATIONS

Representation, Relaxation, and Homogenization

CHAPMAN & HALL/CRC
Monographs and Surveys in
Pure and Applied Mathematics   125

UNBOUNDED FUNCTIONALS

IN THE

CALCULUS OF VARIATIONS

Representation, Relaxation, and Homogenization

LUCIANO CARBONE

RICCARDO DE ARCANGELIS

CRC Press
Taylor & Francis Group
Boca Raton London New York

CRC Press is an imprint of the
Taylor & Francis Group, an **informa** business
A CHAPMAN & HALL BOOK

Published 2002 by Chapman & Hall /CRC Press
Taylor & Francis Group
6000 Broken Sound Parkway NW, Suite 300
Boca Raton, FL 33487-2742

© 2002 by Taylor & Francis Group, LLC
CRC Press is an imprint of Taylor & Francis Group, an Informa business

First issued in paperback 2019

No claim to original U.S. Government works

ISBN-13: 978-0-367-45507-1 (pbk)
ISBN-13: 978-1-58488-235-0 (hbk)

**Visit the Taylor & Francis Web site at
http://www.taylorandfrancis.com**

**and the CRC Press Web site at
http://www.crcpress.com**

Library of Congress Cataloging-in-Publication Data

Carbone, L. (Luciano)
 Unbounded functionals in the calculus of variations : representation, relaxation, and homogenization / Luciano Carbone, Riccardo De Arcangelis.
  p. cm.-- (Monographs & surveys in pure and applied math)
 Includes bibliographical references and index.
 ISBN 1-58488-235-2
 1. Calculus of variations. 2. Functionals. I. Arcangelis, Riccardo de. II. Title. III. Chapman & Hall/CRC monographs and surveys in pure and applied mathematics.
QA315. C274 2001
515'.64—dc21
                       2001053756

Library of Congress Card Number 2001053756

To Gemma and Emilia

Table of contents

Preface

In the last 30 years several problems have been examined in the framework of the study of certain composite materials having the particular feature that they can be described by means of minimizing configurations of energies not necessarily finite on all the "smooth" admissible ones.

Problems involving energies with these features appeared, for example, in the study of elastic-plastic torsion theory, of electrostatic screening, and of the modelling of some rubber-like nonlinear elastomers, and have been generally approached by means of ad hoc, or particular mathematical techniques.

The aim of the present volume is to propose a systematic and unifying mathematical framework, within the calculus of variations, for the treatment of problems of this nature, at least in the stationary case.

From this point of view, the fundamental notion that appears to play a central role is the one of *unbounded functional*. These functionals take nonnegative extended real values, and represent the energies under consideration. They depend, in a classical manner, essentially on two variables: one of set-type nature in which the functional enjoys measure theoretic properties, and one of scalar configuration-type nature in which it enjoys convexity and lower semicontinuity properties. On the other side, the above energies behave also in a "non-classical" way. They turn out to take finite values only on those configurations that are subject to pointwise constraints on the strains, hence not depending on the regularity of the configurations themselves.

The analysis of this notion requires the reconsideration of well-established concepts and techniques. Therefore the book naturally divides into two parts.

In the first part (Chapters 1 to 5), we aim to allow as much as possible a self-contained reading of the volume. The main notions of convex analysis are recalled, together with those of measure theory, and of theory of variational convergences. Then we introduce some function spaces usually considered in calculus of variations, where we study some lower semicontinuity and minimization problems for energy functionals. Such notions are also adapted to the new setting by means of the necessary changes and the required extensions.

At the end of the first part, Chapter 6 plays the role of a hinge chapter. It begins with a brief survey on some aspects of the theory of the standard functionals of the calculus of variations such as unique extension properties, representation as integrals of the calculus of variations, relaxation theory, and homogenization processes. Then, the mathematical aspects of some physical models, which suggest the notion of unbounded functional, are briefly explained.

By unique extension properties, we mean those types of problems in which one tries to extend a function defined in a set to a wider one by preserving some of its characteristic features, and gaining uniqueness of the extension.

The representation as integrals of the calculus of variations problems refers to the identification of sufficient conditions (possibly also necessary) on an abstract functional F implying its description as

$$F(\Omega, u) = \int_\Omega f(x, \nabla u)dx,$$

where Ω is the set-type variable and u the configuration-type one.

Given a function F defined in a topological space, relaxation problems deal with the study of representation formulas for the description of the relaxed function of F, namely of the greatest lower semicontinuous function less than or equal to F, having in mind the qualitative property according to which the greatest lower bound of a function agrees with the minimum of its relaxed function.

By homogenization problems we mean those in which one tries to simulate the behaviour of composite materials finely grained in a "regular" way (somehow comparable to a periodic distribution of two or more components) by means of a homogeneous one, and vice-versa. In this book, we restrict ourselves to the cases where such simulation can occur in the sense that the minimum energy of the homogeneous material turns out to be close, for every admissible external force, to the one of the composite materials, as much as the graining is fine.

In the physical models inspiring unbounded functionals, the energies involved have an integral form on "regular" configurations, but the energy densities f are unbounded.

Thus, in the second part of the volume (Chapters 7 to 13), which is the most original one, a tentative theory of unbounded functionals is developed according to the scheme proposed in Chapter 6, having in mind the described models and focusing mainly on homogenization. This is done, at least in the case of unique extension, integral representation and relaxation, for "translation invariant" functionals, i.e. functions that don't change their values when both the set-type variable and the configuration-type one undergo translations.

Finally, in Chapter 14, the homogenization results obtained are exploited to provide some explicit descriptions of the homogenized materials relative to the unbounded energies proposed in Chapter 6.

In our opinion, the theory developed in such a way allows to obtain deeper results than the already known ones, and to address interesting new problems, including ones in applied mathematics.

In memory of Ennio De Giorgi and Jacques-Louis Lions, we would like to point out that several ideas contained in this book originated from their scientific visions and mathematical concepts.

We are also indebted to Haïm Brezis for his warm encouragements in the preparation of the book and for some deep discussions, and to Sergio Spagnolo for his friendly mathematical teachings.

Finally, we want to remark that the research activities on composite materials can be considered as a common effort, to which a lot of mathematicians contribute with different competencies. So we are also indebted to many colleagues for several comments and discussions.

The book contains both published and new results. It is mainly aimed at graduate students and researchers in mathematics, but we hope that it may be useful to engineers and continuum physicists.

Naples, July 2001 Luciano Carbone
Riccardo De Arcangelis

μονή   πρόοδος
ἐπιστροφή

Authors' addresses:

Luciano Carbone: *Dipartimento di Matematica e Applicazioni "R. Caccioppoli", via Cintia, Complesso Monte S. Angelo, 80126 Napoli, Italy.*
E-mail: *carbone@biol.dgbm.unina.it*

Riccardo De Arcangelis: *Dipartimento di Matematica e Applicazioni "R. Caccioppoli", via Cintia, Complesso Monte S. Angelo, 80126 Napoli, Italy.*
E-mail: *dearcang@unina.it*

Basic Notations and Recalls

The present chapter is devoted to the introduction of the general notations and the basic facts that we are going to use throughout the book.

Basic Notations

Let X be a set. For every $S \subseteq X$ we denote by χ_S the *characteristic function* of S defined by

$$\chi_S(x) = \begin{cases} 1 & \text{if } x \in S \\ 0 & \text{if } x \in X \setminus S, \end{cases}$$

and by I_S the *indicator function* of S given by

$$I_S(x) = \begin{cases} 0 & \text{if } x \in S \\ +\infty & \text{if } x \in X \setminus S. \end{cases}$$

If $f \colon X \to [-\infty, +\infty]$ and $x_0 \in X$, we say that x_0 is a *minimizer* of f if x_0 is a minimum point of f. Given $\{x_h\} \subseteq X$, we say that $\{x_h\}$ is a *minimizing sequence* of f if the limit $\lim_{h \to +\infty} f(x_h)$ exists, and

$$\lim_{h \to +\infty} f(x_h) = \inf\{f(x) : x \in X\}.$$

For every $r \in \mathbf{R}$ we denote by $[r]$ the integer part of r, i.e. $[r] = \max\{m \in \mathbf{Z} : m \leq r\}$.

Let $n \in \mathbf{N}$.

We say that an element of $(\mathbf{N} \cup \{0\})^n$ is a *multiindex*. For every multiindex $\alpha = (\alpha_1, \ldots, \alpha_n)$ the length $|\alpha|$ of α is defined as $|\alpha| = \alpha_1 + \ldots + \alpha_n$.

We denote by $\mathbf{R}^n$ the space of the n-tuples $x = (x_1 \ldots, x_n)$ of real numbers, that we endow with the usual scalar product, euclidean norm, and topology. For every $x, y \in \mathbf{R}^n$ we denote by $x \cdot y$, respectively by $|x|$, the scalar product of x and y, respectively the norm of x. We also denote by $\mathbf{e}_1 = (1,0,\ldots,0),\ldots,\mathbf{e}_n = (0,\ldots,0,1)$ the elements of the canonical basis of $\mathbf{R}^n$, and by 0 both the origin of $\mathbf{R}^n$ and the real number zero, in general the meaning of 0 being clear from the context.

As usual, for every $x_0 \in \mathbf{R}^n$, $S \subseteq \mathbf{R}^n$ and $r \in\,]0,+\infty[$, we denote by $B_r(x_0)$ the open ball of $\mathbf{R}^n$ centred at x_0 and with radius r, by $Q_r(x_0)$ the open cube centred at x_0 having sidelength r, and set

$$\mathrm{dist}(x_0, S) = \inf\{|x_0 - x| : x \in S\},$$

$$S_r^- = \{x \in S : \mathrm{dist}(x, \partial S) > r\} \quad S_r^+ = \{x \in \mathbf{R}^n : \mathrm{dist}(x, S) < r\}.$$

For tradition reasons, we set $Y = \,]0,1[^n$.

We say that a subset of $\mathbf{R}^n$ is a *polyhedral set* if it can be expressed as the intersection of a finite number of closed half-spaces.

By $\mathcal{A}_0$ we denote the set of the bounded open subsets of $\mathbf{R}^n$. For every open subset Ω of $\mathbf{R}^n$, we denote by $\mathcal{A}(\Omega)$ the set of the open subsets of Ω.

We denote by $\mathcal{L}^n$ the Lebesgue measure on $\mathbf{R}^n$. Given a Lebesgue measurable set Ω, we denote by $\mathcal{L}_n(\Omega)$ the family of the Lebesgue measurable subsets of Ω. When considering Lebesgue measure on subsets of $\mathbf{R}^n$, we generally write for simplicity "measurable," "a.e.," and so on in place of "$\mathcal{L}_n$-measurable," "$\mathcal{L}^n$-almost everywhere," and so on. For tradition reasons, we also write dx in place of $d\mathcal{L}^n$ in the integrals of measurable functions.

Finally, we denote by $[-\infty, +\infty]$ the extended real numbers system, that we endow with the usual topological structure that makes it a compact space.

Basic Topological Facts

Let (U, τ) be a topological space.

For every $A \subseteq U$ we denote by $\mathrm{int}(A)$, $\overline{A}$ and ∂A respectively the interior, the closure and the boundary of A.

Given $E \subseteq \mathbf{R}$, $\varepsilon_0 \in \overline{E}$, a family $\{u_\varepsilon\}_{\varepsilon \in E} \subseteq U$ and $u \in U$, we write $u_\varepsilon \to u$ as $\varepsilon \to \varepsilon_0$ to mean that $\{u_\varepsilon\}_{\varepsilon \in E}$ converges to u in τ as $\varepsilon \to \varepsilon_0$. In particular, if $\{u_h\}$ is a sequence of points of U, we write $u_h \to u$ to mean that $\{u_h\}$ converges to u in τ as h goes to $+\infty$.

For every $u \in U$ we denote by $\mathcal{N}(u)$ the set of the open neighborhoods of u in τ.

Let $\{u_h\} \subseteq U$ and $u \in U$. We say that u is a *cluster point* of $\{u_h\}$ if for every $I \in \mathcal{N}(u)$ and every $h \in \mathbf{N}$ there exists $k \geq h$ such that $u_k \in I$.

It is clear that if $\{u_h\}$ converges, or if it has a converging subsequence, then the limit of $\{u_h\}$, as well as the limit of every converging subsequence of $\{u_h\}$, is a cluster point of $\{u_h\}$. The converse is not true in general topological spaces, in the sense that a cluster point of $\{u_h\}$ need not be the limit of a converging subsequence of $\{u_h\}$. It is true if U satisfies the first countability axiom.

We say that $X \subseteq U$ is *sequentially closed* if for every $\{u_h\} \subseteq X$ converging to $u \in U$ it results that $u \in X$.

It is obvious that a closed set is also sequentially closed, the converse being, in general, false.

We say that $K \subseteq U$ is *compact* if every open covering of K has a finite subcovering, we say that K is *sequentially compact* if every $\{u_h\} \subseteq K$ has a subsequence that converges to a point of K. We say that K is *relatively compact* if $\overline{K}$ is compact, and that K is *relatively sequentially compact* if $\overline{K}$ is sequentially compact.

We recall that, in general, the notions of compactness and of sequential compactness are independent (cf. for example [Ro, Chapter 9, Problems 6 and 27]), and that they agree provided U satisfies the first countability axiom.

If U is a metric space and $X \subseteq U$, we say that X is *precompact* if every $\{u_h\} \subseteq X$ has a Cauchy subsequence. It is clear that if X is relatively sequentially compact, then it is also precompact. In general, the converse is not true, but it holds if U is complete.

A topological space Ω is said to be *locally compact* if every point of Ω has a relatively compact neighborhood.

One of the most important topological notions with which the book is concerned is the one of lower semicontinuity, that we recall briefly.

Let (U, τ) be a topological space and $F: U \to [-\infty, +\infty]$.

We say that F is *τ-lower semicontinuous*, or simply lower semicontinuous if no ambiguity occurs, if for every $\lambda \in \mathbf{R}$ the set $\{u \in U : F(u) > \lambda\}$ is open.

For every $u \in U$ we denote by $\liminf_{v \to u} F(v)$ the lower limit of F at u defined by

$$\liminf_{v \to u} F(v) = \sup_{I \in \mathcal{N}(u)} \inf_{v \in I} F(v).$$

Let $u \in U$, we say that F is *lower semicontinuous at u* if

$$F(u) \leq \liminf_{v \to u} F(v).$$

Then it turns out that F is lower semicontinuous if and only if F is lower semicontinuous at u for every $u \in U$. Consequently, lower semicontinuity turns out to be a local property.

It must be remarked that, since it is obviously always true that $F(u) \geq \liminf_{v \to u} F(v)$ for every $u \in U$, it turns out that F is lower semicontinuous if and only if

$$F(u) = \liminf_{v \to u} F(v) \text{ for every } u \in U.$$

It is clear that if $\{F_\theta\}_{\theta \in \mathcal{T}}$ is a collection of lower semicontinuous functions defined on U, then $u \in U \mapsto \sup_{\theta \in \mathcal{T}} F_\theta(u)$ too is lower semicontinuous. Analogously, if $\mathcal{T}$ is finite, then $u \in U \mapsto \inf_{\theta \in \mathcal{T}} F_\theta(u)$ too is lower semicontinuous, but, besides this case, in general the infimum of a family of lower semicontinuous functions need not be lower semicontinuous.

It is clear that a set $X \subseteq U$ is closed if and only if I_X is lower semicontinuous.

We say that F is *sequentially τ-lower semicontinuous*, or simply sequentially lower semicontinuous if no ambiguity occurs, if

$$F(u) \leq \liminf_{h \to +\infty} F(u_h) \text{ for every } u \in U, \text{ and every } \{u_h\} \subseteq U \text{ with } u_h \to u.$$

It is clear that a lower semicontinuous function is also sequentially lower semicontinuous. The converse is, in general, false, but it becomes true if U satisfies the first countability axiom.

Finally, we point out that

$$(0.1) \qquad\qquad F(u) \leq \limsup_{h \to +\infty} F(u_h)$$

whenever $\{u_h\} \subseteq U$, and u is a cluster point of $\{u_h\}$.

Basic Facts about Topological Vector Spaces and Banach Spaces

We now describe the main properties of those spaces in which topological structures interact with vectorial ones.

All the vector spaces that we are going to consider in this book will be real.

A vector space W is said to be a *topological vector space* if W is endowed with a topology for which the functions

$$(u, v) \in W \times W \mapsto u + v \in W, \quad (\lambda, u) \in \mathbf{R} \times W \mapsto \lambda u \in W$$

are continuous.

It is well known that in a topological vector space a set I is a neighborhood of a point u if and only if its translated $I - u = \{x - u : x \in I\}$ is a neighborhood of the origin.

A sequence $\{u_h\}$ of points of a topological vector space is said to be a *Cauchy sequence* if for every neighborhood I of the origin there exists $n_I \in \mathbf{N}$ such that $u_n - u_m \in I$ whenever $n, m > n_I$. A topological vector space is *sequentially complete* if every Cauchy sequence converges to a point of the space. In a metric space the notions of sequential completeness coincides with the one of completeness.

A particular class of topological vector spaces is the one where the topology is generated by a family of seminorms.

Let W be a vector space.

A *seminorm* on W is a function $p \colon W \to [0, +\infty[$ such that

$$p(\lambda v) = |\lambda| p(v) \text{ for every } \lambda \in \mathbf{R} \text{ and } v \in W,$$

and

$$p(u + v) \leq p(u) + p(v) \text{ for every } u, \ v \in W.$$

Of course, a seminorm p on W for which $p(u) = 0$ implies $u = 0$ is a *norm* on W. In this case, W is said to be a *normed* space.

Then, if $\{p_\theta\}_{\theta \in \mathcal{T}}$ is a family of seminorms on W, for every $u \in U$ the family of the finite intersection of sets of the type $\{v \in W : p_\theta(v - u) < \eta\}$, with $\theta \in \mathcal{T}$ and $\eta > 0$, forms a basis of neighborhoods of u, thus generating a topology on W that makes it a topological vector space.

In particular, if $\{p_\theta\}_{\theta \in \mathcal{T}}$ is made up of a single norm, the topology generated by $\{p_\theta\}_{\theta \in \mathcal{T}}$ is nothing more than the one generated by the norm itself.

A complete normed space is said to be a *Banach* space.

As usual, for every topological vector space W, we denote by W' the *dual space* of W, i.e. the set of the real continuous linear functionals on W.

If, in addition, W is also normed with norm $\| \cdot \|$, then W' turns out to be a Banach space, once we endow it with the *dual norm*

$$\| \cdot \|_{W'} \colon L \in W' \mapsto \sup\{L(u) : u \in W, \ \|u\| \leq 1\}.$$

If W is a topological vector space, and for every $\theta \in W'$ we define $p_\theta \colon u \in W \mapsto |\theta(u)|$, then p_θ turns out to be a seminorm on W, and the topology generated by $\{p_\theta\}_{\theta \in W'}$ is the so called *weak topology* on W, and is denoted by weak-W.

Analogously, if W is a topological vector space, and for every $u \in W$ we define $p_u \colon \theta \in W' \mapsto |\theta(u)|$, then p_u turns out to be a seminorm on W', and the topology generated by $\{p_u\}_{u \in W}$ is the so called *weak* topology* on W', and is denoted by weak*-W'.

In a normed space W the norm is weakly lower semicontinuous since

$$\|u\|_W = \sup\{L(u) : \|L\|_{W'} \leq 1\} \text{ for every } u \in W,$$

and, just by definition, in the dual of a normed space the dual norm is weakly* lower semicontinuous.

We now recall the following weak and weak* compactness criteria.

The basic result in this field is a weak* compactness theorem based on Tychonoff's theorem.

Theorem 0.1 (Alaoglu's Theorem). *Let W be a Banach space. Then the strongly closed balls of W' are compact in the weak*-W' topology.*

By using Alaoglu's theorem, it is easy to deduce a weak compactness criterium in reflexive spaces. Actually this criterium turns out to characterize reflexive spaces, and this is the deepest part of the following result.

Theorem 0.2 (Bourbaki-Kakutani-Šmulian Theorem). *Let W be a Banach space. Then W is reflexive if and only if its strongly closed balls are compact in the weak-W topology.*

Bourbaki-Kakutani-Šmulian Theorem describes a weak compactness property. The result below is the key to deduce a similar result for sequential weak compactness.

Theorem 0.3 (Eberlein-Šmulian Theorem). *Let W be a Banach space, and $S \subseteq W$. Then the following facts are equivalent.*
i) S is relatively sequentially compact in the weak-W topology,
ii) for every $\{u_h\} \subseteq S$ the set of the cluster points of $\{u_h\}$ in the weak-W topology in nonempty,
iii) S is relatively compact in the weak-W topology.

By Bourbaki-Kakutani-Šmulian Theorem, and Eberlein-Šmulian Theorem, the result below follows.

Theorem 0.4. *Let W be a Banach space. Then W is reflexive if and only if the strongly closed balls of W are sequentially compact in the weak-W topology.*

We recall also the sequential version of Alaoglu's theorem. It holds under separability assumptions, and follows by exploiting the metrizability of the weak*-W' topology of the strongly closed balls of the dual of a separable Banach space W (cf. for example [Br2, Corollaire III.26]).

Theorem 0.5. *Let W be a separable Banach space. Then the strongly closed balls of W' are sequentially compact in the weak*-W' topology.*

Finally, we recall the following metrizability criterium (cf. for example [Br2, Théorème III.25]).

Theorem 0.6. *Let W be a Banach space. Then W is separable if and only if for every ball B of W' the weak*-W' topology on B is metrizable.*

Basic Function Spaces

If Ω is a topological space, we denote by $C^0(\Omega)$ the set of the continuous real functions on Ω, and with $C^0_b(\Omega)$ the class of the bounded elements of $C^0(\Omega)$. It is clear that, if Ω is compact, then $C^0(\Omega) = C^0_b(\Omega)$.

With a slight abuse of notations, we denote by $\|\cdot\|_{C^0(\Omega)}$ the norm

$$\|\cdot\|_{C^0(\Omega)}\colon u \in C_b^0(\Omega) \mapsto \sup_\Omega |u|,$$

call again with $C_b^0(\Omega)$ the topology on $C_b^0(\Omega)$ induced by this norm, and recall that, once we equip $C_b^0(\Omega)$ with it, $C_b^0(\Omega)$ becomes a Banach space.

If $u\colon \Omega \to [-\infty, +\infty]$, we define the *support* $\mathrm{spt}(u)$ of u as the closure of $\{x \in \Omega : u(x) \neq 0\}$, and set

$$C_0^0(\Omega) = \left\{ u \in C^0(\Omega) : \mathrm{spt}(u) \text{ is compact} \right\}.$$

We also denote by $\widehat{C}_0^0(\Omega)$ the closure of $C_0^0(\Omega)$ in $C_b^0(\Omega)$. Then it is clear that $\widehat{C}_0^0(\Omega)$ is a Banach space with norm $\|\cdot\|_{C^0(\Omega)}$, that $C_0^0(\Omega) \subseteq \widehat{C}_0^0(\Omega) \subseteq C_b^0(\Omega)$, and that, when Ω is compact, $C_0^0(\Omega) = \widehat{C}_0^0(\Omega) = C^0(\Omega)$.

It is easy to prove that $\max_\Omega |u|$ exists for every $u \in \widehat{C}_0^0(\Omega)$.

The space $\widehat{C}_0^0(\Omega)$ is usually called the space of the *continuous functions vanishing at infinity*, since, when Ω is Hausdorff and locally compact, it is easy to verify that $u \in \widehat{C}_0^0(\Omega)$ if and only if for every $\varepsilon > 0$ there exists a compact subset K of Ω such that $\sup_K |u| < \varepsilon$.

Let now Ω be an open subset of $\mathbf{R}^n$.

Given $m \in \mathbf{N}$, we denote by $C^m(\Omega)$ the set of the functions having continuous partial derivatives of order up to m in Ω, and by $C^m(\overline{\Omega})$ the one of the elements in $C^m(\Omega)$ that can be extended, together with all their partial derivatives of order up to m, to continuous functions on $\overline{\Omega}$. If, in addition, Ω is bounded, we endow $C^m(\overline{\Omega})$ with the usual topology induced by the norm

$$\|\cdot\|_{C^m(\Omega)}\colon u \in C^m(\overline{\Omega}) \mapsto \max_{0 \leq |\alpha| \leq m} \left\| \left\| \frac{\partial^{|\alpha|} u}{\partial x^\alpha} \right\| \right\|_{C^0(\Omega)}.$$

In general, we endow $C^m(\Omega)$ with the usual topology generated by the family of seminorms $p_A\colon u \in C^m(\Omega) \mapsto \|u\|_{C^m(A)}$, with A varying in the set of the bounded open subsets of Ω such that $\overline{A} \subseteq \Omega$. We denote again by $C^m(\Omega)$ such topology, and recall that, once endowed with it, $C^m(\Omega)$ becomes a complete metrizable topological vector space.

We set $C^\infty(\Omega) = \cap_{m \in \mathbf{N}} C^m(\Omega)$, and endow it with the usual topology generated by the family of seminorms $p_{m,A}\colon u \in C^\infty(\Omega) \mapsto \|u\|_{C^m(A)}$, with m varying in $\mathbf{N} \cup \{0\}$, and A in the set of the bounded open subsets of Ω such that $\overline{A} \subseteq \Omega$. We denote again by $C^\infty(\Omega)$ such topology, and recall that, once endowed with it, $C^\infty(\Omega)$ becomes a complete metrizable topological vector space.

Finally, for every $m \in \mathbf{N}$, we denote by $C_0^m(\Omega)$ the set of the functions in $C^m(\Omega)$ having compact support in Ω, and set $C_0^\infty(\Omega) = \cap_{m \in \mathbf{N}} C_0^m(\Omega)$.

We will always identify the functions in $C_0^0(\Omega)$, with their null extensions to $\mathbf{R}^n$.

For every $z \in \mathbf{R}^n$ we denote by u_z the linear function with gradient z, i.e.

$$u_z \colon x \in \mathbf{R}^n \mapsto z \cdot x.$$

A function u on $\mathbf{R}^n$ is said to be *piecewise affine* on $\mathbf{R}^n$ if it is continuous, and if

$$u(x) = \sum_{j=1}^{m} \left(u_{z_j}(x) + c_j \right) \chi_{P_j}(x) \text{ for every } x \in \cup_{j=1}^{m} \mathrm{int}(P_j),$$

where $m \in \mathbf{N}$, $z_1, \ldots, z_m \in \mathbf{R}^n$, $c_1, \ldots, c_m \in \mathbf{R}$, and $P_1, \ldots, P_m$ are polyhedral sets with pairwise disjoint nonempty interiors such that $\cup_{j=1}^{m} P_j = \mathbf{R}^n$. We denote by $PA(\mathbf{R}^n)$ the set of the piecewise affine functions on $\mathbf{R}^n$. For every $u = \sum_{j=1}^{m}(u_{z_j} + s_j)\chi_{P_j}$ in $PA(\mathbf{R}^n)$ we set $B_u = \cup_{j=1}^{m}(\overline{P_j} \setminus \mathrm{int}(P_j))$.

We will make use of the following approximation result (cf. for example [ET, Chapter X, Proposition 2.1]).

Theorem 0.8. *Let* $u \in C^1(\mathbf{R}^n)$. *Then there exists* $\{u_h\} \subseteq PA(\mathbf{R}^n)$ *such that* $\lim_{h \to +\infty} \|u_h - u\|_{C^0(K)} = 0$, *and*

$$\lim_{h \to +\infty} \sup\{|\nabla u(x) - \nabla u_h(x)| : x \in K \setminus B_{u_h}\} = 0$$

for every $K \subseteq \mathbf{R}^n$ *compact.*

Chapter 1

Elements of Convex Analysis

The present chapter is concerned with the main notions and results of convex analysis used in the book.

In the first section we present the basics of convex analysis in the abstract setting of locally convex topological vector spaces. Then the treatment goes on to the setting of $\mathbf{R}^n$, even if some of the results are still valid in more general frameworks.

In particular, the convex and the lower semicontinuous envelopes of a function are introduced and described, and their compositions in the two possible different orders are studied and compared. This study is motivated by the deep importance that both these composite operators have in calculus of variations.

For a deeper treatment of convex analysis, we refer, for example, to the books [DuS], [ET], [R], and [RW].

§1.1 Convex Sets and Functions

Let V be a vector space over the reals.

Given $k \in \mathbf{N}$, $x_1, \ldots, x_k \in V$, and $t_1, \ldots, t_k \in [0, +\infty[$ such that $\sum_{j=1}^{k} t_j = 1$, we say that the point $\sum_{j=1}^{k} t_j x_j$ is a *convex combination* of $x_1, \ldots, x_k$.

In particular, if x, $y \in V$, and $t \in [0, 1]$, the point $tx + (1 - t)y$ is a convex combination of x and y. From a geometrical point of view, a convex combination of x and y lies on the line through x and y, but between them, thus the set $\{tx + (1 - t)y : t \in [0, 1]\}$ of the convex combinations of x and y is the closed line segment joining x and y.

We say that $C \subseteq V$ is *convex* if $tx + (1 - t)y \in C$ whenever x, $y \in C$, and $t \in [0, 1]$. In other words, C is convex if C contains the closed line segment joining x and y, whenever x, $y \in C$.

Equivalently, it is possible to say that $C \subseteq V$ is convex if $\sum_{j=1}^{k} t_j x_j$ whenever $k \in \mathbf{N}$, $x_1, \ldots, x_k \in C$, and $t_1, \ldots, t_k \in [0, +\infty[$ are such that $\sum_{j=1}^{k} t_j = 1$. As above, it is possible to say that C is convex if C contains all the convex combinations of finitely many of its points.

It is clear that if $\{C_\theta\}_{\theta \in \mathcal{T}}$ is a collection of convex sets, then $\cap_{\theta \in \mathcal{T}} C_\theta$ too is convex. On the contrary, the union of two convex sets need not be convex.

A fundamental tool for the study of convex analysis is furnished by the separation properties of convex sets.

To describe precisely such argument, we need to recall briefly the notion of locally convex topological vector space.

A topological vector space is said to be *locally convex* if the origin possesses a fundamental family of convex neighborhoods. For example, every topological vector space whose topology is generated by a family of seminorms is locally convex.

It is important to note that the converse is also true. In fact, by using Minkowski functionals, it can be proved that, given a locally convex topological vector space W, a family of seminorms on W can be constructed that generates the topology of W. Thus, locally convex topological vector spaces turn out to place, in some sense, intermediately between topological vector spaces and normed spaces.

A subset H of V is said to be a *hyperplane* if $H = \{x \in V : L(x) = c\}$ for some $L \in V'$ not identically equal to zero, and $c \in \mathbf{R}$.

Given $A, B \subseteq V$, and a hyperplane H of V with $H = \{x \in V : L(x) = c\}$, we say that H *separates* A and B if $L(x) \leq c$ for every $x \in A$ and $L(x) \geq c$ for every $x \in B$. We say that H *strictly separates* A and B if $L(x) < c$ for every $x \in A$ and $L(x) > c$ for every $x \in B$.

If now W is a topological vector space, it is well known that a hyperplane H of W is closed if and only if the linear functional that determines H is continuous.

We can now recall the Separation Theorem.

Theorem 1.1.1 (Separation Theorem). *Let W be a Hausdorff locally convex topological vector space, $C \subseteq W$ be closed and convex, and $x_0 \in W \setminus C$. Then C and x_0 can be strictly separated by a closed hyperplane of W.*

One of the most significant consequences for our purposes of the Separation Theorem is deduced in the following result.

Theorem 1.1.2. *Let W be a Banach space, and $C \subseteq W$ be convex. Then the following conditions*
i) C is closed in the strong topology of W,
ii) C is closed in the weak-W topology,
iii) C is sequentially closed in the weak-W topology
are equivalent.

Proof. It is clear that ii) implies i).

On the contrary, if i) holds, then, by the Separation Theorem, it turns out that C agrees with the intersection of all the strongly closed half-spaces that contain C itself. Let Σ be one of such half-spaces, then $W \backslash \Sigma$ is trivially open in the weak-W topology, and therefore Σ is also closed in the weak-W one. Because of this, C turns out to be closed in the weak-W topology.

To complete the proof, let us assume that iii) holds. Then C is sequentially closed in W, and therefore it is closed in W. Because of this, and by the previous equivalence, C turns out to be closed in the weak-W topology, and ii) holds. ■

In a similar order of ideas, the result below holds when weak* topologies are considered. To prove it, we first need to recall the Krein-Šmulian closedness criterium.

Theorem 1.1.3 (Krein-Šmulian Theorem). *Let W be a Banach space, and $C \subseteq W'$ be convex. Then C is closed in the weak*-W' topology if and only so does $C \cap \{y \in W' : \|y\|_{W'} \leq r\}$ for every $r > 0$.*

Theorem 1.1.4. *Let W be a separable Banach space, and $C \subseteq W'$ be convex. Then C is closed in the weak*-W' topology if and only if C is sequentially closed in the weak*-W' topology.*

Proof. We only have to prove that if C is sequentially closed in the weak*-W' topology, then C is closed in the same one.

To do this, let us assume that C is sequentially closed in the weak*-W' topology.

By virtue of Krein-Šmulian Theorem, to prove the claim it suffices to verify that, for fixed $k \in \mathbf{N}$, $C \cap \{y \in W' : \|y\|_{W'} \leq k\}$ is closed in the weak*-W' topology.

Let $k \subset \mathbf{N}$. Then, by Theorem 0.6, $C \cap \{y \in W' : \|y\|_{W'} \leq k\}$ turns out to be closed in the weak*-W' topology if and only if it is sequentially closed in the same one.

To prove this last condition, we observe that $\| \cdot \|_{W'}$ is weak*-W'-lower semicontinuous, and that, consequently, $\{y \in W' : \|y\|_{W'} \leq k\}$ turns out to be closed in weak*-W'. This, together with the sequential closure in weak*-W' of C, provides the proof. ■

We now confine ourselves to the study of convex subsets of $\mathbf{R}^n$.

Convex subsets of $\mathbf{R}^n$ enjoy the special feature to possess always "interior" points.

To see this, we recall that a subset M of $\mathbf{R}^n$ is said to be *affine* if it is the translate of a vector subspace of $\mathbf{R}^n$.

For a given $S \subseteq \mathbf{R}^n$ we denote by $\mathrm{aff}(S)$ the *affine hull* of S, defined as the intersection of all the affine sets containing S. It is clear that $\mathrm{aff}(S)$ turns out to be the smallest affine set containing S.

If $C \subseteq \mathbf{R}^n$ is convex, we denote by $\mathrm{ri}(C)$ the *relative interior* of C, i.e. the set of the interior points of C, in the topology of $\mathrm{aff}(C)$, once we regard it as a subspace of $\mathrm{aff}(C)$, and by $\mathrm{rb}(C)$ the *relative boundary* of C, i.e. the set $\overline{C} \setminus \mathrm{ri}(C)$. When $\mathrm{aff}(C) = \mathbf{R}^n$ we write as usual $\mathrm{ri}(C) = \mathrm{int}(C)$ and $\mathrm{rb}(C) = \partial C$.

The following result summarizes the main properties of relative interiors, and can be proved by means of standard techniques in convex analysis (cf. for example [R, Section 6]).

Proposition 1.1.5. *Let $C \subseteq \mathbf{R}^n$ be nonempty and convex. Then $\mathrm{ri}(C)$ is nonempty and convex, $\overline{C}$ is convex,*

$$\mathrm{aff}(\mathrm{ri}(C)) = \mathrm{aff}(\overline{C}) = \mathrm{aff}(C),$$

$$\overline{\mathrm{ri}(C)} = \overline{C}, \quad \mathrm{ri}(\overline{C}) = \mathrm{ri}(C),$$

and

$$\overline{x_0 + t(C - x_0)} = x_0 + t(\overline{C} - x_0) \subseteq \mathrm{ri}(C) \text{ for every } x_0 \in \mathrm{ri}(C), \ t \in [0,1[.$$

By the Separation Theorem, we deduce the following representation result.

Proposition 1.1.6. *Let $C \subseteq \mathbf{R}^n$ be closed and convex. Then there exists a sequence of open half-spaces $\{\Sigma_h\}$ such that $C = \cap_{h \in \mathbf{N}} \Sigma_h$, i.e. there exist $\{z_h\} \subseteq \mathbf{R}^n \setminus \{0\}$ and $\{c_h\} \subseteq \mathbf{R}$ such that*

$$x \in C \text{ if and only if } z_h \cdot x < c_h \text{ for every } h \in \mathbf{N}.$$

Proof. Let us first assume that $\mathrm{int}(C) \neq \emptyset$.

Let $\{x_h\}$ be a dense sequence in $\mathbf{R}^n \setminus C$. Then, by the Separation Theorem, for every $h \in \mathbf{N}$ there exists a hyperplane H_h that strictly separates C and x_h. For every $h \in \mathbf{N}$ let Σ_h be the open half-space containing C whose boundary is H_h. Let us prove that $C = \cap_{h \in \mathbf{N}} \Sigma_h$.

It is obvious that

$$C \subseteq \cap_{h \in \mathbf{N}} \Sigma_h,$$

hence we have to prove only that

$$\mathbf{R}^n \setminus C \subseteq \mathbf{R}^n \setminus \cap_{h \in \mathbf{N}} \Sigma_h.$$

To do this, let $x \in \mathbf{R}^n \setminus C$, and set $C_x = \{tx + (1-t)y : y \in \mathrm{int}(C), \ t \in [0,1[\}$. Then, since $\mathrm{int}(C) \neq \emptyset$, C_x turns out to be nonempty and open.

Since $\mathbf{R}^n \setminus C$ is open, and $\{x_h\}$ is dense in $\mathbf{R}^n \setminus C$, we can find $\overline{h} \in \mathbf{N}$ such that $x_{\overline{h}} \in (\mathbf{R}^n \setminus C) \cap C_x$.

It is clear that $C \subseteq \Sigma_{\overline{h}}$, and $x_{\overline{h}} \notin \Sigma_{\overline{h}}$. Moreover it turns out that $x \notin \Sigma_{\overline{h}}$, otherwise, since $\Sigma_{\overline{h}}$ is convex, we would also have that $x_{\overline{h}} \in C_x \subseteq \Sigma_{\overline{h}}$. Consequently, $x \in \mathbf{R}^n \setminus \cap_{h \in \mathbf{N}} \Sigma_h$, and the proof follows under the assumption that $\text{int}(C) \neq \emptyset$.

Finally, if $\text{int}(C) = \emptyset$, we can regard C as a subset of $\text{aff}(C)$, where C has "nonempty interior," and repeat the above considerations by replacing $\mathbf{R}^n$ with $\text{aff}(C)$, and $\text{int}(C)$ with $\text{ri}(C)$, thus obtaining a sequence $\{\Sigma'_h\}$ of half-spaces in $\text{aff}(C)$ such that

$$(1.1.1) \qquad\qquad C = \cap_{h \in \mathbf{N}} \Sigma'_h.$$

For every $h \in \mathbf{N}$, we now take a half-space Σ_h of $\mathbf{R}^n$ satisfying $\Sigma'_h = \Sigma_h \cap \text{aff}(C)$. Then, once we observe that a finite number of half-spaces $\widetilde{\Sigma}_1 \ldots, \widetilde{\Sigma}_m$ of $\mathbf{R}^n$ can be found such that $\text{aff}(C) = \cap_{j=1}^m \widetilde{\Sigma}_j$, by (1.1.1), it follows that

$$C = (\cap_{h \in \mathbf{N}} \Sigma_h) \cap \widetilde{\Sigma}_1 \cap \ldots \cap \widetilde{\Sigma}_m,$$

which proves the theorem. ∎

Let $C \subseteq \mathbf{R}^n$ be convex. A *supporting half-space* to C is a closed half-space containing C and having a point of C in its boundary. A *non-trivial supporting hyperplane* to C is a hyperplane not containing C which is the boundary of a supporting half-space to C.

The following result is well known (cf. for example [R, Theorem 11.6]).

Theorem 1.1.7. *Let C be a convex subset of $\mathbf{R}^n$, and let $x \in C$. Then there exists a non-trivial supporting hyperplane to C containing x if and only if $x \notin \text{ri}(C)$.*

We now define convex functions.

To do this, we first have to specify some rules to properly carry out arithmetic operations in $[-\infty, +\infty]$.

Of course the result of arithmetic operations between real elements of $[-\infty, +\infty]$ is well defined, as well as the one between elements of $[-\infty, +\infty]$ when no reasonable ambiguity may occur. Thus, for example, we naturally accept to define $+\infty$ as the result of expressions like $x + (+\infty)$, and $\lambda \cdot (+\infty)$, when $x \in \,]-\infty, +\infty]$, and $\lambda \in \,]0, +\infty]$. Analogously, we define $-\infty$ as the result of expressions like $x + (-\infty)$, and $\lambda \cdot (-\infty)$, when $x \in [-\infty, +\infty[$ and $\lambda \in [-\infty, +\infty[$.

On the contrary, expressions like $0 \cdot (+\infty)$, $0 \cdot (-\infty)$, $-\infty + (+\infty)$, and $+\infty + (-\infty)$ present a higher degree of ambiguity, and their values may depend on the general context in which they are considered.

In the context of convex analysis, it is customary to set $0 \cdot (+\infty) = (+\infty) \cdot 0 = 0 \cdot (-\infty) = (-\infty) \cdot 0 = 0$, and to adopt the so called inf-addition convention in which $-\infty + (+\infty) = +\infty + (-\infty) = +\infty$.

Under these rules, extended arithmetic obeys associative, commutative, and distributive laws, with the only exception of the equality

$$\lambda \cdot (+\infty + (-\infty)) = \lambda \cdot (+\infty) + \lambda \cdot (-\infty)$$

that no more holds when $\lambda < 0$.

In addition, in the sequel we will take into account last upper bounds and greatest lower bounds of possibly empty sets. In this case, as usual, we set $\inf \emptyset = +\infty$, and $\sup \emptyset = -\infty$.

For every set U, and every $F: U \to [-\infty, +\infty]$ we define the *effective domain* of F as

$$\mathrm{dom} F = \{x \in U : F(x) < +\infty\},$$

and the *epigraph* of F as

$$\mathrm{epi} F = \{(x, \lambda) \in U \times \mathbf{R} : F(x) \le \lambda\}.$$

It is clear that $\mathrm{dom} F$ is the projection of $\mathrm{epi} F$ on U, in the sense that

$$\mathrm{dom} F = \{x \in U : (x, \lambda) \in \mathrm{epi} F \text{ for some } \lambda \in \mathbf{R}\}.$$

Let V be a vector space, and $C \subseteq V$ be convex.
A function $F: C \to [-\infty, +\infty]$ is said to be *convex* if

$$F(tx + (1-t)y) \le tF(x) + (1-t)F(y) \text{ for every } x, \ y \in V, \ t \in [0,1].$$

From a geometrical point of view, we can say that F is convex if F along the convex combinations of two points of its domain lies below the convex combinations of its values.

Equivalently, it is easy to verify that F is convex if and only if

$$F\left(\sum_{j=1}^{k} t_j x_j\right) \le \sum_{j=1}^{k} t_j F(x_j)$$

for every $k \in \mathbf{N}$, $x_1, \ldots, x_k \in C$, $t_1, \ldots, t_k \in [0, +\infty[$ such that $\sum_{j=1}^{k} t_j = 1$.

If $F: C \to [-\infty, +\infty]$, then the function

$$\widehat{F}: x \in V \mapsto \begin{cases} F(x) & \text{if } x \in C \\ +\infty & \text{if } x \in V \setminus C \end{cases}$$

is convex if and only if C is convex, and F is a convex function. Consequently, it is not restrictive to consider functions defined on the whole V.

A convex function $F: V \to [-\infty, +\infty]$ that takes the value $-\infty$ in a point x_0 behaves in a very special way. In fact it is easy to verify that in this case for every $x_1 \in V$ there exists $t_1 \in [0, 1]$ such that $F((1-t)x_0 + tx_1) = -\infty$ for every $t \in [0, t_1[$, $F((1-t)x_0 + tx_1) = +\infty$ for every $t \in]t_1, 1]$, and $F((1-t_1)x_0 + t_1 x_1)$ may be any value in $[-\infty, +\infty]$.

It is clear that if $\{F_\theta\}_{\theta \in T}$ is a collection of convex functions defined on V, then $x \in V \mapsto \sup_{\theta \in T} F_\theta(x)$ too is convex. On the contrary, the minimum of two convex functions need not be convex.

It is also obvious that, given $C \subseteq V$, it results that C is convex if and only if I_C is a convex function.

Proposition 1.1.8. *Let V be a vector space, and $F: V \to [-\infty, +\infty]$. Then F is convex if and only if epi$F$ is convex.*

Proof. Let us first assume that F is convex, and let (x_1, λ_1), $(x_2, \lambda_2) \in$ epi$F$, $t \in [0, 1]$. Then

$$F(tx_1 + (1-t)x_2) \leq tF(x_1) + (1-t)F(x_2) \leq t\lambda_1 + (1-t)\lambda_2,$$

that is $t(x_1, \lambda_1) + (1-t)(x_2, \lambda_2) \in$ epi$F$. Because of this, the convexity of epi$F$ follows.

Conversely, let us assume that epi$F$ is convex, and let x_1, $x_2 \in V$, $t \in [0, 1]$. We can clearly assume that x_1, $x_2 \in$ dom$F$. Let λ_1, $\lambda_2 \in \mathbf{R}$ be such that (x_1, λ_1), $(x_2, \lambda_2) \in$ epi$F$. Then, because of the convexity of epi$F$, we have that $t(x_1, \lambda_1) + (1-t)(x_2, \lambda_2) \in$ epi$F$, that is

$$F(tx_1 + (1-t)x_2) \leq t\lambda_1 + (1-t)\lambda_2.$$

Because of this, the proof follows letting λ_1 decrease to $F(x_1)$, and λ_2 decrease to $F(x_2)$. ∎

Besides convexity, also lower semicontinuity properties can be characterized by means of epigraphs.

Proposition 1.1.9. *Let (U, τ) be a topological space, and $F: U \to [-\infty, +\infty]$. Then F is τ-lower semicontinuous if and only if epi$F$ is closed in the product topology of $U \times \mathbf{R}$.*

Proof. Let us first assume that F is τ-lower semicontinuous. Let us prove that $U \times \mathbf{R} \setminus$ epi$F$ is open.

Let $(x_0, \lambda_0) \in U \times \mathbf{R} \setminus$ epi$F$. Then $\lambda_0 < F(x_0)$, and let $\lambda \in]\lambda_0, F(x_0)[$. By the τ-lower semicontinuity of F, there exists $I_{x_0} \in \mathcal{N}(x_0)$ such that $\lambda < F(x)$ for every $x \in I_{x_0}$, and therefore $I_{x_0} \times]\lambda_0 - 1, \lambda[$ turns out to be a neighborhood of (x_0, λ_0) having empty intersection with epi$F$. Because of this, it follows that $U \times \mathbf{R} \setminus$ epi$F$ is open.

Conversely, let us assume that epi$F$ is closed. Then $\{(x, \lambda) \in U \times \mathbf{R} : F(x) > \lambda\}$ is open in the product topology of $U \times \mathbf{R}$, and, consequently, for every $\lambda \in \mathbf{R}$, $\{x \in U : F(x) > \lambda\}$ is open in U. This yields the τ-lower semicontinuity of F. ∎

We now study some properties of convex, lower semicontinuous functions.

Proposition 1.1.10. *Let W be a topological vector space, and $F: W \to [-\infty, +\infty]$ be convex and lower semicontinuous. Assume that F takes the value $-\infty$. Then $F(W) \subseteq \{-\infty, +\infty\}$.*

Proof. Let $x_0 \in W$ be such that $F(x_0) = -\infty$, and let $x_1 \in W$. Then, by the lower semicontinuity, and the convexity of F, it follows that

$$F(x_1) \leq \liminf_{x \to x_1} F(x) \leq \liminf_{t \to 0^+} F(tx_0 + (1-t)x_1) \leq$$

$$\leq \liminf_{t \to 0^+} \{tF(x_0) + (1-t)F(x_1)\} = -\infty + F(x_1),$$

from which the nonfiniteness of $F(x_1)$ can be deduced.

By the arbitrariness of x_1, the proof follows. ∎

The following result yields a characterization of convex, lower semicontinuous functions.

Theorem 1.1.11. *Let W be a locally convex topological vector space, and $F: W \to [-\infty, +\infty]$. Then F is convex, lower semicontinuous, and identically equal to $-\infty$ provided $F(x) = -\infty$ for at least one $x \in W$ if and only if*

$$F(x) = \sup\{L(x) + c : L \in W', \ c \in \mathbf{R}, \ L + c \leq F \text{ in } W\} \text{ for every } x \in W.$$

Proof. For the sake of simplicity, let us set

$$s: x \in W \mapsto \sup\{L(x) + c : L \in W', \ c \in \mathbf{R}, \ L + c \leq F \text{ in } W\}.$$

Then it is clear that s is convex, lower semicontinuous, and that

$$(1.1.2) \qquad\qquad s(x) \leq F(x) \text{ for every } x \in W.$$

Consequently, if $F(x) = s(x)$ for every $x \in W$, it follows that F is convex and lower semicontinuous. In addition, if $F(x) = -\infty$ for some $x \in W$, then $\{L(x) + c : L \in W', \ c \in \mathbf{R}, \ L + c \leq F \text{ in } W\} = \emptyset$, and F, being the pointwise supremum of the empty set, is identically equal to $-\infty$.

Conversely, let us assume that F is convex, lower semicontinuous, and that, if $F(x) = -\infty$ for some $x \in W$. Then $F(x) = -\infty$ for every $x \in W$.

We can assume that $\text{dom}F \neq \emptyset$, otherwise the theorem is obvious. Then, by Proposition 1.1.8, and Proposition 1.1.9, $\text{epi}F$ turns out to be nonempty, convex, and closed in the product topology of $W \times \mathbf{R}$.

Let $x_0 \in W$. If $F(x_0) = -\infty$, then $F(x) = -\infty$ for every $x \in W$ and $\{L(x) + c : L \in W', \ c \in \mathbf{R}, \ L + c \le F \text{ in } W\} = \emptyset$. Consequently, $s(x) = -\infty$ for every $x \in W$, and the theorem follows.

If $F(x_0) > -\infty$, let $\lambda_0 \in \ ]-\infty, F(x_0)[$, then $(x_0, \lambda_0) \notin \mathrm{epi}F$, and by the Separation Theorem, there exist $(L, a) \in W' \times \mathbf{R} \setminus (0, 0)$ and $c \in \mathbf{R}$ such that

(1.1.3) $\qquad L(x_0) + a\lambda_0 < c < L(x) + a\lambda$ for every $(x, \lambda) \in \mathrm{epi}F$.

Moreover, since (1.1.3) yields

$$a > \frac{c}{\lambda} - \frac{1}{\lambda}L(x) \text{ for every } x \in \mathrm{dom}F, \ \lambda > \max\{F(x), 0\},$$

we obtain that $a \ge 0$.

Let us consider separately the cases $a > 0$, and $a = 0$.

If $a > 0$, we have that

(1.1.4) $$\lambda_0 < \frac{c}{a} - \frac{1}{a}L(x_0),$$

and by the second inequality in (1.1.3) with $(x, \lambda) = (x, F(x))$, that

(1.1.5) $\qquad \dfrac{c}{a} - \dfrac{1}{a}L(x) < F(x)$ for every $x \in \mathrm{dom}F$.

By (1.1.4), and (1.1.5) we deduce that $\frac{c}{a} - \frac{1}{a}L$ is just one of the functionals appearing in the definition of s, thus, by (1.1.4), we conclude that

$$\lambda_0 < s(x_0) \text{ for every } \lambda_0 \in \ ]-\infty, F(x_0)[,$$

and therefore that

(1.1.6) $\qquad F(x_0) \le s(x_0)$ for every $x_0 \in W$, provided $a > 0$.

In particular, since (1.1.3) yields

$$L(x_0) + a\lambda_0 < c < L(x_0) + aF(x_0) \text{ if } x_0 \in \mathrm{dom}F,$$

we conclude that, if $x_0 \in \mathrm{dom}F$, then $a > 0$, and by the previously treated case, that

(1.1.7) $\qquad F(x) \le s(x)$ for every $x \in \mathrm{dom}F$.

If now $a = 0$, let $y_0 \in \mathrm{dom}F$, and $\mu_0 \in \ ]-\infty, F(y_0)[$. Then, by (1.1.7), we get $M \in W'$, and $b \in \mathbf{R}$ such that

(1.1.8) $\qquad \mu_0 < M(y_0) + b \quad M(x) + b \le F(x)$ for every $x \in W$.

Therefore, by (1.1.8), and (1.1.3) with $a = 0$ we conclude that

$$M(x) + b + \gamma(c - L(x)) \leq F(x) \text{ for every } x \in W, \ \gamma > 0,$$

and

(1.1.9) $\lambda_0 < M(x_0) + b + \gamma(c - L(x_0))$ provided γ is large enough.

Consequently, for γ sufficiently large, $M + b + \gamma(c - L)$ is just one of the functionals appearing in the definition of s, thus, by (1.1.9), we conclude that

$$\lambda_0 < s(x_0) \text{ for every } \lambda_0 \in \ ]-\infty, F(x_0)[,$$

and therefore that

(1.1.10) $F(x_0) \leq s(x_0)$ for every $x_0 \in W$, if $a = 0$.

In conclusion, by (1.1.2), (1.1.6), and (1.1.10) the identity between F and s follows. This completes the proof. ∎

When $W = \mathbf{R}^n$ Theorem 1.1.11 can be specified, as shown in the result below.

Proposition 1.1.12. *Let $f : \mathbf{R}^n \to \ ]-\infty, +\infty]$ be convex and lower semi-continuous. Then there exist $\{a_h\} \subseteq \mathbf{R}^n$ and $\{b_h\} \subseteq \mathbf{R}$ such that*

$$f(z) = \sup\{a_h \cdot z + b_h : h \in \mathbf{N}\} \text{ for every } z \in \mathbf{R}^n.$$

Proof. Of course we can assume that f is not identically equal to $+\infty$.

The proof is similar to the one of Proposition 1.1.6 with $C = \text{epi} f$, but by using Theorem 1.1.11 in place of the Separation Theorem.

Let us first assume that $\text{int}(\text{dom} f) \neq \emptyset$, and let us observe that in this case $\text{int}(\text{epi} f) \neq \emptyset$.

Let $\{(z_h, \lambda_h)\}$ be a countable dense sequence in $\mathbf{R}^{n+1} \setminus \text{epi} f$. Then, by Theorem 1.1.11, for every $h \in \mathbf{N}$ there exist $a_h \in \mathbf{R}^n$ and $b_h \in \mathbf{R}$ such that

$$\lambda_h < a_h \cdot z_h + b_h, \quad a_h \cdot z + b_h \leq f(z) \text{ for every } z \in \mathbf{R}^n.$$

It is obvious that

(1.1.11) $\sup\{a_h \cdot z + b_h : h \in \mathbf{N}\} \leq f(z)$ for every $z \in \mathbf{R}^n$.

To prove the reverse inequality, let $z \in \mathbf{R}^n$ and $\lambda < f(z)$, and set $E = \{t(z, \lambda) + (1 - t)y : y \in \text{int}(\text{epi} f), \ t \in [0, 1[\}$, then E turns out to be nonempty and open. Moreover, since $\mathbf{R}^{n+1} \setminus \text{epi} f$ is open, we can find $\overline{h} \in \mathbf{N}$ such that $(z_{\overline{h}}, \lambda_{\overline{h}}) \in (\mathbf{R}^{n+1} \setminus \text{epi} f) \cap E$.

At this point the same arguments of the proof of Proposition 1.1.6 apply, and we deduce that $\lambda < a_{\overline{h}} \cdot z + b_{\overline{h}}$, and hence that

$$(1.1.12) \qquad f(z) \leq \sup\{a_h \cdot z + b_h : h \in \mathbf{N}\} \text{ for every } z \in \mathbf{R}^n.$$

By (1.1.11) and (1.1.12), the proposition follows when $\text{int}(\text{dom} f) \neq \emptyset$.

If now $\text{int}(\text{dom} f) = \emptyset$, we can regard $\text{dom} f$ as a subset of $\text{aff}(\text{dom} f)$, and repeat the above considerations by replacing $\mathbf{R}^n$ with $\text{aff}(\text{dom} f)$, and $\text{int}(\text{dom} f)$ with $\text{ri}(\text{dom} f)$, thus obtaining $\{a'_h\} \subseteq \mathbf{R}^n$ and $\{b'_h\} \subseteq \mathbf{R}$ such that

$$(1.1.13) \qquad f(z) = \sup\{a'_h \cdot z + b'_h : h \in \mathbf{N}\} \text{ for every } z \in \text{aff}(\text{dom} f).$$

In order to complete the proof, let us assume for the moment that $f(z) \geq 0$ for every $z \in \mathbf{R}^n$.

Let us take $\{a''_h\} \subseteq \mathbf{R}^n$ and $\{b''_h\} \subseteq \mathbf{R}$ such that

$$(1.1.14) \qquad a''_h \cdot z + b''_h = 0 \text{ for every } z \in \text{aff}(\text{dom} f),$$

$$(1.1.15) \qquad \sup_{h \in \mathbf{N}}\{a''_h \cdot z + b''_h\} = +\infty \text{ for every } z \in \mathbf{R}^n \setminus \text{aff}(\text{dom} f),$$

then, since $f(z) = +\infty$ for every $z \in \mathbf{R}^n \setminus \text{aff}(\text{dom} f)$, the proposition follows from (1.1.13)÷(1.1.15) with $\{a_h\}$ given by the union of $\{a'_h\}$ and $\{a''_h\}$, and $\{b_h\}$ by the one of $\{b'_h\}$, and $\{b''_h\}$.

Finally, if f changes sign, it suffices to take $a \in \mathbf{R}^n$ and $b \in \mathbf{R}$ such that $a \cdot z + b \leq f(z)$ for every $z \in \mathbf{R}^n$, whose existence is guaranteed by Theorem 1.1.11, and consider $f - a \cdot (\cdot) - b$. ∎

Finally, we discuss on the lower semicontinuity of convex functions in Banach spaces.

Theorem 1.1.13. *Let W be a Banach space, and $F: W \to [-\infty, +\infty]$ be convex. Then the following conditions*
i) F is W-lower semicontinuous,
ii) F is weak-W-lower semicontinuous,
iii) F is sequentially weak-W-lower semicontinuous
are equivalent

Proof. Follows from Theorem 1.1.2, and the obvious remark that, if iii) holds, then for every $\lambda \in \mathbf{R}$ the set $\{x \in W : F(x) \leq \lambda\}$ is sequentially closed in the weak-W topology. ∎

Theorem 1.1.14. *Let W be a separable Banach space, and $F: W' \to [-\infty, +\infty]$ be convex. Then F is weak*-W'-lower semicontinuous if and only if F is sequentially weak*-W'-lower semicontinuous.*

Proof. Follows from Theorem 1.1.4, and the obvious remark that, if F is sequentially weak*-W'-lower semicontinuous, then for every $\lambda \in \mathbf{R}$ the set $\{y \in W' : F(y) \leq \lambda\}$ is sequentially closed in the weak*-W' topology. ∎

Convex functions, even if defined just by means of vectorial properties, naturally enjoy nice continuity properties.

Proposition 1.1.15. *Let W be a topological vector space, and $F: W \to [-\infty, +\infty]$ be convex. Assume that there exists a nonempty open subset A of W such that $\sup_A F < +\infty$. Then $\mathrm{int}(\mathrm{dom}F) \neq \emptyset$, and F is continuous in $\mathrm{int}(\mathrm{dom}F)$.*

Proof. It is clear that $\mathrm{int}(\mathrm{dom}F) \neq \emptyset$.

First of all, let us prove that

$$(1.1.16) \qquad \text{for every } x \in \mathrm{int}(\mathrm{dom}F) \text{ there exists } A_x \in \mathcal{N}(x)$$

$$\text{such that } \sup_{A_x} F < +\infty.$$

To do this, let $x_0 \in A$. Then for every $x \in \mathrm{int}(\mathrm{dom}F)$ there exists $r > 1$ such that $z = x_0 + r(x - x_0) \in \mathrm{int}(\mathrm{dom}F)$.

Let us set $A_x = \frac{1}{r}z + (1 - \frac{1}{r})A$. Then $A_x \in \mathcal{N}(x)$, $\frac{r}{r-1}(y - \frac{1}{r}z) \in A$ for every $y \in A_x$, and by the convexity of F, we have

$$F(y) = F\left(\frac{1}{r}z + \left(1 - \frac{1}{r}\right)\frac{r}{r-1}\left(y - \frac{1}{r}z\right)\right) \leq \frac{1}{r}F(z) + \left(1 - \frac{1}{r}\right)\sup_A F$$

$$\text{for every } y \in A_x,$$

from which (1.1.16) follows.

Let now $x_0 \in \mathrm{int}(\mathrm{dom}F)$. Let us prove that F is continuous in x_0.

It is not restrictive to assume that $x_0 = 0$.

Let A_0 be given by (1.1.16) with $x = 0$, and set $I_0 = A_0 \cap (-A_0)$. Then I_0 is a symmetric neighborhood of 0.

Let $\varepsilon \in \,]0, 1[$, and $x \in \varepsilon I_0$. Then, since $\frac{1}{\varepsilon}x$ and $-\frac{1}{\varepsilon}x \in I_0$, by the convexity of F it follows that

$$(1.1.17) \qquad F(x) \leq (1 - \varepsilon)F(0) + \varepsilon F\left(\frac{1}{\varepsilon}x\right) \leq (1 - \varepsilon)F(0) + \varepsilon \sup_{A_0} F,$$

from which the continuity of F in 0 follows when $F(0) = -\infty$.

On the other side, if $F(0) \in \mathbf{R}$, again the convexity of F yields

$$F(x) \geq (1 + \varepsilon)F(0) - \varepsilon F\left(-\frac{1}{\varepsilon}x\right) \geq (1 + \varepsilon)F(0) - \varepsilon \sup_{A_0} F,$$

from which, together with (1.1.17), it follows that $|F(x) - F(0)| \leq \varepsilon(F(0) + \sup_{A_0} F)$ whenever $x \in \varepsilon I_0$, namely that F is continuous in 0.

Because of this, the proof follows. ∎

The above continuity property of convex functions can be improved under stronger assumption on the topology of W.

Proposition 1.1.16. *Let W be a normed space, and $F: W \to\,]-\infty, +\infty]$ be convex. Assume that there exists a nonempty open subset A of W such that $\sup_A F < +\infty$. Then $\operatorname{int}(\operatorname{dom}F) \neq \emptyset$, and F is locally Lipschitz in $\operatorname{int}(\operatorname{dom}F)$.*

Proof. It is clear that $\operatorname{int}(\operatorname{dom}F) \neq \emptyset$.

Let us prove that for every $x_0 \in \operatorname{int}(\operatorname{dom}F)$ there exist $\delta > 0$ and $M > 0$ such that $B_\delta(x_0) \subseteq \operatorname{int}(\operatorname{dom}F)$, and

$$(1.1.18) \qquad |F(x) - F(y)| \leq M|x - y| \text{ for every } x,\, y \in B_\delta(x_0).$$

Let $x_0 \in \operatorname{int}(\operatorname{dom}F)$. Then, by Proposition 1.1.15, F is continuous in x_0. Consequently there exists $\delta > 0$ such that $\sup_{B_{2\delta}(x_0)} F - \inf_{B_{2\delta}(x_0)} F < +\infty$.

If now $x,\, y \in B_\delta(x_0)$ satisfy $x \neq y$, let us set $z = y + \frac{\delta}{|x-y|}(y - x)$. Then $z \in B_{2\delta}(x_0)$, and, since $y = \frac{|x-y|}{|x-y|+\delta}z + \frac{\delta}{|x-y|+\delta}x$, by the convexity of F we conclude that

$$F(y) - F(x) \leq \frac{|x-y|}{|x-y|+\delta}F(z) + \frac{\delta}{|x-y|+\delta}F(x) - F(x) =$$

$$= \frac{|x-y|}{|x-y|+\delta}(F(z) - F(x)) \leq \frac{1}{\delta}\left(\sup_{B_{2\delta}(x_0)} F - \inf_{B_{2\delta}(x_0)} F\right)|x - y|$$

for every $x,\, y \in B_\delta(x_0)$.

Because of this, up to an interchange of the roles of x and y, (1.1.18) follows with $M = \frac{1}{\delta}(\sup_{B_{2\delta}(x_0)} F - \inf_{B_{2\delta}(x_0)} F)$. ∎

By Proposition 1.1.16 we deduce the following results.

Theorem 1.1.17. *Let $f: \mathbf{R}^n \to [-\infty, +\infty]$ be convex. Then f is continuous in $\operatorname{ri}(\operatorname{dom}f)$.*

If, in addition, $f(z) > -\infty$ for every $z \in \mathbf{R}^n$, then f is locally Lipschitz in $\operatorname{ri}(\operatorname{dom}f)$.

Proof. By considering the restriction of f to $\operatorname{aff}(\operatorname{dom}f)$, it is not restrictive to assume that $\operatorname{int}(\operatorname{dom}f) \neq \emptyset$.

Let $z_0 \in \operatorname{int}(\operatorname{dom}f)$, and $z_1, \ldots, z_{2^n}$ be the vertices of an open cube Q satisfying $z_0 \in Q$, and $Q \subseteq \operatorname{int}(\operatorname{dom}f)$. Then, since every point of Q is a convex combination of $z_1, \ldots, z_{2^n}$, by the convexity of f it follows that

$$f(z) \leq \sum_{j=1}^{2^n} f(z_j) \text{ for every } z \in Q.$$

Because of this, it follows that $\sup_Q f < +\infty$. Consequently, Proposition 1.1.15 and Proposition 1.1.16 apply, and the theorem follows. ∎

Theorem 1.1.18. *Let W be a Banach space, and $F: W \rightarrow]-\infty, +\infty]$ be convex and lower semicontinuous. Assume that $\text{int}(\text{dom}F) \neq \emptyset$. Then F is locally Lipschitz in $\text{int}(\text{dom}F)$.*

Proof. Let $x_0 \in \text{int}(\text{dom}F)$, and, let us set $C = \{x \in W : F(x) \leq F(x_0) + 1\}$. Then C turns out to be convex, and closed.

For every $y \in W$ let us define $f_y: t \in \mathbf{R} \mapsto F(x_0 + t(y - x_0))$. Then, f_y turns out to be convex, and, since $x_0 \in \text{int}(\text{dom}F)$, it results that $0 \in \text{int}(\text{dom}f_y)$. Moreover, by Theorem 1.1.17, f_y turns out to be continuous in 0, and we have proved that

for every $y \in W$ there exists $\varepsilon_y > 0$ such that

$$\{x_0 + t(y - x_0) : t \in]-\varepsilon_y, \varepsilon_y[\} \subseteq C.$$

Because of this, we have that $W = \cup_{h \in \mathbf{N}} x_0 + h(C - x_0)$, where, for every $h \in \mathbf{N}$, $x_0 + h(C - x_0)$ is closed. Consequently, by the Baire Category Theorem, there must be $h_0 \in \mathbf{N}$ such that $\text{int}(x_0 + h_0(C - x_0)) \neq \emptyset$, from which we conclude that $\text{int}(C) \neq \emptyset$.

In conclusion, since obviously $\sup_{\text{int}(C)} F < +\infty$, the proof follows from Proposition 1.1.16. ∎

Finally, we introduce recession functions.

To do this, we first recall that $g: \mathbf{R}^n \rightarrow [-\infty, +\infty]$ is said to be *positively 1-homogeneous* if $g(0) = 0$, and $g(tz) = tg(z)$ for every $z \in \mathbf{R}^n$ and $t > 0$.

Let $f: \mathbf{R}^n \rightarrow]-\infty, +\infty]$ be convex with $\text{dom}f \neq \emptyset$, and $z_0 \in \text{dom}f$. Then it is well known that, due to the convexity of f, for every $z \in \mathbf{R}^n$ the function $t \in]0, +\infty[\mapsto \frac{f(z_0+tz)-f(z_0)}{t}$ is increasing. Consequently, the limit $\lim_{t \to +\infty} \frac{1}{t}f(z_0 + tz)$ exists for every $z \in \mathbf{R}^n$, and we define the *recession function* f^∞ of f by

$$f^\infty: z \in \mathbf{R}^n \mapsto \lim_{t \to +\infty} \frac{f(z_0 + tz) - f(z_0)}{t}.$$

In some sense, the recession function of f describes the growth speed at infinity of f. In particular, it is obvious that if $\lim_{z \to \infty} \frac{f(z_0+z)}{|z|} = +\infty$, then

$$f^\infty(z) = \begin{cases} 0 & \text{if } z = 0 \\ +\infty & \text{if } z \neq 0. \end{cases}$$

In the following result the main properties of recession functions are summarized.

Proposition 1.1.19. *Let $f: \mathbf{R}^n \rightarrow]-\infty, +\infty]$ be convex with $\text{dom}f \neq \emptyset$, $z_0 \in \text{dom}f$, and let f^∞ be the recession function of f. Then f^∞ is convex, and positively 1-homogeneous.*

If, in addition, f is also lower semicontinuous, then f^∞ is independent of z_0, and is lower semicontinuous.

Proof. By the convexity of f it follows that

$$f^\infty(\lambda z_1 + (1-\lambda)z_2) = \lim_{t\to+\infty} \frac{f(\lambda(z_0+tz_1)+(1-\lambda)(z_0+tz_2)) - f(z_0)}{t} \le$$

$$\le \lambda \lim_{t\to+\infty} \frac{f(z_0+tz_1)-f(z_0)}{t} + (1-\lambda) \lim_{t\to+\infty} \frac{f(z_0+tz_2)-f(z_0)}{t} =$$

$$= \lambda f^\infty(z_1) + (1-\lambda)f^\infty(z_2) \text{ for every } z_1,\ z_2 \in \mathbf{R}^n,\ \lambda \in [0,1],$$

and

$$f^\infty(\lambda z) = \lambda \lim_{t\to+\infty} \frac{f(z_0+\lambda tz)-f(z_0)}{\lambda t} = \lambda \lim_{s\to+\infty} \frac{f(z_0+sz)-f(z_0)}{s}$$

$$\text{for every } z \in \mathbf{R}^n,\ \lambda > 0,$$

from which, once we observe that $f^\infty(0) = 0$, the first part of the proposition follows.

If now f is also lower semicontinuous, by Theorem 1.1.11 there exist $\{a_i\}_{i\in\mathcal{I}} \subseteq \mathbf{R}^n$ and $\{b_i\}_{i\in\mathcal{I}} \subseteq \mathbf{R}$ such that $a_i \cdot z + b_i \le f(z)$ for every $i \in \mathcal{I}$ and $z \in \mathbf{R}^n$, and $\sup_{i\in\mathcal{I}} a_i \cdot z + b_i = f(z)$ for every $z \in \mathbf{R}^n$. Consequently, it turns out that

$$f^\infty(z) = \sup_{t>0} \frac{f(z_0+tz)-f(z_0)}{t} = \sup_{t>0}\sup_{i\in\mathcal{I}} \frac{a_i \cdot z_0 + ta_i \cdot z + b_i - f(z_0)}{t} =$$

$$= \sup_{i\in\mathcal{I}} \left\{ a_i \cdot z + \sup_{t>0} \frac{a_i \cdot z_0 + b_i - f(z_0)}{t} \right\} = \sup_{i\in\mathcal{I}} a_i \cdot z \text{ for every } z \in \mathbf{R}^n,$$

from which also the last part of the proposition follows. ∎

§1.2 Convex and Lower Semicontinuous Envelopes in $\mathbf{R}^n$

For every $S \subseteq \mathbf{R}^n$ we denote by co(S) the *convex hull* of S, i.e. the intersection of all the convex subsets of $\mathbf{R}^n$ containing S. It is clear that co(S) is the smallest convex set containing S.

For example, a closed cube of $\mathbf{R}^n$ is the convex hull of its vertices.

If $k \le n$, and $x_0, x_1, \ldots, x_k$ are $k+1$ points in $\mathbf{R}^n$ such that the vectors $x_1 - x_0, \ldots, x_k - x_0$ are linearly independent, then the *k-simplex* with vertices $x_0, x_1, \ldots, x_k$ is the convex hull of the points $x_0, x_1, \ldots, x_k$.

It is easy to prove that every k-simplex is closed, and that every n-simplex has nonempty interior.

Actually, if $S = \text{co}(\{x_0, x_1, \ldots, x_n\})$ is an n-simplex, then a point $x \in \text{int}(S)$ if and only if $x = \sum_{j=0}^n t_j x_j$ where $t_1, \ldots, t_{n+1} \in]0, +\infty[$, and $\sum_{j=0}^n t_j = 1$. Consequently, it is easy to verify that $\partial S = \cup_{j=0}^n \text{co}(\{x_0, \ldots, x_{j-1}, x_{j+1}, \ldots, x_n\})$, and therefore that ∂S is made up by $n+1$ $(n-1)$-simplexes.

The structure of the convex hull of a set is described by Carathéodory's theorem.

Theorem 1.2.1 (Carathéodory's Theorem). *Let $S \subseteq \mathbf{R}^n$ be nonempty. Then every point of $\mathrm{co}(S)$ can be expressed as a convex combination of at most $n+1$ points of S.*

Proof. First of all, let us prove that

$$(1.2.1) \qquad \mathrm{co}(S) = \left\{ \sum_{j=1}^{m} t_j x_j : m \in \mathbf{N}, \right.$$

$$\left. x_j \in S,\ t_j \in [0, +\infty[\text{ for every } j \in \{1, \dots, m\},\ \sum_{j=1}^{m} t_j = 1 \right\}.$$

To see this, let us denote by Σ the right-hand side of (1.2.1). Then it is easy to verify that Σ is convex, and that $S \subseteq \Sigma$, from which it follows that $\mathrm{co}(S) \subseteq \Sigma$.

Conversely, again by the convexity of Σ, it follows that every convex subset of $\mathbf{R}^n$ containing S must necessarily contain Σ too, that is $\Sigma \subseteq \mathrm{co}(S)$. This concludes the proof of (1.2.1).

Let now $x \in \mathrm{co}(S)$. Then (1.2.1) yields $m \in \mathbf{N}$, $x_1 \dots, x_m \in S$, and $t_1 \dots, t_m \in [0, +\infty[$ satisfying $\sum_{j=1}^{m} t_j = 1$ such that $x = \sum_{j=1}^{m} t_j x_j$.

If $m = n+1$ the proof is complete.

If $m < n+1$ the theorem follows by choosing additional arbitrary points $x_{m+1}, \dots, x_{n+1} \in S$, and $t_{m+1} = \dots = t_{n+1} = 0$ to get that $x = \sum_{j=1}^{n+1} t_j x_j$.

If $m > n+1$, the points $x_2 - x_1, \dots, x_m - x_1$ are linearly dependent, and we can find $s_2', \dots, s_m' \in \mathbf{R}$, not all equal to 0, such that $s_2'(x_2 - x_1) + \dots + s_m'(x_m - x_1) = 0$. Consequently, there exist $s_1, \dots, s_m \in \mathbf{R}$, not all equal to 0 and verifying $\sum_{j=1}^{m} s_j = 0$, such that $\sum_{j=1}^{m} s_j x_j = 0$, and

$$(1.2.2) \qquad x = \sum_{j=1}^{m} t_j x_j = \sum_{j=1}^{m} t_j x_j - c \sum_{j=1}^{m} s_j x_j = \sum_{j=1}^{m} (t_j - c s_j) x_j$$

for every $c \in \mathbf{R}$.

In particular, since $s_i \neq 0$ for some $i \in \{1, \dots, m\}$ and $\sum_{j=1}^{m} s_j = 0$, there exists $i \in \{1, \dots, m\}$ such that $s_i > 0$. Therefore, by taking $c = \min\{\frac{t_j}{s_j} : j \in \{1, \dots, m\} \text{ such that } s_j > 0\}$ in (1.2.2), say for example $c = \frac{t_1}{s_1}$, it follows that $t_j - c s_j \in [0, +\infty[$ for every $j \in \{1 \dots, m\}$, $\sum_{j=2}^{m}(t_j - c s_j) = 1$, and $x = \sum_{j=2}^{m}(t_j - c s_j)x_j$. We have thus expressed x as a convex combinations of $m - 1$ points of S.

By iterating such argument $m - n - 1$ times, we arrive to express x as a convex combinations of $n + 1$ points of S, thus getting the theorem. ∎

Remark 1.2.2. Carathéodory's theorem can be improved if $S \subseteq \mathbf{R}^n$ is nonempty and connected. In fact it can be proved that in this case the elements of co(S) can be expressed as convex combinations of n points of S (cf. for example [RW, 2.29 Theorem]).

We now introduce some types of envelopes of functions.

For every $f: \mathbf{R}^n \to [-\infty, +\infty]$ we denote by $\mathrm{co}f$ the *convex envelope* of f, i.e. the function

$$\mathrm{co}f: z \in \mathbf{R}^n \mapsto \sup\{\phi(z) : \phi: \mathbf{R}^n \to [-\infty, +\infty] \text{ convex}, \ \phi \le f \text{ in } \mathbf{R}^n\}$$

It is clear that $\mathrm{co}f$ turns out to be convex, and that

$$(1.2.3) \qquad\qquad \mathrm{co}f(z) \le f(z) \text{ for every } z \in \mathbf{R}^n.$$

It is clear that, if $S \subseteq \mathbf{R}^n$, then $\mathrm{co}I_S = I_{\mathrm{co}(S)}$.

Proposition 1.2.3. *Let* $f: \mathbf{R}^n \to [-\infty, +\infty]$. *Then*

$$\mathrm{co}f(z) = \inf\{\lambda \in \mathbf{R} : (z, \lambda) \in \mathrm{co}(\mathrm{epi}f)\} \text{ for every } z \in \mathbf{R}^n.$$

Proof. For the sake of simplicity, let us set

$$i: z \in \mathbf{R}^n \mapsto \inf\{\lambda \in \mathbf{R} : (z, \lambda) \in \mathrm{co}(\mathrm{epi}f)\}.$$

By exploiting the convexity of co(epi$f$), it is easy to verify that i is convex. Moreover, since obviously $i(z) \le f(z)$ for every $z \in \mathrm{dom}f$, we immediately deduce that $i(z) \le f(z)$ for every $z \in \mathbf{R}^n$. Consequently

$$(1.2.4) \qquad\qquad i(z) \le \mathrm{co}f(z) \text{ for every } z \in \mathbf{R}^n.$$

Conversely, if $\phi: \mathbf{R}^n \to [-\infty, +\infty]$ is convex, and $\phi \le f$ in $\mathbf{R}^n$, then epi$f \subseteq$ epiϕ and, being this last set convex, co(epi$f$) $\subseteq$ epiϕ. Consequently,

$$\phi(z) = \inf\{\lambda \in \mathbf{R} : (z, \lambda) \in \mathrm{epi}\phi\} \le i(z) \text{ for every } z \in \mathrm{dom}\phi,$$

from which, once we observe that dom$i \subseteq$ domϕ, it follows that

$$(1.2.5) \qquad\qquad \mathrm{co}f(z) \le i(z) \text{ for every } z \in \mathbf{R}^n.$$

By (1.2.4), and (1.2.5) the proof follows. ∎

By Proposition 1.2.3 it follows that for every $f: \mathbf{R}^n \to [-\infty, +\infty]$ it results

$$(1.2.6) \qquad\qquad \mathrm{dom}(\mathrm{co}f) = \mathrm{co}(\mathrm{dom}f).$$

Proposition 1.2.4 below yields also information about epigraphs of convex envelopes.

Proposition 1.2.4. *Let* $f : \mathbf{R}^n \to [-\infty, +\infty]$. *Then*

$$\mathrm{co}(\mathrm{epi} f) \subseteq \mathrm{epi}(\mathrm{co} f) \subseteq \overline{\mathrm{co}(\mathrm{epi} f)}.$$

Proof. Since $\mathrm{epi}(\mathrm{co} f)$ is convex and contains $\mathrm{epi} f$, it turns out that $\mathrm{co}(\mathrm{epi} f) \subseteq \mathrm{epi}(\mathrm{co} f)$.

Let now $(z, \lambda) \in \mathrm{epi}(\mathrm{co} f)$. Then, by Proposition 1.2.3, it follows that for every $\varepsilon > 0$, there exists $\lambda_\varepsilon \in \,]\mathrm{co} f(z), \lambda + \varepsilon[$ such that $(z, \lambda_\varepsilon) \in \mathrm{co}(\mathrm{epi} f)$, that is $(z, \lambda) \in \overline{\mathrm{co}(\mathrm{epi} f)}$. Consequently, $\mathrm{epi}(\mathrm{co} f) \subseteq \overline{\mathrm{co}(\mathrm{epi} f)}$, and the proof follows. ∎

Remark 1.2.5. We remark that, in spite of Proposition 1.2.4, it is not true, in general, that for a given $f : \mathbf{R}^n \to [-\infty, +\infty]$, $\mathrm{epi}(\mathrm{co} f) = \mathrm{co}(\mathrm{epi} f)$, as it can be easily checked by considering $f : z \in \mathbf{R}^n \mapsto \begin{cases} |z| & \text{if } z \neq 0 \\ 1 & \text{if } z = 0 \end{cases}$, for which $\mathrm{epi}(\mathrm{co} f) = \{(z, \lambda) \in \mathbf{R}^2 : \lambda \geq |z|\}$, whilst $\mathrm{co}(\mathrm{epi} f) = \{(z, \lambda) \in \mathbf{R}^2 : \lambda \geq |z|\} \setminus \{(0, 0)\}$.

By Carathéodory's theorem we infer the following representation result for convex envelopes.

Theorem 1.2.6. *Let* $f : \mathbf{R}^n \to [-\infty, +\infty]$. *Then*

$$\mathrm{co} f(z) =$$

$$= \inf \left\{ \sum_{j=1}^{n+1} t_j f(z_j) : z_j \in \mathbf{R}^n, \ t_j \in [0, +\infty[\text{ for every } j \in \{1, \ldots, n+1\}, \right.$$

$$\left. \sum_{j=1}^{n+1} t_j = 1, \ \sum_{j=1}^{n+1} t_j z_j = z \right\} \text{ for every } z \in \mathbf{R}^n.$$

Proof. Let $z \in \mathrm{co}(\mathrm{dom} f)$. Then (1.2.6) yields $\mathrm{co} f(z) < +\infty$, from which, by using also Proposition 1.2.3, it follows that $\{\lambda \in \mathbf{R} : (z, \lambda) \in \mathrm{epi} f\} \neq \emptyset$. Let $\lambda \in \mathbf{R}$ be such that $(z, \lambda) \in \mathrm{co}(\mathrm{epi} f)$. Let us prove that there exist $(z_1, \lambda_1), \ldots, (z_{n+1}, \lambda_{n+1}) \in \mathrm{co}(\mathrm{epi} f)$, and $s_1, \ldots, s_{n+1} \in [0, +\infty[$ with $\sum_{j=1}^{n+1} s_j = 1$ such that $z = \sum_{j=1}^{n+1} s_j z_j$, and $\lambda \geq \sum_{j=1}^{n+1} s_j \lambda_j$.

By Carathéodory's theorem applied to $\mathrm{epi} f$ we get that (z, λ) can be expressed as a convex combination of $n + 2$ points of $\mathrm{epi} f$, say $(z_1, \lambda_1), \ldots, (z_{n+2}, \lambda_{n+2})$. Let S be the convex hull of $\{(z_1, \lambda_1), \ldots, (z_{n+2}, \lambda_{n+2})\}$. Then it may occur that $(z, \lambda) \in \partial S$, or that $\mathrm{int}(S) \neq \emptyset$ and $(z, \lambda) \in \mathrm{int}(S)$.

If $(z, \lambda) \in \partial S$, and S is an $(n + 1)$-simplex, then, once we recall that ∂S is made up by $n + 2$ n-simplexes, we obtain that (z, λ) belongs to one of these. Consequently, (z, λ) turns out to be a convex combination of at most $n + 1$ points of $\{(z_1, \lambda_1), \ldots, (z_{n+2}, \lambda_{n+2})\}$, say for example $(z_1, \lambda_1), \ldots, (z_{n+1}, \lambda_{n+1})$, from which we deduce the existence of

$s_1, \ldots, s_{n+1} \in [0, +\infty[$ with $\sum_{j=1}^{n+1} s_j = 1$ such that $z = \sum_{j=1}^{n+1} s_j z_j$, and $\lambda = \sum_{j=1}^{n+1} s_j \lambda_j$.

If $(z, \lambda) \in \partial S$, and S is not an $(n+1)$-simplex, then the vectors $(z_2, \lambda_2) - (z_1, \lambda_1), \ldots, (z_{n+2}, \lambda_{n+2}) - (z_1, \lambda_1)$ are not linearly independent. Therefore, by using an argument similar to the one exploited in the proof of Carathéodory's theorem, we infer that (z, λ) can be expressed as a convex combination of $k + 1$ vectors of $\{(z_1, \lambda_1), \ldots, (z_{n+2}, \lambda_{n+2})\}$, where k is the dimension of $\mathrm{aff}(\{(z_1, \lambda_1), \ldots, (z_{n+2}, \lambda_{n+2})\})$, and $k < n+1$. Because of this, the same above conclusion holds also in this case.

On the other side, if $\mathrm{int}(S) \neq \emptyset$ and $(z, \lambda) \in \mathrm{int}(S)$, the line (in $\mathbf{R}^{n+1}$) through (z, λ) orthogonal to the hyperplane $\lambda = 0$ meets ∂S in two points $(z, \overline{\lambda}_1), (z, \overline{\lambda}_2)$ with $\overline{\lambda}_1 < \lambda < \overline{\lambda}_2$. Consequently, since $(z, \overline{\lambda}_1) \in \partial S$, by the previously considered case there exist $n+1$ points of $\{(z_1, \lambda_1), \ldots, (z_{n+2}, \lambda_{n+2})\}$, say for example $(z_1, \lambda_1), \ldots, (z_{n+1}, \lambda_{n+1})$, and $s_1, \ldots, s_{n+1} \in [0, +\infty[$ with $\sum_{j=1}^{n+1} s_j = 1$ such that $z = \sum_{j=1}^{n+1} s_j z_j$ and $\lambda > \overline{\lambda}_1 = \sum_{j=1}^{n+1} s_j \lambda_j$.

In conclusion, from what we have already proved, and (1.2.3) we get that for every $\lambda \in \mathbf{R}$ such that $(z, \lambda) \in \mathrm{co}(\mathrm{epi} f)$ it results

$$\lambda \geq \sum_{j=1}^{n+1} s_j \lambda_j > \sum_{j=1}^{n+1} s_j f(z_j) \geq$$

$$\geq \inf \left\{ \sum_{j=1}^{n+1} t_j f(z_j) : t_j \in [0, +\infty[\text{ for every } j \in \{1, \ldots, n+1\}, \right.$$

$$\left. \sum_{j=1}^{n+1} t_j = 1, \ \sum_{j=1}^{n+1} t_j z_j = z \right\} \geq$$

$$\geq \inf \left\{ \sum_{j=1}^{n+1} t_j \mathrm{co} f(z_j) : t_j \in [0, +\infty[\text{ for every } j \in \{1, \ldots, n+1\}, \right.$$

$$\left. \sum_{j=1}^{n+1} t_j = 1, \ \sum_{j=1}^{n+1} t_j z_j = z \right\} \geq \mathrm{co} f(z),$$

from which, together with Proposition 1.2.3, the proof follows when $z \in \mathrm{co}(\mathrm{dom} f)$.

If now $z \notin \mathrm{co}(\mathrm{dom} f)$, then (1.2.6) implies that $\mathrm{co} f(z) = +\infty$. On the other side, let us observe that, for every $z_1, \ldots, z_{n+1} \in \mathbf{R}^n$, $t_1, \ldots, t_{n+1} \in [0, +\infty[$ such that $\sum_{j=1}^{n+1} t_j = 1$, $\sum_{j=1}^{n+1} t_j z_j = z$, it cannot be $f(z_j) < +\infty$ for every $j \in \{1, \ldots, n+1\}$, otherwise z would be in $\mathrm{co}(\mathrm{dom} f)$. Consequently $z_j \notin \mathrm{dom} f$ for some $j \in \{1, \ldots, n+1\}$, and the proof follows also in this case. ∎

We now introduce lower semicontinuous envelopes.

For every $f: \mathbf{R}^n \to [-\infty, +\infty]$ we denote by $\mathrm{sc}^- f$ the *lower semicontinuous envelope* of f, i.e. the function

$$\mathrm{sc}^- f: z \in \mathbf{R}^n \mapsto$$

$$\sup\{\phi(z) : \phi: \mathbf{R}^n \to [-\infty, +\infty] \text{ lower semicontinuous, } \phi \leq f \text{ in } \mathbf{R}^n\}.$$

It is clear that $\mathrm{sc}^- f$ turns out to be lower semicontinuous, and that

$$(1.2.7) \qquad \mathrm{sc}^- f(z) \leq f(z) \text{ for every } z \in \mathbf{R}^n.$$

Moreover, it is easy to verify that

$$(1.2.8) \qquad \mathrm{sc}^- f(z) = \liminf_{y \to z} f(y) \text{ for every } z \in \mathbf{R}^n,$$

from which it follows that

$$(1.2.9) \qquad \mathrm{dom} f \subseteq \mathrm{dom}(\mathrm{sc}^- f) \subseteq \overline{\mathrm{dom} f}.$$

Given $S \subseteq \mathbf{R}^n$, it results that $\mathrm{sc}^- I_S = I_{\overline{S}}$.

Proposition 1.2.7. *Let $f: \mathbf{R}^n \to [-\infty, +\infty]$. Then*

$$\mathrm{sc}^- f(z) = \inf\{\lambda \in \mathbf{R} : (z, \lambda) \in \overline{\mathrm{epi} f}\} \text{ for every } z \in \mathbf{R}^n.$$

Proof. For the sake of simplicity, let us set

$$j: z \in \mathbf{R}^n \mapsto \inf\{\lambda \in \mathbf{R} : (z, \lambda) \in \overline{\mathrm{epi} f}\}.$$

If $\phi: \mathbf{R}^n \to [-\infty, +\infty]$ is lower semicontinuous, and $\phi \leq f$ in $\mathbf{R}^n$, then $\mathrm{epi} f \subseteq \mathrm{epi} \phi$ and, being this last set closed by Proposition 1.1.9, $\overline{\mathrm{epi} f} \subseteq \mathrm{epi} \phi$. Consequently,

$$\phi(z) = \inf\{\lambda \in \mathbf{R} : (z, \lambda) \in \mathrm{epi} \phi\} \leq j(z) \text{ for every } z \in \mathrm{dom} \phi,$$

from which, once we observe that $\mathrm{dom} j \subseteq \mathrm{dom} \phi$, it follows that

$$(1.2.10) \qquad \mathrm{sc}^- f(z) \leq j(z) \text{ for every } z \in \mathbf{R}^n.$$

Let us now prove that j is lower semicontinuous. To do this, we take $z \in \mathbf{R}^n$, $\{z_h\} \subseteq \mathbf{R}^n$ with $z_h \to z$, and observe that, possibly passing to subsequences, it is not restrictive to assume that the limit $\lim_{h \to +\infty} j(z_h)$ exists and is in $[-\infty, +\infty[$. Call λ such limit and let, for every $h \in \mathbf{N}$, $j_h \in \mathbf{R}$ be such that $j(z_h) < j_h$, and $\lim_{h \to +\infty} j_h = \lambda$. Then, for every $h \in \mathbf{N}$, there exists $\lambda_h \in \mathbf{R}$ such that $(z_h, \lambda_h) \in \mathrm{epi} f$, and $j(z_h) \leq \lambda_h < j_h$. Consequently $(z, \lambda) \in \overline{\mathrm{epi} f}$, and $j(z) \leq \lambda$, i.e. j is lower semicontinuous.

In addition, since clearly $j(z) \leq f(z)$ for every $z \in \mathrm{dom} f$, we conclude that $j(z) \leq f(z)$ for every $z \in \mathbf{R}^n$, and therefore that

$$(1.2.11) \qquad j(z) \leq \mathrm{sc}^- f(z) \text{ for every } z \in \mathbf{R}^n.$$

By (1.2.10), and (1.2.11) the proof follows. ∎

By Proposition 1.2.7 we deduce the following corollary.

Proposition 1.2.8. Let $f: \mathbf{R}^n \to [-\infty, +\infty]$. Then

$$\text{epi}(\text{sc}^- f) = \overline{\text{epi} f}.$$

Proof. Let $(z, \lambda) \in \overline{\text{epi} f}$. Then $\lambda \geq \inf\{\mu \in \mathbf{R} : (z, \mu) \in \overline{\text{epi} f}\}$ and by Proposition 1.2.7, it turns out that $\lambda \geq \text{sc}^- f(z)$, i.e.

$$(1.2.12) \qquad\qquad \overline{\text{epi} f} \subseteq \text{epi}(\text{sc}^- f).$$

Conversely, let $(z, \lambda) \in \text{epi}(\text{sc}^- f)$. Then $\text{sc}^- f(z) \leq \lambda$ and, for every $\mu > \lambda$ and $\varepsilon > 0$ there exists $z_{\mu,\varepsilon} \in \mathbf{R}^n$ such that $|z_{\mu,\varepsilon} - z| < \varepsilon$, and $f(z_{\mu,\varepsilon}) \leq \mu$. This yields $(z_{\mu,\varepsilon}, \mu) \in \text{epi} f$, from which we conclude that $(z, \lambda) \in \overline{\text{epi} f}$, i.e. that

$$(1.2.13) \qquad\qquad \text{epi}(\text{sc}^- f) \subseteq \overline{\text{epi} f}.$$

By (1.2.12), and (1.2.13) the proof follows. ∎

§1.3 Lower Semicontinuous Envelopes of Convex Envelopes

In the present section we start the study of the composition of convex and lower semicontinuous operators.

First of all we observe that, by using (1.2.8), it is easy to deduce that the lower semicontinuous envelope of a convex function is again convex.

For every $f: \mathbf{R}^n \to [-\infty, +\infty]$ we denote by f^{**} the function defined by

$$(1.3.1) \qquad\qquad f^{**}: z \in \mathbf{R}^n \mapsto$$

$$\sup\{a \cdot z + c : a \in \mathbf{R}^n, \ c \in \mathbf{R}, \ \alpha \cdot \zeta + c \leq f(\zeta) \text{ for every } \zeta \in \mathbf{R}^n\}.$$

It is clear that f^{**} turns out to be convex and lower semicontinuous, and that

$$(1.3.2) \qquad f^{**}(z) \leq \text{co} f(z) \leq f(z) \text{ for every } z \in \mathbf{R}^n.$$

Moreover, by using Theorem 1.1.11, it is easy to prove that

$$(1.3.3) \qquad\qquad f^{**}(z) = \sup\{\phi(z) :$$

$$\phi: \mathbf{R}^n \to [-\infty, +\infty] \text{ convex and lower semicontinuous, } \phi \leq f \text{ in } \mathbf{R}^n\}$$

$$\text{for every } z \in \mathbf{R}^n.$$

Finally, we remark that f^{**} agrees with the bipolar of f (cf. for example [ET, Chapter I, Proposition 4.1]).

The following result provides a description of the structure of the function defined in (1.3.1).

Proposition 1.3.1. *Let $f: \mathbf{R}^n \to [-\infty, +\infty]$. Then*

$$\mathrm{sc}^-(\mathrm{co} f)(z) = f^{**}(z) \text{ for every } z \in \mathbf{R}^n.$$

Proof. Since f^{**} is convex, lower semicontinuous, and $f^{**} \leq f$, it is clear that

$$(1.3.4) \qquad\qquad f^{**}(z) \leq \mathrm{sc}^-(\mathrm{co} f)(z) \text{ for every } z \in \mathbf{R}^n.$$

Analogously, since $\mathrm{sc}^-(\mathrm{co} f)$ too is convex, lower semicontinuous, and $\mathrm{sc}^-(\mathrm{co} f) \leq f$, by (1.3.3) it follows that

$$(1.3.5) \qquad\qquad \mathrm{sc}^-(\mathrm{co} f)(z) \leq f^{**}(z) \text{ for every } z \in \mathbf{R}^n.$$

By (1.3.4), and (1.3.5) the proof follows. ∎

The following result collects some elementary properties of the function defined in (1.3.1).

Proposition 1.3.2. *Let $f: \mathbf{R}^n \to [-\infty, +\infty]$. Then*

$$(1.3.6) \qquad \begin{cases} \mathrm{ri}(\mathrm{dom} f^{**}) = \mathrm{ri}(\mathrm{dom}(\mathrm{co} f)) = \mathrm{ri}(\mathrm{co}(\mathrm{dom} f)), \\ \mathrm{rb}(\mathrm{dom} f^{**}) = \mathrm{rb}(\mathrm{dom}(\mathrm{co} f)) = \mathrm{rb}(\mathrm{co}(\mathrm{dom} f)) \end{cases}$$

and

$$(1.3.7) \qquad f^{**}(z) = \mathrm{co} f(z) \text{ for every } z \in \mathbf{R}^n \setminus \mathrm{rb}(\mathrm{co}(\mathrm{dom} f)),$$

$$(1.3.8) \qquad\qquad f^{**}(z) = \lim_{t \to 1^-} \mathrm{co} f(tz + (1-t)z_0)$$

for every $z \in \mathbf{R}^n$, $z_0 \in \mathrm{ri}(\mathrm{co}(\mathrm{dom} f))$.

Proof. By Proposition 1.3.1, and (1.2.9) we obtain that

$$\mathrm{dom}(\mathrm{co} f) \subseteq \mathrm{dom} f^{**} \subseteq \overline{\mathrm{dom}(\mathrm{co} f)},$$

from which, together with Proposition 1.1.5 and (1.2.6), equalities in (1.3.6) follow.

Equality in (1.3.7) comes from Proposition 1.3.1, and the continuity properties of convex functions (cf. Theorem 1.1.17) from which it follows that $\mathrm{sc}^-(\mathrm{co} f)$ agrees with $\mathrm{co} f$ except perhaps in $\mathrm{rb}(\mathrm{dom}(\mathrm{co} f))$, and from (1.3.6).

Finally, since by the lower semicontinuity and the convexity of f^{**}, and by (1.3.6) it follows that

$$f^{**}(z) \leq \liminf_{t \to 1^-} f^{**}(tz + (1-t)z_0) \leq \limsup_{t \to 1^-} f^{**}(tz + (1-t)z_0) \leq$$

$$\leq \limsup_{t \to 1^-} \{ t f^{**}(z) + (1-t) f^{**}(z_0) \} = f^{**}(z)$$

for every $z \in \mathbf{R}^n$, $z_0 \in \mathrm{ri}(\mathrm{co}(\mathrm{dom} f))$,

equality (1.3.8) follows from (1.3.7), once we observe that for every $z \in \mathbf{R}^n$ and $z_0 \in \mathrm{ri}(\mathrm{co}(\mathrm{dom} f))$, $tz+(1-t)z_0 \in \mathbf{R}^n \setminus \mathrm{rb}(\mathrm{co}(\mathrm{dom} f))$ for every $t \in [0, 1[$ sufficiently close to 1. ∎

In particular, given $f: \mathbf{R}^n \to [-\infty, +\infty]$, by (1.2.6), (1.3.2), and Proposition 1.3.2 we deduce that

(1.3.9) $$\mathrm{co}(\mathrm{dom} f) \subseteq \mathrm{dom} f^{**} \subseteq \overline{\mathrm{co}(\mathrm{dom} f)}.$$

By using Carathéodory's theorem we can prove a representation result for the function defined in (1.3.1), in the same order of ideas of Theorem 1.2.6.

Lemma 1.3.3. *Let* $f: \mathbf{R}^n \to [0, +\infty]$, *and assume that* $\lim_{z \to \infty} \frac{f(z)}{|z|} = +\infty$. *Then there exists* $\vartheta: [0, +\infty[\to [0, +\infty[$ *increasing, convex, and satisfying* $\lim_{t \to +\infty} \vartheta(t)/t = +\infty$ *such that*

$$\vartheta(|z|) \leq f(z) \text{ for every } z \in \mathbf{R}^n.$$

Proof. The assumptions on f yield that for every $k \in \mathbf{N} \cup \{0\}$ we can find $r_k \in [0, +\infty[$ such that $\frac{f(z)}{|z|} \geq k$ for every $z \in \mathbf{R}^n \setminus B_{r_k}(0)$. Moreover, it is not restrictive to assume that $\{r_k\}$ is strictly increasing and diverging.

Because of this, the function

$$\psi: t \in [0, +\infty[\mapsto \sum_{k=1}^{+\infty} (k-1) \chi_{[r_{k-1}, r_k[}(t)$$

turns out to be increasing, finite, and satisfying $\lim_{t \to +\infty} \psi(t) = +\infty$ and

(1.3.10) $$f(z) \geq |z| \psi(|z|) \text{ for every } z \in \mathbf{R}^n.$$

Let

$$\vartheta: t \in [0, +\infty[\mapsto \int_0^t \psi(s) ds.$$

Then ϑ is increasing, finite, convex, and satisfies

(1.3.11) $$\vartheta(t) \leq t\psi(t) \text{ for every } t \in [0, +\infty[.$$

Moreover, since the monotonicity of ψ implies that

$$\liminf_{t \to +\infty} \frac{\vartheta(t)}{t} \geq \lim_{t \to +\infty} \frac{1}{t} \left(\int_0^r \psi(s) ds + (t-r)\psi(r) \right) = \psi(r)$$

for every $r \in [0, +\infty[$,

and $\lim_{t \to +\infty} \psi(t) = +\infty$, we immediately obtain that $\lim_{t \to +\infty} \vartheta(t)/t = +\infty$.

Finally, from (1.3.10), and (1.3.11) we conclude that

$$\vartheta(|z|) \leq f(z) \text{ for every } z \in \mathbf{R}^n,$$

that completes the proof. ∎

Theorem 1.3.4. *Let* $f \colon \mathbf{R}^n \to]-\infty, +\infty]$. *Assume that* f *is bounded from below, and that* $\lim_{z \to \infty} \frac{f(z)}{|z|} = +\infty$. *Then*

$$f^{**}(z) = \min \left\{ \sum_{j=1}^{n+1} t_j \mathrm{sc}^- f(z_j) : z_j \in \mathbf{R}^n, \ t_j \in [0, +\infty[\right.$$

for every $j \in \{1, \ldots, n+1\}$, $\displaystyle\sum_{j=1}^{n+1} t_j = 1$, $\left. \displaystyle\sum_{j=1}^{n+1} t_j z_j = z \right\}$ *for every* $z \in \mathbf{R}^n$.

Proof. First of all, let us observe that, possibly considering $f - \inf_{\mathbf{R}^n} f$, it is not restrictive to assume that $f(z) \geq 0$ for every $z \in \mathbf{R}^n$.

Let us preliminarily prove the theorem under the additional assumption that f is lower semicontinuous, i.e. $f = \mathrm{sc}^- f$.

Let us prove that $\mathrm{co} f$ is lower semicontinuous.

To do this, let $z \in \mathbf{R}^n$, $\{z_h\} \subseteq \mathbf{R}^n$ be such that $z_h \to z$. Let us observe that, since $\mathrm{co} f(\xi) \geq 0$ for every $\xi \in \mathbf{R}^n$, possibly passing to subsequences we can assume that the limit $\lim_{h \to +\infty} \mathrm{co} f(z_h)$ exists and is finite.

By Theorem 1.2.6, for every $h \in \mathbf{N}$ there exist $s_{h,1}, \ldots, s_{h,n+1} \in [0, +\infty[$, with $\sum_{j=1}^{n+1} s_{h,j} = 1$, and $z_{h,1} \ldots, z_{h,n+1} \in \mathbf{R}^n$ satisfying $\sum_{j=1}^{n+1} s_{h,j} z_{h,j} = z_h$ such that the limit $\lim_{h \to +\infty} \sum_{j=1}^{n+1} s_{h,j} f(z_{h,j})$ exists, and

$$(1.3.12) \qquad \lim_{h \to +\infty} \sum_{j=1}^{n+1} s_{h,j} f(z_{h,j}) = \lim_{h \to +\infty} \mathrm{co} f(z_h).$$

Let ϑ be given by Lemma 1.3.3. Then, by (1.3.12), and the finiteness of $\lim_{h \to +\infty} \mathrm{co} f(z_h)$, we infer that

$$(1.3.13) \qquad \limsup_{h \to +\infty} \sum_{j=1}^{n+1} s_{h,j} \vartheta(|z_{h,j}|) < +\infty.$$

Possibly passing to subsequences, we have that for every $j \in \{1, \ldots, n+1\}$, there exists $s_j \in [0, +\infty[$ such that $s_{h,j} \to s_j$, and $\sum_{j=1}^{n+1} s_j = 1$. Moreover, again possibly passing to subsequences, and by setting $I = \{j \in$

$\{1, \ldots, n+1\} : z_{h,j} \to \zeta_j$ for some $\zeta_j \in \mathbf{R}^n\}$ and $J = \{j \in \{1, \ldots, n+1\} :$
$\liminf_{h \to +\infty} |z_{h,j}| = +\infty\}$, we can assume that $I \cup J = \{1, \ldots, n+1\}$.
Therefore, by (1.3.13), and the growth properties of ϑ, it turns out that

$$\limsup_{h \to +\infty} s_{h,i}|z_{h,i}| = \limsup_{h \to +\infty} s_{h,i}\vartheta(|z_{h,i}|) \lim_{h \to +\infty} \frac{|z_{h,i}|}{\vartheta(|z_{h,i}|)} \leq$$

$$\leq \limsup_{h \to +\infty} \sum_{j=1}^{n+1} s_{h,j}\vartheta(|z_{h,j}|) \lim_{h \to +\infty} \frac{|z_{h,i}|}{\vartheta(|z_{h,i}|)} = 0 \text{ for every } i \in J,$$

from which we also conclude that $s_i = 0$ for every $i \in J$.

Because of this, and by setting $\zeta_i = 0$ for every $i \in J$, we deduce that $\sum_{j=1}^{n+1} s_j\zeta_j = z$, from which, together with (1.3.12), the lower semicontinuity of f, and Theorem 1.2.6, we obtain that

$$(1.3.14) \qquad \liminf_{h \to +\infty} \mathrm{co}f(z_h) \geq \liminf_{h \to +\infty} \sum_{j \in I} s_{h,j}f(z_{h,j}) \geq$$

$$\geq \sum_{j \in I} \liminf_{h \to +\infty} s_{h,j}f(z_{h,j}) \geq \sum_{j \in I} s_j f(\zeta_j) = \sum_{j=1}^{n+1} s_j f(\zeta_j) \geq \mathrm{co}f(z),$$

that is the lower semicontinuity of $\mathrm{co}f$.

In particular, (1.3.14) with $z_h = z$ for every $h \in \mathbf{N}$, and Theorem 1.2.6, yield that

$$\inf\left\{ \sum_{j=1}^{n+1} t_j f(z_j) : z_j \in \mathbf{R}^n,\ t_j \in [0, +\infty[\text{ for every } j \in \{1, \ldots, n+1\}, \right.$$

$$\left. \sum_{j=1}^{n+1} t_j = 1,\ \sum_{j=1}^{n+1} t_j z_j = z \right\} = \sum_{j=1}^{n+1} s_j f(\zeta_j),$$

from which it follows that for every $z \in \mathbf{R}^n$ the minimum $\min\{\sum_{j=1}^{n+1} t_j f(z_j)$
$: z_j \in \mathbf{R}^n,\ t_j \in [0, +\infty[\text{ for every } j \in \{1, \ldots, n+1\},\ \sum_{j=1}^{n+1} t_j = 1,$
$\sum_{j=1}^{n+1} t_j z_j = z\}$ is attained.

In conclusion, from what was just proved, Proposition 1.3.1 and Theorem 1.2.6 we obtain that

$$f^{**}(z) = \mathrm{sc}^-(\mathrm{co}f)(z) = \mathrm{co}f(z) = \min\left\{ \sum_{j=1}^{n+1} t_j f(z_j) : z_j \in \mathbf{R}^n,\ t_j \in [0, +\infty[\right.$$

$$\left. \text{for every } j \in \{1, \ldots, n+1\},\ \sum_{j=1}^{n+1} t_j = 1,\ \sum_{j=1}^{n+1} t_j z_j = z \right\} \text{ for every } z \in \mathbf{R}^n,$$

that proves the theorem when f is lower semicontinuous.

In order to treat the general case, let us preliminarily observe that $f^{**} \leq \mathrm{sc}^- f \leq f$ from which, since clearly $(f^{**})^{**} = f^{**}$, it follows that

(1.3.15) $f^{**}(z) = (\mathrm{sc}^- f)^{**}(z)$ for every $z \in \mathbf{R}^n$.

Let us now prove that

(1.3.16) $$\lim_{z \to \infty} \frac{\mathrm{sc}^- f(z)}{|z|} = +\infty.$$

To do this, we know that for every $k \in \mathbf{N}$ there exists $r_k \in [0, +\infty[$ such that $f(z) \geq k|z|$ for every $z \in \mathbf{R}^n \setminus B_{r_k}(0)$. Consequently, since

$$f(z) \geq \begin{cases} 0 & \text{if } z \in B_{r_k}(0) \\ k|z| & \text{if } z \in \mathbf{R}^n \setminus B_{r_k}(0) \end{cases} \quad \text{for every } k \in \mathbf{N},$$

we conclude that

$$\mathrm{sc}^- f(z) \geq \max\{0, k|z| - kr_k\} \text{ for every } k \in \mathbf{N}.$$

Because of this, (1.3.16) follows.

In conclusion, by (1.3.15), (1.3.16), and the previously treated case applied to $\mathrm{sc}^- f$, we obtain that

$$f^{**}(z) = (\mathrm{sc}^- f)^{**}(z) = \min\left\{ \sum_{j=1}^{n+1} t_j \mathrm{sc}^- f(z_j) : z_j \in \mathbf{R}^n, \ t_j \in [0, +\infty[\right.$$

$$\left. \text{for every } j \in \{1,\ldots, n+1\}, \ \sum_{j=1}^{n+1} t_j = 1, \ \sum_{j=1}^{n+1} t_j z_j = z \right\} \text{ for every } z \in \mathbf{R}^n,$$

which proves the theorem. ■

Remark 1.3.5. We observe that Theorem 1.3.4 can be no more true if the boundedness from below condition on f is dropped, as it is verified by considering $f : z \in \mathbf{R} \mapsto \begin{cases} 0 & \text{if } z = 0 \\ \ln|z| & \text{if } 0 < |z| \leq 1. \\ +\infty & \text{if } 1 < |z| \end{cases}$ In this case it turns out that $f^{**}(z) = -\infty$ for every $z \in \mathbf{R}$, whilst $\min\{t\mathrm{sc}^- f(z_1) + (1-t)\mathrm{sc}^- f(z_2) : t \in [0,1], \ z_1, z_2 \in \mathbf{R}, \ tz_1 + (1-t)z_2 = z\} = \begin{cases} -\infty & \text{if } |z| < 1 \\ +\infty & \text{if } 1 \leq |z| \end{cases}$ for every $z \in \mathbf{R}$.

On the other side, Theorem 1.3.4 becomes trivially true if f cannot take the value $+\infty$, and is not bounded from below in the sense that $\mathrm{sc}^- f(z_0) = -\infty$ for some $z_0 \in \mathbf{R}^n$. In fact, in this case, there can be no $a \in \mathbf{R}^n$,

$c \in \mathbf{R}$ such that $\alpha \cdot \zeta + c \leq f(\zeta)$ for every $\zeta \in \mathbf{R}^n$, otherwise it would be $\alpha \cdot \zeta + c \leq \mathrm{sc}^- f(\zeta)$ for every $\zeta \in \mathbf{R}^n$ too. Therefore $f^{**}(z) = -\infty$ for every $z \in \mathbf{R}^n$. Moreover, since for every $z \in \mathbf{R}^n$ we can always take z_0 as one of the vectors z_j in the right-hand side of the claim of Theorem 1.3.4, it turns out that the minimum described there is attained, and equal to $-\infty$.

Remark 1.3.6. We point out that, in particular, the claim of Theorem 1.3.4 holds provided $f: \mathbf{R}^n \to]-\infty, +\infty]$ satisfies $\lim_{z \to \infty} \frac{f(z)}{|z|} = +\infty$, and is lower semicontinuous.

§1.4 Convex Envelopes of Lower Semicontinuous Envelopes

In the present section, given $f: \mathbf{R}^n \to [-\infty, +\infty]$, we carry out the study of $\mathrm{co}(\mathrm{sc}^- f)$ and, in particular, of its relationships with f^{**}.

The following result collects some elementary properties of convex envelopes of the lower semicontinuous envelopes.

Proposition 1.4.1. *Let* $f: \mathbf{R}^n \to [-\infty, +\infty]$. *Then* $\mathrm{co}(\mathrm{sc}^- f)$ *is convex, and*

$$(1.4.1) \qquad f^{**}(z) \leq \mathrm{co}(\mathrm{sc}^- f)(z) \leq \mathrm{co} f(z) \text{ for every } z \subset \mathbf{R}^n,$$

$$(1.4.2) \quad \begin{cases} \mathrm{ri}(\mathrm{dom}(\mathrm{co}(\mathrm{sc}^- f))) = \mathrm{ri}(\mathrm{dom} f^{**}) = \\ \qquad\qquad = \mathrm{ri}(\mathrm{dom}(\mathrm{co} f)) = \mathrm{ri}(\mathrm{co}(\mathrm{dom} f)), \\ \mathrm{rb}(\mathrm{dom}(\mathrm{co}(\mathrm{sc}^- f))) = \mathrm{rb}(\mathrm{dom} f^{**}) = \\ \qquad\qquad = \mathrm{rb}(\mathrm{dom}(\mathrm{co} f)) = \mathrm{rb}(\mathrm{co}(\mathrm{dom} f)), \end{cases}$$

$$(1.4.3) \quad \mathrm{co}(\mathrm{sc}^- f)(z) = f^{**}(z) = \mathrm{co} f(z) \text{ for every } z \in \mathbf{R}^n \setminus \mathrm{rb}(\mathrm{co}(\mathrm{dom} f)).$$

Proof. It is clear that $\mathrm{co}(\mathrm{sc}^- f)$ is convex.

Since obviously $\mathrm{sc}^- f \leq f$, we immediately obtain that

$$(1.4.4) \qquad \mathrm{co}(\mathrm{sc}^- f)(z) \leq \mathrm{co} f(z) \text{ for every } z \in \mathbf{R}^n.$$

On the other side, being f^{**} lower semicontinuous, we have that $f^{**} \leq \mathrm{sc}^- f$, from which, taking into account the convexity of f^{**}, we conclude that

$$(1.4.5) \qquad f^{**}(z) \leq \mathrm{co}(\mathrm{sc}^- f)(z) \text{ for every } z \in \mathbf{R}^n.$$

By (1.4.4) and (1.4.5), inequalities in (1.4.1) follow.

Conditions (1.4.2), and (1.4.3) follow from (1.4.1), and Proposition 1.3.2. ∎

In spite of (1.4.3), the following examples prove that, in general, for a given function f, $co(sc^- f)$ and f^{**} may be different.

Example 1.4.2. Let $n = 2$, and let f be defined by

$$f \colon (z_1, z_2) \in \mathbf{R}^2 \mapsto \begin{cases} z_2 - z_1 e^{z_2} & \text{if } z_2 \geq 0 \text{ and } 0 < z_1 \leq z_2 e^{-z_2} \\ 0 & \text{if } z_2 \geq 0 \text{ and } z_2 e^{-z_2} < z_1 \\ +\infty & \text{otherwise.} \end{cases}$$

Then $\mathrm{dom} f$ is convex, f is upper semicontinuous in $\mathbf{R}^2$ and locally Lipschitz in $\mathrm{dom} f$. Moreover, it is clear that

$$f^{**}(z_1, z_2) = \begin{cases} 0 & \text{if } z_1 \geq 0 \text{ and } z_2 \geq 0 \\ +\infty & \text{otherwise} \end{cases} \quad \text{for every } (z_1, z_2) \in \mathbf{R}^2,$$

whilst it is easy to see that

$$co(sc^- f)(z_1, z_2) = \begin{cases} z_2 & \text{if } z_2 \geq 0 \text{ and } z_1 = 0 \\ 0 & \text{if } z_2 \geq 0 \text{ and } z_1 > 0 \text{ for every } (z_1, z_2) \in \mathbf{R}^2. \\ +\infty & \text{otherwise} \end{cases}$$

Note that in this case $co(sc^- f)$ is not lower semicontinuous.

In the example below we observe that $co(sc^- f)$ and f^{**} can be different also when f is bounded in $\mathrm{dom} f$, and $\mathrm{dom} f$ is very regular.

Example 1.4.3. Let $n = 2$, and let f be defined by

$$f \colon (z_1, z_2) \in \mathbf{R}^2 \mapsto \begin{cases} +\infty & \text{if } z_1 \leq 0 \\ 1 - z_1 e^{z_2^2} & \text{if } 0 < z_1 \leq e^{-z_2^2} \\ 0 & \text{if } z_1 > e^{-z_2^2}, \end{cases}$$

then $\mathrm{dom} f$ is convex, f is bounded and upper semicontinuous in $\mathbf{R}^2$, and locally Lipschitz in $\mathrm{dom} f$. Moreover it is clear that

$$f^{**}(z_1, z_2) = \begin{cases} +\infty & \text{if } z_1 < 0 \\ 0 & \text{if } z_1 \geq 0 \end{cases} \quad \text{for every } (z_1, z_2) \in \mathbf{R}^2,$$

whilst $co(sc^- f)$ is given by

$$co(sc^- f)(z_1, z_2) = \begin{cases} +\infty & \text{if } z_1 < 0 \\ 1 & \text{if } z_1 = 0 \text{ for every } (z_1, z_2) \in \mathbf{R}^2. \\ 0 & \text{if } z_1 > 0 \end{cases}$$

Also in this case $co(sc^- f)$ is not lower semicontinuous.

In spite of the above examples, for a given $f \colon \mathbf{R}^n \to \ ] - \infty, +\infty]$, $co(sc^- f)$ can be constructed from f by means of a suitable use of the ** operator.

To do this, we say that $f \colon \mathbf{R}^n \to \ ] - \infty, +\infty]$ is *locally bounded from below* if for every compact set $K \subseteq \mathbf{R}^n$ there exists $c_K \in \mathbf{R}$ such that $f(z) \geq c_K$ for every $z \in K$.

Proposition 1.4.4. *Let* $f: \mathbf{R}^n \to]-\infty, +\infty]$. *Assume that* f *is locally bounded from below. Then*

$$\mathrm{co}(\mathrm{sc}^- f)(z) = \inf_{m \in \mathbf{N}} (f + I_{Q_m(0)})^{**}(z) \text{ for every } z \in \mathbf{R}^n.$$

Proof. It is clear that

$$(1.4.6) \qquad \inf_{m \in \mathbf{N}} (f + I_{Q_m(0)})^{**}(z) \leq (f + I_{Q_k(0)})^{**}(z) \leq$$

$$\leq \mathrm{sc}^-(f + I_{Q_k(0)})(z) = \mathrm{sc}^- f(z) \text{ for every } z \in \mathbf{R}^n, \ k \in \mathbf{N} \text{ with } z \in Q_k(0).$$

Moreover, being $\{(f + I_{Q_m(0)})^{**}(z)\}$ decreasing for every $z \in \mathbf{R}^n$, the function $\inf_{m \in \mathbf{N}} (f + I_{Q_m(0)})^{**}$ turns out to be convex. Consequently, by (1.4.6) we deduce that

$$(1.4.7) \qquad \inf_{m \in \mathbf{N}} (f + I_{Q_m(0)})^{**}(z) \leq \mathrm{co}(\mathrm{sc}^- f)(z) \text{ for every } z \in \mathbf{R}^n.$$

In order to prove the reverse inequality, we fix $z \in \mathbf{R}^n$ and $m \in \mathbf{N}$, and observe that $f + I_{Q_m(0)}$ is bounded from below. Then by Theorem 1.3.4 applied to $f + I_{Q_m(0)}$, and Theorem 1.2.6, we get $z_1^m, \ldots, z_{n+1}^m \in \mathbf{R}^n$, $t_1^m, \ldots, t_{n+1}^m \in [0, +\infty[$ with $\sum_{j=1}^{n+1} t_j^m = 1$, $\sum_{i=1}^{n+1} t_i^m z_i^m = z$ such that

$$(f + I_{Q_m(0)})^{**}(z) = \sum_{j=1}^{n+1} t_j^m \mathrm{sc}^-(f + I_{Q_m(0)})(z_j^m) \geq \sum_{j=1}^{n+1} t_j^m \mathrm{sc}^- f(z_j^m) \geq$$

$$\geq \mathrm{co}(\mathrm{sc}^- f)(z) \text{ for every } m \in \mathbf{N}.$$

Therefore, as m diverges, we conclude that

$$(1.4.8) \qquad \inf_{m \in \mathbf{N}} (f + I_{Q_m(0)})^{**}(z) \geq \mathrm{co}(\mathrm{sc}^- f)(z) \text{ for every } z \in \mathbf{R}^n.$$

By (1.4.7) and (1.4.8) the proof follows. ∎

Remark 1.4.5. We observe explicitly that Proposition 1.4.4 continues to hold if we replace the cubes considered there with another increasing sequence of sets covering $\mathbf{R}^n$. In particular, it is easy to see that

$$\inf_{m \in \mathbf{N}} (f + I_{Q_m(0)})^{**} = \inf_{m \in \mathbf{N}} (f + I_{z_0 + m(A - z_0)})^{**}$$

whenever $f: \mathbf{R}^n \to [-\infty, +\infty]$, $A \in \mathcal{A}_0$, $z_0 \in A$.

Let $f: \mathbf{R}^n \to [-\infty, +\infty]$. We now propose some conditions in order to have identity between the $\mathrm{co}(\mathrm{sc}^- f)$ and f^{**}.

Remark 1.4.6. Let $f: \mathbf{R}^n \to [-\infty, +\infty]$. Then, by using the convexity of $\text{co}(\text{sc}^- f)$, (1.3.3), and (1.4.1), we deduce that the following conditions are equivalent

$$\text{co}(\text{sc}^- f)(z) = f^{**}(z) \text{ for every } z \in \mathbf{R}^n,$$

$$\text{co}(\text{sc}^- f) \text{ is lower semicontinuous.}$$

Proposition 1.4.7. Let $f: \mathbf{R}^n \to [-\infty, +\infty]$. Assume that $\text{co}(\text{dom} f)$ is an affine set. Then $\text{co}(\text{sc}^- f) = f^{**}$.
  In particular, $\text{co}(\text{sc}^- f) = f^{**}$ if $\text{co}(\text{dom} f) = \mathbf{R}^n$, or if $\text{dom} f = \mathbf{R}^n$.

Proof. By (1.2.6), $\text{co} f$ turns out to be convex and finite in $\text{co}(\text{dom} f)$. Therefore, since our assumptions imply that $\text{co}(\text{dom} f) = \text{ri}(\text{co}(\text{dom} f))$, by using Theorem 1.1.17, $\text{co} f$ turns out to be continuous in $\text{co}(\text{dom} f)$.
  On the other side, our assumptions imply also that $\text{co}(\text{dom} f)$ is closed. This, together with the continuity of $\text{co} f$ in $\text{co}(\text{dom} f)$, yields the lower semicontinuity of $\text{co} f$ on the whole $\mathbf{R}^n$, and hence that $\text{co} f(z) \leq f^{**}(z)$ for every $z \in \mathbf{R}^n$.
  Because of this, and (1.2.7) the first part of the proposition follows.
  The second part follows from the first one, since, by (1.2.3), $\text{dom} f \subseteq \text{co}(\text{dom} f)$. ∎

Proposition 1.4.8. Let $f: \mathbf{R}^n \to]-\infty, +\infty]$. Assume that f is bounded from below, and that $\lim_{z \to \infty} \frac{f(z)}{|z|} = +\infty$. Then $\text{co}(\text{sc}^- f) = f^{**}$.

Proof. By Theorem 1.3.4, and Theorem 1.2.6 we obtain that

$$f^{**}(z) = \min \left\{ \sum_{j=1}^{n+1} t_j \text{sc}^- f(z_j) : z_j \in \mathbf{R}^n, \ t_j \in [0, +\infty[\right.$$

$$\left. \text{for every } j \in \{1, \ldots, n+1\}, \ \sum_{j=1}^{n+1} t_j = 1, \ \sum_{j=1}^{n+1} t_j z_j = z \right\} = \text{co}(\text{sc}^- f)(z)$$

$$\text{for every } z \in \mathbf{R}^n,$$

which proves the proposition. ∎

Remark 1.4.9. Let $f: \mathbf{R}^n \to [-\infty, +\infty]$. Then, by (1.4.3) of Proposition 1.4.1, we deduce that

$$f^{**}(z) = \text{co}(\text{sc}^- f)(z) = \text{co} f(z) \text{ for every } z \in \mathbf{R}^n \setminus \text{rb}(\text{co}(\text{dom} f)),$$

therefore, to prove identity between $\text{co}(\text{sc}^- f)$ and f^{**}, we have to prove only their coincidence in $\text{rb}(\text{co}(\text{dom} f))$.

In the following results, given $f: \mathbf{R}^n \to [-\infty, +\infty]$, we prove that coincidence of $\mathrm{co}(\mathrm{sc}^- f)$ with f^{**} depends, in some cases, only on some geometric properties of $\mathrm{dom} f$. We also characterize the convex subsets of $\mathbf{R}^n$ that are convex hulls of effective domains of functions for which such coincidence holds.

We start with some results of local nature.

Proposition 1.4.10. *Let* $f: \mathbf{R}^n \to \,]-\infty, +\infty]$, *and* $z_0 \in \mathrm{rb}(\mathrm{co}(\mathrm{dom} f))$. *Assume that* f *is locally bounded from below, and that there exists a non-trivial supporting hyperplane to* $\mathrm{co}(\mathrm{dom} f)$ *having bounded intersection with* $\mathrm{rb}(\mathrm{co}(\mathrm{dom} f))$ *and containing* z_0. *Then*

$$\mathrm{co}(\mathrm{sc}^- f)(z_0) = f^{**}(z_0).$$

Proof. Let H be the non-trivial supporting hyperplane to $\mathrm{co}(\mathrm{dom} f)$ having bounded intersection with $\mathrm{rb}(\mathrm{co}(\mathrm{dom} f))$ and containing z_0, Σ be the closed half-space containing $\mathrm{co}(\mathrm{dom} f)$ whose boundary is H, and $r > 0$ be such that

$$(1.4.9) \qquad\qquad H \cap \mathrm{rb}(\mathrm{co}(\mathrm{dom} f)) \subseteq B_r(z_0).$$

Let $m \in \mathbf{N}$ be such that $B_{2r}(z_0) \subseteq Q_m(0)$.

By using the local boundedness from below assumption, let us take an affine function α with $\alpha(z) \leq (f + I_{Q_m(0)})(z)$ for every $z \in \mathbf{R}^n$. Moreover, let $\eta \in \mathbf{R}$ with $\eta < \min\{\alpha(z_0), 0\}$, and, for every $\tau > 0$, let α_τ be an affine function verifying

$$(1.4.10) \quad \begin{cases} \alpha_{\tau_2}(z) < \alpha_{\tau_1}(z) < \alpha(z) \\ \qquad \text{for every } \tau_1, \, \tau_2 \in \,]0, +\infty[\text{ with } \tau_1 < \tau_2, \, z \in \mathrm{int}(\Sigma), \\ \lim_{\tau \to +\infty} \alpha_\tau(z) = -\infty \text{ for every } z \in \mathrm{int}(\Sigma), \\ \alpha_\tau(z) = \alpha(z) \text{ for every } \tau > 0, \, z \in H. \end{cases}$$

Finally, for every $\tau > 0$, let us set $P_\tau = \{z \in \mathbf{R}^n : \alpha_\tau(z) = \eta\}$, and denote by Σ_τ the closed half-space containing z_0 whose boundary is P_τ.

Let us prove that

$$(1.4.11) \quad \text{there exists } \tau_0 > 0 \text{ such that } \Sigma \cap \Sigma_{\tau_0} \cap \overline{\mathrm{co}(\mathrm{dom} f)} \subseteq B_{2r}(z_0).$$

To do this we argue by contradiction. We assume that for every $h \in \mathbf{N}$ there exists $\underline{z_h} \in \Sigma \cap \Sigma_h \cap \overline{\mathrm{co}(\mathrm{dom} f)}$ with $|z_h - z_0| \geq 2r$. Then, by the convexity of $\overline{\mathrm{co}(\mathrm{dom} f)}$, we get that

$$(1.4.12) \qquad \xi_h = z_0 + 2r \frac{z_h - z_0}{|z_h - z_0|} \in \overline{\mathrm{co}(\mathrm{dom} f)} \text{ for every } h \in \mathbf{N}.$$

It is clear that $|\xi_h - z_0| = 2r$, that by (1.4.10) $\lim_{h \to +\infty} \mathrm{dist}(\xi_h, H) = 0$, and that there exist $\{\xi_{h_k}\} \subseteq \{\xi_h\}$ and $\xi \in \mathbf{R}^n$ such that $\lim_{k \to +\infty} \xi_{h_k} = \xi$.

Then, once we observe that $\overline{\text{co}(\text{dom}f)} \cap H = \text{rb}(\text{co}(\text{dom}f)) \cap H$, by (1.4.12) it follows that $\xi \in \text{rb}(\text{co}(\text{dom}f)) \cap H$ and $|\xi - z_0| = 2r$, contrary to (1.4.9).

Let τ_0 be given by (1.4.11). Then, since $f(z) = +\infty$ for every $z \in \mathbf{R}^n \setminus \Sigma$ and $f(z) \geq 0 > \eta$ for every $z \in \Sigma$, it turns out that

$$(1.4.13) \qquad \alpha_{\tau_0}(z) \leq f(z) \text{ for every } z \in (\mathbf{R}^n \setminus \Sigma) \cup (\mathbf{R}^n \setminus \Sigma_{\tau_0}).$$

Moreover, since $B_{2r}(z_0) \subseteq Q_m(0)$, by (1.4.11) we get that $f(z) = +\infty$ for every $z \in (\Sigma \cap \Sigma_{\tau_0}) \setminus Q_m(0)$, and hence, taking into account also (1.4.10), that

$$(1.4.14) \qquad \alpha_{\tau_0}(z) \leq f(z) \text{ for every } z \in \Sigma \cup \Sigma_{\tau_0}.$$

In conclusion, by (1.4.13) and (1.4.14), we have that $\alpha_{\tau_0}(z) \leq f(z)$ for every $z \in \mathbf{R}^n$, from which, together with (1.4.10), we infer that

$$(1.4.15) \qquad \alpha(z_0) = a_{\tau_0}(z_0) \leq f^{**}(z_0).$$

By (1.4.15), since α is a generic affine function with $\alpha \leq f + I_{Q_m(0)}$ on $\mathbf{R}^n$, we conclude that $(f + I_{Q_m(0)})^{**}(z_0) \leq f^{**}(z_0)$ and, by (1.4.1) of Proposition 1.4.1, that

$$f^{**}(z_0) \leq \text{co}(\text{sc}^- f)(z_0) \leq (f + I_{Q_m(0)})^{**}(z_0) \leq f^{**}(z_0),$$

which proves the proposition. ∎

Proposition 1.4.10 can be inverted. To do this, let us first prove the following result.

Lemma 1.4.11. *Let C be a convex subset of $\mathbf{R}^n$, and H be a non-trivial supporting hyperplane to C. Then $H \cap \text{rb}(C)$ is unbounded if and only if $H \cap \text{rb}(C)$ contains a half-line.*

Proof. It is clear that, if $H \cap \text{rb}(C)$ contains a half-line, then $H \cap \text{rb}(C)$ is unbounded.

Conversely, let us assume that $H \cap \text{rb}(C)$ is unbounded, let $z_0 \in H \cap \text{rb}(C)$, and observe that it is not restrictive to assume that $z_0 = 0$.

For every $h \in \mathbf{N}$ there exists $z_h \in H \cap \text{rb}(C)$ with $|z_h| > h$, and set $\xi_h = z_h/|z_h|$. Then, since $0 \in H \cap \overline{C}$, by the convexity of $H \cap \overline{C}$ we deduce that $\xi_h \in H \cap \overline{C}$ for every $h \in \mathbf{N}$. Let $\xi_0 \in \mathbf{R}^n$ be such that $|\xi_0| = 1$ and, up to subsequences, $\xi_h \to \xi_0$. Then, being $H \cap \overline{C}$ closed, we get also that $\xi_0 \in H \cap \overline{C}$.

Let us prove that the half-line $\{t\xi_0 : t \geq 0\}$ is contained in $H \cap \overline{C}$, this will conclude the proof since $H \cap \overline{C} = H \cap \text{rb}(C)$.

Let $t > 0$. Then it is clear that $t\xi_0 \in H$, so we only have to prove that $t\xi_0 \in \overline{C}$.

Let $r > 0$, and take $h \in \mathbf{N}$ be such that $|z_h| > t$, and $\xi_h \in B_{r/(2t)}(\xi_0)$. Then, since $0 \in \overline{C}$, by the convexity of $\overline{C}$ we conclude that $t\xi_h = \frac{t}{|z_h|} z_h \in \overline{C}$, and that $t\xi_h \in B_{r/2}(t\xi_0)$. Because of this, we infer that $B_r(t\xi_0) \cap \mathrm{ri}(C) \neq \emptyset$ for every $r > 0$, i.e. $t\xi_0 \in \overline{C}$. ∎

Proposition 1.4.12. *Let C be a convex subset of $\mathbf{R}^n$, H be a non-trivial supporting hyperplane to C, and assume that $\mathrm{co}(\mathrm{sc}^- f)(z) = f^{**}(z)$ for every $f \colon \mathbf{R}^n \to [0, +\infty]$ with $\mathrm{co}(\mathrm{dom} f) = C$ and every $z \in H \cap \mathrm{rb}(C)$. Then $H \cap \mathrm{rb}(C)$ is bounded.*

Proof. If $n = 1$ the proposition is certainly true since $\mathrm{rb}(C)$ is empty or bounded.

If $n > 1$ let us prove that if $H \cap \mathrm{rb}(C)$ is unbounded, then

(1.4.16)   there exist $f \colon \mathbf{R}^n \to [0, +\infty]$ with $\mathrm{co}(\mathrm{dom} f) = C$,

and $\overline{z} \in H \cap \mathrm{rb}(C)$ such that $\mathrm{co}(\mathrm{sc}^- f)(\overline{z}) \neq f^{**}(\overline{z})$.

To do this let l be the half-line with $l \subseteq H \cap \mathrm{rb}(C)$ given by Lemma 1.4.11, and assume for the moment that $H = \{z \in \mathbf{R}^n : z_1 = 0\}$, $\{z \in \mathbf{R}^n : z_1 = z_2 = \ldots = z_{n-1} = 0, \ z_n \geq -1\} \subseteq l$, and that $C \subseteq \{z \in \mathbf{R}^n : z_1 \geq 0\}$.

As in Example 1.4.2, let f_0 be given by

$$f_0 \colon (y_1, y_2) \in \mathbf{R}^2 \mapsto \begin{cases} y_2 - y_1 e^{y_2} & \text{if } y_2 > 0 \text{ and } 0 \leq y_1 < y_2 e^{-y_2} \\ 0 & \text{if } y_1 \geq \max\{y_2 e^{-y_2}, 0\} \\ +\infty & \text{if } y_1 < 0, \end{cases}$$

and set

$$f \colon (z_1, \ldots, z_n) \in \mathbf{R}^n \mapsto f_0(z_1, z_n) + I_C(z_1, \ldots, z_n).$$

Then $\mathrm{co}(\mathrm{dom} f) = \mathrm{dom} f = C$.

Let $\tilde{z} \in \mathrm{ri}(C)$ with $\tilde{z}_n = 0$, and set $S = \{t\tilde{z} + (1-t)z : z \in l \text{ with } z_n \geq 0, \ t \in \ ]0, 1]\}$. Then it is clear that $S \subseteq \mathrm{ri}(C)$, and hence that

(1.4.17)   for every $z \in S$ there exist $\xi_1, \ \xi_2 \in S$, $\tau \in [0, 1]$ such that

$$z = (1 - \tau)\xi_1 + \tau\xi_2 \text{ and } f(\xi_1) = f(\xi_2) = 0.$$

Therefore, by the convexity of $\mathrm{co} f$, (1.2.3), and (1.4.17), we conclude that

(1.4.18) $\mathrm{co} f(z) \leq (1 - \tau)\mathrm{co} f(\xi_1) + \tau \mathrm{co} f(\xi_2) \leq (1 - \tau)f(\xi_1) + \tau f(\xi_2) = 0,$

$$\text{for every } z \in S$$

and, by Proposition 1.3.1 and (1.4.18), that

(1.4.19) $f^{**}(0, 0, \ldots, 0, z_n) = 0$ for every $z_n > 0$.

Let now $m \in \mathbf{N}$, and observe that the affine function $\alpha_m \colon (z_1, \ldots, z_n) \in \mathbf{R}^n \mapsto z_n - e^{m/2} z_1$ is such that $\alpha_m \leq f_0 + I_{Q_m(0)} \leq f + I_{Q_m(0)}$ on $\mathbf{R}^n$, and that this yields

$$(1.4.20) \qquad \qquad \overline{z}_n = \alpha_m(\overline{z}) \leq (f + I_{Q_m(0)})^{**}(\overline{z})$$

for every $m \in \mathbf{N}$, $\overline{z} \in \{z \in \mathbf{R}^n : z_1 = z_2 = \ldots = z_{n-1} = 0,\ 0 \leq z_n \leq m/2\}$.

In conclusion, by (1.4.19), (1.4.20), and Proposition 1.4.4 we obtain that

$$f^{**}(\overline{z}) < \mathrm{co}(\mathrm{sc}^- f)(\overline{z})$$

for every $\overline{z} \in \{z \in \mathbf{R}^n : z_1 = z_2 = \ldots = z_{n-1} = 0,\ z_n > 0\}$

provided that $H = \{z \in \mathbf{R}^n : z_1 = 0\}$, $\{z \in \mathbf{R}^n : z_1 = z_2 = \ldots = z_{n-1} = 0,\ z_n \geq -1\} \subseteq l$, and that $C \subseteq \{z \in \mathbf{R}^n : z_1 \geq 0\}$.

In order to prove (1.4.16) in the general case, let $A \colon \mathbf{R}^n \to \mathbf{R}^n$ be a one-to-one affine mapping such that $A(H) = \{\zeta \in \mathbf{R}^n : \zeta_1 = 0\}$, $A(l) \supseteq \{\zeta \in \mathbf{R}^n : \zeta_1 = \zeta_2 = \ldots = \zeta_{n-1} = 0,\ \zeta_n \geq -1\}$, and $A(C) \subseteq \{\zeta \in \mathbf{R}^n : \zeta_1 \geq 0\}$. Then, by (1.4.16) in the just considered particular case, we deduce the existence of a function $g \colon \mathbf{R}^n \to [0, +\infty]$ with $\mathrm{co}(\mathrm{dom}\, g) = A(C)$ such that

$$(1.4.21) \qquad \mathrm{co}(\mathrm{sc}^- g)(\overline{\zeta}) > g^{**}(\overline{\zeta}) \text{ for some } \overline{\zeta} \in A(H) \cap A(C),$$

and set $f = g(A(\cdot))$.

By using Theorem 1.2.6 it is not difficult to verify that $\mathrm{co} f(z) = \mathrm{co} g(A(z))$ for every $z \in \mathbf{R}^n$, from which, together with Proposition 1.3.1, we conclude that

$$(1.4.22) \qquad \qquad f^{**}(z) = g^{**}(A(z)) \text{ for every } z \in \mathbf{R}^n.$$

Analogously, for every $m \in \mathbf{N}$, we have that $f + I_{Q_m(0)} = g(A(\cdot)) + I_{A(Q_m(0))}(A(\cdot))$, and therefore that $(f + I_{Q_m(0)})^{**} = (g + I_{A(Q_m(0))})^{**}(A(\cdot))$. Therefore by Proposition 1.4.4, and Remark 1.4.5, we infer that

$$(1.4.23) \qquad \mathrm{co}(\mathrm{sc}^- f)(z) = \inf_{m \in \mathbf{N}} (f + I_{Q_m(0)})^{**}(z) =$$

$$= \inf_{m \in \mathbf{N}} (g + I_{A(Q_m(0))})^{**}(A(z)) = \inf_{m \in \mathbf{N}} (g + I_{Q_m(0)})^{**}(A(z)) = \mathrm{co}(\mathrm{sc}^- g)(A(z))$$

for every $z \in \mathbf{R}^n$.

By (1.4.23), (1.4.21), and (1.4.22) we obtain that

$$\mathrm{co}(\mathrm{sc}^- f)(\overline{z}) > f^{**}(\overline{z}) \text{ for some } \overline{z} \in H \cap C,$$

from which (1.4.16) follows. This completes the proof. ∎

By the previous results we deduce the following characterization of global nature.

Theorem 1.4.13. *Let C be a convex subset of $\mathbf{R}^n$. Then the following conditions are equivalent*

(1.4.24)   *for every $z_0 \in \mathrm{rb}(C)$ there exists a non-trivial supporting*

hyperplane H to C containing z_0 such that $H \cap \mathrm{rb}(C)$ is bounded,

(1.4.25) $\mathrm{co}(\mathrm{sc}^- f) = f^{**}$ *for every $f : \mathbf{R}^n \to]-\infty, +\infty]$*

locally bounded from below, with $\mathrm{co}(\mathrm{dom} f) = C$,

(1.4.26)   *for every non-trivial supporting hyperplane H to C,*

$$H \cap \mathrm{rb}(C) \text{ is bounded.}$$

Proof. Let us prove that $(1.4.24) \Rightarrow (1.4.25) \Rightarrow (1.4.26) \Rightarrow (1.4.24)$.

It is clear that (1.4.24), together with Remark 1.4.9, and Proposition 1.4.10, implies (1.4.25), and that, by Proposition 1.4.12, (1.4.26) follows from (1.4.25).

Finally let $z_0 \in \mathrm{rb}(C)$, and let H be the non-trivial supporting hyperplane to C containing z_0 given by Theorem 1.1.7. Then (1.4.26) yields (1.4.24). ∎

By Theorem 1.4.13 we deduce the following corollaries.

Corollary 1.4.14. *Let $f : \mathbf{R}^n \to]-\infty, +\infty]$ be bounded from below, and assume that $\mathrm{dom} f$ is bounded. Then $\mathrm{co}(\mathrm{sc}^- f) = f^{**}$.*

Proof. Follows by Theorem 1.4.13 once we observe that, if $\mathrm{dom} f$ is bounded, so is also $\mathrm{rb}(\mathrm{co}(\mathrm{dom} f))$. ∎

Let C be a convex set, we recall that C is said to be *strictly convex* if for every $z_1, z_2 \in \mathrm{rb}(C)$ with $z_1 \neq z_2$ and $t \in]0,1[$, it results $tz_1 + (1-t)z_2 \in \mathrm{ri}(C)$ (or, equivalently, if every point of $\mathrm{rb}(C)$ is an extreme point of $\overline{C}$; cf. for example [R, Chapter 18]).

Corollary 1.4.15. *Let $f : \mathbf{R}^n \to]-\infty, +\infty]$ be locally bounded from below. Assume that $\mathrm{co}(\mathrm{dom} f)$ is strictly convex. Then $\mathrm{co}(\mathrm{sc}^- f) = f^{**}$.*

Proof. Follows from Theorem 1.4.13, once we observe that if $\mathrm{co}(\mathrm{dom} f)$ is strictly convex, then for every non-trivial supporting hyperplane H to $\mathrm{co}(\mathrm{dom} f)$, $H \cap \mathrm{rb}(\mathrm{co}(\mathrm{dom} f))$ consists of only one point. ∎

Corollary 1.4.16. *Let $f : \mathbf{R} \to]-\infty, +\infty]$ be locally bounded from below. Then $\mathrm{co}(\mathrm{sc}^- f) = f^{**}$.*

Proof. Let us observe that in one dimension $\mathrm{rb}(\mathrm{co}(\mathrm{dom} f))$ can be empty, or made up by one or two points.

If it is empty, then $\mathrm{co}(\mathrm{dom} f) = \mathbf{R}$, and the corollary follows from Proposition 1.4.7. Otherwise $\mathrm{rb}(\mathrm{co}(\mathrm{dom} f))$ is bounded, and Theorem 1.4.13 applies. ∎

Chapter 2

Elements of Measure
and Increasing Set Functions
Theories

The present chapter is devoted to the treatment of set functions in a measure theoretic framework (cf. for example [Co], [DuS], and [Ru] for general references on the subject).

In the first sections we recall the main concepts and results from measure theory needed in the book, together with the basics of L^p spaces.

The final sections deal mainly with increasing set functions, that are introduced, and whose main properties are established. In particular the notion of inner regular envelope is recalled, and some abstract criteria ensuring the identity of an increasing set function with its inner regular envelope are established. The link between increasing set functions and measure theory is furnished by the De Giorgi-Letta Extension Theorem (cf. [DM2], [DGL]), which is also proved in our setting.

Applications are made to functionals, depending on open sets and functions, that are increasing when the second variable is fixed.

§2.1 Measures and Integrals

Let Ω be a nonempty set. We say that a collection $\mathcal{E}$ of subsets of Ω is a σ-algebra on Ω if

$$\emptyset \in \mathcal{E},$$

$$\Omega \setminus A \in \mathcal{E} \text{ whenever } A \in \mathcal{E},$$

$$\cup_{h \in \mathbf{N}} A_h \in \mathcal{E} \text{ whenever } A_h \in \mathcal{E} \text{ for every } h \in \mathbf{N}.$$

Given a σ-algebra on Ω, we say that the couple $(\Omega, \mathcal{E})$ is a *measure space*.

If Ω is a topological space, we denote by $\mathcal{B}(\Omega)$ the intersection of all the σ-algebras on Ω containing the open subsets of Ω. It turns out that $\mathcal{B}(\Omega)$ is actually the smallest σ-algebra on Ω containing the open subsets of Ω, and is called the *σ-algebra of the Borel subsets of Ω*, and its elements are called *Borel sets*. In this way, $(\Omega, \mathcal{B}(\Omega))$ becomes a measure space, called *Borel measure space*.

Let $(\Omega, \mathcal{E})$ be a measure space. In order to define what we are going to call measures, we introduce the two different notions of positive measure and of real or vector measure, that, even if similar, enjoy different peculiarities, and play different roles.

If $\mu: \mathcal{E} \to [0, +\infty]$, we say that μ is a *positive measure* on $\mathcal{E}$ (or simply a measure if no confusion may occur) if $\mu(\emptyset) = 0$, and μ is *countably additive* in the sense that

$$(2.1.1) \qquad \mu\left(\cup_{h=1}^{+\infty} A_h\right) = \sum_{h=1}^{+\infty} \mu(A_h)$$

whenever $A_1, \ldots, A_h, \ldots \in \mathcal{E}$ are pairwise disjoint.

If $m \in \mathbf{N}$, and $\mu: \mathcal{E} \to \mathbf{R}^m$, we say that μ is a *measure* on $\mathcal{E}$ (or simply a measure if no confusion may occur) if $\mu(\emptyset) = 0$, and μ is countably additive in the sense of (2.1.1). When $m = 1$, we say that μ is a *real measure*, when $m > 1$ we say that μ is a *vector measure*.

We observe that, in the case of measures, the series in (2.1.1) must necessarily converge absolutely since the union in the left-hand side of (2.1.1) does not depend on the order in which the sets $A_1, \ldots, A_h, \ldots$ are listed.

For every (real or vector) measure μ on $\mathcal{E}$, we define the *total variation* $|\mu|$ of μ as the set function defined by

$$|\mu|: A \in \mathcal{E} \mapsto \sup\left\{ \sum_{h=1}^{+\infty} |\mu(A_h)| : A_h \in \mathcal{E} \text{ for every } h \in \mathbf{N}, \right.$$

$$\left. A_1,\ A_2, \ldots,\ A_h, \ldots \text{ pairwise disjoint},\ \cup_{h=1}^{+\infty} A_h = A \right\}.$$

Moreover, if μ is a real measure, we define the *positive part* μ^+ of μ, and the *negative part* μ^- of μ as the set functions defined by

$$\mu^+: A \in \mathcal{E} \mapsto \frac{|\mu|(A) + \mu(A)}{2}, \quad \mu^-: A \in \mathcal{E} \mapsto \frac{|\mu|(A) - \mu(A)}{2}.$$

Then it is well known that the total variation of a measure turns out to be a positive measure taking only finite values, and

$$|\mu(A)| \leq |\mu|(A) \text{ for every } A \in \mathcal{E}.$$

Consequently, so do the positive and negative parts of a real measure, and

$$0 \leq \mu^+(A) \leq |\mu|(A), \ 0 \leq \mu^-(A) \leq |\mu|(A) \text{ for every } A \in \mathcal{E}.$$

It is also well known that the total variation is a norm on the set of the measures on $\mathcal{E}$, and that, once we endow it with the topology induced by $|\cdot|$, this set actually becomes a Banach space.

We say that a positive measure μ is *σ-finite* if $\Omega = \cup_{h \in \mathbf{N}} A_h$, where, for every $h \in \mathbf{N}$, $A_h \in \mathcal{E}$, and $\mu(A_h) < +\infty$.

If $(\Omega_1, \mathcal{E}_1)$, $(\Omega_2, \mathcal{E}_2)$ are measure spaces, the intersection of all the σ-algebras on $\Omega_1 \times \Omega_2$ containing $\{A_1 \times A_2 : A_1 \in \mathcal{E}_1, A_2 \in \mathcal{E}_2\}$ is denoted, with an abuse of notation, by $\mathcal{E}_1 \times \mathcal{E}_2$. It turns out that $\mathcal{E}_1 \times \mathcal{E}_2$ is actually the smallest σ-algebra on $\Omega_1 \times \Omega_2$ containing $\{A_1 \times A_2 : A_1 \in \mathcal{E}_1, A_2 \in \mathcal{E}_2\}$, and is called the *product σ-algebra* of $\mathcal{E}_1$ and $\mathcal{E}_2$. In this way, $(\Omega_1 \times \Omega_2, \mathcal{E}_1 \times \mathcal{E}_2)$ becomes a measure space, called *product measure space* of $(\Omega_1, \mathcal{E}_1)$ and $(\Omega_2, \mathcal{E}_2)$.

If Ω is a topological space, a positive measure (respectively a measure) μ on $\mathcal{B}(\Omega)$ is said to be a *Borel positive measure* (respectively a *Borel measure*) on Ω. A Borel positive measure on Ω that is finite on each compact subset of Ω is said to be a *Radon positive measure* on Ω.

The restriction of Lebesgue measure to $\mathcal{B}(\mathbf{R}^n)$ is the classical example of Radon positive measure on $\mathbf{R}^n$.

For every $E \subseteq \mathbf{R}^n$, and $\delta > 0$ let us set

$$\mathcal{H}_\delta^{n-1}(E) = \frac{\omega_{n-1}}{2^{n-1}} \inf \left\{ \sum_{j=1}^{+\infty} (\mathrm{diam}(E_j))^{n-1} : E \subseteq \cup_{j=1}^{+\infty} E_j, \right.$$

$$\left. \mathrm{diam}(E_j) < \delta \text{ for every } j \in \mathbf{N} \right\},$$

where ω_{n-1} denotes the Lebesgue measure of the unit ball in $\mathbf{R}^{n-1}$.

Then, fixed $E \subseteq \mathbf{R}^n$, it is easy to verify that $\delta \in]0, +\infty[\mapsto \mathcal{H}_\delta^{n-1}(E)$ is increasing, consequently the limit

$$\mathcal{H}^{n-1}(E) = \lim_{\delta \to 0^+} \mathcal{H}_\delta^{n-1}(E)$$

exists and is in $[0, +\infty]$. The value $\mathcal{H}^{n-1}(E)$ is called the $(n-1)$-*dimensional Hausdorff outer measure* of E, and the set function $E \in \mathcal{B}(\mathbf{R}^n) \mapsto \mathcal{H}^{n-1}(E)$ is a Borel positive measure called $(n-1)$-*dimensional Hausdorff measure*, and denoted by $\mathcal{H}^{n-1}$.

Roughly speaking, $\mathcal{H}^{n-1}$ measures "$(n-1)$-dimensional" sets. The meaning, and the analysis of such property is quite elaborate, and goes beyond the scopes of the present book. Nevertheless we recall that $\mathcal{H}^{n-1}(E) =$

$+\infty$ for every $E \in \mathcal{B}(\mathbf{R}^n)$ such that $\text{int}(E) \neq \emptyset$, and that $\mathcal{H}^{n-1}(E)$ agrees with the classical surface area of E provided $E \in \mathcal{B}(\mathbf{R}^n)$ is regular smooth surface.

Because of this, $\mathcal{H}^{n-1}$ turns out to be a Borel positive measure on $\mathbf{R}^n$, but not a Radon positive measure $\mathbf{R}^n$.

We say that a subset S of a topological space Ω is σ-compact if $E = \cup_{h=1}^{+\infty} K_h$ where K_h is compact for every $h \in \mathbf{N}$.

We denote by $\mathcal{M}(\Omega)$ the set of the Borel real measures on Ω, and, consequently, by $(\mathcal{M}(\Omega))^m$ the one of the Borel vector measures on Ω (with values in $\mathbf{R}^m$). Analogously, we define $\mathcal{M}_{\text{loc}}(\Omega) = \cap_{K \text{compact}, K \subseteq \Omega} \mathcal{M}(K)$, i.e. the set of the real valued functions defined in $\mathcal{B}(K)$ and that are in $\mathcal{M}(K)$ for every compact subset K of Ω. The meaning of $(\mathcal{M}_{\text{loc}}(\Omega))^m$ is now obvious. The elements of $(\mathcal{M}_{\text{loc}}(\Omega))^m$ are usually called *Radon measures on Ω*, to be more precise *Radon real measures* if $m = 1$, or *Radon vector measures* if $m > 1$.

We emphasize that $|\mu|(\Omega) < +\infty$ whenever $\mu \in (\mathcal{M}(\Omega))^m$, and that $|\mu|(K) < +\infty$ for every compact subset K of Ω whenever $\mu \in (\mathcal{M}_{\text{loc}}(\Omega))^m$.

It is worth while to remark that Radon measures are not, in general, measures in the sense of the above definition, at least because they are defined on $\cup_{K \text{compact}, K \subseteq \Omega} \mathcal{B}(K)$, that can also not be a σ-algebra. Nevertheless, the following result can be proved.

Proposition 2.1.1. *Let $\mu \in (\mathcal{M}_{\text{loc}}(\Omega))^m$. Then $|\mu|$ can be extended from $\cup_{K \text{compact}, K \subseteq \Omega} \mathcal{B}(K)$ to $\mathcal{B}(\Omega)$, and the resulting set function is a Radon positive measure on Ω.*

If in addition $\sup\{|\mu|(K) : K \text{ compact subset of } \Omega\} < +\infty$, then μ can be extended from $\cup_{K \text{compact}, K \subseteq \Omega} \mathcal{B}(K)$ to $\mathcal{B}(\Omega)$, and the resulting set function is a Borel measure.

Proof. For every $B \in \mathcal{B}(\Omega)$ let us set

$$|\mu|_e(B) = \sup\{|\mu|(B \cap K) : K \text{ compact subset of } \Omega\}.$$

Then it is clear that $|\mu|_e(B) = |\mu|(B)$ for every $B \in \mathcal{B}(\Omega)$ such that $B \subseteq K$ for some compact set K, and that $|\mu|_e(K) < +\infty$ for every compact set K.

If now $\{B_h\} \subseteq \mathcal{B}(\Omega)$ are pairwise disjoint, we have that

$$|\mu|(K \cap \cup_{h=1}^{+\infty} B_h) = \sum_{h=1}^{+\infty} |\mu|(K \cap B_h) \leq \sum_{h=1}^{+\infty} |\mu|_e(B_h)$$

for every compact subset K of Ω,

from which we obtain that

$$(2.1.2) \qquad |\mu|_e(\cup_{h=1}^{+\infty} B_h) \leq \sum_{h=1}^{+\infty} |\mu|_e(B_h).$$

On the other side, it turns out that for every $h \in \mathbf{N}$, and every $\lambda_h \in \mathbf{R}$ with $\lambda_h < |\mu|_e(B_h)$ there exists a compact set K_h such that $\lambda_h < |\mu|(K_h \cap B_h)$. Consequently, for every $m \in \mathbf{N}$, we deduce that

$$\sum_{h=1}^{m} \lambda_h \le \sum_{h=1}^{m} |\mu|(K_h \cap B_h) = |\mu|\left((\cup_{h=1}^{m} K_h) \cap (\cup_{h=1}^{m} B_h)\right) \le |\mu|_e(\cup_{h=1}^{+\infty} B_h),$$

from which, letting first λ_h increase to $|\mu|(B_h)$ for every $h \in \{1, \dots, m\}$, and then m increase to $+\infty$, it follows that

(2.1.3) $$\sum_{h=1}^{+\infty} |\mu|_e(B_h) \le |\mu|_e(\cup_{h=1}^{+\infty} B_h).$$

From (2.1.2), and (2.1.3) we conclude that $|\mu|_e$ is also countably additive.

If now $\sup\{|\mu|(K) : K \text{ compact subset of } \Omega\} < +\infty$, arguments similar to the one above exposed imply that the set functions

$$\mu_e^+ : B \in \mathcal{B}(\Omega) \mapsto \sup\{\mu^+(B \cap K) : K \text{ compact subset of } \Omega\},$$

and

$$\mu_e^- : B \in \mathcal{B}(\Omega) \mapsto \sup\{\mu^-(B \cap K) : K \text{ compact subset of } \Omega\}$$

are in $\mathcal{M}(\Omega)$, and extend respectively μ^+ and μ^-. Therefore $\mu_e^+ - \mu_e^-$ is the desired extension of μ. ∎

If $\mu \in (\mathcal{M}_{\mathrm{loc}}(\Omega))^m$, we will always perform the extension process described in Proposition 2.1.1, and continue to denote with the same symbols $|\mu|$, and μ the Radon positive measure, and the Borel measure given there as extensions of $|\mu|$, and μ. Consequently, given $\mu \in (\mathcal{M}_{\mathrm{loc}}(\Omega))^m$, we can think to $|\mu|$ as to a Radon positive measure, and, provided $\sup\{|\mu|(K) : K \text{ compact subset of } \Omega\} < +\infty$, to μ as to a Borel measure.

We now define integrals.

Let $(\Omega, \mathcal{E})$ be a measure space.

A function $u : \Omega \to \mathbf{R}$ is said to be *simple* if there exist $m \in \mathbf{N}$, $c_1, \dots, c_m \in \mathbf{R}$, and $S_1, \dots, S_m \subseteq \Omega$ pairwise disjoint such that $u(x) = \sum_{j=1}^{m} c_j \chi_{S_j}(x)$ for every $x \in \Omega$. A simple function $u = \sum_{j=1}^{m} c_j \chi_{S_j}$ is said to be *simple $\mathcal{E}$-measurable* if $S_j \in \mathcal{E}$ for every $j \in \{1, \dots, m\}$.

Let now μ be a positive measure on $\mathcal{E}$.

In order to properly define integrals, we assume, as usual in measure theory, that $0 \cdot (+\infty) = 0$.

For every simple $\mathcal{E}$-measurable function $u = \sum_{j=1}^{m} c_j \chi_{S_j}$ we define the integral $\int_\Omega u d\mu$ of u over Ω as

$$\int_\Omega u d\mu = \sum_{j=1}^{m} c_j \mu(S_j).$$

It is well known that such definition is well posed, in the sense that it does not depend on the particular choice of the values c_j and of the sets S_j used to represent u.

If now $u\colon \Omega \to [0,+\infty]$, we define the *integral* $\int_\Omega u d\mu$ of u over Ω as

$$\int_\Omega u d\mu =$$

$$= \sup\left\{\int_\Omega s d\mu : s \text{ simple } \mathcal{E}\text{-measurable}, \ s(x) \le u(x) \text{ for every } x \in \Omega\right\}.$$

When $u\colon \Omega \to [-\infty,+\infty]$, we say that u is *μ-summable* on Ω if $\int_\Omega |u| d\mu < +\infty$.

Again when $u\colon \Omega \to [-\infty,+\infty]$, we say that u is *μ-integrable* on Ω if $\int_\Omega u^+ d\mu < +\infty$ or $\int_\Omega u^- d\mu < +\infty$, where u^+ and u^- are respectively the positive and the negative part of u defined by

$$u^+\colon x \in \Omega \mapsto \max\{u(x),0\} \quad u^-\colon x \in \Omega \mapsto -\min\{u(x),0\}.$$

In this case, we define the *integral* $\int_\Omega u d\mu$ of u over Ω as

$$\int_\Omega u d\mu = \int_\Omega u^+ d\mu - \int_\Omega u^- d\mu.$$

If $u_1,\ldots,u_m$ are μ-integrable, and $u = (u_1,\ldots,u_m)$, we set $\int_\Omega u d\mu = (\int_\Omega u_1 d\mu,\ldots,\int_\Omega u_m d\mu)$.

If now μ is a real measure, and u is $|\mu|$-summable on Ω, we have obviously that $\int_\Omega u d\mu^+ \le \int_\Omega |u| d|\mu| < +\infty$, and $\int_\Omega u d\mu^- \le \int_\Omega |u| d|\mu| < +\infty$. Consequently we can define the *integral* $\int_\Omega u d\mu$ of u over Ω as

$$\int_\Omega u d\mu = \int_\Omega u d\mu^+ - \int_\Omega u d\mu^-.$$

Finally, if $\mu = (\mu_1,\ldots,\mu_m)$ is a vector measure, and u is $|\mu|$-summable on Ω, we set $\int_\Omega u d\mu = (\int_\Omega u d\mu_1,\ldots,\int_\Omega u d\mu_m)$. If $\mu = (\mu_1,\ldots,\mu_m)$ is a measure, and $u = (u_1,\ldots,u_m)$ is such that $|u|$ is $|\mu|$-summable on Ω, we set $\int_\Omega u d\mu = \sum_{j=1}^m \int_\Omega u_j d\mu_j$.

In conclusion, we observe that if μ is a positive measure on $\mathcal{E}$ and u is μ-integrable on Ω, or if μ is a measure on $\mathcal{E}$ and u is $|\mu|$-summable on Ω, then the integral $\int_A u d\mu$ is well defined for every $A \in \mathcal{E}$, and

$$\int_A u d\mu = \int_\Omega u \chi_A d\mu.$$

We point out that, simply by looking at the above definitions, no assumption on the functions involved seems to be needed, and in this setting even some elementary properties of the integral can be proved. For example, the result below follows directly from the definition of integral.

Theorem 2.1.2 (Monotonicity and Additivity of the Integral). *Let*
$(\Omega, \mathcal{E})$ *be a measure space. Then,*
i) if μ is a positive measure on $\mathcal{E}$, it results that

$$\int_\Omega u d\mu \le \int_\Omega v d\mu$$

whenever u, v are μ-integrable on Ω, and $u(x) \le v(x)$ for every $x \in \Omega$,

$$\int_A u d\mu \le \int_B u d\mu$$

whenever $u \colon \Omega \to [0, +\infty]$, and A, $B \in \mathcal{E}$, satisfy $A \subseteq B$,

$$\int_{A \cup B} u d\mu = \int_A u d\mu + \int_B u d\mu$$

whenever $u \colon \Omega \to [0, +\infty]$, and A, $B \in \mathcal{E}$ are disjoint.

ii) if μ is a measure on $\mathcal{E}$, and u is $|\mu|$-summable on Ω, it results that

$$\left| \int_\Omega u d\mu \right| \le \int_\Omega |u| d|\mu|,$$

$$\int_\Omega c u d\mu = c \int_\Omega u d\mu \text{ for every } c \in \mathbf{R},$$

$$\int_{A \cup B} u d\mu = \int_A u d\mu + \int_B u d\mu \text{ whenever } A, B \in \mathcal{E} \text{ are disjoint.}$$

In spite of Theorem 2.1.2, other basic properties of the integral needed in order to deal with a reasonable theory fail to be true if no additional hypotheses are assumed on the functions to be integrated. For example this happens for the linearity property, as it can be easily checked by means of simple examples.

To overcome such difficulties, the notion of measurability of a function is introduced. It provides a quite natural and general tool, that allows the development of a complete and flexible theory of integration provided it is concerned with measurable functions.

Because of this, we will deal mainly with integrals of measurable functions, even if occasionally the integral of non-necessarily measurable ones might be taken into account.

Let $(\Omega, \mathcal{E})$ be a measure space. A function $u \colon \Omega \to [-\infty, +\infty]$ is said to be $\mathcal{E}$-measurable if $u^{-1}(A) \in \mathcal{E}$ for every open set $A \subseteq [-\infty, +\infty]$. It is well known that the measurability property is equivalent to the requirement that $u^{-1}(]\lambda, +\infty]) \in \mathcal{E}$ for every $\lambda \in \mathbf{R}$, as well as $u^{-1}(B) \in \mathcal{E}$ for every $B \in \mathcal{B}([-\infty, +\infty])$. If Ω is a topological space, a $\mathcal{B}(\Omega)$-measurable function is called a *Borel function*.

It is easy to verify that, if $u, v: \Omega \to [-\infty, +\infty]$ are $\mathcal{E}$-measurable, and $f: [-\infty, +\infty] \to [-\infty, +\infty]$ is Borel, then $u + v$, $u \cdot v$, $\frac{u}{v}$, when defined, are $\mathcal{E}$-measurable, as well as $f(u)$, $\max\{u, v\}$, and $\min\{u, v\}$. In particular so is $f(u)$ when f is continuous, and therefore so are $|u|$, $|u|^p$ with $p > 0$, u^+, and u^-.

If $\{u_h\}$ is a sequence of $\mathcal{E}$-measurable functions on Ω, then $\inf_{h \in \mathbf{N}} u_h$, $\sup_{h \in \mathbf{N}} u_h$, $\liminf_{h \to +\infty} u_h$, and $\limsup_{h \to +\infty} u_h$ too are $\mathcal{E}$-measurable. In addition, it is easy to verify that a simple function $u = \sum_{j=1}^m c_j \chi_{S_j}$ is $\mathcal{E}$-measurable if and only if it is simple $\mathcal{E}$-measurable.

Especially when in connection with integration theory, given a positive measure μ on $\mathcal{E}$, *equivalence classes of $\mathcal{E}$-measurable functions* are considered rather than $\mathcal{E}$-measurable ones, being two $\mathcal{E}$-measurable functions u_1 and u_2 defined on Ω equivalent if $\mu(\{x \in \Omega : u_1(x) \neq u_2(x)\}) = 0$. As usual in this setting, equivalence classes of $\mathcal{E}$-measurable functions are then thought as functions defined in Ω up to sets of zero measure.

Such feature suggests the introduction of the expression *μ-almost everywhere* (μ-a.e.) *in Ω*, to express that a given pointwise property holds for every point in $\Omega \setminus N$ with $\mu(N) = 0$.

So, given a sequence $\{u_h\}$ of $\mathcal{E}$-measurable functions on Ω, and a $\mathcal{E}$-measurable function u on Ω, if $\lim_{h \to +\infty} u_h(x) = u(x)$ μ-a.e. in Ω, then we say that $\{u_h\}$ converges to u *μ-almost everywhere in Ω* (μ-a.e. in Ω).

The set of $\mathcal{E}$-measurable functions on Ω can be endowed with a topology that makes it a metric space, and, given a sequence $\{u_h\}$ of $\mathcal{E}$-measurable functions on Ω, and a $\mathcal{E}$-measurable function u on Ω, it turns out that $u_h \to u$ in such topology if and only if

$$\lim_{h \to +\infty} \mu(\{x \in \Omega : |u_h(x) - u(x)| > \varepsilon\}) = 0 \text{ for every } \varepsilon > 0.$$

When this happens, we say that $\{u_h\}$ converges to u *in μ-measure*, or *in measure* if no ambiguity occurs.

Convergence in measure of a sequence of $\mathcal{E}$-measurable functions is strictly linked to its almost everywhere convergence.

Proposition 2.1.3. *Let $(\Omega, \mathcal{E})$ be a measure space, and μ a positive measure on $\mathcal{E}$. Let $u_1, \ldots, u_h, \ldots, u$ be $\mathcal{E}$-measurable functions on Ω. Then,*
i) if $\mu(\Omega) < +\infty$, and $u_h \to u$ μ-a.e. in Ω, it turns out that $u_h \to u$ in μ-measure,
ii) if $u_h \to u$ in μ-measure, it turns out that there exists $\{u_{h_k}\} \subseteq \{u_h\}$ such that $u_h \to u$ μ-a.e. in Ω.

The main properties of the integral of $\mathcal{E}$-measurable functions are recalled in the results below.

Theorem 2.1.4. *Let $(\Omega, \mathcal{E})$ be a measure space, μ a positive measure on $\mathcal{E}$, and $u: \Omega \to [0, +\infty]$ be $\mathcal{E}$-measurable. Then there exists a sequence $\{s_h\}$ of*

$\mathcal{E}$-measurable, simple functions such that $0 \le s_1(x) \le s_2(x) \le \ldots \le u(x)$, $\lim_{h \to +\infty} s_h(x) = u(x)$ for every $x \in \Omega$, and

$$\lim_{h \to +\infty} \int_\Omega s_h d\mu = \int_\Omega u d\mu.$$

Proposition 2.1.5 (Linearity of the Integral). *Let* $(\Omega, \mathcal{E})$ *be a measure space, and* μ *a measure on* $\mathcal{E}$. *Then*

$$\int_\Omega (au + bv) d\mu = a \int_\Omega u d\mu + b \int_\Omega v d\mu$$

whenever u, v *are* $\mathcal{E}$-*measurable and* $|\mu|$-*summable on* Ω, *and* a, $b \in \mathbf{R}$.

Theorem 2.1.6 (Monotone Convergence Theorem). *Let* $(\Omega, \mathcal{E})$ *be a measure space, and* μ *a positive measure on* $\mathcal{E}$. *For every* $h \in \mathbf{N}$ *let* $u_h : \Omega \to [0, +\infty]$ *be* $\mathcal{E}$-*measurable, and such that* $u_1(x) \le u_2(x) \le \ldots \le u_h(x) \ldots$ *for every* $x \in \Omega$. *Then the limit* $\lim_{h \to +\infty} \int_\Omega u_h d\mu$ *exists, and*

$$\lim_{h \to +\infty} \int_\Omega u_h d\mu = \int_\Omega \sup_{h \in \mathbf{N}} u_h d\mu.$$

Theorem 2.1.7 (Fatou's Lemma). *Let* $(\Omega, \mathcal{E})$ *be a measure space, and* μ *a positive measure on* $\mathcal{E}$. *For every* $h \in \mathbf{N}$ *let* $u_h : \Omega \to [0, +\infty]$ *be* $\mathcal{E}$-*measurable. Then*

$$\int_\Omega \liminf_{h \to +\infty} u_h d\mu \le \liminf_{h \to +\infty} \int_\Omega u_h d\mu.$$

Theorem 2.1.8 (Lebesgue Dominated Convergence Theorem). *Let* $(\Omega, \mathcal{E})$ *be a measure space, and* μ *a positive measure on* $\mathcal{E}$. *For every* $h \in \mathbf{N}$ *let* $u_h : \Omega \to [-\infty, +\infty]$ *be* $\mathcal{E}$-*measurable such that the limit* $u(x) = \lim_{h \to +\infty} u_h(x)$ *exists for* μ-*a.e.* $x \in \Omega$, *and* $\sup_{h \in \mathbf{N}} |u_h|$ *is* μ-*summable in* Ω. *Then*

$$\lim_{h \to +\infty} \int_\Omega |u_h - u| d\mu = 0.$$

Consequently, the limit $\lim_{h \to +\infty} \int_\Omega u_h d\mu$ *exists, and*

$$\lim_{h \to +\infty} \int_\Omega u_h d\mu = \int_\Omega u d\mu.$$

If now $(\Omega_1, \mathcal{E}_1)$, $(\Omega_2, \mathcal{E}_2)$ are measure spaces, and μ_1, μ_2 are σ-finite positive measures respectively on $\mathcal{E}_1$ and $\mathcal{E}_2$, it turns out that, for every $E \in$

$\mathcal{E}_1 \times \mathcal{E}_2$, $\varphi_{E,1} \colon x_1 \in \Omega_1 \mapsto \mu_2(\{x_2 \in \Omega_2 : (x_1, x_2) \in E\})$ is $\mathcal{E}_1$-measurable, $\varphi_{E,2} \colon x_2 \in \Omega_2 \mapsto \mu_1(\{x_1 \in \Omega_1 : (x_1, x_2) \in E\})$ is $\mathcal{E}_2$-measurable, and

$$\int_{\Omega_1} \varphi_{E,1} d\mu_1 = \int_{\Omega_2} \varphi_{E,2} d\mu_2.$$

The above equality allows the definition of a measure on $\mathcal{E}_1 \times \mathcal{E}_2$, called *product measure* of μ_1 and μ_2, and denoted by $\mu_1 \times \mu_2$, as

$$\mu_1 \times \mu_2 \colon E \in \mathcal{E}_1 \times \mathcal{E}_2 \mapsto \int_{\Omega_1} \varphi_{E,1} d\mu_1 = \int_{\Omega_2} \varphi_{E,2} d\mu_2.$$

Of course the above definition implies that

$$\mu_1 \times \mu_2(E_1 \times E_2) = \mu_1(E_1)\mu_2(E_2) \text{ for every } E_1 \in \mathcal{E}_1, \ E_2 \in \mathcal{E}_2.$$

The following result describes integration in product measure spaces.

Theorem 2.1.9 (Fubini's Theorem). *Let $(\Omega_1, \mathcal{E}_1)$, $(\Omega_2, \mathcal{E}_2)$ be measure spaces, μ_1, μ_2 be σ-finite positive measures respectively on $\mathcal{E}_1$ and $\mathcal{E}_2$, and u be $(\mathcal{E}_1 \times \mathcal{E}_2)$-measurable. Then, the following facts hold:*
i) if u takes its values in $[0, +\infty]$, it turns out that the functions

$$x_1 \in \Omega_1 \mapsto \int_{\Omega_2} u(x_1, x_2) d\mu_2(x_2), \quad x_2 \in \Omega_2 \mapsto \int_{\Omega_1} u(x_1, x_2) d\mu_1(x_1)$$

are respectively $\mathcal{E}_1$-measurable and $\mathcal{E}_2$-measurable, and that

$$(2.1.4) \qquad \int_{\Omega_1} \left(\int_{\Omega_2} u(x_1, x_2) d\mu_2(x_2) \right) d\mu_1(x_1) = \int_{\Omega_1 \times \Omega_2} u \, d\mu_1 \times \mu_2 =$$

$$= \int_{\Omega_2} \left(\int_{\Omega_1} u(x_1, x_2) d\mu_1(x_1) \right) d\mu_2(x_2),$$

ii) if u takes its values in $\mathbf{R}^m$, and $\int_{\Omega_1} (\int_{\Omega_2} |u(x_1, x_2)| d\mu_2(x_2)) d\mu_1(x_1) < +\infty$ or $\int_{\Omega_2} (\int_{\Omega_1} |u(x_1, x_2)| d\mu_1(x_1)) d\mu_2(x_2) < +\infty$, it turns out that

$$\int_{\Omega_1 \times \Omega_2} |u| d\mu_1 \times \mu_2 < +\infty,$$

iii) if $\int_{\Omega_1 \times \Omega_2} |u| d\mu_1 \times \mu_2 < +\infty$, it turns out that $\int_{\Omega_2} |u(x_1, x_2)| d\mu_2 < +\infty$ for μ_1-a.e. $x_1 \in \Omega_1$ and $\int_{\Omega_1} |u(x_1, x_2)| d\mu_1 < +\infty$ for μ_2-a.e. $x_2 \in \Omega_2$, that $\int_{\Omega_2} (\int_{\Omega_1} |u(x_1, x_2)| d\mu_1(x_1)) d\mu_2(x_2) + \int_{\Omega_1} (\int_{\Omega_2} |u(x_1, x_2)| d\mu_2(x_2)) d\mu_1(x_1) < +\infty$, and that (2.1.4) holds.

Finally, we recall the notions of translation of a function and of a measure on a subset of $\mathbf{R}^n$.

For every $E \subseteq \mathbf{R}^n$, every function u on E, and $x_0 \in \mathbf{R}^n$ we define the *translated* of u as

$$T[x_0]u\colon x \in E - x_0 \mapsto u(x + x_0).$$

For every $\Omega \in \mathcal{B}(\mathbf{R}^n)$, $\nu \in (\mathcal{M}(\Omega))^m$, and $x_0 \in \mathbf{R}^n$ we define the *translated* of ν as

$$T[x_0]\nu\colon A \in \mathcal{B}(\Omega - x_0) \mapsto \nu(x_0 + A).$$

Then it is easy to prove that for every $\nu \in (\mathcal{M}(\Omega))^m$ and $x_0 \in \mathbf{R}^n$, $T[x_0]\nu$ turns out to be in $(\mathcal{M}(\Omega - x_0))^m$. Moreover, for every Borel positive measure λ on Ω and every λ-summable function u on Ω, it results that $T[x_0]u$ is $T[x_0]\lambda$-summable on $\Omega - x_0$, and by using standard approximation results by means of measurable simple functions, that

(2.1.5)
$$\int_{\Omega - x_0} T[x_0]u\,d(T[x_0]\lambda) - \int_\Omega u\,d\lambda.$$

§2.2 Basics on L^p Spaces

This section provides a brief recall of the theory of L^p spaces.

Let $(\Omega, \mathcal{E})$ be a measure space, μ a positive measure on $\mathcal{E}$, and $p \in [1, +\infty]$.

If $p \in [1, +\infty[$, we denote by $L^p(\Omega, \mu)$ the set of the (equivalence classes of) $\mathcal{E}$-measurable functions u on Ω for which $\int_\Omega |u|^p d\mu < +\infty$. If $p = +\infty$, $L^\infty(\Omega, \mu)$ is the set of the (equivalence classes of) $\mathcal{E}$-measurable functions u on Ω such that $\operatorname{ess\,sup}_\Omega |u| < +\infty$. As usual, we think to the elements of $L^p(\Omega, \mu)$ as to functions defined μ-a.e. in Ω.

Once equipped with the norm

$$\| \cdot \|_{L^p(\Omega,\mu)}\colon u \in L^p(\Omega, \mu) \mapsto \begin{cases} \left(\int_\Omega |u|^p dx\right)^{1/p} & \text{if } p \in [1, +\infty[\\ \operatorname{ess\,sup}_\Omega |u| & \text{if } p = +\infty, \end{cases}$$

$L^p(\Omega, \mu)$ turns out to be a Banach space. With an abuse of notation, for every $p \in [1, +\infty]$ we denote again by $L^p(\Omega, \mu)$ the topology of $L^p(\Omega, \mu)$.

For every $p \in [1, +\infty]$, we denote by p' the conjugate of p defined as

$$p' = \begin{cases} +\infty & \text{if } p = 1 \\ \frac{p}{p-1} & \text{if } 1 < p < +\infty \\ 1 & \text{if } p = +\infty. \end{cases}$$

Fundamental in the study of L^p spaces is Hölder's inequality.

Theorem 2.2.1 (Hölder's Inequality). *Let $(\Omega, \mathcal{E})$ be a measure space, μ be a positive measure on $\mathcal{E}$, and u, $v \colon \Omega \to [-\infty, +\infty]$ be $\mathcal{E}$-measurable. Then*

$$\int_\Omega |uv| d\mu \le \left(\int_\Omega |u|^p d\mu \right)^{1/p} \left(\int_\Omega |v|^{p'} d\mu \right)^{1/p'}.$$

If $\mu(\Omega) < +\infty$, from Hölder's inequality it follows that $L^p(\Omega, \mu) \subseteq L^q(\Omega, \mu)$ provided $1 \le q \le p \le +\infty$. In this case, given $p \in]1, +\infty]$, we denote again with $\cap_{q \in [1,p[} L^q(\Omega, \mu)$ the topology on $\cap_{q \in [1,p[} L^q(\Omega, \mu)$ generated by the family of seminorms $u \in \cap_{q \in [1,p[} L^q(\Omega, \mu) \mapsto \|u\|_{L^q(\Omega, \mu)}$, as q varies in $[1, p[$. Once endowed with the $\cap_{q \in [1,p[} L^q(\Omega, \mu)$ topology, $\cap_{q \in [1,p[} L^q(\Omega, \mu)$ turns out to be a complete metrizable topological vector space.

Convergence in L^p is linked to μ-a.e. convergence, as shown by the following result (cf. for example [Br2, Théorème IV.9]).

Proposition 2.2.2. *Let $(\Omega, \mathcal{E})$ be a measure space, μ be a positive measure on $\mathcal{E}$, $\{u_h\} \subseteq L^p(\Omega, \mu)$, and $u \in L^p(\Omega, \mu)$. Assume that $u_h \to u$ in $L^p(\Omega, \mu)$. Then there exist $\{u_{h_k}\} \subseteq \{u_h\}$, and $g \in L^p(\Omega, \mu)$ such that $u_{h_k} \to u$ μ-a.e. in Ω, and $\sup_{k \in \mathbf{N}} |u_{h_k}(x)| \le g(x)$ for μ-a.e. $x \in \Omega$.*

We recall that, for every $p \in]1, +\infty[$, the dual space of $L^p(\Omega, \mu)$ can be identified with $L^{p'}(\Omega, \mu)$. The same property holds also when $p = 1$, provided μ is σ-finite. Therefore, given $\{u_h\} \subseteq L^p(\Omega, \mu)$ and $u \in L^p(\Omega, \mu)$, it turns out that, when $p \in [1 + \infty[$, $u_h \to u$ in weak-$L^p(\Omega, \mu)$, (respectively, when μ is σ-finite and $p = \infty$, $u_h \to u$ in weak*-$L^\infty(\Omega, \mu)$) if and only if

$$\int_\Omega u_h v dx \to \int_\Omega uv dx \text{ for every } v \in L^{p'}(\Omega, \mu).$$

If Ω is a topological space and μ is a Borel positive measure, for every $p \in [1, +\infty]$ we denote by $L^p_{\text{loc}}(\Omega, \mu)$ the set of the $\mathcal{B}(\Omega)$-measurable functions u on Ω such that $u \in L^p(K, \mu)$ for every compact subset K of Ω.

We endow $L^p_{\text{loc}}(\Omega, \mu)$ with its usual topology, denoted again by $L^p_{\text{loc}}(\Omega, \mu)$, that is with the one generated by the family of seminorms $u \in L^p_{\text{loc}}(\Omega, \mu) \mapsto \|u\|_{L^p(K,\mu)}$ with K varying among the compact subsets of Ω, that makes it a sequentially complete Hausdorff locally convex topological vector space. In addition, if Ω is σ-compact, then $L^p_{\text{loc}}(\Omega, \mu)$ turns out to be metrizable for every $p \in [1, +\infty]$.

Finally, given $p \in]1, +\infty]$ we also denote with $\cap_{q \in [1,p[} L^q_{\text{loc}}(\Omega, \mu)$ the topology on $\cap_{q \in [1,p[} L^q_{\text{loc}}(\Omega, \mu)$ generated by the family of seminorms $u \in \cap_{q \in [1,p[} L^q_{\text{loc}}(\Omega, \mu) \mapsto \|u\|_{L^q(K,\mu)}$, with K varying among the compact subsets of Ω, and q in $[1, p[$. Once endowed with the $\cap_{q \in [1,p[} L^q_{\text{loc}}(\Omega, \mu)$ topology, $\cap_{q \in [1,p[} L^q_{\text{loc}}(\Omega, \mu)$ turns out to be a sequentially complete Hausdorff locally convex topological vector space, metrizable if Ω is σ-compact.

We recall the following relative weak compactness criterion in L^1.

Theorem 2.2.3 (Dunford-Pettis-de la Vallée Poussin Theorem).
Let $(\Omega, \mathcal{E})$ be a measure space, μ be a finite positive measure on $\mathcal{E}$, $m \in \mathbf{N}$,
and $X \subseteq (L^1(\Omega, \mu))^m$. Then the following conditions are equivalent:
i) X is weak-$(L^1(\Omega, \mu))^m$ relatively compact,
ii) X is weak-$(L^1(\Omega, \mu))^m$ relatively sequentially compact,
iii) X is bounded, and for every $\varepsilon > 0$ there exists $\delta > 0$ such that

$$\sup_{u \in X} \int_A |u| d\mu < \varepsilon \text{ for every } A \in \mathcal{E} \text{ with } \mu(A) < \delta,$$

iv) there exists $\vartheta \colon [0, +\infty[\to [0, +\infty]$ Borel, and satisfying $\lim_{t \to +\infty} \vartheta(t)/t = +\infty$, such that

$$\sup_{u \in X} \int_\Omega \vartheta(|u|) d\mu < +\infty.$$

Remark 2.2.4. If $(\Omega, \mathcal{E})$, μ are as in Theorem 2.2.3, and X satisfies conditions i) or ii), then, by Lemma 1.3.3, it can be assumed that the function ϑ produced by Theorem 2.2.3 takes its values in $[0, +\infty[$, is increasing, and convex.

In particular, from Theorem 2.2.3 and Remark 2.2.4 the following result holds.

Corollary 2.2.5. Let $(\Omega, \mathcal{E})$ be a measure space, μ be a finite positive measure on $\mathcal{E}$, $m \in \mathbf{N}$, and $u \in (L^1(\Omega, \mu))^m$. Then there exists $\vartheta \colon [0, +\infty[\to [0, +\infty[$ convex, and satisfying $\lim_{t \to +\infty} \vartheta(t)/t = +\infty$, such that $\vartheta(|u|) \in L^1(\Omega, \mu)$.

Weak compactness in L^p spaces when $p \in]1, +\infty]$ is less involved than the one in L^1, as described in the result below.

Theorem 2.2.6. Let $(\Omega, \mathcal{E})$ be a measure space, μ be a finite positive measure on $\mathcal{E}$, $p \in]1, +\infty]$, $m \in \mathbf{N}$, and $X \subseteq (L^p(\Omega, \mu))^m$. Then the following conditions are equivalent:
i) X is bounded,
ii) X is weak-$(L^p(\Omega, \mu))^m$ (weak*-$(L^\infty(\Omega, \mu))^m$ if $p = +\infty$) relatively compact,
iii) X is weak-$(L^p(\Omega, \mu))^m$ (weak*-$(L^\infty(\Omega, \mu))^m$ if $p = +\infty$ and $(L^p(\Omega, \mu))^m$ is separable) relatively sequentially compact.

If $\Omega \in \mathcal{L}_n(\mathbf{R}^n)$ and $\mu = \mathcal{L}^n$, we simply write $L^p(\Omega)$, $\| \cdot \|_{L^p(\Omega)}$, and $L^p_{\mathrm{loc}}(\Omega)$ in place of $L^p(\Omega, \mathcal{L}^n)$, $\| \cdot \|_{L^p(\Omega, \mathcal{L}^n)}$, and $L^p_{\mathrm{loc}}(\Omega, \mathcal{L}^n)$.

Theorem 2.2.7 (Continuity of Translations in L^p). Let $p \in [1, +\infty[$, and $u \in L^p_{\mathrm{loc}}(\mathbf{R}^n)$. Then the function

$$y \in \mathbf{R}^n \mapsto T[y]u \in L^p_{\mathrm{loc}}(\mathbf{R}^n)$$

is continuous.

The uniform validity of the condition expressed by Theorem 2.2.7 is the main tool to characterize strong compactness in $L^p(\mathbf{R}^n)$ spaces.

Theorem 2.2.8. *Let $p \in [1, +\infty[$, and $X \subseteq L^p(\mathbf{R}^n)$. Then X is relatively compact in $L^p(\mathbf{R}^n)$ if and only if*
i) X is bounded,
ii) for every $\varepsilon > 0$ there exists $\delta > 0$ such that

$$\int_{\mathbf{R}^n} |T[h]u - u|^p dx < \varepsilon \text{ whenever } u \in X, \ h \in \mathbf{R}^n \text{ satisfies } |h| < \delta,$$

iii) for every $\varepsilon > 0$ there exists $r_\varepsilon > 0$ such that

$$\int_{\mathbf{R}^n \setminus B_{r_\varepsilon}(0)} |u|^p dx < \varepsilon \text{ whenever } u \in X.$$

Let $u \in L^1_{\text{loc}}(\mathbf{R}^n)$. We say that u is Y-periodic if $\int_{x+Y} u\,dy = \int_Y u\,dy$ for every $x \in \mathbf{R}^n$.

The result below analyzes the asymptotic behaviour of oscillating periodic functions as the frequency increases. It is classical, and we prove it, in the form that we need, because of the importance of the role that it plays in homogenization theory.

Theorem 2.2.9. *Let $p \in [1, +\infty]$, $u \in L^p_{\text{loc}}(\mathbf{R}^n)$ be Y-periodic, and set, for every $s > 0$, $u_s \colon x \in \mathbf{R}^n \mapsto u(sx)$. Then, for every bounded $\Omega \in \mathcal{L}_n(\mathbf{R}^n)$,*

$$u_s \rightarrow \int_Y u\,dy$$

in weak-$L^p(\Omega)$ if $p \in [1, +\infty[$, in weak*-$L^\infty(\Omega)$ if $p = +\infty$, as $s \rightarrow +\infty$.

Proof. To prove the theorem it suffices to verify that for every $l > 0$, and every $\{s_h\} \subseteq]0, +\infty[$ strictly increasing there exists $\{s_{h_k}\} \subseteq \{s_h\}$ such that $u_{s_{h_k}} \rightarrow \int_Y u\,dy$ weakly in $L^p(Q_l(0))$ if $p \in [1, +\infty[$, weakly* in $L^\infty(Q_l(0))$ if $p = +\infty$.

To do this, let $l > 0$. Let us preliminarily prove that $\{\|u_s\|_{L^p(Q_l(0))}\}_{s>0}$ is bounded if $p \in]1, +\infty]$, or that there exists $\vartheta \colon [0, +\infty[\rightarrow [0, +\infty[$ increasing, convex, and satisfying $\lim_{t \rightarrow +\infty} \vartheta(t)/t = +\infty$, such that the family $\{\int_{Q_l(0)} \vartheta(|u_s|)dx\}_{s>0}$ is bounded if $p = 1$.

This is obvious if $p = +\infty$. If $p \in [1, +\infty[$, we treat only the case in which $p = 1$, the others being similar.

By Corollary 2.2.5 there exists $\vartheta \colon [0, +\infty[\rightarrow [0, +\infty[$ increasing, convex, and satisfying $\lim_{t \rightarrow +\infty} \vartheta(t)/t = +\infty$, such that $\int_{Q_l(0)} \vartheta(|u|)dx < +\infty$. Then, the Y-periodicity assumption yields

(2.2.1) $$\int_{Q_l(0)} \vartheta(|u_s|)dx = \frac{1}{s^n} \int_{Q_{sl}(0)} \vartheta(|u|)dy \le$$

$$\leq \frac{1}{s^n} \int_{\cup_{\{\varsigma \in \mathbf{Z}^n : \varsigma + Y \cap Q_{sl}(0) \neq \emptyset\}} \varsigma + Y} \vartheta(|u|) dy =$$

$$= \frac{1}{s^n} \sum_{\{\varsigma \in \mathbf{Z}^n : \varsigma + Y \cap Q_{sl}(0) \neq \emptyset\}} \int_{\varsigma + Y} \vartheta(|u|) dy \leq \frac{([sl] + 1)^n}{s^n} \int_Y \vartheta(|u|) dy,$$

from which the desired boundedness follows.

Let now $\{s_h\} \subseteq]0, +\infty[$ be strictly increasing. Then, by Theorem 2.2.3 if $p = 1$, or Theorem 2.2.6 if $p \in]1, +\infty]$ there exist $\{s_{h_k}\} \subseteq \{s_h\}$ and $u_\infty \in L^p(Q_l(0))$ such that $u_{s_{h_k}} \to u_\infty$ in weak-$L^p(Q_l(0))$ if $p \in [1, +\infty[$, in weak*-$L^\infty(Q_l(0))$ if $p = +\infty$. Therefore it only remains to prove that $u_\infty = \int_Y u \, dy$.

To do this, let us preliminarily observe that, possibly considering u^+ and u^-, it is not restrictive to assume that $u \geq 0$ a.e. in $Q_l(0)$.

Let now $x_0, y_0 \in Q_l(0)$, and $r > 0$ such that $Q_r(x_0) \cup Q_r(y_0) \subseteq Q_l(0)$. Then we have that

$$\int_{Q_r(x_0)} u_s \, dx = \frac{1}{s^n} \int_{Q_{sr}(sx_0)} u \, dx \leq \frac{1}{s^n} \int_{\cup_{\{\varsigma \in \mathbf{Z}^n : \varsigma + Y \cap Q_{sr}(sx_0) \neq \emptyset\}} \varsigma + Y} u \, dy =$$

$$= \frac{1}{s^n} \sum_{\{\varsigma \in \mathbf{Z}^n : \varsigma + Y \cap Q_{sr}(sx_0) \neq \emptyset\}} \int_{\varsigma + Y} u \, dy =$$

$$= \frac{1}{s^n} \sum_{\{\varsigma \in \mathbf{Z}^n : \varsigma + Y \cap Q_{sr}(sx_0) \neq \emptyset\}} \int_{sy_0 - sx_0 + \varsigma + Y} u \, dy =$$

$$= \frac{1}{s^n} \int_{\cup_{\{\varsigma \in \mathbf{Z}^n : \varsigma + Y \cap Q_{sr}(sy_0) \neq \emptyset\}} \varsigma + Y} u \, dy \leq \frac{1}{s^n} \int_{Q_{[sr]+4}(sy_0)} u \, dx =$$

$$= \int_{Q_{\frac{[sr]+4}{s}}(y_0)} u_s \, dx \quad \text{for every } s > 0,$$

from which we conclude that

$$(2.2.2) \qquad \int_{Q_r(x_0)} u_\infty \, dx = \lim_{k \to +\infty} \int_{Q_r(x_0)} u_{s_{h_k}} \, dx \leq$$

$$\leq \limsup_{k \to +\infty} \int_{Q_{\frac{[s_{h_k} r]+4}{s_{h_k}}}(y_0)} u_{s_{h_k}} \, dx =$$

$$= \int_{Q_r(y_0)} u_\infty \, dx + \limsup_{k \to +\infty} \int_{Q_{\frac{[s_{h_k} r]+4}{s_{h_k}}}(y_0) \setminus Q_r(y_0)} u_{s_{h_k}} \, dx.$$

We now observe that an argument similar to the one used in (2.2.1) yields

$$\int_{Q_{\frac{[s_{h_k}r]+4}{s_{h_k}}}(y_0)\backslash Q_r(y_0)} u_s dx = \frac{1}{s_{h_k}^n}\int_{Q_{[s_{h_k}r]+4}(s_{h_k}y_0)\backslash Q_{s_{h_k}r}(s_{h_k}y_0)} u dx \le$$

$$\le \frac{([s_{h_k}r]+4)^{n-1}}{s_{h_k}^n}\int_Y u dy,$$

from which it follows that

$$(2.2.3) \qquad \limsup_{k\to+\infty}\int_{Q_{\frac{[s_{h_k}r]+4}{s_{h_k}}}(y_0)\backslash Q_r(y_0)} u_{s_{h_k}} dx = 0.$$

In conclusion, by (2.2.2), and (2.2.3) it results that $\int_{Q_r(x_0)} u_\infty dx \le \int_{Q_r(y_0)} u_\infty dx$, and consequently, by replacing the roles of x_0 and y_0, that

$$(2.2.4) \qquad \int_{Q_r(x_0)} u_\infty dx = \int_{Q_r(y_0)} u_\infty dx$$

for every x_0, $y_0 \in Q_l(0)$, $r > 0$ sufficiently small.

Finally, by (2.2.4), and by Theorem 2.2.9 of the next section, we conclude that u_∞ is a.e. constant in $Q_l(0)$.

In order to determine such constant value, we observe that, again using an argument similar to the one in (2.2.1), it follows that

$$u_\infty = \int_Y u_\infty dy = \lim_{k\to+\infty}\int_Y u_{s_{h_k}} dx = \frac{1}{s_{h_k}^n}\int_{s_{h_k}Y} u dx = \int_Y u dy,$$

that completes the proof. ∎

§2.3 Derivation of Measures

Let $(\Omega, \mathcal{E})$ be a measure space, μ be a positive measure, and ν a (real or vector) measure on $\mathcal{E}$. We say that ν is *absolutely continuous* with respect to μ if

$$|\nu|(A) = 0 \text{ whenever } A \in \mathcal{E} \text{ satisfies } \mu(A) = 0.$$

We say that ν is *singular* with respect to μ if there exists $N_0 \in \mathcal{E}$ such that $\mu(N_0) = 0$, and $|\nu|(\Omega \backslash N_0) = 0$.

It is clear that every measure is absolutely continuous with respect to its total variation. Moreover, if μ be a positive measure, and $u \in (L^1(\Omega, \mu))^m$, the measure $A \in \mathcal{E} \mapsto \int_A u d\mu \in \mathbf{R}^m$ is absolutely continuous with respect to μ, and is usually denoted by $u\mu$.

Theorem 2.3.1 (Radon-Nikodym Theorem). *Let $(\Omega, \mathcal{E})$ be a measure space, μ be a σ-finite positive measure on $\mathcal{E}$, and $\nu \colon \mathcal{E} \to \mathbf{R}^m$ be a measure on $\mathcal{E}$. Assume that ν is absolutely continuous with respect to μ. Then there is a unique $u \in (L^1(\Omega, \mu))^m$ such that*

$$\nu(A) = \int_A u d\mu \text{ for every } A \in \mathcal{E}.$$

The function u in Theorem 2.3.1 is called the *Radon-Nikodym deriva-tive* of ν with respect to μ, and is denoted by $\frac{d\nu}{d\mu}$.

By using Radon-Nikodym Theorem, the following results hold.

Theorem 2.3.2. *Let $(\Omega, \mathcal{E})$ be a measure space, μ be a σ-finite positive measure on $\mathcal{E}$, and $\nu \colon \mathcal{E} \to \mathbf{R}$ be a measure on $\mathcal{E}$. Assume that ν is absolutely continuous with respect to μ. Then for every $u \in L^1(\Omega, \nu)$ it results that $u\frac{d\nu}{d\mu} \in L^1(\Omega, \mu)$, and*

$$\int_\Omega u d\nu = \int_\Omega u \frac{d\nu}{d\mu} d\mu.$$

Proof. If $u = \chi_A$ for $A \in \mathcal{E}$ the theorem follows from Radon-Nikodym Theorem.

Consequently, the theorem follows when u is a $\mathcal{E}$-measurable simple function, and by using Theorem 2.1.4 and the Monotone Convergence The-orem, also when ν is positive and $u \in L^1(\Omega, \nu)$ is such that $u(x) \geq 0$ for ν-a.e. $x \in \Omega$.

Because of this, the theorem follows also when $u \in L^1(\Omega, |\nu|)$ by con-sidering separately ν^+, ν^-, u^+, and u^-. ∎

Theorem 2.3.3. *Let $(\Omega, \mathcal{E})$ be a measure space, μ be a σ-finite positive measure on $\mathcal{E}$, and $\nu \colon \mathcal{E} \to \mathbf{R}^m$ be a measure on $\mathcal{E}$. Assume that ν is absolutely continuous with respect to μ. Then*

$$|\nu|(A) = \int_A \left| \frac{d\nu}{d\mu} \right| d\mu \text{ for every } A \in \mathcal{E}.$$

Proof. By the Radon-Nikodym Theorem, it follows that for every $A \in \mathcal{E}$, and every sequence $\{A_h\} \subseteq \mathcal{E}$ of pairwise disjoint sets whose union is A it results that

$$\sum_{h=1}^{+\infty} |\nu(A_h)| = \sum_{h=1}^{+\infty} \left| \int_{A_h} \frac{d\nu}{d\mu} d\mu \right| \leq \sum_{h=1}^{+\infty} \int_{A_h} \left| \frac{d\nu}{d\mu} \right| d\mu = \int_A \left| \frac{d\nu}{d\mu} \right| d\mu,$$

and, consequently, that

(2.3.1) $$|\nu|(A) \leq \int_A \left| \frac{d\nu}{d\mu} \right| d\mu \text{ for every } A \in \mathcal{E}.$$

Let now $A \in \mathcal{E}$, $\{z_h\} \subseteq \mathbf{R}^n$ be dense in $\{z \in \mathbf{R}^n : |z| = 1\}$, and $\varepsilon > 0$. Let $A_1 = \{x \in A : (1 - \varepsilon)|\frac{d\nu}{d\mu}(x)| \leq \frac{d\nu}{d\mu}(x) \cdot z_1\}$, and set, for every $h \in \mathbf{N} \setminus \{1\}$, $A_h = \{x \in A : (1 - \varepsilon)|\frac{d\nu}{d\mu}(x)| \leq \frac{d\nu}{d\mu}(x) \cdot z_h\} \setminus \cup_{j=1}^{h-1} A_j$. Then $A_h \in \mathcal{E}$ for every $h \in \mathbf{N}$, $\cup_{h=1}^{+\infty} A_h = A$, and

$$(1 - \varepsilon) \int_A \left|\frac{d\nu}{d\mu}\right| d\mu = (1 - \varepsilon) \sum_{h=1}^{+\infty} \int_{A_h} \left|\frac{d\nu}{d\mu}\right| d\mu \leq \sum_{h=1}^{+\infty} \int_{A_h} \frac{d\nu}{d\mu} \cdot z_h d\mu =$$

$$= \sum_{h=1}^{+\infty} \nu(A_h) \cdot z_h \leq \sum_{h=1}^{+\infty} |\nu(A_h)| \leq |\nu|(A) \text{ for every } \varepsilon > 0,$$

from which, together with (2.3.1), the proof follows. ∎

The following decomposition theorem is classical in measure theory.

Theorem 2.3.4 (Lebesgue Decomposition Theorem). *Let $(\Omega, \mathcal{E})$ be a measure space, let μ be a σ-finite positive measure on $\mathcal{E}$, and ν a measure on $\mathcal{E}$. Then there is a unique measure ν^{a} on $\mathcal{E}$ absolutely continuous with respect to μ, and a unique measure ν^{s} on $\mathcal{E}$ singular with respect to μ such that*

$$\nu = \nu^{\mathrm{a}} + \nu^{\mathrm{s}}.$$

Formula in Theorem 2.3.4 yields the Lebesgue decomposition of ν with respect to μ.

Finally, the result below provides an interpretation, at least when $\Omega \in \mathcal{A}(\mathbf{R}^n)$, of the Radon-Nikodym derivatives as limits of ratios of measures.

Theorem 2.3.5. *Let $\Omega \in \mathcal{A}(\mathbf{R}^n)$ and $m \in \mathbf{N}$. Then, for every $\nu \in (\mathcal{M}(\Omega))^m$ and for $\mathcal{L}^n$-a.e. $x \in \Omega$ the limit $\lim_{r \to 0^+} \frac{\nu(Q_r(x))}{r^n}$ exists, and*

$$\lim_{r \to 0^+} \frac{\nu(Q_r(x))}{r^n} = \frac{d\nu^{\mathrm{a}}}{d\mathcal{L}^n}(x).$$

By using Theorem 2.3.5 it can be proved that if $\Omega \subseteq \mathbf{R}^n$ is open, $m \in \mathbf{N}$, and $u \in (L^1(\Omega))^m$, then

$$\lim_{r \to 0^+} \frac{1}{r^n} \int_{Q_r(x)} |u(y) - u(x)| dy = 0 \text{ for } \mathcal{L}^n\text{-a.e. } x \in \Omega.$$

A point in which this occurs is called a *Lebesgue point* of u.

Remark 2.3.6. It is important to observe that the Radon-Nikodym Theorem allows the identification of L^p spaces with suitable spaces of measures.

More precisely, if $(\Omega, \mathcal{E})$ and μ are as in the Radon-Nikodym Theorem, and $p \in [1, +\infty]$, then the mapping

$$u \to u\mu$$

turns out to be an isomorphism between $L^p(\Omega, \mu)$ and the set of the measures $\nu \colon \mathcal{E} \to \mathbf{R}$ that are absolutely continuous with respect to μ and such that $|\frac{d\nu}{d\mu}|^p$ is μ-summable on Ω if $p \in [1, +\infty[$, or such that $\operatorname{ess\,sup}_\Omega |\frac{d\nu}{d\mu}|$ is finite if $p = +\infty$.

This interpretations become more shrinking and expressive if $p = 1$. In fact, in this case the above mapping becomes an Banach space isomorphism between $L^1(\Omega, \mu)$ and the space of the real measures on Ω that are absolutely continuous with respect to μ.

Analogously, if in addition Ω is a topological space and μ is a Borel positive measure, then $L^p_{\mathrm{loc}}(\Omega, \mu)$ can be regarded as the space of the Radon measures $\nu \colon \mathcal{E} \to \mathbf{R}$ that are absolutely continuous with respect to μ and such that $|\frac{d\nu}{d\mu}|^p$ is μ-summable on every compact subset of Ω if $p \in [1, +\infty[$, or such that $\operatorname{ess\,sup}_K |\frac{d\nu}{d\mu}|$ is finite for every compact subset K of Ω if $p = +\infty$.

In this order of ideas, also continuous functions on Ω can be thought as Radon measures on $\mathcal{B}(\Omega)$, and in particular, when $\Omega \in \mathcal{A}(\mathbf{R}^n)$, so do the elements of $C^\infty(\Omega)$ by means of the mapping

$$u \to u\mathcal{L}^n.$$

We will come back to this approach in §7.2.

§2.4 Abstract Measure Theory in Topological Settings

Let Ω be a topological space. In the present section we describe how the requirement of slight additional assumptions on the topology allows a deeper description of the structure of Borel functions and measures.

To do this, we first need to select some special classes of Borel measures, called regular measures.

Definition 2.4.1. *Let Ω be a topological space, μ be a Borel positive measure on Ω, and $B \in \mathcal{B}(\Omega)$. We say that μ is*
i) inner regular in B if

$$\mu(B) = \sup\{\mu(K) : K \text{ compact, } K \subseteq B\},$$

ii) outer regular in B if

$$\mu(B) = \inf\{\mu(A) : A \text{ open, } A \supseteq B\},$$

iii) regular in B if it is both inner and outer regular in B.

We also say that μ is inner regular, outer regular, regular if so it is in B for every $B \in \mathcal{B}(\Omega)$.

Regular Borel measures form a subclass, in general proper, of the one of Borel measures. We refer to [Co, Chapter 7] for an example in this direction.

The result below shows that non-regularity of a measure seems to be due to the lack of some properties of the space, rather than of the measure itself.

Theorem 2.4.2. *Let Ω be a Hausdorff locally compact space in which every open set is σ-compact. Let μ be a Radon positive measure on Ω. Then μ is regular.*

It is worth while to deduce from Proposition 2.1.1, and Theorem 2.4.2 the following corollary.

Corollary 2.4.3. *Let Ω be a Hausdorff locally compact space in which every open set is σ-compact, and $\mu \in (\mathcal{M}_{\mathrm{loc}}(\Omega))^m$. Then $|\mu|$ is regular.*

In particular, if $\mu \in (\mathcal{M}(\Omega))^m$, then $|\mu|$ is regular.

We observe explicitly that Theorem 2.4.2, and Corollary 2.4.3 hold when $\Omega = \mathbf{R}^n$.

Theorem 2.4.4 (Lusin's Theorem). *Let Ω be a Hausdorff locally compact space, μ a regular Radon positive measure on Ω, and u be Borel and equal to 0 outside a set with finite measure. Then for every $\varepsilon > 0$ there exists $v_\varepsilon \in C_0^0(\Omega)$ such that $\|v_\varepsilon\|_{C^0(\Omega)} \leq \mathrm{ess\,sup}_\Omega |u|$, and $\mu(\{x \in \Omega : v_\varepsilon(x) \neq u(x)\}) < \varepsilon$.*

As consequence, the following approximation result in L^p spaces holds.

Theorem 2.4.5. *Let Ω be a Hausdorff locally compact space, μ a regular Radon positive measure on Ω, and $p \in [1, +\infty[$. Then $C_0^0(\Omega)$ is dense in $L^p(\Omega, \mu)$.*

We now pass to the study of the structure of Borel measures, that is described by the Riesz Representation Theorem (cf. for example [Ru, 6.19 Theorem]).

Theorem 2.4.6 (Riesz Representation Theorem). *Let Ω be a Hausdorff locally compact space, $m \in \mathbf{N}$, and let $L\colon (\widehat{C}_0^0(\Omega))^m \to \mathbf{R}$ be linear and continuous. Then there exists a unique $\mu \in (\mathcal{M}(\Omega))^m$, with $|\mu|$ regular, such that*

$$L(u) = \int_\Omega u\,d\mu \quad \text{for every } u \in (\widehat{C}_0^0(\Omega))^m.$$

Moreover

$$\sup\left\{L(u) : u \in (\widehat{C}_0^0(\Omega))^m, \ \|u\|_{C^0(\Omega)} \le 1\right\} = |\mu|(\Omega).$$

As corollary, from Riesz Representation Theorem we deduce the following result that we prove for sake of completeness.

Corollary 2.4.7. *Let Ω be a Hausdorff locally compact space, and $\nu \in \mathcal{M}(\Omega)$ be such that $|\nu|$ is inner regular. Then*

$$\nu^+(\Omega) = \sup\left\{\int_\Omega \varphi d\nu : \varphi \in C_0^0(\Omega), \ 0 \le \varphi(x) \le 1 \text{ for every } x \in \Omega\right\},$$

$$\nu^-(\Omega) = \sup\left\{\int_\Omega \varphi d\nu : \varphi \in C_0^0(\Omega), \ -1 \le \varphi(x) \le 0 \text{ for every } x \in \Omega\right\}.$$

In particular, if μ is an inner regular Borel positive measure on Ω, and $u \in L^1(\Omega, \mu)$, then

$$\int_\Omega u^+ d\mu = \sup\left\{\int_\Omega \varphi u d\mu : \varphi \in C_0^0(\Omega), \ 0 \le \varphi(x) \le 1 \text{ for every } x \in \Omega\right\},$$

$$\int_\Omega u^- d\mu = \sup\left\{\int_\Omega \varphi u d\mu : \varphi \in C_0^0(\Omega), \ -1 \le \varphi(x) \le 0 \text{ for every } x \in \Omega\right\}.$$

Proof. We prove only the formulas for ν^+ and u^+, the proof of the one for ν^- and u^- being similar.

It is clear that

$$\int_\Omega \varphi d\nu = \int_\Omega \varphi d\nu^+ - \int_\Omega \varphi d\nu^- \le \int_\Omega \varphi d\nu^+ \le \nu^+(\Omega)$$

for every $\varphi \in C_0^0(\Omega)$ satisfying $0 \le \varphi(x) \le 1$ for every $x \in \Omega$,

from which we deduce that

$$(2.4.1) \quad \sup\left\{\int_\Omega \varphi d\nu : \varphi \in C_0^0(\Omega), \ 0 \le \varphi(x) \le 1 \text{ for every } x \in \Omega\right\} \le$$

$$\le \nu^+(\Omega).$$

On the other side, because of Riesz representation theorem, it turns out that

$$(2.4.2) \qquad \nu^+(\Omega) = \frac{|\nu|(\Omega) + \nu(\Omega)}{2} =$$

$$= \frac{1}{2}\left\{ \sup\left\{ \int_\Omega \phi d\nu : \phi \in \widehat{C}_0^0(\Omega), \ |\phi(x)| \leq 1 \text{ for every } x \in \Omega \right\} + \int_\Omega d\nu \right\} =$$

$$= \sup\left\{ \int_\Omega \frac{\phi+1}{2} d\nu : \phi \in \widehat{C}_0^0(\Omega), \ |\phi(x)| \leq 1 \text{ for every } x \in \Omega \right\}.$$

Let now $\{\psi_h\} \subseteq C_0^0(\Omega)$ be given by Theorem 2.4.5 with $p = 1$ such that $0 \leq \psi_h(x) \leq 1$ for every $h \in \mathbf{N}$ and $x \in \Omega$, and $\psi_h(x) \to 1$ for $|\nu|$-a.e. $x \in \Omega$. Then, Lebesgue Dominated Convergence Theorem provides that

$$\int_\Omega \frac{\phi+1}{2} d\nu = \lim_{h\to+\infty} \int_\Omega \frac{\phi+1}{2}\psi_h d\nu \leq$$

$$\leq \sup\left\{ \int_\Omega \varphi d\nu : \varphi \in C_0^0(\Omega), \ |\varphi(x)| \leq 1 \text{ for every } x \in \Omega \right\}$$

for every $\phi \in \widehat{C}_0^0(\Omega)$ satisfying $|\phi(x)| \leq 1$ for every $x \in \Omega$,

from which, together with (2.4.2) and (2.4.1), the formula for $\nu^+(\Omega)$ follows.

In particular, if $\nu = u\mu$, then Theorem 2.3.3 and the above formula for $\nu^+(\Omega)$ yield

$$\int_\Omega u^+ d\mu = \int_\Omega \frac{|u|+u}{2} d\mu = \frac{|\nu|(\Omega)+\nu(\Omega)}{2} = \nu^+(\Omega) =$$

$$= \sup\left\{ \int_\Omega \varphi u d\mu : \varphi \in C_0^0(\Omega), \ |\varphi(x)| \leq 1 \text{ for every } x \in \Omega \right\},$$

that completes the proof of the corollary. ∎

When Ω is a Hausdorff locally compact space, and $m \in \mathbf{N}$, the Riesz Representation Theorem allows the identification of $(\mathcal{M}(\Omega))^m$ with the dual of the Banach space $(\widehat{C}_0^0(\Omega))^m$. Consequently, a weak* topology turns out to be canonically defined on $(\mathcal{M}(\Omega))^m$. As usual, we denote it by weak*-$(\mathcal{M}(\Omega))^m$.

We recall that, given $\{\mu_h\} \subseteq (\mathcal{M}(\Omega))^m$, and $\mu \in (\mathcal{M}(\Omega))^m$, it results that $\mu_h \to \mu$ in weak*-$(\mathcal{M}(\Omega))^m$ if and only if $\int_\Omega \varphi d\mu_h \to \int_\Omega \varphi d\mu$ for every $\varphi \in (\widehat{C}_0^0(\Omega))^m$. We also observe that, by the Banach-Steinhaus Theorem, if $\mu_h \to \mu$ in weak*-$(\mathcal{M}(\Omega))^m$, then $\{|\mu_h|(\Omega)\}$ turns out to be bounded.

In particular, by the Riesz Representation Theorem, the following lower semicontinuity follows.

Proposition 2.4.8. *Let Ω be a Hausdorff locally compact space, and $m \in \mathbf{N}$. Then the functional $\nu \in (\mathcal{M}(\Omega))^m \mapsto |\nu|(\Omega)$ is weak*-$(\mathcal{M}(\Omega))^m$ lower semicontinuous.*

Proof. Follows from Riesz Representation Theorem, once we observe that for every $\nu \in (\mathcal{M}(\Omega))^m$, $|\nu|(\Omega)$ is the pointwise supremum of a family of weak*-$(\mathcal{M}(\Omega))^m$ continuous functionals. ∎

By using Alaoglu's theorem, the following compactness result holds.

Theorem 2.4.9. *Let* Ω *be a Hausdorff locally compact space, and* $m \in$ **N**. *Then the bounded subsets of* $(\mathcal{M}(\Omega))^m$ *are weak*-*$(\mathcal{M}(\Omega))^m$ *relatively compact.*

In particular, under separability assumptions, the following sequential version of Theorem 2.4.9 follows from Theorem 0.5.

Theorem 2.4.10. *Let* Ω *be a Hausdorff locally compact space, and* $m \in$ **N**. *Assume that* $(\widehat{C}_0^0(\Omega))^m$ *is separable. Then the strongly bounded subsets of* $(\mathcal{M}(\Omega))^m$ *are weak*-*$(\mathcal{M}(\Omega))^m$ *relatively sequentially compact.*

We remark that, in general, the assumptions of Theorem 2.4.10 are fulfilled provided suitable hypotheses on Ω are assumed. For example, Theorem 2.4.10 holds if $\Omega \subseteq \mathbf{R}^n$.

If Ω is a Hausdorff locally compact space, $m \in \mathbf{N}$, μ is a Borel positive measure on Ω, and $\{u_h\}$ is a bounded sequence in $(L^1(\Omega, \mu))^m$, then Theorem 2.4.9, and Theorem 2.4.10 applied with $\mu_h = u_h \mu$ yield the relative compactness of $\{\mu_h\}$ only in the weak*-$(\mathcal{M}(\Omega))^m$ topology. Consequently, in general, its cluster points need not be in $(L^1(\Omega, \mu))^m$.

On the other side, if $\{\int_\Omega \vartheta(|u_h|)d\mu\}$ is bounded for some $\vartheta \colon [0, +\infty[\to [0, +\infty]$ Borel, and satisfying $\lim_{t \to +\infty} \vartheta(t)/t = +\infty$, Theorem 2.2.3 applies, and the existence of a weak-$(L^1(\Omega, \mu))^m$ converging subsequence of $\{u_h\}$ follows. In particular this holds if $\vartheta(t) = t^p$ for every $t \in [0, +\infty[$, and some $p \in]1, +\infty[$, case in which Theorem 2.2.6 applies, and $\{u_h\}$ turns out to have a weak-$L^p(\Omega, \mu)$ converging subsequence.

§2.5 Local Properties of Boundaries of Open Subsets of **R**^{n}

The present section is devoted to a discussion on some types of convexity properties of certain classes of open subsets of $\mathbf{R}^n$ that we will use in this book.

We say that $\Omega \in \mathcal{A}(\mathbf{R}^n)$ has *Lipschitz boundary* if for every $x \in \partial\Omega$ there exists a neighborhood I_x of x such that $I_x \cap \partial\Omega$ is the graph, in a suitable coordinate system, of a Lipschitz continuous function whose epigraph contains $I_x \cap \Omega$.

If $\Omega \in \mathcal{A}(\mathbf{R}^n)$ has Lipschitz boundary, then for $\mathcal{H}^{n-1}$-a.e. $x \in \partial\Omega$ there exists the outward unit vector normal to $\partial\Omega$, that we denote by $\mathbf{n}_\Omega$.

Proposition 2.5.1. *Let* $\Omega \in \mathcal{A}(\mathbf{R}^n)$ *be convex. Then* Ω *has Lipschitz boundary.*

Proof. Let $x_0 \in \Omega$. Then, being Ω open, let $r > 0$ be such that $B_r(x_0) \subseteq \Omega$.

Let $x \in \partial\Omega$. Let us prove that there exists $I_x \in \mathcal{N}(x)$ such that $I_x \cap \partial\Omega$ is the graph, in a suitable coordinate system, of a finite convex function whose epigraph contains $I_x \cap \Omega$.

To do this, let us consider the half-line $l = \{x_0 + t(x - x_0) : t \in [0, +\infty[\}$, take $y_0 \in l$ such that the hyperplane H_0 containing y_0 and orthogonal to l has empty intersection with $\overline{\Omega}$, and set $B'_r = \{y \in H_0 : |y - y_0| < r\}$.

Let us set $I_x = \{ty + (1-t)(x_0 - y_0 + y) : y \in B'_r, \; t \in \;]0,1[\}$. Then it is easy to see that $I_x \in \mathcal{N}(x)$.

Let us fix $y \in B'_r$, and denote by S_y the open line segment joining y to $x_0 - y_0 + y$. Then, since S_y has one endpoint in Ω and the other in $\mathbf{R}^n \setminus \overline{\Omega}$, it must result $S_y \cap \partial\Omega \neq \emptyset$. Moreover it is clear that $S_y \cap \partial\Omega$ is made up by a single point, otherwise, taken $x_1, \; x_2 \in S_y \cap \partial\Omega$ with $x_1 \neq x_2$, it would necessarily occur that $x_1 = t_0 x_2 + (1 - t_0)(x_0 + y - y_0)$ or $x_2 = t_0 x_1 + (1 - t_0)(x_0 + y - y_0)$ for some $t_0 \in \;]0,1[$. To fix ideas, let us assume that $x_1 = t_0 x_2 + (1 - t_0)(x_0 + y - y_0)$ for some $t_0 \in \;]0,1[$. Then, by the last item of Proposition 1.1.5, it would result that $x_1 \in \Omega$, contrary to the fact that $x_1 \in \partial\Omega$.

Consequently, the application that to every $y \in B'_r$ associates the only element of $S_y \cap \partial\Omega$ defines, in the coordinate system centred in y_0, with H_0 equal to the hyperplane of first $n-1$ coordinates, and with the line through y_0 and x_0 equal to the n-th coordinate axis, a finite function whose graph is contained in $I_x \cap \partial\Omega$.

On the other side, since every point in $I_x \cap \partial\Omega$ is in $S_y \cap \partial\Omega$ for a suitable $y \in B'_r$, it turns out that the graph of the above defined function actually agrees with $I_x \cap \partial\Omega$. Moreover, it is immediately verified that such function is convex, and that, just by construction, its epigraph contains $I_x \cap \Omega$.

Finally, by possibly considering a smaller neighborhood of x compactly contained in I_x, and by Theorem 1.1.17, the proof follows. ∎

We now introduce the class of the strongly star shaped open sets that will play a crucial role in the proof of some regularity results of measure type functions.

Definition 2.5.2. *Let $\Omega \in \mathcal{A}(\mathbf{R}^n)$, and $x_0 \in \Omega$. We say that Ω is strongly star shaped with respect to x_0 if it is star shaped with respect to x_0, and if for every $x \in \overline{\Omega}$ the half open line segment joining x_0 and x, and not containing x, is contained in Ω.*

We say that an open set Ω is strongly star shaped if there exists $x_0 \in \Omega$ such that Ω is strongly star shaped with respect to x_0.

In the following result some elementary properties of strongly star shaped open sets are collected.

Proposition 2.5.3. *Let $\Omega \in \mathcal{A}(\mathbf{R}^n)$, $x_0 \in \Omega$ be such that Ω is strongly star shaped with respect to x_0. Then*

$$(2.5.1) \qquad x_0 + t(\Omega - x_0) \text{ is strongly star shaped with respect to } x_0$$

$$\text{for every } t \in \,]0, +\infty[,$$

(2.5.2) $$\overline{x_0 + r(\Omega - x_0)} \subseteq \Omega, \ \overline{\Omega} \subseteq x_0 + s(\Omega - x_0)$$

for every $r, \ s \in [0, +\infty[$ with $r < 1 < s$.

Proof. We preliminarily observe that

(2.5.3) $$\overline{x_0 + r(\Omega - x_0)} = x_0 + r(\overline{\Omega} - x_0) \text{ for every } r \in [0, +\infty[.$$

To prove (2.5.1) we observe that, by (2.5.3), for every $t \in \,]0, +\infty[$, and $y \in x_0 + t(\Omega - x_0)$ the half open line segment S joining x_0 and y, and not containing y, agrees with $x_0 + t((x_0 + \frac{S - x_0}{t}) - x_0)$, that $x_0 + \frac{S - x_0}{t}$ is the half open line segment joining x_0 and $x_0 + \frac{y - x_0}{t}$, and that $x_0 + \frac{y - x_0}{t} \in \overline{\Omega}$. Because of this, it turns out that $x_0 + \frac{S - x_0}{t} \subseteq \Omega$, and therefore that $S \subseteq x_0 + t(\Omega - x_0)$, from which the star shapedness of $x_0 + t(\Omega - x_0)$ follows.

Let now $r \in [0, 1[$, and let $y \in \overline{x_0 + r(\Omega - x_0)}$. Then, by (2.5.3), we conclude that $y = x_0 + r(z - x_0)$ for some $z \in \overline{\Omega}$. Consequently, y belongs to the half open line segment joining x_0 and z, and not containing z, which is a subset of Ω. Therefore the left-hand side of (2.5.2) follows.

In conclusion, if $s \in \,]1, +\infty[$, the right-hand side of (2.5.2) follows from the left-hand side one, once we observe that $\Omega = x_0 + \frac{1}{s}(\{x_0 + s(\Omega - x_0)\} - x_0)$, and that $x_0 + s(\Omega - x_0)$ is strongly star shaped by (2.5.1). ∎

Moreover, by using Proposition 1.1.5, it is easy to verify that

(2.5.4) Ω is strongly star shaped with respect to each of its points

whenever $\Omega \in \mathcal{A}(\mathbf{R}^n)$ is convex.

The class of the strongly star shaped open sets is sufficiently wide to provide the following covering result.

Proposition 2.5.4. *Let* $\Omega \in \mathcal{A}_0$ *have Lipschitz boundary. Then there exists a finite open covering* $\{\Omega_j\}_{j \in \{1,\dots,m\}}$ *of* $\overline{\Omega}$ *such that, for every* $j \in \{1,\dots,m\}$, $\Omega_j \cap \Omega$ *is strongly star shaped with Lipschitz boundary.*

Proof. Let $x \in \partial\Omega$, and let $I_x \in \mathcal{N}(x)$ such that $I_x \cap \partial\Omega$ is the graph, in a suitable coordinate system, of a Lipschitz continuous function whose epigraph contains $I_x \cap \Omega$. It is clear that it is not restrictive to assume that $I_x = \tilde{B} \times] - \varepsilon, \varepsilon[$, where $\tilde{B}$ is an open ball of $\mathbf{R}^{n-1}$ centred in the origin, and $\varepsilon > 0$.

For every $y \in \mathbf{R}^n$ let us set $\hat{y} = (y_1, \dots, y_{n-1})$, and denote again by $|\hat{y}|$ its norm. Then we can assume that

$$I_x \cap \overline{\Omega} = \{y \in \mathbf{R}^n : -\varepsilon < y_n \leq \vartheta(\hat{y}), \ \hat{y} \in \tilde{B}\}$$

for some $\vartheta \colon \mathbf{R}^{n-1} \to \mathbf{R}$ Lipschitz continuous, and that $x = (\hat{0}, \vartheta(\hat{0}))$, with $\vartheta(\hat{0}) > 0$. Moreover, if c is the Lipschitz constant of ϑ, we can assume that the radius of $\widetilde{B}$ is strictly smaller than $\frac{\vartheta(\hat{0})}{2c}$.

Let us prove that $I_x \cap \Omega$ is strongly star shaped with respect to 0.

To do this, let us first observe that $\overline{I_x \cap \Omega} = (\overline{I_x} \cap \partial\Omega) \cup (\overline{I_x} \cap \Omega)$, and take $y \in \overline{I_x} \cap \partial\Omega$. Then $y = (\hat{y}, \vartheta(\hat{y}))$, and, since $|\hat{y}| < \frac{\vartheta(\hat{0})}{2c}$, we have that

$$0 < \vartheta(\hat{0}) - 2ct|\hat{y}| \leq \vartheta(t\hat{y}) - ct|\hat{y}| \text{ for every } t \in [0,1],$$

from which we deduce that

$$(2.5.5) \qquad t\vartheta(\hat{y}) = t(\vartheta(\hat{y}) - \vartheta(t\hat{y})) + t\vartheta(t\hat{y}) \leq ct(1-t)|\hat{y}| + t\vartheta(t\hat{y}) <$$

$$< (1-t)\vartheta(t\hat{y}) + t\vartheta(t\hat{y}) = \vartheta(t\hat{y}) \text{ for every } t \in [0,1[.$$

By (2.5.5), once we observe that $-\varepsilon < \vartheta(\hat{y})$, we conclude that the half open line segment joining 0 and y, but not containing this last point, is contained in $I_x \cap \Omega$.

Let now $y \in \overline{I_x} \cap \Omega$. Then $-\varepsilon \leq y_n < \vartheta(\hat{y})$. Consequently, the half open line segment joining 0 and y lies between the ones joining 0 and $(\hat{y}, -\varepsilon)$, and 0 and $(\hat{y}, \vartheta(\hat{y}))$, from which we conclude that, also in this case, the half open line segment joining 0 and y, but not containing this last point, is contained in $I_x \cap \Omega$.

We have thus proved that $I_x \cap \Omega$ is strongly star shaped with respect to 0.

Let us now prove that $I_x \cap \Omega$ has Lipschitz boundary.

To do this, we observe that $\partial(I_x \cap \Omega) = (I_x \cap \partial\Omega) \cup (\partial I_x \cap \Omega) \cup (\partial I_x \cap \partial\Omega)$, and let $y \in \partial(I_x \cap \Omega)$.

It is clear that, if $y \in (I_x \cap \partial\Omega) \cup (\partial I_x \cap \Omega)$, it is possible to find $J_y \in \mathcal{N}(y)$ such that $J_y \cap \partial(I_x \cap \Omega)$ is the graph, in a suitable coordinate system, of a Lipschitz continuous function whose epigraph contains $J_y \cap (I_x \cap \Omega)$.

On the other side, if $y \in \partial I_x \cap \partial\Omega$, by carrying out a slight space rotation, it is possible to find again $J_y \in \mathcal{N}(y)$ such that $J_y \cap \partial(I_x \cap \Omega)$ is the graph, in the new coordinate system, of the minimum between two Lipschitz continuous functions, that is again Lipschitz continuous, and whose epigraph contains $J_y \cap (I_x \cap \Omega)$.

In conclusion, we have proved that for every $x \in \partial\Omega$ there exists $I_x \in \mathcal{N}(x)$ such that $I_x \cap \Omega$ is strongly star shaped, and with Lipschitz boundary. Because of this, the proof follows once we observe that for every $x \in \Omega$ there exists a ball centred in x and contained in Ω, that is certainly strongly star shaped and with Lipschitz boundary, and by taking into account the compactness of $\overline{\Omega}$. ∎

§2.6 Increasing Set Functions

For every $A, B \in \mathcal{A}(\mathbf{R}^n)$, we write $A \subset\subset B$ if $\overline{A}$ is a compact subset of B.

Definition 2.6.1. Let $\mathcal{O} \subseteq \mathcal{A}(\mathbf{R}^n)$, and $\alpha\colon \mathcal{O} \to [0, +\infty]$. We say that α is increasing if

$$\alpha(\Omega_1) \leq \alpha(\Omega_2) \text{ for every } \Omega_1,\ \Omega_2 \in \mathcal{O} \text{ such that } \Omega_1 \subseteq \Omega_2.$$

Definition 2.6.2. Let $\mathcal{O} \subseteq \mathcal{A}(\mathbf{R}^n)$, and $\alpha\colon \mathcal{O} \to [0, +\infty]$. For every $\mathcal{E} \subseteq \mathcal{O}$, we define the $\mathcal{E}$-inner regular envelope $\alpha_{\mathcal{E}-}$ of α as

$$\alpha_{\mathcal{E}-}\colon \Omega \in \mathcal{A}(\mathbf{R}^n) \mapsto$$

$$\begin{cases} 0 & \text{if } \{A \in \mathcal{E} : A \subset\subset \Omega\} = \emptyset \\ \sup\{\alpha(A) : A \in \mathcal{E},\ A \subset\subset \Omega\} & \text{if } \{A \in \mathcal{E} : A \subset\subset \Omega\} \neq \emptyset, \end{cases}$$

and say that α is $\mathcal{E}$-inner regular, or simply inner regular when $\mathcal{E} = \mathcal{O}$, if

$$\alpha(\Omega) = \alpha_{\mathcal{E}-}(\Omega) \text{ for every } \Omega \in \mathcal{O}.$$

When $\mathcal{E} = \mathcal{O}$ we write α_- in place of $\alpha_{\mathcal{O}-}$.

Remark 2.6.3. It is clear that, if $\mathcal{O} \subseteq \mathcal{A}(\mathbf{R}^n)$ and $\alpha\colon \mathcal{O} \to [0, +\infty]$, then, for every $\mathcal{E} \subseteq \mathcal{O}$, $\alpha_{\mathcal{E}-}$ is increasing. Moreover, if α is increasing, then

$$\alpha_{\mathcal{E}-}(\Omega) \leq \alpha_-(\Omega) \leq \alpha(\Omega) \text{ for every } \Omega \in \mathcal{O}.$$

Inner regular envelopes are inner regular, as proved by the following result.

Proposition 2.6.4. Let $\mathcal{O} \subseteq \mathcal{A}(\mathbf{R}^n)$, and $\alpha\colon \mathcal{O} \to [0, +\infty]$ be increasing. Then α_- is inner regular, i.e.

$$\alpha_-(\Omega) = (\alpha_-)_-(\Omega) = (\alpha_-)_{\mathcal{A}(\mathbf{R}^n)-}(\Omega) = (\alpha_-)_{\mathcal{A}_0-}(\Omega)$$

$$\text{for every } \Omega \in \mathcal{A}(\mathbf{R}^n).$$

Proof. The second and the third equalities are nothing more than the definition of α_-, therefore we have to prove only the first one.

Since for every $\Omega \in \mathcal{A}(\mathbf{R}^n)$, and $A \in \mathcal{O}$ we can find $B \in \mathcal{A}_0$ satisfying $A \subset\subset B \subset\subset \Omega$, we have that

$$\alpha_-(\Omega) = \sup\{\alpha(A) : A \in \mathcal{O},\ A \subset\subset \Omega\} \leq$$

$$\leq \sup\{\alpha_-(B) : B \in \mathcal{A}_0,\ B \subset\subset \Omega\} = (\alpha_-)_-(\Omega) \text{ for every } \Omega \in \mathcal{A}(\mathbf{R}^n).$$

Because of this, and by Remark 2.6.3, the proof follows. ∎

In order to establish some inner regularity criteria, we need to give some definitions.

Definition 2.6.5. *Let $\mathcal{O} \subseteq \mathcal{A}(\mathbf{R}^n)$, and $\alpha \colon \mathcal{O} \to [0, +\infty]$. We say that α is*
i) weakly superadditive if

$$\alpha(\Omega_1) + \alpha(\Omega_2) \leq \alpha(\Omega)$$

for every Ω_1, Ω_2, $\Omega \in \mathcal{O}$ with $\Omega_1 \cap \Omega_2 = \emptyset$, $\Omega_1 \cup \Omega_2 \subset\subset \Omega$,

ii) weakly subadditive if

$$\alpha(\Omega) \leq \alpha(\Omega_1) + \alpha(\Omega_2) \text{ for every } \Omega, \Omega_1, \Omega_2 \in \mathcal{O} \text{ with } \Omega \subset\subset \Omega_1 \cup \Omega_2,$$

iii) superadditive if
$$\alpha(\Omega_1) + \alpha(\Omega_2) \leq \alpha(\Omega)$$

for every Ω_1, Ω_2, $\Omega \in \mathcal{O}$ with $\Omega_1 \cap \Omega_2 = \emptyset$, $\Omega_1 \cup \Omega_2 \subseteq \Omega$,

iv) subadditive if

$$\alpha(\Omega) \leq \alpha(\Omega_1) + \alpha(\Omega_2) \text{ for every } \Omega, \Omega_1, \Omega_2 \in \mathcal{O} \text{ with } \Omega \subseteq \Omega_1 \cup \Omega_2.$$

If in addition $\mathcal{O}$ fulfils the following assumption

(2.6.1) $\Omega \setminus \overline{A} \in \mathcal{O}$ *for every Ω, $A \in \mathcal{O}$ such that $A \subset\subset \Omega$,*

we say that α is
v) boundary superadditive if

$$\alpha(A) + \alpha(\Omega \setminus \overline{B}) \leq \alpha(\Omega) \text{ for every } \Omega, A, B \in \mathcal{O} \text{ such that } A \subset\subset B \subset\subset \Omega,$$

vi) boundary subadditive if

$$\alpha(\Omega) \leq \alpha(B) + \alpha(\Omega \setminus \overline{A}) \text{ for every } \Omega, A, B \in \mathcal{O} \text{ such that } A \subset\subset B \subset\subset \Omega.$$

Remark 2.6.6. It is obvious that, if $\mathcal{O} \subseteq \mathcal{A}(\mathbf{R}^n)$ and $\alpha \colon \mathcal{O} \to [0, +\infty]$ is superadditive, then it is also weakly superadditive. Analogously, if α is subadditive, then it is also weakly subadditive.

It is also clear that, if $\mathcal{O}$ satisfies (2.6.1), and α is superadditive, then it is also boundary superadditive. Analogously, if $\mathcal{O}$ satisfies (2.6.1), and α is subadditive, then it is also boundary subadditive.

Definition 2.6.7. *Let $\mathcal{O} \subseteq \mathcal{A}(\mathbf{R}^n)$. A family $\mathcal{D} \subseteq \mathcal{A}(\mathbf{R}^n)$ is said to be dense in $\mathcal{O}$ if for every Ω_1, $\Omega_2 \in \mathcal{O}$ with $\Omega_1 \subset\subset \Omega_2$ there exists $D \in \mathcal{D}$ satisfying $\Omega_1 \subset\subset D \subset\subset \Omega_2$.*

A family $\mathcal{P} \subseteq \mathcal{A}(\mathbf{R}^n)$ is said to be perfect in $\mathcal{O}$ if for every $\Omega \in \mathcal{P}$, $A \in \mathcal{O}$ with $A \subset\subset \Omega$ there exists $P \in \mathcal{P}$ such that $A \subset\subset P \subset\subset \Omega$.

It is clear that if $\mathcal{D}$ is dense in $\mathcal{O}$, and $\mathcal{D} \subseteq \mathcal{O}$, then $\mathcal{D}$ is also perfect in $\mathcal{O}$.

Proposition 2.6.8. *Let $\mathcal{O} \subseteq \mathcal{A}(\mathbf{R}^n)$ be dense in $\mathcal{A}(\mathbf{R}^n)$, and $\alpha \colon \mathcal{O} \to [0, +\infty]$. If α is weakly superadditive, then α_- is superadditive. Analogously, if α is weakly subadditive, then α_- is subadditive.*

Proof. Let us assume that α is weakly superadditive.

Let Ω_1, Ω_2, $\Omega \in \mathcal{A}(\mathbf{R}^n)$ with $\Omega_1 \cap \Omega_2 = \emptyset$, $\Omega_1 \cup \Omega_2 \subseteq \Omega$, and let A_1, $A_2 \in \mathcal{O}$ be such that $A_1 \subset\subset \Omega_1$, and $A_2 \subset\subset \Omega_2$. Then, by the density of $\mathcal{O}$ in $\mathcal{A}(\mathbf{R}^n)$, there exists $A \in \mathcal{O}$ satisfying $A_1 \cup A_2 \subset\subset A \subset\subset \Omega$, from which, together with the weak superadditivity of α, we conclude that

$$(2.6.2) \qquad \alpha(A_1) + \alpha(A_2) \le \alpha(A) \le \alpha_-(\Omega)$$

for every A_1, $A_2 \in \mathcal{O}$ with $A_1 \subset\subset \Omega_1$, $A_2 \subset\subset \Omega_2$.

By (2.6.2) the superadditivity of α_- follows.

Let us assume now that α is weakly subadditive.

Let Ω, Ω_1, $\Omega_2 \in \mathcal{A}(\mathbf{R}^n)$ with $\Omega \subseteq \Omega_1 \cup \Omega_2$, and let $A \in \mathcal{O}$ be such that $A \subset\subset \Omega$. By using the density of $\mathcal{O}$ in $\mathcal{A}(\mathbf{R}^n)$ it is easy to prove the existence of two increasing sequences $\{A'_h\} \subseteq \mathcal{O}$, and $\{A''_h\} \subseteq \mathcal{O}$ such that $A'_h \subset\subset \Omega_1$ and $A''_h \subset\subset \Omega_2$ for every $h \in \mathbf{N}$, $\Omega_1 = \cup_{h=1}^{+\infty} A'_h$, and $\Omega_2 = \cup_{h=1}^{+\infty} A''_h$. Consequently, by using the compactness of $\overline{A}$, we conclude that there exist A', $A'' \in \mathcal{O}$ satisfying $A' \subset\subset \Omega_1$, $A'' \subset\subset \Omega_2$, and $A \subset\subset A' \cup A''$.

Because of this, and by the weak subadditivity of α, we conclude that

$$(2.6.3) \qquad \alpha(A) \le \alpha(A_1) + \alpha(A_2) \le \alpha_-(\Omega_1) + \alpha_-(\Omega_2)$$

for every $A \in \mathcal{O}$ with $A \subset\subset \Omega$.

By (2.6.3) the subadditivity of α_- follows. ∎

Proposition 2.6.9. *Let $\mathcal{O} \subseteq \mathcal{A}(\mathbf{R}^n)$, and $\alpha \colon \mathcal{O} \to [0, +\infty]$. Then,*
i) *if $\mathcal{P} \subseteq \mathcal{A}(\mathbf{R}^n)$ is perfect in $\mathcal{O}$,*

$$(\alpha_{\mathcal{O}-})_{\mathcal{P}-}(\Omega) = \alpha_{\mathcal{O}-}(\Omega) \text{ for every } \Omega \in \mathcal{P},$$

ii) *if α is increasing, $\mathcal{D} \subseteq \mathcal{O}$, and $\mathcal{D}$ is dense in $\mathcal{O}$,*

$$\alpha_{\mathcal{O}-}(\Omega) = \alpha_{\mathcal{D}-}(\Omega) \text{ for every } \Omega \in \mathcal{O},$$

iii) *if α is increasing, $\mathcal{P} \subseteq \mathcal{O}$, and $\mathcal{P}$ is perfect in $\mathcal{O}$,*

$$\alpha_{\mathcal{O}-}(\Omega) = \alpha_{\mathcal{P}-}(\Omega) \text{ for every } \Omega \in \mathcal{P}.$$

Proof. Let us prove i).

Being $\alpha_{\mathcal{O}-}$ increasing, it is clear that

$$(2.6.4) \qquad (\alpha_{\mathcal{O}-})_{\mathcal{P}-}(\Omega) \le \alpha_{\mathcal{O}-}(\Omega) \text{ for every } \Omega \in \mathcal{A}(\mathbf{R}^n).$$

On the other side, let $\Omega \in \mathcal{P}$, and $A \in \mathcal{O}$ with $A \subset\subset \Omega$. Then, being $\mathcal{P}$ perfect in $\mathcal{O}$, there exists $B \in \mathcal{P}$ such that $A \subset\subset B \subset\subset \Omega$. Therefore we have

$$\alpha(A) \leq \alpha_{\mathcal{O}-}(B) \leq (\alpha_{\mathcal{O}-})_{\mathcal{P}-}(\Omega) \text{ for every } \Omega \in \mathcal{P},$$

from which, together with (2.6.4), condition i) follows.

Let us prove ii).

Since $\mathcal{D}$ is dense in $\mathcal{O}$, and α is increasing, it is easy to deduce that

$$\alpha(A) \leq \alpha_{\mathcal{D}-}(\Omega) \text{ for every } \Omega, \ A \in \mathcal{O} \text{ with } A \subset\subset \Omega,$$

from which it follows that

(2.6.5) $$\alpha_{\mathcal{O}-}(\Omega) \leq \alpha_{\mathcal{D}-}(\Omega) \text{ for every } \Omega \in \mathcal{O}.$$

By (2.6.5), since $\mathcal{D} \subseteq \mathcal{O}$ and consequently

$$\alpha_{\mathcal{D}-}(\Omega) \leq \alpha_{\mathcal{O}-}(\Omega) \text{ for every } \Omega \in \mathcal{A}(\mathbf{R}^n),$$

condition ii) follows.

Finally the proof of iii) is similar to the one of ii), by taking $\Omega \in \mathcal{P}$. ∎

Given $\{A_h\} \subseteq \mathcal{O}$, and $\Omega \in \mathcal{O}$ such that $A_h \subseteq \Omega$ for every $n \in \mathbf{N}$, we say that $\{A_h\}$ is *well increasing* to Ω if $A_h \subset\subset A_{h+1}$ for every $h \in \mathbf{N}$, and $\cup_{h=1}^{+\infty} A_h = \Omega$. We say that $\{A_h\}$ is *well decreasing to the empty set with respect to* Ω if $\{\Omega \setminus A_h\}$ is well increasing to Ω.

We can now prove a first characterization of inner regular functions.

Proposition 2.6.10. *Let $\mathcal{O} \subseteq \mathcal{A}(\mathbf{R}^n)$ satisfy (2.6.1), and $\alpha \colon \mathcal{O} \to [0, +\infty]$. Assume that α is inner regular and boundary superadditive. Then*
i) for every $\Omega \in \mathcal{O}$ for which $\alpha(\Omega) < +\infty$, α is vanishing along the sequences in $\mathcal{O}$ that are well decreasing to the empty set with respect to Ω,
ii) for every $\Omega \in \mathcal{O}$ for which $\alpha(\Omega) = +\infty$, α is diverging along the sequences in $\mathcal{O}$ that are well increasing to Ω.

Conversely, assume that $\mathcal{O}$ is perfect in $\mathcal{A}_0$, that α is increasing, boundary subadditive, and that i) and ii) hold. Then α is inner regular.

Proof. We prove the first part of the proposition.

Let $\Omega \in \mathcal{O}$ be such that $\alpha(\Omega) < +\infty$, and let $\{A_h\}$ be a sequence in $\mathcal{O}$ well decreasing to the empty set with respect to Ω. Then by (2.6.1), and the boundary superadditivity of α it follows that

$$\alpha(A_{n+1}) \leq \alpha(\Omega) - \alpha(\Omega \setminus \overline{A_h}),$$

from which, together with the inner regularity of α, i) follows.

Moreover, the inner regularity of α implies condition ii), and the first part of the proposition.

Let us now prove the second part of the proposition.

Since α is increasing, from Remark 2.6.3 it follows that

(2.6.6) $$\alpha_-(\Omega) \leq \alpha(\Omega) \text{ for every } \Omega \in \mathcal{O}.$$

Let now $\Omega \in \mathcal{O}$, and assume for the moment that $\alpha(\Omega) < +\infty$. Let $K \in \mathcal{A}_0$ with $K \subset\subset \Omega$. Then, being $\mathcal{O}$ perfect in $\mathcal{A}_0$, there exist $A, B \in \mathcal{O}$ such that $K \subset\subset A \subset\subset B \subset\subset \Omega$.

Because of this, (2.6.1), the boundary subadditivity of α, and being α increasing, we conclude that

$$\alpha(\Omega) \leq \alpha(B) + \alpha(\Omega \setminus \overline{A}) \leq \alpha_-(\Omega) + \alpha(\Omega \setminus \overline{A}),$$

from which, together with assumption i), the opposite inequality to (2.6.6) and the inner regularity of α at Ω when $\alpha(\Omega) < +\infty$ follow.

In conclusion, being by assumption ii) α inner regular at Ω also when $\alpha(\Omega) = +\infty$, the inner regularity of α follows. ∎

As corollary, we deduce the following result.

Proposition 2.6.11. *Let $\mathcal{O} \subseteq \mathcal{A}(\mathbf{R}^n)$ satisfy (2.6.1), and $\alpha \colon \mathcal{O} \to [0, +\infty[$. Assume that $\mathcal{O}$ is perfect in $\mathcal{A}_0$, and that α is increasing, boundary subadditive, and such that for every $\Omega \in \mathcal{O}$ there exists a Borel positive measure μ_Ω on Ω satisfying*

$$\alpha(A) \leq \mu_\Omega(A) < +\infty \text{ for every } A \in \mathcal{O} \cap \mathcal{A}(\Omega).$$

Then α is inner regular.

Proof. Follows from Proposition 2.6.10. ∎

The following result is a variant of the De Giorgi-Letta Extension Theorem in our setting (cf. [DGL, Proposition 5.5 and Théorème 5.6], [DM2, Theorem 14.23]).

Theorem 2.6.12. *Let $\mathcal{O} \subseteq \mathcal{A}(\mathbf{R}^n)$ be dense in $\mathcal{A}(\mathbf{R}^n)$, and $\alpha \colon \mathcal{O} \to [0, +\infty]$ be increasing, weakly superadditive, and weakly subadditive. For every $E \subseteq \mathbf{R}^n$ let us set*

$$\alpha^*(E) = \inf\{\alpha_-(A) : A \in \mathcal{A}(\mathbf{R}^n), \ E \subseteq A\}.$$

Then the restriction of α^* to $\mathcal{B}(\mathbf{R}^n)$ is a Borel positive measure that agrees with α_- on $\mathcal{O}$.

If, in addition, α is also inner regular, then α^* agrees with α on $\mathcal{O}$.

Proof. First of all we observe that, being α increasing, it is easy to verify that α^* agrees with α_- on $\mathcal{O}$. In addition, if α is also inner regular, the coincidence of α^* with α on $\mathcal{O}$ follows.

Let us prove that the restriction of α^* to $\mathcal{B}(\mathbf{R}^n)$ is a Borel positive measure.

Since α is increasing, Proposition 2.6.4 yields the inner regularity of α_-. Moreover, since α is weakly superadditive and weakly subadditive, and $\mathcal{O}$ is dense in $\mathcal{A}(\mathbf{R}^n)$, Proposition 2.6.8 yields the superadditivity and the subadditivity of α_-.

If $\alpha_-(\emptyset) \neq 0$, by using the superadditivity and subadditivity properties of α_-, it must necessarily result $\alpha_-(\emptyset) = +\infty$. Consequently, α^* turns out to agree with the Borel positive measure identically equal to $+\infty$.

Because of this, we can assume that $\alpha_-(\emptyset) = 0$, and, consequently, that $\alpha^*(\emptyset) = 0$.

Let us first prove that α_- is countably subadditive, i.e.

$$(2.6.7) \qquad \alpha_-(\Omega) \leq \sum_{h=1}^{+\infty} \alpha_-(\Omega_h)$$

whenever $\Omega, \Omega_1, \ldots, \Omega_h, \ldots \in \mathcal{A}(\mathbf{R}^n)$ satisfy $\Omega \subseteq \cup_{h=1}^{+\infty}\Omega_h$.

To do this, let $\Omega, \Omega_1, \ldots, \Omega_h, \ldots$ be as in (2.6.7), and let $A \in \mathcal{A}(\mathbf{R}^n)$ be such that $A \subset\subset \Omega$. Then, by using the compactness of $\overline{A}$, and the subadditivity of α_- it is easy to verify that

$$\alpha_-(A) \leq \sum_{h=1}^{+\infty} \alpha_-(\Omega_h) \text{ for every } A \in \mathcal{A}(\mathbf{R}^n) \text{ such that } A \subset\subset \Omega,$$

from which, together with the inner regularity of α_-, (2.6.7) follows.

By (2.6.7) it follows that

$$(2.6.8) \qquad \alpha^*(S) \leq \sum_{h=1}^{+\infty} \alpha^*(S_h)$$

whenever $S, S_1, \ldots, S_h, \ldots \subseteq \mathbf{R}^n$ satisfy $S \subseteq \cup_{h=1}^{+\infty}S_h$.

In fact, let $S, S_1, \ldots, S_h, \ldots \subseteq \mathbf{R}^n$ be as in (2.6.8). We can clearly assume that $\sum_{h=1}^{+\infty} \alpha^*(S_h) < +\infty$, so that, given $\varepsilon > 0$, for every $h \in \mathbf{N}$ we can find $A_h \in \mathcal{A}(\mathbf{R}^n)$ with $S_h \subseteq A_h$, and $\alpha_-(A_h) < \alpha^*(S_h) + \frac{\varepsilon}{2^h}$. Because of this, and (2.6.7) we conclude that

$$\alpha^*(S) \leq \alpha_-\left(\cup_{h=1}^{+\infty}A_h\right) \leq \sum_{h=1}^{+\infty} \alpha_-(A_h) \leq \sum_{h=1}^{+\infty} \alpha^*(S_h) + \varepsilon \text{ for every } \varepsilon > 0,$$

from which (2.6.8) follows.

Since $\alpha^*(\emptyset) = 0$, by (2.6.8) it turns out that α^* is an outer measure, and consequently (cf., for example [Co, Chapter 1]) that the set $\mathcal{S}_{\alpha^*}(\mathbf{R}^n) =$

$\{B \subseteq \mathbf{R}^n : \alpha^*(B) = \alpha^*(B \cap S) + \alpha^*(B \setminus S)$ for every $S \subseteq \mathbf{R}^n\}$ is a σ-algebra on $\mathbf{R}^n$, and that the restriction of α^* to $\mathcal{S}_{\alpha^*}(\mathbf{R}^n)$ is a positive measure. Therefore, in order to complete the proof, we only have to prove that $\mathcal{B}(\mathbf{R}^n) \subseteq \mathcal{S}_{\alpha^*}(\mathbf{R}^n)$.

To do this, being $\mathcal{S}_{\alpha^*}(\mathbf{R}^n)$ a σ-algebra on $\mathbf{R}^n$, it suffices to prove that $\mathcal{A}(\mathbf{R}^n) \subseteq \mathcal{S}_{\alpha^*}(\mathbf{R}^n)$.

Let $\Omega \in \mathcal{A}(\mathbf{R}^n)$. Then, by (2.6.8) we only have to prove that

$$(2.6.9) \qquad \alpha^*(S \cap \Omega) + \alpha^*(S \setminus \Omega) \le \alpha^*(S) \text{ for every } S \subseteq \mathbf{R}^n.$$

If this is not the case, let $S \subseteq \mathbf{R}^n$ be such that

$$\alpha^*(S) < \alpha^*(S \cap \Omega) + \alpha^*(S \setminus \Omega),$$

and let $A \in \mathcal{A}(\mathbf{R}^n)$ be such that $S \subseteq A$, and

$$\alpha_-(A) < \alpha_-(A \cap \Omega) + \alpha^*(A \setminus \Omega).$$

Moreover, by exploiting the inner regularity of α_-, let $B \in \mathcal{A}(\mathbf{R}^n)$ be such that $B \subset\subset A \cap \Omega$, and

$$(2.6.10) \qquad \alpha_-(A) < \alpha_-(B) + \alpha^*(A \setminus \Omega).$$

In conclusion, by the superadditivity of α_-, we deduce that

$$\alpha_-(B) + \alpha^*(A \setminus \Omega) \le \alpha_-(B) + \alpha_-(A \setminus \overline{B}) \le \alpha_-(A),$$

contrary to (2.6.10).

Because of this, (2.6.9) holds, and $\mathcal{A}(\mathbf{R}^n) \subseteq \mathcal{S}_{\alpha^*}(\mathbf{R}^n)$. This concludes the proof. ∎

From Theorem 2.6.12 we deduce the following result.

Proposition 2.6.13. *Let* $\alpha \colon \mathcal{A}_0 \to [0, +\infty]$ *be increasing, weakly superadditive, and weakly subadditive. Then the limit* $\lambda(x) = \lim_{r \to 0+} \frac{1}{r^n} \alpha(Q_r(x))$ *exists for* $\mathcal{L}^n$*-a.e.* $x \in \mathbf{R}^n$, λ *is* $\mathcal{L}_n$*-measurable, and*

$$\alpha_-(\Omega) \ge \int_\Omega \lambda(x)dx \text{ for every } \Omega \in \mathcal{A}_0.$$

Proof. Let us preliminarily observe that, being α increasing, we have

$$\limsup_{r \to 0+} \frac{1}{r^n} \alpha(Q_r(x)) \ge \limsup_{r \to 0+} \frac{1}{r^n} \alpha_-(Q_r(x)) \ge t^n \limsup_{r \to 0+} \frac{1}{t^n r^n} \alpha(Q_{tr}(x)) =$$

$$= t^n \limsup_{s \to 0+} \frac{1}{s^n} \alpha(Q_s(x)) \text{ for every } x \in \mathbf{R}^n, \ t \in]0, 1[,$$

from which we deduce that

$$(2.6.11) \quad \limsup_{r \to 0^+} \frac{1}{r^n} a(Q_r(x)) = \limsup_{r \to 0^+} \frac{1}{r^n} \alpha_-(Q_r(x)) \text{ for every } x_0 \in \mathbf{R}^n.$$

Since α is increasing, weakly superadditive, and weakly subadditive, by Theorem 2.6.12 we deduce the existence of a Borel positive measure α^* on $\mathbf{R}^n$ that agrees with α_- on $\mathcal{A}_0$.

Let now $\Omega \in \mathcal{A}_0$, and observe that we can assume that $\alpha_-(\Omega) < +\infty$.

Let $\nu: A \in \mathcal{B}(\Omega) \mapsto \alpha^*(A)$. Then ν is a Borel real measure on Ω, therefore, by the Lebesgue Decomposition Theorem, we can decompose ν into the sum of its absolutely continuous part with respect to Lebesgue measure ν^{a} and of its singular part ν^{s}.

By Theorem 2.3.5, (2.6.11), and Remark 2.6.3 we obtain that for $\mathcal{L}^n$-a.e. $x \in \mathbf{R}^n$ the limit $\lambda(x) = \lim_{r \to 0^+} \frac{1}{r^n} \alpha(Q_r(x))$ exists, and $\lambda(x) = \frac{d\nu^{\mathrm{a}}}{d\mathcal{L}^n}(x)$. In fact we have that

$$\frac{d\nu^{\mathrm{a}}}{d\mathcal{L}^n}(x) = \limsup_{r \to 0^+} \frac{1}{r^n} \alpha_-(Q_r(x)) = \limsup_{r \to 0^+} \frac{1}{r^n} \alpha(Q_r(x)) \geq$$

$$\geq \liminf_{r \to 0^+} \frac{1}{r^n} \alpha(Q_r(x)) \geq \liminf_{r \to 0^+} \frac{1}{r^n} \alpha_-(Q_r(x)) = \frac{d\nu^{\mathrm{a}}}{d\mathcal{L}^n}(x)$$

$$\text{for } \mathcal{L}^n\text{-a.e. } x \in \mathbf{R}^n.$$

In conclusion, taking into account that Ω is open, by the Lebesgue Decomposition Theorem, and the Radon-Nikodym Theorem we conclude that

$$\alpha_-(\Omega) = \alpha^*(\Omega) \geq \nu^{\mathrm{a}}(\Omega) = \int_\Omega \frac{d\nu^{\mathrm{a}}}{d\mathcal{L}^n}(x) dx = \int_\Omega \lambda(x) dx,$$

which proves the proposition. ∎

Finally, we make some remarks about *translation invariant set functions*, i.e. functions of the type $\alpha: \mathcal{A}(\mathbf{R}^n) \to [0, +\infty]$ satisfying

$$\alpha(x_0 + A) = \alpha(A) \text{ for every } A \in \mathcal{A}(\mathbf{R}^n), \ x_0 \in \mathbf{R}^n.$$

Proposition 2.6.14. *Let $\alpha: \mathcal{A}(\mathbf{R}^n) \to [0, +\infty]$ be increasing, weakly superadditive, weakly subadditive, and translation invariant. Then*

$$(2.6.12) \qquad \alpha_-(A) = \alpha_-(Q_1(0))\mathcal{L}^n(A) \text{ for every } A \in \mathcal{A}(\mathbf{R}^n).$$

Proof. Let us first recall that, by Theorem 2.6.12, α_- turns out to be the restriction to $\mathcal{A}(\mathbf{R}^n)$ of a Borel positive measure on $\mathbf{R}^n$ that results to be translation invariant.

In order to identify α_- let us consider separately the cases in which $\alpha_-(Q_1(0)) < +\infty$ and $\alpha_-(Q_1(0)) = +\infty$.

If $\alpha_-(Q_1(0)) < +\infty$ we observe that, since every bounded open set can be covered by a finite number of translated of $Q_1(0)$ and α_- is translation invariant, α_- turns out to be locally finite on $\mathbf{R}^n$. Because of this, and well known properties of translation invariant measures (see for example [Co, Proposition 1.4.5]), equality (2.6.12) follows.

If $\alpha_-(Q_1(0)) = +\infty$ we observe that, since for every $r > 0$ $Q_1(0)$ can be covered by a finite number of translated of $Q_r(0)$ and α_- is translation invariant, it turns out that $\alpha_-(Q_r(0)) = +\infty$. Because of this, and again the translation invariance of α_- we deduce that $\alpha_-(A) = +\infty$ for every $A \in \mathcal{A}(\mathbf{R}^n)$ from which equality (2.6.12) follows. ■

By Proposition 2.6.14 we trivially deduce the following result.

Proposition 2.6.15. *Let* $\alpha\colon \mathcal{A}(\mathbf{R}^n) \to [0, +\infty]$ *be increasing, weakly superadditive, weakly subadditive, and translation invariant. Then*

$$\alpha(A) \le \alpha(Q_1(0))\mathcal{L}^n(\overline{A}) \text{ for every } A \in \mathcal{A}(\mathbf{R}^n),$$

$$\alpha(A) \ge \alpha(Q_1(0))\mathcal{L}^n(A) \text{ for every } A \in \mathcal{A}(\mathbf{R}^n),$$

(2.6.13) $$\alpha(A) = \alpha_-(A) = \alpha(Q_1(0))\mathcal{L}^n(A)$$

for every $A \in \mathcal{A}(\mathbf{R}^n)$ *with* $\mathcal{L}^n(\partial A) = 0$.

Remark 2.6.16. It is clear that (2.6.13) of Proposition 2.6.15 cannot hold for every $A \in \mathcal{A}(\mathbf{R}^n)$. To see this let us set, for every $A \in \mathcal{A}(\mathbf{R}^n)$, $\alpha(A) = \mathcal{L}^n(\overline{A})$. Then α fulfils the assumptions of Proposition 2.6.15 but clearly (2.6.13) does not hold if $\mathcal{L}^n(\partial A) \neq 0$.

§2.7 Increasing Set Functionals

Definition 2.7.1. *Let* $\mathcal{O} \subseteq \mathcal{A}(\mathbf{R}^n)$, U *be a set, and* $\Phi\colon \mathcal{O} \times U \to [0, +\infty]$. *We say that* Φ *is increasing if for every* $u \in U$, $\Phi(\cdot, u)$ *is increasing.*

Definition 2.7.2. *Let* $\mathcal{O} \subseteq \mathcal{A}(\mathbf{R}^n)$, U *be a set, and* $\Phi\colon \mathcal{O} \times U \to [0, +\infty]$. *For every* $\mathcal{E} \subseteq \mathcal{O}$, *we introduce the* $\mathcal{E}$-inner regular envelope $\Phi_{\mathcal{E}-}$ *of* Φ *as the function defined by*

$$\Phi_{\mathcal{E}-}\colon (\Omega, u) \in \mathcal{A}(\mathbf{R}^n) \times U \mapsto \Phi(\cdot, u)_{\mathcal{E}-}(\Omega),$$

and say that Φ *is* $\mathcal{E}$-inner regular, *or simply inner regular when* $\mathcal{E} = \mathcal{O}$, *if*

$$\Phi(\Omega, u) = \Phi_{\mathcal{E}-}(\Omega, u) \text{ for every } (\Omega, u) \in \mathcal{O} \times U.$$

When $\mathcal{E} = \mathcal{O}$ we write Φ_- in place of $\Phi_{\mathcal{O}-}$.

Definition 2.7.3. Let $\mathcal{O} \subseteq \mathcal{A}(\mathbf{R}^n)$, U be a set, and $\Phi: \mathcal{O} \times U \to [0, +\infty]$. We say that Φ is
i) *weakly superadditive if for every* $u \in U$, $\Phi(\cdot, u)$ *is weakly superadditive.*
ii) *weakly subadditive if for every* $u \in U$, $\Phi(\cdot, u)$ *is weakly subadditive.*

For every $E \subseteq \mathbf{R}^n$, every function u on E, and $t \in]0, +\infty[$ we define the *rescaled homothety* of u as

$$O_t u: x \in \frac{1}{t}E \mapsto \frac{1}{t}u(tx).$$

Let U be a set of functions on $\mathbf{R}^n$ such that

(2.7.1) $T[-x_0]O_tT[x_0]u \in U$ whenever $u \in U$, $x_0 \in \mathbf{R}^n$, $t \in]0, 1]$,

let $\mathcal{O} \subseteq \mathcal{A}(\mathbf{R}^n)$, and let $\Phi: \mathcal{O} \times U \to [0, +\infty]$ satisfy

(2.7.2) $$\liminf_{t \to 1^-} \Phi(\Omega, T[-x_0]O_tT[x_0]u) \geq \Phi(\Omega, u)$$

for every $\Omega \in \mathcal{O}$ strongly star shaped with respect to x_0, $u \in U$

and

(2.7.3) $$\limsup_{t \to 1^+} \Phi_-(x_0 + t(\Omega - x_0), T[-x_0]O_{1/t}T[x_0]u) \leq \Phi_-(\Omega, u)$$

for every $\Omega \in \mathcal{O}$ strongly star shaped with respect to x_0, $u \in U$.

Then the following inner regularity result holds.

Proposition 2.7.4. *Let* $\mathcal{O} \subseteq \mathcal{A}_0$, U *be a set of functions on* $\mathbf{R}^n$ *satisfying* (2.7.1), *and let* $\Phi: \mathcal{O} \times U \to [0, +\infty]$ *be increasing, and satisfying* (2.7.2), (2.7.3). *Then*

(2.7.4) $\Phi(\Omega, u) = \Phi_-(\Omega, u)$ *for every* $\Omega \in \mathcal{O}$ *strongly star shaped,* $u \in U$.

Proof. Let Ω, u be as in (2.7.4), $x_0 \in \Omega$ be such that Ω is strongly star shaped with respect to x_0, and $t \in]1, +\infty[$. Then, since Proposition 2.5.3 yields $\Omega \subset\subset x_0 + t(\Omega - x_0)$, we have that

(2.7.5) $\Phi(\Omega, T[-x_0]O_{1/t}T[x_0]u) \leq \Phi_-(x_0 + t(\Omega - x_0), T[-x_0]O_{1/t}T[x_0]u),$

hence as t decreases to 1, by (2.7.5), (2.7.2), (2.7.3), and Remark 2.6.3 we deduce (2.7.4). ∎

In order to extend Proposition 2.7.4 to wider classes of open sets let us consider a set U, and $\Phi: \mathcal{A}_0 \times U \to [0, +\infty]$. Let us introduce the following assumptions

(2.7.6) $$\Phi(\Omega, u) \leq \Phi(\Omega \cap \Omega_1, u) + \Phi(\Omega \cap \Omega_2, u)$$

whenever Ω, Ω_1, $\Omega_2 \in \mathcal{A}_0$ satisfy with $\Omega \subset\subset \Omega_1 \cup \Omega_2$, $u \in U$,

(2.7.7) for every $\Omega \in \mathcal{A}_0$, $u \in U$ with $\Phi(\Omega, u) < +\infty$

there exists a Borel positive measure $\mu_{\Omega,u}$ on Ω satisfying

$$\Phi(A, u) \leq \mu_{\Omega,u}(A) < +\infty \text{ for every } A \in \mathcal{A}(\Omega).$$

Lemma 2.7.5. *Let U be a set, and let $\Phi: \mathcal{A}_0 \times U \to [0, +\infty]$ be increasing, and satisfying (2.7.6). Then*

(2.7.8) $$\Phi(\Omega, u) \leq \sum_{j=1}^{m} \Phi(\Omega \cap \Omega_j, u)$$

whenever Ω, $\Omega_1, \ldots, \Omega_m \in \mathcal{A}_0$ satisfy $\Omega \subset\subset \cup_{j=1}^{m} \Omega_j$, $u \in U$.

Proof. We argue by induction on m.

If $m = 2$ (2.7.8) follows from (2.7.6).

If $m > 2$ let us assume that (2.7.8) holds with m replaced by $m - 1$, and prove it with m. To do this, we first take Ω, $\Omega_1, \ldots, \Omega_m$ as in (2.7.8), and, for every $j \in \{1, \ldots, m\}$, an open set A_j with $A_j \subset\subset \Omega_j$ such that $\Omega \subset\subset \cup_{j=1}^{m} A_j$. Then, by (2.7.6), we have

(2.7.9) $$\Phi(\Omega, u) \leq \Phi\left(\Omega \cap \cup_{j=1}^{m-1} A_j, u\right) + \Phi(\Omega \cap A_m, u).$$

Let us now observe that $\Omega \cap \cup_{j=1}^{m-1} A_j \subset\subset \cup_{j=1}^{m-1} \Omega_j$, hence by (2.7.9), the induction assumption, and (2.6.11) we get

$$\Phi(\Omega, u) \leq \sum_{j=1}^{m-1} \Phi\left((\Omega \cap \cup_{i=1}^{m-1} A_i) \cap \Omega_j, u\right) + \Phi(\Omega \cap A_m, u) \leq$$

$$\leq \sum_{j=1}^{m-1} \Phi(\Omega \cap \Omega_j, u) + \Phi(\Omega \cap \Omega_m, u),$$

which proves the lemma. ∎

Theorem 2.7.6. *Let U be a set of functions on $\mathbf{R}^n$ satisfying (2.7.1), and let $\Phi: \mathcal{A}_0 \times U \to [0, +\infty]$ be increasing, and satisfying (2.7.2), (2.7.3), (2.7.6), and (2.7.7). Then*

(2.7.10) $$\Phi(\Omega, u) = \Phi_-(\Omega, u)$$

for every $\Omega \in \mathcal{A}_0$ with Lipschitz boundary, $u \in U$.

Proof. Let Ω, u be as in (2.7.10).

If $\Phi(\Omega, u) < +\infty$ let us set $\alpha \colon A \in \mathcal{A}(\Omega) \mapsto \Phi(A, u) \in [0, +\infty[$. Let us observe that (2.7.6) implies the boundary subadditivity of α. In fact, let Ω', A, $B \in \mathcal{A}(\Omega)$ be such that $A \subset\subset B \subset\subset \Omega'$, and let $\Omega_1 \in \mathcal{A}(\Omega)$, $\Omega_2 \in \mathcal{A}_0$ satisfy $A \subset\subset \Omega_1 \subset\subset B$, and $\Omega' \setminus \overline{\Omega_1} \subset\subset \Omega_2$. Then $\Omega' \subset\subset \Omega_1 \cup \Omega_2$, and (2.7.6), once we recall that α is increasing, yields

$$\alpha(\Omega') \leq \alpha(\Omega' \cap \Omega_1) + \alpha(\Omega' \cap \Omega_2) \leq \alpha(B) + \alpha(\Omega' \setminus \overline{A}).$$

Because of this, and of (2.7.7), Proposition 2.6.11 applies, and (2.7.10) follows.

If $\Phi(\Omega, u) = +\infty$ let us prove that also $\Phi_-(\Omega, u) = +\infty$. If this is not the case, being Ω with Lipschitz boundary, let $\{\Omega_j\}_{j=1,\ldots,m}$ be the open covering of $\overline{\Omega}$ given by Proposition 2.5.4. Then by Lemma 2.7.5, Proposition 2.7.4, and the increasing character of Φ we have

$$\Phi(\Omega, u) \leq \sum_{j=1}^m \Phi(\Omega_j \cap \Omega, u) = \sum_{j=1}^m \Phi_-(\Omega_j \cap \Omega, u) \leq m\Phi_-(\Omega, u) < +\infty,$$

that contradicts our assumption. ∎

Chapter 3

Minimization Methods
and Variational Convergences

In the present chapter we recall the notion and the main properties of De Giorgi's Γ-convergence, introduced in the seventies to propose a framework in which settle the study of the asymptotic behaviour of families of variational problems.

In this chapter we describe the abstract features of Γ-convergence, and refer to Chapters 6 and those from 10 onwards for its applications to more concrete situations.

To properly introduce the subject, in the first section, we recall the abstract framework in which settle the study of minimization of variational problems. Then we introduce Γ-convergence theory and describe its applications to the calculus of variations. The last section is devoted to the study of a particular case of Γ-convergence: the one of relaxation.

We refer to [DG7], [DGF1], [DGF2], [DM2], and [DG6] for a more complete exposition on the subject.

§3.1 The Direct Methods in the Calculus of Variations

In this section we briefly recall the the main notions needed in order to treat the abstract problem of the minimization of a functional over a set.

As usual, such notions will be of topological nature, and the final result will be a variant of the well celebrated Weierstrass Theorem based on the lower semicontinuity properties of the functional, and on the compactness of the set. Nevertheless, in view of applications to the study of minimization problems in Γ-convergence theory, it seems to be more natural to follow an approach based on the weaker notion of countable compactness in place of compactness.

Let (U, τ) be a topological space.

Definition 3.1.1. *We say that a subset $K \subseteq U$ is countably compact if every countable open covering of K has a finite subcovering.*

We say that K is relatively countably compact if $\overline{K}$ is countably compact.

It is clear that a compact set is countably compact, and that, in general, the converse is false. Nevertheless, it becomes true if U satisfies the second countability axiom.

In general, countably compact sets, even Hausdorff, need not be closed, and the closure of a countably compact set need not be countably compact. On the other side, closed subsets of countably compact sets are again countably compact, and, provided U satisfies the first countability axiom, countably compact sets are closed.

Countably compact spaces have the nice feature to enjoy the Bolzano-Weierstrass property, as explained in the following result (cf. for example [Ro, Chapter 9, Proposition 7]).

Theorem 3.1.2. *A subset K of U is countably compact if and only if for every $\{u_h\} \subseteq K$ the set of the cluster points of $\{u_h\}$ in K is nonempty.*

Proof. Let us first assume that K is countably compact, and let $\{u_h\} \subseteq K$. For every $k \in \mathbf{N}$ let us set $A_k = U \setminus \overline{\{u_h : h \geq k\}}$. Then, for every $k \in \mathbf{N}$, A_k is open, and $A_k \subseteq A_{k+1}$. It is clear that $\{A_k\}$ cannot be a covering of K, otherwise, by the countable compactness of K, it would be $K \subseteq A_{k_0}$ for some $k_0 \in \mathbf{N}$, contrary to the fact that $u_{k_0+1} \in K \setminus A_{k_0}$. Because of this, $K \setminus \cup_{k=1}^{+\infty} A_k \neq \emptyset$. Let $u \in K \setminus \cup_{k=1}^{+\infty} A_k \neq \emptyset$. Then $u \in \cap_{k=1}^{+\infty} \overline{\{u_h : h \geq k\}}$ and, consequently, is a cluster point of $\{u_h\}$.

Let us assume now that for every sequence in K the set of the cluster points in K of the sequence is nonempty, and let $\{A_h\}_{h \in \mathbf{N}}$ be a countable covering of K. If $\{A_h\}_{h \in \mathbf{N}}$ has no finite subcoverings, then $K \setminus \cup_{h=1}^{k} A_h \neq \emptyset$ for every $k \in \mathbf{N}$. For every $k \in \mathbf{N}$ let $u_k \in K \setminus \cup_{h=1}^{k} A_h$, and let $u \in K$ be a cluster point of $\{u_k\}$. Then, being for every $k \in \mathbf{N}$, $K \setminus \cup_{h=1}^{k} A_h$ closed, and $K \setminus \cup_{h=1}^{k+1} A_h \subseteq K \setminus \cup_{h=1}^{k} A_h$, it turns out that $u \in \cap_{h=1}^{+\infty} (K \setminus \cup_{h=1}^{k} A_h) = K \setminus \cup_{h=1}^{+\infty} A_h$, contrary to the fact that $K \setminus \cup_{h=1}^{+\infty} A_h = \emptyset$. This yields that $\{A_h\}_{h \in \mathbf{N}}$ has a finite subcovering, and therefore that K is countably compact. ∎

By using Theorem 3.1.2, it is easy to see that a relatively sequentially compact set is countably compact. The converse in false in general topological spaces, but it becomes true if U satisfies the first countability axiom, or, by virtue of the Eberlein-Šmulian Theorem, if U is a Banach space equipped with its weak topology.

We now come to the problem of the minimization of a function.

Definition 3.1.3. *Let $F: U \to [-\infty, +\infty]$. We say that F is*
i) coercive if for every $\lambda \in \mathbf{R}$ there exists a compact subset K_λ of U such

that $\{u \in U : F(u) \leq \lambda\} \subseteq K_\lambda$,

ii) countably coercive if for every $\lambda \in \mathbf{R}$ there exists a countably compact subset K_λ of U such that $\{u \in U : F(u) \leq \lambda\} \subseteq K_\lambda$,

iii) strongly countably coercive if for every $\lambda \in \mathbf{R}$ there exists a closed and countably compact subset K_λ of U such that $\{u \in U : F(u) \leq \lambda\} \subseteq K_\lambda$,

iv) sequentially coercive if for every $\lambda \in \mathbf{R}$ there exists a sequentially compact subset K_λ of U such that $\{u \in U : F(u) \leq \lambda\} \subseteq K_\lambda$

v) strongly sequentially coercive if for every $\lambda \in \mathbf{R}$ there exists a closed and sequentially compact subset K_λ of U such that $\{u \in U : F(u) \leq \lambda\} \subseteq K_\lambda$.

We can state now the main result on the minimization of a functional over a set.

Theorem 3.1.4. Let $F:U \to [-\infty, +\infty]$ be lower semicontinuous and countably coercive (respectively lower semicontinuous and coercive, sequentially lower semicontinuous and sequentially coercive). Then F has a minimum in U.

Proof. We deal only with the non-sequential case, the proof for the others being similar with the obvious changes.

If F is identically equal to $+\infty$, the proof is obvious.

If not, let $\{\lambda_h\} \subseteq \mathbf{R}$ be strictly decreasing and such that $\lim_{h \to +\infty} \lambda_h = \inf_U F$, and let $\{u_h\} \subseteq U$ satisfy $F(u_h) \leq \lambda_h$ for every $h \in \mathbf{N}$.

Since $\{u_h\} \subseteq \{v \in U : F(v) \leq \lambda_1\}$, the countable coerciveness of F and Theorem 3.1.2 yield a cluster point $u \in U$ of $\{u_h\}$. Therefore (0.1) applies, and we get

$$\inf_U F \leq F(u) \leq \limsup_{h \to +\infty} F(u_h) \leq \lim_{h \to +\infty} \lambda_h = \inf_U F,$$

from which we conclude that u is a minimizer of F. ∎

Remark 3.1.5. It is worth while to remark that the part of Theorem 3.1.4 dealing with countable coerciveness still holds by replacing the lower semicontinuity assumption on F with the condition expressed by (0.1). It is easy to verify that the lower semicontinuity of F implies (0.1), and that (0.1) implies the sequential lower semicontinuity of F.

From Theorem 3.1.4 we deduce the following corollaries in the case of Banach spaces.

Theorem 3.1.6. Let W be a reflexive Banach space, $X \subseteq W$, and $F:X \to [-\infty, +\infty]$. Assume that X is convex and closed, that F is convex and W-lower semicontinuous, and that, if X is not bounded,

$$\lim_{\|u\| \to +\infty} F(u) = +\infty.$$

Then F has a minimum in X.

Proof. Let $\widehat{F}$ be defined as

$$\widehat{F} : u \in W \mapsto \begin{cases} F(u) & \text{if } u \in X \\ +\infty & \text{if } u \in W \setminus X, \end{cases}$$

then $\widehat{F}$ turns out to be convex and lower semicontinuous. Consequently, by Theorem 1.1.13, it results to be also weak-W-lower semicontinuous.

Then, by using such property, and the assumption on the behaviour of F at infinity if X is not bounded, it follows that for every $\lambda \in \mathbf{R}$ the set $\{u \in W : \widehat{F}(u) \leq \lambda\}$ is bounded and closed in the weak-W topology. Consequently, by the Bourbaki-Kakutani-Šmulian Theorem, it is also compact in the same topology, and the coerciveness of $\widehat{F}$ in the weak-W topology too follows.

Because of this, Theorem 3.1.4 applies to $\widehat{F}$, and we conclude that $\widehat{F}$ has a minimum in W. This trivially implies that F has a minimum in X, and concludes the proof. ∎

The results below deal with the case in which the functionals are defined in a subset of a dual space.

Theorem 3.1.7. *Let W be a Banach space, $X \subseteq W'$, and $F : X \to [-\infty, +\infty]$. Assume that X is closed in the weak*-W' topology, that F is weak*-W'-lower semicontinuous, and that, if X is not bounded,*

$$\lim_{\|y\| \to +\infty} F(y) = +\infty.$$

Then F has a minimum in X.

Proof. Let $\widehat{F}$ be defined as

$$\widehat{F} : y \in W' \mapsto \begin{cases} F(y) & \text{if } y \in X \\ +\infty & \text{if } y \in W' \setminus X. \end{cases}$$

Then, as in the proof of Theorem 3.1.6, by using the closure properties of X, and the lower semicontinuity ones of F, the lower semicontinuity of $\widehat{F}$ in the weak*-W' topology follows. Moreover, by using the assumption on the behaviour of F at infinity if X is not bounded, it follows that for every $\lambda \in \mathbf{R}$ the set $\{y \in W' : \widehat{F}(y) \leq \lambda\}$ is bounded and closed in the weak*-W' topology. Consequently, by Alaoglu's theorem, it is also compact in the same topology, and the coerciveness of $\widehat{F}$ in the weak*-W' topology follows.

Because of this, Theorem 3.1.4 applies to $\widehat{F}$, and we conclude that $\widehat{F}$ has a minimum in W'. This trivially implies that F has a minimum in X, and concludes the proof. ∎

Theorem 3.1.8. *Let W be a separable Banach space, $X \subseteq W'$, and $F: X \to [-\infty, +\infty]$. Assume that X is convex and sequentially closed in the weak*-W' topology, that F is convex and sequentially weak*-W'-lower semicontinuous, and that, if X is not bounded,*

$$\lim_{\|y\| \to +\infty} F(y) = +\infty.$$

Then F has a minimum in X.

Proof. First of all, we observe that Theorem 1.1.4 yields the closure of X in the weak*-W' topology.

Let $\widehat{F}$ be defined as in the proof of Theorem 3.1.7, then, by the properties of X and F, and by Theorem 1.1.14, it follows that $\widehat{F}$, and consequently F, are weak*-W'-lower semicontinuous.

Therefore, Theorem 3.1.7 applies, and the proof follows. ■

§3.2 Γ-Convergence

In the following, specially in view of the applications that we are going to develop, we will need to utilize a notion of variational convergence slightly more general of the usual one of Γ-convergence for sequences of functionals. It is the notion of multiple Γ-limit introduced in [DG5], and [DG6].

We need it since we are going to work with families of functionals depending on parameters that can be also real numbers varying in an interval.

Let (U, τ) be a topological space.

Definition 3.2.1. *Let $E \subseteq [-\infty, +\infty]$, $\varepsilon_0 \in \overline{E}$, and let, for every $\varepsilon \in E$, $F_\varepsilon: U \to [-\infty, +\infty]$. We define the $\Gamma^-(\tau)$-lower limit, and the $\Gamma^-(\tau)$-upper limit of $\{F_\varepsilon\}_{\varepsilon \in E}$ as ε goes to ε_0 as the functionals defined by*

$$\Gamma^-(\tau) \liminf_{\varepsilon \to \varepsilon_0} F_\varepsilon : u \in U \mapsto \sup_{I \in \mathcal{N}(u)} \liminf_{\varepsilon \to \varepsilon_0} \inf_{v \in I} F_\varepsilon(v),$$

$$\Gamma^-(\tau) \limsup_{\varepsilon \to \varepsilon_0} F_\varepsilon : u \in U \mapsto \sup_{I \in \mathcal{N}(u)} \limsup_{\varepsilon \to \varepsilon_0} \inf_{v \in I} F_\varepsilon(v).$$

If in $u \in U$ it results

$$\Gamma^-(\tau) \liminf_{\varepsilon \to \varepsilon_0} F_\varepsilon(u) = \Gamma^-(\tau) \limsup_{\varepsilon \to \varepsilon_0} F_\varepsilon(u),$$

we say that the family $\{F_\varepsilon\}_{\varepsilon \in E}$ $\Gamma^-(\tau)$-converges in u as ε goes to ε_0, and we define the $\Gamma^-(\tau)$-limit in u of $\{F_\varepsilon\}_{\varepsilon \in E}$ as ε goes to ε_0 by

$$\Gamma^-(\tau) \lim_{\varepsilon \to \varepsilon_0} F_\varepsilon(u) = \Gamma^-(\tau) \liminf_{\varepsilon \to \varepsilon_0} F_\varepsilon(u) = \Gamma^-(\tau) \limsup_{\varepsilon \to \varepsilon_0} F_\varepsilon(u).$$

When $E = \mathbf{N}$ we always take $\varepsilon_0 = +\infty$. In this case the above definitions reduce to the usual ones of Γ-upper limit, Γ-lower limit, and Γ-limit of a sequence of functionals proposed in [DGF1]. As usual in this case, we write "$\Gamma^-(\tau) \liminf_{h \to +\infty} F_h$" in place of "$\Gamma^-(\tau) \liminf_{\varepsilon \to +\infty} F_\varepsilon$," and use analogous notations for the remaining limits.

We observe explicitly that the $\Gamma^-(\tau)$-lower limit, and the $\Gamma^-(\tau)$-upper limit of $\{F_\varepsilon\}_{\varepsilon \in E}$ as ε goes to ε_0 exist for every $u \in U$.

It is clear that

$$(3.2.1) \qquad \Gamma^-(\tau) \liminf_{\varepsilon \to \varepsilon_0} F_\varepsilon(u) \leq \Gamma^-(\tau) \limsup_{\varepsilon \to \varepsilon_0} F_\varepsilon(u) \text{ for every } u \in U,$$

and that, if τ' is another topology on U, finer than τ, it results that

$$\Gamma^-(\tau) \liminf_{\varepsilon \to \varepsilon_0} F_\varepsilon(u) \leq \Gamma^-(\tau') \liminf_{\varepsilon \to \varepsilon_0} F_\varepsilon(u),$$

$$\Gamma^-(\tau) \limsup_{\varepsilon \to \varepsilon_0} F_\varepsilon(u) \leq \Gamma^-(\tau') \limsup_{\varepsilon \to \varepsilon_0} F_\varepsilon(u)$$

for every $u \in U$.

Γ-limits turn out to be stable with respect to continuous perturbations, as proved by the following result.

Proposition 3.2.2. *Let* $E \subseteq [-\infty, +\infty]$, $\varepsilon_0 \in \overline{E}$, *and let, for every* $\varepsilon \in E$, $F_\varepsilon : U \to [-\infty, +\infty]$, *and* $G : U \to \mathbf{R}$. *Assume that* G *is continuous. Then*

$$\Gamma^-(\tau) \liminf_{\varepsilon \to \varepsilon_0} (F_\varepsilon + G)(u) = \Gamma^-(\tau) \liminf_{\varepsilon \to \varepsilon_0} F_\varepsilon(u) + G(u)$$

$$\Gamma^-(\tau) \limsup_{\varepsilon \to \varepsilon_0} (F_\varepsilon + G)(u) = \Gamma^-(\tau) \limsup_{\varepsilon \to \varepsilon_0} F_\varepsilon(u) + G(u)$$

for every $u \in U$.

Proof. We prove only the second equality, the proof of the first one being analogous.

Let $u \in U$. Let us fix $I_0 \in \mathcal{N}(u)$, and let us observe that

$$\Gamma^-(\tau) \limsup_{\varepsilon \to \varepsilon_0} (F_\varepsilon + G)(u) = \sup_{I \in \mathcal{N}(u),\, I \subseteq I_0} \limsup_{\varepsilon \to \varepsilon_0} \inf_{v \in I} (F_\varepsilon + G)(v),$$

$$\Gamma^-(\tau) \limsup_{\varepsilon \to \varepsilon_0} F_\varepsilon(u) = \sup_{I \in \mathcal{N}(u),\, I \subseteq I_0} \limsup_{\varepsilon \to \varepsilon_0} \inf_{v \in I} F_\varepsilon(v).$$

Then we have that

$$\Gamma^-(\tau) \limsup_{\varepsilon \to \varepsilon_0} (F_\varepsilon + G)(u) \geq \sup_{I \in \mathcal{N}(u),\, I \subseteq I_0} \limsup_{\varepsilon \to \varepsilon_0} \left\{ \inf_{v \in I} F_\varepsilon(v) + \inf_{v \in I} G(v) \right\} \geq$$

$$\geq \sup_{I \in \mathcal{N}(u), \, I \subseteq I_0} \limsup_{\varepsilon \to \varepsilon_0} \left\{ \inf_{v \in I} F_\varepsilon(v) + \inf_{v \in I_0} G(v) \right\} =$$

$$= \Gamma^-(\tau) \limsup_{\varepsilon \to \varepsilon_0} F_\varepsilon(u) + \inf_{v \in I_0} G(v) \text{ for every } I_0 \in \mathcal{N}(u).$$

Because of this, and by taking into account also the continuity of G, we conclude that

$$(3.2.2) \qquad \Gamma^-(\tau) \limsup_{\varepsilon \to \varepsilon_0} (F_\varepsilon + G)(u) \geq$$

$$\geq \Gamma^-(\tau) \limsup_{\varepsilon \to \varepsilon_0} F_\varepsilon(u) + \sup_{I_0 \in \mathcal{N}(u)} \inf_{v \in I_0} G(v) = \Gamma^-(\tau) \limsup_{\varepsilon \to \varepsilon_0} F_\varepsilon(u) + G(u).$$

In order to prove the reverse inequality, let $u \in U$. Let us take $I_0 \in \mathcal{N}(u)$. Then

$$\Gamma^-(\tau) \limsup_{\varepsilon \to \varepsilon_0} (F_\varepsilon + G)(u) \leq \sup_{I \in \mathcal{N}(u), \, I \subseteq I_0} \limsup_{\varepsilon \to \varepsilon_0} \left\{ \inf_{v \in I} F_\varepsilon(v) + \sup_{v \in I} G(v) \right\} \leq$$

$$\leq \sup_{I \in \mathcal{N}(u), \, I \subseteq I_0} \limsup_{\varepsilon \to \varepsilon_0} \left\{ \inf_{v \in I} F_\varepsilon(v) + \sup_{v \in I_0} G(v) \right\} =$$

$$= \Gamma^-(\tau) \limsup_{\varepsilon \to \varepsilon_0} F_\varepsilon(u) + \sup_{v \in I_0} G(v) \text{ for every } I_0 \in \mathcal{N}(u),$$

from which, taking into account also the continuity of G, we conclude that

$$(3.2.3) \qquad \Gamma^-(\tau) \limsup_{\varepsilon \to \varepsilon_0} (F_\varepsilon + G)(u) \leq$$

$$\leq \Gamma^-(\tau) \limsup_{\varepsilon \to \varepsilon_0} F_\varepsilon(u) + \inf_{I_0 \in \mathcal{N}(u)} \sup_{v \in I_0} G(v) = \Gamma^-(\tau) \limsup_{\varepsilon \to \varepsilon_0} F_\varepsilon(u) + G(u).$$

By (3.2.2), and (3.2.3) the proof follows. ∎

It is clear that, if $\{\varepsilon_h\} \subseteq E$ is such that $\varepsilon_h \to \varepsilon_0$, then, by using also (3.2.1), it follows that

$$(3.2.4) \qquad \Gamma^-(\tau) \liminf_{\varepsilon \to \varepsilon_0} F_\varepsilon(u) \leq \Gamma^-(\tau) \liminf_{h \to +\infty} F_{\varepsilon_h}(u) \leq$$

$$\leq \Gamma^-(\tau) \limsup_{h \to +\infty} F_{\varepsilon_h}(u) \leq \Gamma^-(\tau) \limsup_{\varepsilon \to \varepsilon_0} F_\varepsilon(u) \text{ for every } u \in U.$$

In particular, when $E = \mathbf{N}$, $\varepsilon_0 = +\infty$, and $\{h_k\} \subseteq \mathbf{N}$ diverges, it results

$$(3.2.5) \qquad \Gamma^-(\tau) \liminf_{h \to +\infty} F_h(u) \leq \Gamma^-(\tau) \liminf_{k \to +\infty} F_{h_k}(u) \leq$$

$$\leq \Gamma^-(\tau) \limsup_{k \to +\infty} F_{h_k}(u) \leq \Gamma^-(\tau) \limsup_{h \to +\infty} F_h(u) \text{ for every } u \in U.$$

The Γ^--upper limit of a family of functionals $\{F_\varepsilon\}_{\varepsilon \in E}$ can be characterized by means of the Γ^--upper limits of sequences in $\{F_\varepsilon\}_{\varepsilon \in E}$.

Proposition 3.2.3. *Let* $E \subseteq [-\infty, +\infty]$, $\varepsilon_0 \in \overline{E}$, *and let, for every* $\varepsilon \in E$, $F_\varepsilon : U \to [-\infty, +\infty]$. *Then*

$$\Gamma^-(\tau) \limsup_{\varepsilon \to \varepsilon_0} F_\varepsilon(u) =$$

$$= \sup \left\{ \Gamma^-(\tau) \limsup_{h \to +\infty} F_{\varepsilon_h}(u) : \{\varepsilon_h\} \subseteq E,\ \varepsilon_h \to \varepsilon_0 \right\} \text{ for every } u \in U.$$

Proof. Let $u \in U$. Then obviously

(3.2.6) $$\Gamma^-(\tau) \limsup_{\varepsilon \to \varepsilon_0} F_\varepsilon(u) \geq$$

$$= \sup \left\{ \Gamma^-(\tau) \limsup_{h \to +\infty} F_{\varepsilon_h}(u) : \{\varepsilon_h\} \subseteq E,\ \varepsilon_h \to \varepsilon_0 \right\}.$$

Let now $I_0 \in \mathcal{N}(u)$, and $\{\varepsilon_{0,h}\} \subseteq E$ with $\varepsilon_{0,h} \to \varepsilon_0$ be such that $\limsup_{h \to +\infty} \inf_{v \in I_0} F_{\varepsilon_{0,h}}(v) = \limsup_{\varepsilon \to \varepsilon_0} \inf_{v \in I_0} F_\varepsilon(v)$. Then it is clear that

$$\limsup_{\varepsilon \to \varepsilon_0} \inf_{v \in I_0} F_\varepsilon(v) \leq \sup_{I \in \mathcal{N}(u)} \limsup_{h \to +\infty} \inf_{v \in I} F_{\varepsilon_{0,h}}(v) \leq$$

$$\leq \sup \left\{ \Gamma^-(\tau) \limsup_{h \to +\infty} F_{\varepsilon_h}(u) : \{\varepsilon_h\} \subseteq E,\ \varepsilon_h \to \varepsilon_0 \right\},$$

from which, together with (3.2.6), the proof follows. ∎

When (U, τ) satisfies the first countability axiom, the following sequential characterization of Γ-limits holds.

Proposition 3.2.4. *Assume that* (U, τ) *satisfies the first countability axiom. For every* $h \in \mathbf{N}$ *let* $F_h : U \to [-\infty, +\infty]$. *Then*

$$\Gamma^-(\tau) \liminf_{h \to +\infty} F_h(u) = \min \left\{ \liminf_{h \to +\infty} F_h(v_h) : v_h \to u \right\},$$

$$\Gamma^-(\tau) \limsup_{h \to +\infty} F_h(u) = \min \left\{ \limsup_{h \to +\infty} F_h(v_h) : v_h \to u \right\}$$

for every $u \in U$.

Proof. Let $u \in U$. Let us preliminarily observe that, since for every $I \in \mathcal{N}(u)$, and every $\{v_h\} \subseteq U$ such that $v_h \to u$ it results that $v_h \in I$ definitively, it turns out that

(3.2.7) $$\Gamma^-(\tau) \liminf_{h \to +\infty} F_h(u) \leq \inf \left\{ \liminf_{h \to +\infty} F_h(v_h) : v_h \to u \right\}.$$

In order to prove the reverse inequality, it is not restrictive to assume that $\Gamma^-(\tau)\liminf_{h\to+\infty} F_h(u) < +\infty$. Let $\{\lambda_k\} \subseteq \mathbf{R}$ be strictly decreasing, and such that $\lim_{k\to+\infty}\lambda_k = \Gamma^-(\tau)\liminf_{h\to+\infty} F_h(u)$, and let $I_1 \supseteq I_2 \supseteq \ldots \supseteq I_k \supseteq \ldots$ be a countable basis of neighborhoods at u. Then, since clearly

$$\Gamma^-(\tau)\liminf_{h\to+\infty} F_h(u) \geq \liminf_{h\to+\infty}\inf_{v\in I_k} F_h(v) \text{ for every } k \in \mathbf{N},$$

we can find $\{h_k\} \subseteq \mathbf{N}$ strictly increasing satisfying $\lambda_k > \inf_{v\in I_k} F_{h_k}(v)$ for every $k \in \mathbf{N}$ and, consequently, $\{v_k\} \subseteq U$, such that $v_k \in I_k$ and $\lambda_k > F_{h_k}(v_k)$ for every $k \in \mathbf{N}$.

We now set $h_0 = 0$, and define a sequence $\{u_m\}$ by setting $u_m = v_k$ whenever $h_{k-1} < m \leq h_k$ for some $k \in \mathbf{N}$. Then $u_m \to u$, $u_{h_k} = v_k$ for every $k \in \mathbf{N}$, and

$$(3.2.8) \qquad \Gamma^-(\tau)\liminf_{h\to+\infty} F_h(u) \geq \liminf_{k\to+\infty} F_{h_k}(u_{h_k}) \geq \liminf_{h\to+\infty} F_h(u_h).$$

By (3.2.7), and (3.2.8) the first part of the proposition follows.

In order to prove the remaining one, we take $u \in U$ and, as before, we observe that

$$(3.2.9) \qquad \Gamma^-(\tau)\limsup_{h\to+\infty} F_h(u) \leq \inf\left\{\limsup_{h\to+\infty} F_h(v_h) : v_h \to u\right\}.$$

To prove the reverse inequality, it is not restrictive to assume that $\Gamma^-(\tau)\limsup_{h\to+\infty} F_h(u) < +\infty$. Let $\{\lambda_k\} \subseteq \mathbf{R}$ be strictly decreasing, and such that $\lim_{k\to+\infty}\lambda_k = \Gamma^-(\tau)\limsup_{h\to+\infty} F_h(u)$, and let $\{I_k\}$ be as before. Then, since clearly

$$\Gamma^-(\tau)\limsup_{h\to+\infty} F_h(u) \geq \limsup_{h\to+\infty}\inf_{v\in I_k} F_h(v) \text{ for every } k \in \mathbf{N},$$

we can find $\{h_k\} \subseteq \mathbf{N}$ strictly increasing satisfying $\lambda_k > \inf_{v\in I_k} F_h(v)$ for every $k \in \mathbf{N}$ and $h \geq h_k$. Because of this, for every $k \in \mathbf{N}$ we can find $\{v_{k,h}\}_{h\in\mathbf{N}} \subseteq I_k$ such that $\lambda_k > F_h(v_{k,h})$ for every $h \geq h_k$.

We now take $u_1,\ldots,u_{h_1-1} \in U$, and set $u_m = v_{k,m}$ whenever $h_k \leq m < h_{k+1}$ for some $k \in \mathbf{N}$. Then $u_m \to u$, $\lambda_k > F_h(u_h)$ for every $k \in \mathbf{N}$ and $h_k \leq h < h_{k+1}$, and

$$(3.2.10) \qquad \Gamma^-(\tau)\limsup_{h\to+\infty} F_h(u) \geq \limsup_{k\to+\infty} F_h(u_h).$$

By (3.2.9), and (3.2.10) also the last part of the proposition follows. ∎

It is worth while to observe explicitly that Proposition 3.2.4, when (U,τ) satisfies the first countability axiom, and, for every $h \in \mathbf{N}$, $F_h: U \to$

$[-\infty, +\infty]$, yields that for every $u \in U$, and for every $\{v_h\} \subseteq U$ such that $v_h \to u$ it results

$$\Gamma^-(\tau) \liminf_{h \to +\infty} F_h(u) \leq \liminf_{h \to +\infty} F_h(v_h),$$

and that for every $u \in U$ there exists $\{u_h\} \subseteq U$ such that $u_h \to u$, and

$$\Gamma^-(\tau) \liminf_{h \to +\infty} F_h(u) \geq \liminf_{h \to +\infty} F_h(u_h).$$

Analogously, for every $u \in U$, and for every $\{v_h\} \subseteq U$ such that $v_h \to u$ it results

$$\Gamma^-(\tau) \limsup_{h \to +\infty} F_h(u) \leq \limsup_{h \to +\infty} F_h(v_h),$$

and for every $u \in U$ there exists $\{u_h\} \subseteq U$ such that $u_h \to u$, and

$$\Gamma^-(\tau) \limsup_{h \to +\infty} F_h(u) \geq \limsup_{h \to +\infty} F_h(u_h).$$

Proposition 3.2.5. *Assume that (U, τ) satisfies the first countability axiom. For every $h \in \mathbf{N}$ let $F_h: U \to [-\infty, +\infty]$, $u \in U$, and $\lambda \in [-\infty, +\infty]$. Then*

(3.2.11) $$\lambda = \Gamma^-(\tau) \lim_{h \to +\infty} F_h(u)$$

if and only if

(3.2.12)   *for every $\{h_k\} \subseteq \mathbf{N}$ strictly increasing*

there exists $\{h_{k_j}\} \subseteq \{h_k\}$ such that $\lambda = \Gamma^-(\tau) \lim_{j \to +\infty} F_{h_{k_j}}(u)$.

Proof. By (3.2.5) it immediately follows that (3.2.11) implies (3.2.12) (actually with $\{h_{k_j}\} = \{h_k\}$).

Let us assume now that (3.2.12) is fulfilled. Let us prove that (3.2.11) holds.

By Proposition 3.2.4, we can find $\{u_h\} \subseteq U$ such that $u_h \to u$ and $\Gamma^-(\tau) \liminf_{h \to +\infty} F_h(u) = \liminf_{h \to +\infty} F_h(u_h)$, and let $\{h_k\} \subseteq \mathbf{N}$ strictly increasing satisfy $\Gamma^-(\tau) \liminf_{h \to +\infty} F_h(u) = \limsup_{k \to +\infty} F_{h_k}(u_{h_k})$. Let $\{h_{k_j}\} \subseteq \{h_k\}$ be given by (3.2.12). Then

(3.2.13) $$\Gamma^-(\tau) \liminf_{h \to +\infty} F_h(u) \geq \Gamma^-(\tau) \limsup_{k \to +\infty} F_{h_k}(u) \geq$$

$$\geq \Gamma^-(\tau) \limsup_{j \to +\infty} F_{h_{k_j}}(u) = \lambda.$$

If we now assume by contradiction that $\lambda < \Gamma^-(\tau) \limsup_{h\to+\infty} F_h(u)$, let $I \in \mathcal{N}(u)$ be such that $\lambda < \limsup_{h\to+\infty} \inf_{v\in I} F_h(v)$, and take $\{h_k\} \subseteq \mathbf{N}$ strictly increasing satisfying $\limsup_{h\to+\infty} \inf_{v\in I} F_h(v) = \liminf_{k\to+\infty} \inf_{v\in I} F_{h_k}(v)$. Let $\{h_{k_j}\} \subseteq \{h_k\}$ be given by (3.2.12). Then

$$\lambda < \Gamma^-(\tau) \liminf_{k\to+\infty} F_{h_k}(u) \leq \Gamma^-(\tau) \liminf_{j\to+\infty} F_{h_{k_j}}(u) = \lambda,$$

thus getting a contradiction. Therefore

(3.2.14) $$\Gamma^-(\tau) \limsup_{h\to+\infty} F_h(u) \leq \lambda.$$

By (3.2.13), and (3.2.14) equality (3.2.11) follows. ∎

Proposition 3.2.6. *Let (U, τ) satisfy the first countability axiom. Let $E \subseteq [-\infty, +\infty]$, $\varepsilon_0 \in \overline{E}$, and let, for every $\varepsilon \in E$, $F_\varepsilon : U \to [-\infty, +\infty]$. Then*

$$\Gamma^-(\tau) \liminf_{\varepsilon\to\varepsilon_0} F_\varepsilon(u) =$$

$$= \min\left\{ \Gamma^-(\tau) \liminf_{h\to+\infty} F_{\varepsilon_h}(u) : \{\varepsilon_h\} \subseteq E, \varepsilon_h \to \varepsilon_0 \right\} \text{ for every } u \in U.$$

Proof. Let $u \in U$. Then, by (3.2.4), it is clear that

(3.2.15) $$\Gamma^-(\tau) \liminf_{\varepsilon\to\varepsilon_0} F_\varepsilon(u) \leq$$

$$\leq \inf\left\{ \Gamma^-(\tau) \liminf_{h\to+\infty} F_{\varepsilon_h}(u) : \{\varepsilon_h\} \subseteq E, \varepsilon_h \to \varepsilon_0 \right\}.$$

To prove the reverse inequality, we assume that $\Gamma^-(\tau) \liminf_{\varepsilon\to\varepsilon_0} F_\varepsilon(u) < +\infty$. Let $\{\lambda_k\} \subseteq \mathbf{R}$ be strictly decreasing, and such that $\lim_{k\to+\infty} \lambda_k = \Gamma^-(\tau) \liminf_{\varepsilon\to\varepsilon_0} F_\varepsilon(u)$, and let $\{I_k\}$ be as in the proof of Proposition 3.2.4. Then, since clearly

$$\Gamma^-(\tau) \liminf_{\varepsilon\to\varepsilon_0} F_\varepsilon(u) \geq \liminf_{\varepsilon\to\varepsilon_0} \inf_{v\in I_k} F_\varepsilon(v) \text{ for every } k \in \mathbf{N},$$

we can find $\{\varepsilon_k\} \subseteq E$ satisfying $\varepsilon_k \to \varepsilon_0$, and $\lambda_k > \inf_{v\in I_k} F_{\varepsilon_k}(v)$ for every $k \in \mathbf{N}$. Consequently, there exists $\{v_k\} \subseteq U$, such that $v_k \in I_k$ and $\lambda_k > F_{\varepsilon_k}(v_k)$ for every $k \in \mathbf{N}$.

It is clear that $u_k \to u$, therefore, by Proposition 3.2.4, it follows that

(3.2.16) $$\Gamma^-(\tau) \liminf_{\varepsilon\to\varepsilon_0} F_\varepsilon(u) \geq \liminf_{k\to+\infty} F_{\varepsilon_k}(v_k) \geq \Gamma^- \liminf_{h\to+\infty} F_{\varepsilon_k}(u).$$

By (3.2.15), and (3.2.16) the proof follows. ∎

Finally, we prove that Γ-convergence has nice compactness properties.

Theorem 3.2.7. *Assume that (U, τ) has a countable base of open sets. For every $h \in \mathbf{N}$ let $F_h : U \to [-\infty, +\infty]$. Then there exists $\{h_k\} \subseteq \mathbf{N}$ strictly increasing for which the limit $\Gamma^-(\tau) \lim_{k \to +\infty} F_{h_k}(u)$ exists for every $u \in U$.*

Proof. Let $\{A_m\}$ be a countable base of open sets for τ.

Let $\{h_k^1\} \subseteq \mathbf{N}$ be strictly increasing and such that the limit $\lim_{k \to +\infty} \inf_{v \in A_1} F_{h_k^1}(v)$ exists. For every $m \in \mathbf{N}$ let us choose $\{h_k^{m+1}\} \subseteq \{h_k^m\}$ for which the limit $\lim_{k \to +\infty} \inf_{v \in A_{m+1}} F_{h_k^{m+1}}(v)$ exists.

We now apply the classical diagonalization argument, and set, for every $k \in \mathbf{N}$, $h_k = h_k^k$. Then it turns out that the limit $\lim_{k \to +\infty} \inf_{v \in A_m} F_{h_k}(v)$ exists for every $m \in \mathbf{N}$. Because of this, we conclude that

$$\Gamma^-(\tau) \liminf_{k \to +\infty} F_{h_k}(u) = \sup_{I \in \mathcal{N}(u) \cap \{A_m\}} \liminf_{k \to +\infty} \inf_{v \in I} F_{h_k}(v) =$$

$$= \sup_{I \in \mathcal{N}(u) \cap \{A_m\}} \limsup_{k \to +\infty} \inf_{v \in I} F_{h_k}(v) = \Gamma^-(\tau) \limsup_{k \to +\infty} F_{h_k}(u) \text{ for every } u \in U,$$

from which the proof follows. ∎

§3.3 Applications to the Calculus of Variations

In the present section we establish the results on the asymptotic behaviour of minima, and of minimum values of families of functionals defined on the same space.

Let (U, τ) be a topological space.

Lemma 3.3.1. *Let $\alpha \colon \tau \to [-\infty, +\infty]$. Then the function*

$$u \in U \mapsto \sup_{I \in \mathcal{N}(u)} \alpha(I)$$

is lower semicontinuous.

Proof. Let us observe that for every $c \in \mathbf{R}$, and every $u \in \{v \in U : \sup_{I \in \mathcal{N}(v)} \alpha(I) > c\}$ there exists $I_c \in \mathcal{N}(u)$ such that $\alpha(I_c) > c$. Consequently $\sup_{I \in \mathcal{N}(v)} \alpha(I) > c$ for every $v \in I_c$, and the set $\{v \in U : \sup_{I \in \mathcal{N}(v)} \alpha(I) > c\}$ turns out to be open. ∎

Proposition 3.3.2. *Let $E \subseteq [-\infty, +\infty]$, $\varepsilon_0 \in \overline{E}$, and let, for every $\varepsilon \in E$, $F_\varepsilon \colon U \to [-\infty, +\infty]$. Then the functionals $\Gamma^-(\tau) \liminf_{\varepsilon \to \varepsilon_0} F_\varepsilon$ and $\Gamma^-(\tau) \limsup_{\varepsilon \to \varepsilon_0} F_\varepsilon$ are lower semicontinuous on U.*

Proof. Follows from Lemma 3.3.1 applied to $\alpha \colon A \in \tau \mapsto \liminf_{\varepsilon \to \varepsilon_0} \inf_{v \in A} F_\varepsilon(v)$, and to $\alpha \colon A \in \tau \mapsto \limsup_{\varepsilon \to \varepsilon_0} \inf_{v \in A} F_\varepsilon(v)$. ∎

Lemma 3.3.3. Let $E \subseteq [-\infty, +\infty]$, $\varepsilon_0 \in \overline{E}$, and let, for every $\varepsilon \in E$, $F_\varepsilon: U \to [-\infty, +\infty]$. Let $A \in \tau$. Then

$$\inf_{u \in A} \Gamma^-(\tau) \liminf_{\varepsilon \to \varepsilon_0} F_\varepsilon(u) \geq \liminf_{\varepsilon \to \varepsilon_0} \inf_{u \in A} F_\varepsilon(u),$$

$$\inf_{u \in A} \Gamma^-(\tau) \limsup_{\varepsilon \to \varepsilon_0} F_\varepsilon(u) \geq \limsup_{\varepsilon \to \varepsilon_0} \inf_{u \in A} F_\varepsilon(u)$$

for every $u \in U$.

Proof. Let $u \in U$. Then we have that

$$\inf_{u \in A} \Gamma^-(\tau) \liminf_{\varepsilon \to \varepsilon_0} F_\varepsilon(u) = \inf_{u \in A} \sup_{I \in \mathcal{N}(u)} \liminf_{\varepsilon \to \varepsilon_0} \inf_{v \in I} F_\varepsilon(v) \geq$$

$$\geq \inf_{u \in A} \liminf_{\varepsilon \to \varepsilon_0} \inf_{v \in A} F_\varepsilon(v) = \liminf_{\varepsilon \to \varepsilon_0} \inf_{u \in A} F_\varepsilon(u).$$

The proof of the second inequality is similar. ∎

Let $E \subseteq [-\infty, +\infty]$, $\varepsilon_0 \in \overline{E}$, $\{u_\varepsilon\}_{\varepsilon \in E} \subseteq U$, and $u \in U$. We say that u is a *cluster point* of $\{u_\varepsilon\}_{\varepsilon \in E}$ as $\varepsilon \to \varepsilon_0$ if for every $I \in \mathcal{N}(u)$, and every neighborhood O of ε_0 there exists $\varepsilon \in E \cap O$ such that $u_\varepsilon \in I$.

Lemma 3.3.4. Let $E \subseteq [-\infty, +\infty]$, $\varepsilon_0 \in \overline{E}$, and let, for every $\varepsilon \in E$, $F_\varepsilon: U \to [-\infty, +\infty]$. Let $\{u_\varepsilon\}_{\varepsilon \in E} \subseteq U$, and u be a cluster point of $\{u_\varepsilon\}_{\varepsilon \in E}$ as $\varepsilon \to \varepsilon_0$. Then

$$\Gamma^-(\tau) \liminf_{\varepsilon \to \varepsilon_0} F_\varepsilon(u) \leq \limsup_{\varepsilon \to \varepsilon_0} F_\varepsilon(u_\varepsilon).$$

Proof. Let $I \in \mathcal{N}(u)$. Let us prove that

$$\liminf_{\varepsilon \to \varepsilon_0} \inf_{v \in I} F_\varepsilon(u) \leq \limsup_{\varepsilon \to \varepsilon_0} F_\varepsilon(u_\varepsilon).$$

This will imply the lemma.

To do this, it is sufficient to note that for every neighborhood O of ε_0 there exists $\eta \in E \cap O$ such that $u_\eta \in I$, and, consequently, such that $\inf_{v \in I} F_\eta(v) \leq F_\eta(u_\eta)$. ∎

Proposition 3.3.5. Let $E \subseteq [-\infty, +\infty]$, $\varepsilon_0 \in \overline{E}$, and let, for every $\varepsilon \in E$, $F_\varepsilon: U \to [-\infty, +\infty]$. Let $K \subseteq U$ be countably compact. Then $\Gamma^-(\tau) \liminf_{\varepsilon \to \varepsilon_0} F_\varepsilon$ attains its minimum on K, and

$$\min_{u \in K} \Gamma^-(\tau) \liminf_{\varepsilon \to \varepsilon_0} F_\varepsilon(u) \leq \liminf_{\varepsilon \to \varepsilon_0} \inf_{u \in K} F_\varepsilon(u).$$

Proof. The existence of the minimum in K of $\Gamma^-(\tau) \liminf_{\varepsilon \to \varepsilon_0} F_\varepsilon$ follows from Proposition 3.3.2, and Theorem 3.1.4.

Let $\{\varepsilon_h\} \subseteq E$ be such that

$$(3.3.1) \qquad \lim_{h \to +\infty} \inf_{u \in K} F_{\varepsilon_h}(u) = \liminf_{\varepsilon \to \varepsilon_0} \inf_{u \in K} F_\varepsilon(u),$$

and let $\{u_h\} \subseteq K$ satisfy

$$(3.3.2) \qquad \lim_{h \to +\infty} F_{\varepsilon_h}(u_h) = \lim_{h \to +\infty} \inf_{u \in K} F_{\varepsilon_h}(u).$$

Since K is countably compact, Theorem 3.1.2 yields the existence of a cluster point of $\{u_h\}$, say u, in K. Consequently, by (3.2.4), Lemma 3.3.4, (3.3.2), and (3.3.1) we conclude that

$$\min_{v \in K} \Gamma^-(\tau) \liminf_{\varepsilon \to \varepsilon_0} F_\varepsilon(v) \le \Gamma^-(\tau) \liminf_{\varepsilon \to \varepsilon_0} F_\varepsilon(u) \le \Gamma^-(\tau) \liminf_{h \to +\infty} F_{\varepsilon_h}(u) \le$$

$$\le \limsup_{h \to +\infty} F_{\varepsilon_h}(u_h) = \lim_{h \to +\infty} \inf_{u \in K} F_{\varepsilon_h}(u) = \liminf_{\varepsilon \to \varepsilon_0} \inf_{u \in K} F_\varepsilon(u),$$

which proves the proposition. ∎

We can now prove the results on the convergence of minima and of minimizers.

Definition 3.3.6. *Let $E \subseteq [-\infty, +\infty]$, and let, for every $\varepsilon \in E$, $F_\varepsilon : U \to [-\infty, +\infty]$. We say that the functionals $\{F_\varepsilon\}_{\varepsilon \in E}$ are*
i) equi-coercive if for every $\lambda \in \mathbf{R}$ there exists a compact subset K_λ of U such that $\{u \in U : F_\varepsilon(u) \le \lambda\} \subseteq K_\lambda$ for every $\varepsilon \in E$,
ii) equi-strongly countably coercive if for every $\lambda \in \mathbf{R}$ there exists a closed and countably compact subset K_λ of U such that $\{u \in U : F_\varepsilon(u) \le \lambda\} \subseteq K_\lambda$ for every $\varepsilon \in E$,
iii) equi-strongly sequentially coercive if for every $\lambda \in \mathbf{R}$ there exists a closed and sequentially compact subset K_λ of U such that $\{u \in U : F_\varepsilon(u) \le \lambda\} \subseteq K_\lambda$ for every $\varepsilon \in E$.

Theorem 3.3.7. *Let $E \subseteq [-\infty, +\infty]$, $\varepsilon_0 \in \overline{E}$, and let, for every $\varepsilon \in E$, $F_\varepsilon : U \to [-\infty, +\infty]$. Assume that the functionals $\{F_\varepsilon\}_{\varepsilon \in E}$ are equi-strongly countably coercive. Then $\Gamma^-(\tau) \liminf_{\varepsilon \to \varepsilon_0} F_\varepsilon$ and $\Gamma^-(\tau) \limsup_{\varepsilon \to \varepsilon_0} F_\varepsilon$ are strongly countably coercive.*

If in addition the limit $\Gamma^-(\tau) \lim_{\varepsilon \to \varepsilon_0} F_\varepsilon(u)$ exists for every $u \in U$, it results that $\Gamma^-(\tau) \lim_{\varepsilon \to \varepsilon_0} F_\varepsilon$ has a minimum on U, that the limit $\lim_{\varepsilon \to \varepsilon_0} \inf_{v \in U} F_\varepsilon(v)$ exists, and that

$$\min_{v \in U} \Gamma^-(\tau) \lim_{\varepsilon \to \varepsilon_0} F_\varepsilon(v) = \lim_{\varepsilon \to \varepsilon_0} \inf_{v \in U} F_\varepsilon(v).$$

Finally, if $\lim_{\varepsilon \to \varepsilon_0} \inf_{v \in U} F_\varepsilon(v) < +\infty$, and if $\{u_\varepsilon\}_{\varepsilon \in E} \subseteq U$ is such that $\lim_{\varepsilon \to \varepsilon_0} F_\varepsilon(u_\varepsilon) = \lim_{\varepsilon \to \varepsilon_0} \inf_{v \in U} F_\varepsilon(v)$, then the set of the cluster points of

$\{u_\varepsilon\}_{\varepsilon \in E}$ as $\varepsilon \to \varepsilon_0$ is nonempty, and every such point is a solution of $\min_{v \in U} \Gamma^-(\tau) \lim_{\varepsilon \to \varepsilon_0} F_\varepsilon(v)$.

Proof. Let us first prove that the functionals $\Gamma^-(\tau) \liminf_{\varepsilon \to \varepsilon_0} F_\varepsilon$ and $\Gamma^-(\tau) \limsup_{\varepsilon \to \varepsilon_0} F_\varepsilon$ are strongly countably coercive. To do this, we treat only the case of the $\Gamma^-(\tau) \limsup_{\varepsilon \to \varepsilon_0} F_\varepsilon$, the other being similar.

For every $\lambda \in \mathbf{R}$ let K_λ be the closed countably compact set given by the equi-countable coerciveness of $\{F_\varepsilon\}_{\varepsilon \in E}$.

Let $\lambda \in \mathbf{R}$, and $u \in \{v \in U : \Gamma^-(\tau) \limsup_{\varepsilon \to \varepsilon_0} F_\varepsilon(v) \leq \lambda\}$. Then, for every $I \in \mathcal{N}(u)$, it turns out that $\limsup_{\varepsilon \to \varepsilon_0} \inf_{v \in I} F_\varepsilon(v) \leq \lambda$ and, by the equi-countable coerciveness of $\{F_\varepsilon\}_{\varepsilon \in E}$, that $I \cap K_{\lambda+\theta} \neq \emptyset$ for every $\theta > 0$, namely, taking also into account the closedness of $K_{\lambda+\theta}$ for every $\theta > 0$, that $u \in \cap_{\theta>0} K_{\lambda+\theta}$.

We now observe that $\cap_{\theta>0} K_{\lambda+\theta}$ is a closed subset of a countably compact space, hence it is itself countably compact.

Because of this, the countable coerciveness of $\Gamma^-(\tau) \limsup_{\varepsilon \to \varepsilon_0} F_\varepsilon$ follows.

We now assume that the limit $\Gamma^-(\tau) \lim_{\varepsilon \to \varepsilon_0} F_\varepsilon(u)$ exists for every $u \in U$. Then, by Proposition 3.3.2, the countable coerciveness of $\Gamma^-(\tau) \limsup_{\varepsilon \to \varepsilon_0} F_\varepsilon$, and Theorem 3.1.4 it follows that $\Gamma^-(\tau) \lim_{\varepsilon \to \varepsilon_0} F_\varepsilon$ has a minimum on U.

If $\liminf_{\varepsilon \to \varepsilon_0} \inf_{v \in U} F_\varepsilon(v) = +\infty$, then Lemma 3.3.3 yields $\Gamma^-(\tau) \lim_{\varepsilon \to \varepsilon_0} F_\varepsilon(u) = +\infty$ for every $u \in U$, and, consequently, that

$$\min_{v \in U} \Gamma^-(\tau) \lim_{\varepsilon \to \varepsilon_0} F_\varepsilon(v) = +\infty = \lim_{\varepsilon \to \varepsilon_0} \inf_{v \in U} F_\varepsilon(v).$$

On the contrary, if $\liminf_{\varepsilon \to \varepsilon_0} \inf_{v \in U} F_\varepsilon(v) < +\infty$, let $\lambda \in \mathbf{R}$ satisfy $\liminf_{\varepsilon \to \varepsilon_0} \inf_{v \in U} F_\varepsilon(v) < \lambda$, and $\{\varepsilon_h\} \subseteq E$ be such that $\varepsilon_h \to \varepsilon_0$, and

$$(3.3.3) \qquad \lim_{h \to +\infty} \inf_{v \in U} F_{\varepsilon_h}(v) = \liminf_{\varepsilon \to \varepsilon_0} \inf_{v \in U} F_\varepsilon(v).$$

Let K_λ be the countably compact set given by the equi-strongly countable coerciveness property. Then it is straightforward to verify that

$$(3.3.4) \qquad \inf_{v \in U} F_{\varepsilon_h}(v) = \inf_{v \in K_\lambda} F_{\varepsilon_h}(v) \text{ for every } h \in \mathbf{N} \text{ sufficiently large.}$$

Finally, by (3.2.4), Proposition 3.3.5, (3.3.4), (3.3.3), and Lemma 3.3.3, we conclude that

$$\min_{v \in U} \Gamma^-(\tau) \lim_{\varepsilon \to \varepsilon_0} F_\varepsilon(v) \leq \min_{v \in U} \Gamma^-(\tau) \lim_{h \to +\infty} F_{\varepsilon_h}(v) \leq$$

$$\leq \min_{v \in K_\lambda} \Gamma^-(\tau) \lim_{h \to +\infty} F_{\varepsilon_h}(v) \leq \lim_{h \to +\infty} \inf_{v \in K_\lambda} F_{\varepsilon_h}(v) = \lim_{h \to +\infty} \inf_{v \in U} F_{\varepsilon_h}(v) =$$

$$= \liminf_{\varepsilon \to \varepsilon_0} \inf_{v \in U} F_\varepsilon(v) \leq \limsup_{\varepsilon \to \varepsilon_0} \inf_{v \in U} F_\varepsilon(v) \leq \inf_{v \in U} \Gamma^-(\tau) \lim_{\varepsilon \to \varepsilon_0} F_\varepsilon(v),$$

from which the part of the theorem concerning the asymptotic behaviour of $\{\inf_{v \in U} F_\varepsilon(v)\}_{\varepsilon \in E}$ follows.

If now $\lim_{\varepsilon \to \varepsilon_0} \inf_{v \in U} F_\varepsilon(v) < +\infty$, and $\{u_\varepsilon\}_{\varepsilon \in E}$ is as above, let $\lambda \in \mathbf{R}$ be such that $\lim_{\varepsilon \to \varepsilon_0} \inf_{v \in U} F_\varepsilon(v) < \lambda$. Then, for every $\varepsilon \in E$ sufficiently close to ε_0, it results that $F_\varepsilon(u_\varepsilon) < \lambda$, and, consequently, by the equi-strongly countable coerciveness of $\{F_\varepsilon\}_{\varepsilon \in E}$, that $u_\varepsilon \in K_\lambda$.

Because of this, Theorem 3.1.2 yields the existence of at least one cluster point of $\{u_\varepsilon\}_{\varepsilon \in E}$ as $\varepsilon \to \varepsilon_0$. Let u be one of such points. Then, by Lemma 3.3.4, Lemma 3.3.3, and the first part of the present theorem, we conclude that

$$\min_{v \in U} \Gamma^-(\tau) \lim_{\varepsilon \to \varepsilon_0} F_\varepsilon(v) \leq \Gamma^-(\tau) \lim_{\varepsilon \to \varepsilon_0} F_\varepsilon(u) \leq$$

$$\leq \limsup_{\varepsilon \to \varepsilon_0} F_\varepsilon(u_\varepsilon) = \lim_{\varepsilon \to \varepsilon_0} \inf_{v \in U} F_\varepsilon(v) = \min_{v \in U} \Gamma^-(\tau) \lim_{\varepsilon \to \varepsilon_0} F_\varepsilon(v),$$

that completes the proof. ∎

From Theorem 3.3.7 we deduce the following corollary in the case of coerciveness, or of strongly sequential coerciveness.

Theorem 3.3.8. *Let $E \subseteq [-\infty, +\infty]$, $\varepsilon_0 \in \overline{E}$, and let, for every $\varepsilon \in E$, $F_\varepsilon : U \to [-\infty, +\infty]$. Assume that the functionals $\{F_\varepsilon\}_{\varepsilon \in E}$ are equi-coercive (respectively equi-strongly sequentially coercive). Then $\Gamma^-(\tau) \liminf_{\varepsilon \to \varepsilon_0} F_\varepsilon$ and $\Gamma^-(\tau) \limsup_{\varepsilon \to \varepsilon_0} F_\varepsilon$ are coercive (respectively strongly sequentially coercive).*

If in addition the limit $\Gamma^-(\tau) \lim_{\varepsilon \to \varepsilon_0} F_\varepsilon(u)$ exists for every $u \in U$, it results that $\Gamma^-(\tau) \lim_{\varepsilon \to \varepsilon_0} F_\varepsilon$ has a minimum on U, that the limit $\lim_{\varepsilon \to \varepsilon_0} \inf_{v \in U} F_\varepsilon(v)$ exists, and that

$$\min_{v \in U} \Gamma^-(\tau) \lim_{\varepsilon \to \varepsilon_0} F_\varepsilon(v) = \lim_{\varepsilon \to \varepsilon_0} \inf_{v \in U} F_\varepsilon(v).$$

Finally, if $\lim_{\varepsilon \to \varepsilon_0} \inf_{v \in U} F_\varepsilon(v) < +\infty$, and if $\{u_\varepsilon\}_{\varepsilon \in E} \subseteq U$ is such that $\lim_{\varepsilon \to \varepsilon_0} F_\varepsilon(u_\varepsilon) = \lim_{\varepsilon \to \varepsilon_0} \inf_{v \in U} F_\varepsilon(v)$, then $\{u_\varepsilon\}_{\varepsilon \in E}$ has cluster points as $\varepsilon \to \varepsilon_0$ (respectively there exists $\{\varepsilon_h\} \subseteq E$ with $\varepsilon_h \to \varepsilon_0$ such that $\{u_{\varepsilon_h}\}$ converges), and every such point (respectively the limit of $\{u_{\varepsilon_h}\}$) is a solution of $\min_{v \in U} \Gamma^-(\tau) \lim_{\varepsilon \to \varepsilon_0} F_\varepsilon(v)$.

Proof. Let $\{K_\lambda\}_{\lambda \in \mathbf{R}}$ be the family of the compact (respectively closed and sequentially compact) sets given by the equi-coerciveness (respectively equi-strongly sequential coerciveness) of $\{F_\varepsilon\}_{\varepsilon \in E}$.

The coerciveness (respectively strongly sequential coerciveness) of $\Gamma^-(\tau) \liminf_{\varepsilon \to \varepsilon_0} F_\varepsilon$ and of $\Gamma^-(\tau) \limsup_{\varepsilon \to \varepsilon_0} F_\varepsilon$ follows exactly as in the corresponding part of the proof of Theorem 3.3.7, once we observe that a closed subset of a compact (respectively sequentially compact) space is itself compact (respectively closed and sequentially compact).

The part of the theorem dealing with the asymptotic behaviours of $\{\inf_{v \in U} F_\varepsilon(v)\}_{\varepsilon \in E}$, and of $\{u_\varepsilon\}_{\varepsilon \in E}$ as $\varepsilon \to \varepsilon_0$ follows from Theorem 3.3.7, once we recall that a compact (respectively sequentially compact) set is also countably compact.

Finally, if $\lim_{\varepsilon \to \varepsilon_0} \inf_{v \in U} F_\varepsilon(v) < +\infty$, and $\{u_\varepsilon\}_{\varepsilon \in E}$ is as above, let $\lambda \in \mathbf{R}$ be such that $\lim_{\varepsilon \to \varepsilon_0} \inf_{v \in U} F_\varepsilon(v) < \lambda$. Then, for every $\varepsilon \in E$ sufficiently close to ε_0, it results that $F_\varepsilon(u_\varepsilon) < \lambda$, and, consequently, by the equi-strongly sequential coerciveness of $\{F_\varepsilon\}_{\varepsilon \in E}$, that $u_\varepsilon \in K_\lambda$.

Let $\{\varepsilon_h\} \subseteq E$ with $\varepsilon_h \to \varepsilon_0$. Then, by the sequential compactness of K_λ, the existence of a converging subsequence of $\{u_{\varepsilon_h}\}$, still denoted by $\{u_{\varepsilon_h}\}$, follows. Let u be the limit of $\{u_{\varepsilon_h}\}$. Then, by (3.2.4), Lemma 3.3.4, Lemma 3.3.3 and the first part of the present theorem, we conclude that

$$\min_{v \in U} \Gamma^-(\tau) \lim_{\varepsilon \to \varepsilon_0} F_\varepsilon(v) \leq \min_{v \in U} \Gamma^-(\tau) \liminf_{h \to +\infty} F_{\varepsilon_h}(v) \leq \Gamma^-(\tau) \liminf_{h \to +\infty} F_{\varepsilon_h}(u) \leq$$

$$\leq \limsup_{h \to +\infty} F_{\varepsilon_h}(u_{\varepsilon_h}) = \lim_{\varepsilon \to \varepsilon_0} \inf_{v \in U} F_\varepsilon(v) = \min_{v \in U} \Gamma^-(\tau) \lim_{h \to +\infty} F_h(v),$$

that completes the proof. ∎

§3.4 Γ-Convergence in Topological Vector Spaces and of Increasing Set Functionals

We conclude this chapter with some remarks on Γ-convergence in the framework of topological vector spaces, and in the framework of increasing set functionals.

Proposition 3.4.1. *Let U be a topological vector space, $E \subseteq [-\infty, +\infty]$, $\varepsilon_0 \in \overline{E}$, and let, for every $\varepsilon \in E$, $F_\varepsilon: U \to [-\infty, +\infty]$. Assume that, for every $\varepsilon \in E$, F_ε is convex. Then $\Gamma^-(\tau) \limsup_{\varepsilon \to \varepsilon_0} F_\varepsilon$ is convex.*

Proof. Let $u_1, u_2 \in U$, $t \in [0,1]$, and let $I \in \mathcal{N}(tu_1 + (1-t)u_2)$.

Since U is a topological vector space, the function $(u,v) \in U \times U \mapsto tu + (1-t)v \in U$ is continuous, consequently there exist $I_1 \in \mathcal{N}(u_1)$ and $I_2 \in \mathcal{N}(u_2)$ such that $tI_1 + (1-t)I_2 \subseteq I$.

Because of this, and by the convexity of each F_ε, we have that

$$\inf_{v \in I} F_\varepsilon(v) \leq \inf_{v \in tI_1 + (1-t)I_2} F_\varepsilon(v) = \inf_{v_1 \in I_1, \, v_2 \in I_2} F_\varepsilon(tv_1 + (1-t)v_2) \leq$$

$$\leq t \inf_{v_1 \in I_1} F_\varepsilon(v_1) + (1-t) \inf_{v_2 \in I_2} F_\varepsilon(v_2) \text{ for every } \varepsilon \in E,$$

from which we deduce that

$$\limsup_{\varepsilon \to \varepsilon_0} \inf_{v \in I} F_\varepsilon(v) \leq t \limsup_{\varepsilon \to \varepsilon_0} \inf_{v \in I_1} F_\varepsilon(v) + (1-t) \limsup_{\varepsilon \to \varepsilon_0} \inf_{v \in I_2} F_\varepsilon(v) \leq$$

$$\le t\Gamma^-(\tau)\limsup_{\varepsilon\to\varepsilon_0} F_\varepsilon(u_1) + (1-t)\Gamma^-(\tau)\limsup_{\varepsilon\to\varepsilon_0} F_\varepsilon(u_2)$$

for every $I \in \mathcal{N}(tu_1 + (1-t)u_2)$.

The convexity of $\Gamma^-(\tau)\limsup_{\varepsilon\to\varepsilon_0} F_\varepsilon$ now follows from the above inequality. ∎

Let $\Omega_0 \in \mathcal{A}(\mathbf{R}^n)$, and (U,τ) be a topological space. For every $h \in \mathbf{N}$ let $F_h\colon \mathcal{A}(\Omega_0) \times U \to [0,+\infty]$ be increasing. Then, it is clear that $\Gamma^-(\tau)\liminf_{h\to+\infty} F_h$, and $\Gamma^-(\tau)\limsup_{h\to+\infty} F_h$ too are increasing.

Proposition 3.4.2. *Let $\Omega_0 \in \mathcal{A}(\mathbf{R}^n)$, and (U,τ) be a topological space satisfying the first countability axiom. For every $h \in \mathbf{N}$ let $F_h\colon \mathcal{A}(\Omega_0) \times U \to [0,+\infty]$ be increasing, and let $F\colon \mathcal{A}(\Omega_0) \times U \to [0,+\infty]$. Then*

$$(3.4.1) \qquad F(\Omega, u) = \sup\left\{\Gamma^-(\tau)\liminf_{h\to+\infty} F_h(A, u) : A \subset\subset \Omega\right\} =$$

$$= \sup\left\{\Gamma^-(\tau)\limsup_{h\to+\infty} F_h(A, u) : A \subset\subset \Omega\right\} \text{ for every } \Omega \in \mathcal{A}(\Omega_0),\ u \in U$$

if and only if

(3.4.2) *for every $\{h_k\} \subseteq \mathbf{N}$ strictly increasing there exists $\{h_{k_j}\} \subseteq \{h_k\}$*

such that $F(\Omega, u) = \sup\left\{\Gamma^-(\tau)\liminf_{j\to+\infty} F_{h_{k_j}}(A, u) : A \subset\subset \Omega\right\} =$

$$= \sup\left\{\Gamma^-(\tau)\limsup_{j\to+\infty} F_{h_{k_j}}(A, u) : A \subset\subset \Omega\right\} \text{ for every } \Omega \in \mathcal{A}(\Omega_0),\ u \in U.$$

Proof. It is clear that (3.4.1) implies (3.4.2).

Conversely, let us assume that (3.4.2) holds. For the sake of simplicity, let us set

$$F'\colon (\Omega, u) \in \mathcal{A}(\Omega_0) \times U \mapsto \Gamma^-(\tau)\liminf_{h\to+\infty} F_h(\Omega, u),$$

$$F''\colon (\Omega, u) \in \mathcal{A}(\Omega_0) \times U \mapsto \Gamma^-(\tau)\limsup_{h\to+\infty} F_h(\Omega, u).$$

Then it is clear that F' and F'' are increasing.

Let $(\Omega, u) \in \mathcal{A}(\Omega_0) \times U$. Then, since F'_- is inner regular, let $\{\Omega_k\} \subseteq \mathcal{A}(\Omega)$ satisfy $\Omega_k \subset\subset \Omega_{k+1} \subset\subset \Omega$ for every $k \in \mathbf{N}$, $\cup_{k=1}^{+\infty}\Omega_k = \Omega$, and

$$(3.4.3) \qquad\qquad F'_-(\Omega, u) = \lim_{k\to+\infty} F'(\Omega_k, u),$$

and let, for every $k \in \mathbf{N}$, $\{u_h^k\} \subseteq U$ be such that $u_h^k \to u$ as h diverges, and

$$(3.4.4) \qquad\qquad F'(\Omega_k, u) = \liminf_{h\to+\infty} F_h(\Omega_k, u_h^k).$$

By using (3.4.3), and (3.4.4), we can construct $\{h_k\} \subseteq \mathbf{N}$ strictly increasing such that $u_{h_k}^k \to u$, and

$$F'_-(\Omega, u) = \limsup_{k \to +\infty} F_{h_k}(\Omega_k, u_{h_k}^k).$$

Consequently, by taking into account the properties of $\{\Omega_k\}$, we obtain that

$$F'_-(\Omega, u) \geq \limsup_{k \to +\infty} F_{h_k}(A, u_{h_k}^k) \geq \Gamma^-(\tau) \limsup_{k \to +\infty} F_{h_k}(A, u)$$

for every $A \subset\subset \Omega$,

from which, together with (3.4.2), we conclude that

(3.4.5) $\qquad F'_-(\Omega, u) \geq \sup\{\Gamma^-(\tau) \limsup_{k \to +\infty} F_{h_k}(A, u) : A \subset\subset \Omega\} \geq$

$$\geq \sup\{\Gamma^-(\tau) \limsup_{j \to +\infty} F_{h_{k_j}}(A, u) : A \subset\subset \Omega\} = F(\Omega, u)$$

for every $(\Omega, u) \in \mathcal{A}(\Omega_0) \times U$.

In order to prove the reverse inequality for F''_-, let, by contradiction, $(\Omega, u) \in \mathcal{A}(\Omega_0) \times U$ be such that $F(\Omega, u) < F''_-(\Omega, u)$, from which it follows that there exists $A \subset\subset \Omega$ satisfying $F(\Omega, u) < F''(A, u)$. Let $I \in \mathcal{N}(u)$, and $\{h_k\} \subseteq \mathbf{N}$ strictly increasing such that

$$F(\Omega, u) < \limsup_{h \to +\infty} \inf_{v \in I} F_h(A, v) = \liminf_{k \to +\infty} \inf_{v \in I} F_{h_k}(A, v).$$

Then, by (3.4.2), we have that

$$F(\Omega, u) < \Gamma^-(\tau) \liminf_{k \to +\infty} F_{h_k}(A, u) \leq \Gamma^-(\tau) \liminf_{j \to +\infty} F_{h_{k_j}}(A, u) \leq$$

$$\leq \sup\{\Gamma^-(\tau) \liminf_{j \to +\infty} F_{h_{k_j}}(A, u) : A \subset\subset \Omega\} = F(\Omega, u),$$

thus getting a contradiction.

Hence, it occurs that

(3.4.6) $\qquad F''_-(\Omega, u) \leq F(\Omega, u)$ for every $(\Omega, u) \in \mathcal{A}(\Omega_0) \times U$.

By (3.4.5), (3.4.6), and (3.2.1), the proof follows. ∎

Finally, we prove the following abstract compactness result for sequences of increasing functionals.

Proposition 3.4.3. *Let* $\Omega_0 \in \mathcal{A}(\mathbf{R}^n)$, *and* (U, τ) *be a topological space satisfying the second countability axiom. For every* $h \in \mathbf{N}$ *let* $F_h \colon \mathcal{A}(\Omega_0) \times U \to [0, +\infty]$ *be increasing. Then there exists* $\{h_k\} \subseteq \mathbf{N}$ *strictly increasing for which*

$$\sup\left\{\Gamma^-(\tau)\liminf_{k\to+\infty} F_{h_k}(A, u) : A \subset\subset \Omega\right\} =$$

$$= \sup\left\{\Gamma^-(\tau)\limsup_{k\to+\infty} F_{h_k}(A, u) : A \subset\subset \Omega\right\} \text{ for every } \Omega \in \mathcal{A}(\Omega_0), \ u \in U$$

Proof. Let $\{D_j\}_{j\in\mathbf{N}} \subseteq \mathcal{A}(\Omega_0)$ be dense in $\mathcal{A}(\Omega_0)$. Then, by Theorem 3.2.7, and an iteration argument, for every $j \in \mathbf{N}$ there exists $\{h_k^j\} \subseteq \mathbf{N}$ strictly increasing, satisfying $\{h_k^{j+1}\} \subseteq \{h_k^j\}$ for every $j \in \mathbf{N}$, for which the limit $\Gamma^-(\tau)\lim_{k\to+\infty} F_{h_k^j}(D_j, u)$ exists for every $u \in U$.

At this point, a diagonalization argument, and (3.2.5) provides the existence of $\{h_k\}$ strictly increasing for which the limit $\Gamma^-(\tau)\lim_{k\to+\infty} F_{h_k}(D_j, u)$ exists for every $j \in \mathbf{N}$, $u \in U$.

Because of this, the proof follows, since

$$\sup\left\{\Gamma^-(\tau)\liminf_{k\to+\infty} F_{h_k}(A, u) : A \subset\subset \Omega\right\} =$$

$$= \sup\left\{\Gamma^-(\tau)\liminf_{k\to+\infty} F_{h_k}(A, u) : A \in \{D_j\}, \ A \subset\subset \Omega\right\} =$$

$$= \sup\left\{\Gamma^-(\tau)\limsup_{k\to+\infty} F_{h_k}(A, u) : A \in \{D_j\}, \ A \subset\subset \Omega\right\} =$$

$$= \sup\left\{\Gamma^-(\tau)\limsup_{k\to+\infty} F_{h_k}(A, u) : A \subset\subset \Omega\right\}$$

for every $\Omega \in \mathcal{A}(\Omega_0)$, $u \in U$. ∎

§3.5 Relaxation

Let (U, τ) be a topological space.

In this section we analyze more deeply the particular case of the Γ-convergence of a constant family of functionals.

Definition 3.5.1. *Let* $F \colon U \to [-\infty, +\infty]$. *We define the relaxed functional* $\mathrm{sc}^-(\tau)F$ *of* F *as*

$$\mathrm{sc}^-(\tau)F \colon u \in U \mapsto \liminf_{v\to u} F(v) = \sup_{I\in\mathcal{N}(u)} \inf_{v\in I} F(v).$$

When U agrees with $\mathbf{R}^n$ endowed with the usual topology, the relaxed function of $f \colon \mathbf{R}^n \to [-\infty, +\infty]$ is denoted simply by $\mathrm{sc}^- f$. By (1.2.8), it agrees with the lower semicontinuous envelope of f already introduced in §1.2.

It is clear that

(3.5.1) $\text{sc}^-(\tau)F(u) \le F(u)$ for every $u \in U$,

and that, by defining $F_h: u \in U \mapsto F(u)$ for every $h \in \mathbf{N}$, the limit $\Gamma^-(\tau)\lim_{h \to +\infty} F_h(u)$ exists for every $u \in U$, and

(3.5.2) $\Gamma^-(\tau)\lim_{h \to +\infty} F_h(u) = \text{sc}^-(\tau)F(u)$ for every $u \in U$.

Because of (3.5.2), many properties of relaxed functionals follow from the corresponding ones of Γ-limits. Thus, if τ' is another topology on U, finer than τ, it results that

$$\text{sc}^-(\tau)F(u) \le \text{sc}^-(\tau')F(u) \text{ for every } u \in U.$$

The results below follows from the corresponding ones of the previous section.

Proposition 3.5.2. *Let* $F: U \to [-\infty, +\infty]$, *and* $G: U \to \mathbf{R}$. *Assume that* G *is continuous. Then*

$$\text{sc}^-(\tau)(F+G)(u) = \text{sc}^-(\tau)F(u) + G(u) \text{ for every } u \in U.$$

When (U, τ) satisfies the first countability axiom, the following sequential characterization of relaxed functionals holds.

Proposition 3.5.3. *Let* $F: U \to [-\infty, +\infty]$. *Assume that* (U, τ) *satisfies the first countability axiom. Then*

$$\text{sc}^-(\tau)F(u) = \min\left\{\liminf_{h \to +\infty} F(v_h) : v_h \to u\right\} \text{ for every } u \in U.$$

Proposition 3.5.4. *Let* $F: U \to [-\infty, +\infty]$. *Then the functional* $\text{sc}^-(\tau)F$ *is lower semicontinuous on* U.

In particular, from (3.5.1), and Proposition 3.5.4 it follows that

(3.5.3) F is lower semicontinuous at u if and only if $F(u) = \text{sc}^-(\tau)F(u)$.

Theorem 3.5.5. *Let* $F: U \to [-\infty, +\infty]$. *Assume that* F *is strongly countably coercive. Then* $\text{sc}^-(\tau)F$ *is strongly countably coercive, has a minimum on* U, *and*

$$\min_{v \in U} \text{sc}^-(\tau)F(v) = \inf_{v \in U} F(v).$$

Moreover, if $\inf_{v \in U} F(v) < +\infty$, *and if* $\{u_h\} \subseteq U$ *is such that* $\lim_{h \to +\infty} F(u_h) = \inf_{v \in U} F(v)$, *then the set of the cluster points of* $\{u_h\}$ *is nonempty, and every such point is a solution of* $\min_{v \in U} \text{sc}^-(\tau)F(v)$.

From Theorem 3.5.5 we deduce the following corollary in the sequential coerciveness case.

Theorem 3.5.6. *Let* $F: U \to [-\infty, +\infty]$. *Assume that* F *is coercive (respectively strongly sequentially coercive). Then* $\mathrm{sc}^-(\tau)F$ *is coercive (respectively strongly sequentially coercive), has a minimum on* U, *and*

$$\min_{v \in U} \mathrm{sc}^-(\tau)F(v) = \inf_{v \in U} F(v).$$

Moreover, if $\inf_{v \in U} F(v) < +\infty$, *and if* $\{u_h\} \subseteq U$ *is such that* $\lim_{h \to +\infty} F(u_h) = \inf_{v \in U} F(v)$, *then* $\{u_h\}$ *has at least a converging subsequence, and every limit point of its converging subsequences is a solution of* $\min_{v \in U} \mathrm{sc}^-(\tau)F(v)$.

The following result shows that, as happens when $U = \mathbf{R}^n$, the relaxed functional of F actually agrees with its lower semicontinuous envelope.

Proposition 3.5.7. *Let* $F: U \to [-\infty, +\infty]$. *Then*

$$\mathrm{sc}^-(\tau)F(u) =$$

$$= \sup\{\Phi(u) : \Phi: U \to [-\infty, +\infty] \text{ lower semicontinuous, } \Phi \leq F \text{ in } U\}$$

for every $u \in U$.

Proof. By Proposition 3.5.4, and (3.5.1) it turns out that $\mathrm{sc}^-(\tau)F$ is lower semicontinuous, and that $\mathrm{sc}^-(\tau)F \leq F$ in U. Consequently,

$$(3.5.4) \qquad\qquad \mathrm{sc}^-(\tau)F(u) \leq$$

$$\leq \sup\{\Phi(u) : \Phi: U \to [-\infty, +\infty] \text{ lower semicontinuous, } \Phi \leq F \text{ in } U\}$$

for every $u \in U$.

On the other side, if $\Phi: U \to [-\infty, +\infty]$ is lower semicontinuous, and $\Phi \leq F$ in U, then, by using (3.5.3), it obviously turns out that

$$\Phi(u) = \mathrm{sc}^-(\tau)\Phi(u) \leq \mathrm{sc}^-(\tau)F(u) \text{ for every } u \in U,$$

from which, together with (3.5.4), the proof follows. ∎

Finally, we point out that the relaxation and the Γ-limit operators commute.

Proposition 3.5.8. *Let* $E \subseteq [-\infty, +\infty]$, $\varepsilon_0 \in \overline{E}$, *and let, for every* $\varepsilon \in E$, $F_\varepsilon: U \to [-\infty, +\infty]$. *Then*

$$\Gamma^-(\tau) \liminf_{\varepsilon \to \varepsilon_0} F_\varepsilon(u) = \Gamma^-(\tau) \liminf_{\varepsilon \to \varepsilon_0} \mathrm{sc}^-(\tau)F_\varepsilon(u)$$

$$\Gamma^-(\tau) \limsup_{\varepsilon \to \varepsilon_0} F_\varepsilon(u) = \Gamma^-(\tau) \limsup_{\varepsilon \to \varepsilon_0} \mathrm{sc}^-(\tau)F_\varepsilon(u)$$

for every $u \in U$.

Proof. Let $\varepsilon \in E$, $u \in U$, and $I \in \mathcal{N}(u)$. Then the constant functional $w \in U \mapsto \inf_{v \in I} F_\varepsilon(v)$ is lower semicontinuous and less that F_ε in I. Consequently, by Proposition 3.5.7 applied to I in place of U, we obtain that

$$\inf_{v \in I} F_\varepsilon(v) \leq \mathrm{sc}^-(\tau) F_\varepsilon(u) \text{ for every } \varepsilon \in E, \ u \in U.$$

Because of this, and by using also (3.5.1) the proof follows. ∎

Finally, given $F: U \to [-\infty, +\infty]$, we deduced from F another functional, that we call *sequential lower value of F* and denote by $\mathrm{sq}^-(\tau)F$, that enjoys intermediate properties between those of $\mathrm{sc}^-(\tau)F$ and of F, in the sense that it has some features of lower semicontinuity type, but inherits the properties of F more directly than what $\mathrm{sc}^-(\tau)F$ does. It is defined as

$$\mathrm{sq}^-(\tau)F: u \in U \mapsto \inf \left\{ \liminf_{h \to +\infty} F(u_h) : u_h \to u \text{ in } \tau \right\}.$$

It is clear that

$$\mathrm{sc}^-(\tau) F(u) \leq \mathrm{sq}^-(\tau) F(u) \leq F(u) \text{ for every } u \in U.$$

We remark that, in general, $\mathrm{sq}^-(\tau)F$ need not be either sequentially lower semicontinuous. An example showing this will be given, in the framework of variational integrals, in §10.9.

On the contrary, if U satisfies the first countability axiom, Proposition 3.5.3 yields the identity between $\mathrm{sc}^- F$ and $\mathrm{sq}^- F$.

Chapter 4
BV and Sobolev Spaces

In this book Sobolev and *BV* spaces are widely used as domains in which variational problems are settled. In the present chapter we briefly introduce them, at least in the case of derivatives of order one, together with their main properties. We refer to [A], [AFP], [EG], [Gu], [Z] for general references of the matter.

Nevertheless, the presentation that we are going to propose differs slightly from those usually described in literature, in which Sobolev spaces are firstly introduced, and *BV* ones are then studied as a generalization of them. On the contrary, we follow an opposite scheme by placing ourselves in the framework of Borel measures, and firstly looking at *BV* spaces as special subsets of a particular space of Borel measures, and then deducing Sobolev spaces by means of successive restrictions.

This unifying approach has the advantage of clarifying the relationships between these spaces, and, in particular, allows a better understanding of the structure certain weak type topologies they are equipped with.

§4.1 Regularization of Measures and of Summable Functions

To carry out the above program, we recall in this section the notion of regularization of a Borel measure and, in particular, of a locally summable Lebesgue measurable function, together with their main properties. We will come back on the notion of regularization in Chapter 7, in a more general context.

We say that $\rho \in C_0^\infty(B_1(0))$ is a *symmetric mollifier* if $\rho(y) \geq 0$ and $\rho(-y) = \rho(y)$ for every $y \in \mathbf{R}^n$, and $\int_{\mathbf{R}^n} \rho(y)dy = 1$.

Let ρ be a symmetric mollifier. Then for every $\Omega \in \mathcal{A}(\mathbf{R}^n)$, $\mu \in \mathcal{M}_{\mathrm{loc}}(\Omega)$, and $\varepsilon > 0$ we define the *regularization* μ_ε of μ as

$$(4.1.1) \qquad \mu_\varepsilon : x \in \Omega_\varepsilon^- \mapsto \frac{1}{\varepsilon^n} \int_\Omega \rho\left(\frac{x-y}{\varepsilon}\right) d\mu(y).$$

We observe that the definition in (4.1.1) is well posed since, being ρ with compact support in $B_1(0)$, for every $x \in \Omega_\varepsilon^-$ the integral in (4.1.1) is actually extended only over $B_\varepsilon(x)$, whose closure is a compact subset of Ω.

In particular, if $u \in L^1_{\text{loc}}(\Omega)$, then (4.1.1) applied with $\mu = u\mathcal{L}^n$, defines the regularization u_ε of u, that, by using Theorem 2.3.2, turns out to be given by

$$(4.1.2) \qquad u_\varepsilon : x \in \Omega_\varepsilon^- \mapsto \frac{1}{\varepsilon^n} \int_\Omega \rho\left(\frac{x-y}{\varepsilon}\right) u(y)\,dy.$$

Because of this, it is clear that the regularization of a function enjoys all the properties of the regularization of a measure.

We list now the main properties of the regularization of a function.

If $\Omega \in \mathcal{A}(\mathbf{R}^n)$, it turns out that

$$w_\varepsilon \in C^\infty(\Omega_\varepsilon^-) \text{ for every } w \in L^1_{\text{loc}}(\Omega), \text{ and } \varepsilon > 0,$$

$$(4.1.3) \qquad \int_{\Omega_\varepsilon^-} |w_\varepsilon|\,dx \leq \int_\Omega |w|\,dx \text{ for every } w \in L^1_{\text{loc}}(\Omega), \text{ and } \varepsilon > 0,$$

and that

$$(4.1.4) \qquad \text{spt}(w_\varepsilon) \subseteq \{x \in \Omega_\varepsilon^- : \text{dist}(x, S) < \varepsilon\} \text{ for every } w \in L^1_{\text{loc}}(\Omega),$$

$$S \subseteq \Omega \text{ with } w = 0 \text{ a.e. in } \Omega \setminus S, \ \varepsilon \in \,]0, \text{dist}(\text{spt}(w), \partial\Omega)[.$$

Moreover, it is easy to verify that

$$(4.1.5) \qquad \lim_{\varepsilon \to 0} \max_{x \in K} |w_\varepsilon(x) - w(x)| = 0$$

for every $w \in C^0(\Omega)$, and every compact subset K of Ω,

from which, by exploiting Theorem 2.4.5, and (4.1.3), it is possible to prove that

$$\lim_{\varepsilon \to 0} \int_K |w_\varepsilon - w|\,dx = 0$$

for every $w \in L^1_{\text{loc}}(\Omega)$, and every compact subset K of Ω.

Finally, it is well known that

$$(4.1.6) \qquad \lim_{\varepsilon \to 0} w_\varepsilon(x) = w(x) \text{ for every } w \in L^1_{\text{loc}}(\Omega), \text{ and a.e. } x \in \Omega.$$

Regularization processes provide a powerful tool to approximate measures by smooth functions, as proved in the following result.

Theorem 4.1.1. Let $\Omega \in \mathcal{A}(\mathbf{R}^n)$, and $\mu \in \mathcal{M}_{\mathrm{loc}}(\Omega)$. Then, for every $\varepsilon > 0$, $\mu_\varepsilon \in C^\infty(\Omega_\varepsilon^-)$,

$$\frac{\partial^{|\alpha|}\mu_\varepsilon}{\partial x^\alpha}(x) = \frac{1}{\varepsilon^{n+|\alpha|}} \int_\Omega \frac{\partial^{|\alpha|}\rho}{\partial z^\alpha}\left(\frac{x-y}{\varepsilon}\right) d\mu(y) \text{ for every } x \in \Omega_\varepsilon^-, \ \alpha \in \mathbf{N}^n,$$

and

$$\int_{\Omega_\varepsilon^-} |\mu_\varepsilon| dx \leq |\mu|(\Omega).$$

Moreover

$$\lim_{\varepsilon \to 0} \int_{\Omega_\varepsilon^-} \varphi \mu_\varepsilon dx = \int_\Omega \varphi d\mu \text{ for every } \varphi \in C_0^0(\Omega),$$

and

$$\lim_{\varepsilon \to 0} \mu_\varepsilon(x) = \frac{d\mu^a}{d\mathcal{L}^n}(x) \text{ for } \mathcal{L}^n\text{-a.e. } x \in \Omega.$$

Proof. The part of the theorem concerning the smoothness properties of the regularizations of μ follows easily by induction on the length of the multiindex α, by directly considering difference quotients, and by using Lebesgue Dominated Convergence Theorem.

In addition, by ii) of Theorem 2.1.2, and by Fubini's theorem, it comes that

$$\int_{\Omega_\varepsilon^-} |\mu_\varepsilon| dx = \frac{1}{\varepsilon^n} \int_{\Omega_\varepsilon^-} \left| \int_\Omega \rho\left(\frac{x-y}{\varepsilon}\right) d\mu(y) \right| dx \leq$$

$$\leq \frac{1}{\varepsilon^n} \int_{\Omega_\varepsilon^-} \int_\Omega \rho\left(\frac{x-y}{\varepsilon}\right) d|\mu|(y) dx \leq$$

$$\leq \frac{1}{\varepsilon^n} \int_\Omega \int_\Omega \rho\left(\frac{x-y}{\varepsilon}\right) dx d|\mu|(y) = \int_\Omega d|\mu|(y) = |\mu|(\Omega)$$

for every $\varepsilon > 0$.

Let now $\varphi \in C_0^0(\Omega)$. Let us preliminarily observe that (4.1.4) implies that $\mathrm{spt}(\varphi_\varepsilon) \subseteq \Omega_\varepsilon^-$ provided $\varepsilon \in \frac{1}{2}]0, \mathrm{dist}(\mathrm{spt}(\varphi), \partial\Omega)[$. Therefore, by Fubini's theorem applied to the positive and negative part of μ, and the symmetry of ρ, we get that

$$\int_{\Omega_\varepsilon^-} \varphi(x)\mu_\varepsilon(x) dx = \int_{\mathrm{spt}(\varphi)} \varphi(x)\mu_\varepsilon(x) dx =$$

$$= \frac{1}{\varepsilon^n} \int_{\mathrm{spt}(\varphi)} \varphi(x) \int_\Omega \rho\left(\frac{x-y}{\varepsilon}\right) d\mu(y) dx =$$

$$= \frac{1}{\varepsilon^n} \int_\Omega \int_{\mathrm{spt}(\varphi)} \rho\left(\frac{x-y}{\varepsilon}\right) \varphi(x) dx d\mu(y) =$$

$$= \int_\Omega \frac{1}{\varepsilon^n} \int_\Omega \rho\left(\frac{y-x}{\varepsilon}\right) \varphi(x) dx d\mu(y) = \int_\Omega \varphi_\varepsilon(y) d\mu(y)$$

for every $\varepsilon \in \frac{1}{2}\,]0, \operatorname{dist}(\operatorname{spt}(\varphi), \partial\Omega)[\,$.

Because of this, and by (4.1.5) we conclude that

$$\lim_{\varepsilon \to 0} \int_{\Omega_\varepsilon^-} \varphi \mu_\varepsilon dx = \int_\Omega \varphi d\mu,$$

once we recall that, by (4.1.4), there exists a compact subset K of Ω such that $\operatorname{spt}(\varphi_\varepsilon) \subseteq K$ for every $\varepsilon > 0$ sufficiently small.

Finally, once we observe that

$$\frac{1}{\varepsilon^n} \int_{B_\varepsilon(x)} \rho\left(\frac{x-y}{\varepsilon}\right) d|\mu^{\mathrm{s}}|(y) \le \max_{\mathbf{R}^n} \rho \frac{1}{\varepsilon^n} |\mu^{\mathrm{s}}|(B_\varepsilon(x))$$

for every $x \in \Omega$, and $\varepsilon \in \,]0, \operatorname{dist}(x, \partial\Omega)[\,$,

and that, by Theorem 2.3.5, $\limsup_{\varepsilon\to 0} \frac{1}{\varepsilon^n}|\mu^{\mathrm{s}}|(B_\varepsilon(x)) = 0$ for $\mathcal{L}^n$-a.e. $x \in \Omega$, we conclude that

$$\limsup_{\varepsilon \to 0} \frac{1}{\varepsilon^n} \left| \int_{B_\varepsilon(x)} \rho\left(\frac{x-y}{\varepsilon}\right) d\mu^{\mathrm{s}}(y) \right| = 0 \quad \text{for } \mathcal{L}^n\text{-a.e. } x \in \Omega.$$

Because of this, and by the Lebesgue Decomposition Theorem, we thus infer that

$$= \liminf_{\varepsilon\to 0} \left(\frac{d\mu^{\mathrm{a}}}{d\mathcal{L}^n}\right)_\varepsilon (x) =$$

$$= \liminf_{\varepsilon\to 0} \left(\frac{d\mu^{\mathrm{a}}}{d\mathcal{L}^n}\right)_\varepsilon (x) - \limsup_{\varepsilon\to 0} \frac{1}{\varepsilon^n} \int_{B_\varepsilon(x)} \rho\left(\frac{x-y}{\varepsilon}\right) d\mu^{\mathrm{s}}(y) \le$$

$$\le \liminf_{\varepsilon\to 0} \left\{ \frac{1}{\varepsilon^n} \int_{B_\varepsilon(x)} \rho\left(\frac{x-y}{\varepsilon}\right) d\mu^{\mathrm{a}}(y) + \frac{1}{\varepsilon^n} \int_{B_\varepsilon(x)} \rho\left(\frac{x-y}{\varepsilon}\right) d\mu^{\mathrm{s}}(y) \right\} =$$

$$= \liminf_{\varepsilon\to 0} \mu_\varepsilon(x) \le \limsup_{\varepsilon\to 0} \mu_\varepsilon(x) =$$

$$= \limsup_{\varepsilon\to 0} \left\{ \frac{1}{\varepsilon^n} \int_{B_\varepsilon(x)} \rho\left(\frac{x-y}{\varepsilon}\right) d\mu^{\mathrm{a}}(y) + \frac{1}{\varepsilon^n} \int_{B_\varepsilon(x)} \rho\left(\frac{x-y}{\varepsilon}\right) d\mu^{\mathrm{s}}(y) \right\} =$$

$$= \lim_{\varepsilon\to 0} \left(\frac{d\mu^{\mathrm{a}}}{d\mathcal{L}^n}\right)_\varepsilon (x) + \limsup_{\varepsilon\to 0} \frac{1}{\varepsilon^n} \int_{B_\varepsilon(x)} \rho\left(\frac{x-y}{\varepsilon}\right) d\mu^{\mathrm{s}}(y) =$$

$$= \limsup_{\varepsilon\to 0} \left(\frac{d\mu^{\mathrm{a}}}{d\mathcal{L}^n}\right)_\varepsilon (x) \quad \text{for } \mathcal{L}^n\text{-a.e. } x \in \Omega,$$

from which, together with (4.1.6), the pointwise convergence result follows. ∎

The above approximation process can be specified by the following result.

Proposition 4.1.2. Let $\Omega \in \mathcal{A}(\mathbf{R}^n)$, and $\mu \in (\mathcal{M}_{\mathrm{loc}}(\Omega))^m$. Then the limit $\lim_{\varepsilon \to 0} \int_{\Omega_\varepsilon^-} |\mu_\varepsilon| dx$ exists, and

$$\lim_{\varepsilon \to 0} \int_{\Omega_\varepsilon^-} |\mu_\varepsilon| dx = |\mu|(\Omega).$$

Proof. By Theorem 4.1.1 we conclude that

$$(4.1.7) \qquad\qquad \limsup_{\varepsilon \to 0} \int_{\Omega_\varepsilon^-} |\mu_\varepsilon| dx \leq |\mu|(\Omega).$$

On the other side, if $A \in \mathcal{A}(\Omega)$ is such that $A \subset\subset \Omega$, then Theorem 4.1.1 yields that $\mu_\varepsilon \mathcal{L}^n \to \mu$ in weak*-$\mathcal{M}(A)$, therefore, by Proposition 2.4.8, it follows that

$$(4.1.8) \quad |\mu|(A) \leq \liminf_{\varepsilon \to 0} \int_A |\mu_\varepsilon| dx \leq \liminf_{\varepsilon \to 0} \int_{\Omega_\varepsilon^-} |\mu_\varepsilon| dx \text{ for every } A \subset\subset \Omega.$$

Therefore, once we recall that Theorem 2.4.2 yields the inner regularity of $|\mu|$, by (4.1.8), and (4.1.7) we obtain that

$$|\mu|(\Omega) = \sup\{|\mu|(A) : A \subset\subset \Omega\} \leq \liminf_{\varepsilon \to 0} \int_{\Omega_\varepsilon^-} |\mu_\varepsilon| dx \leq$$

$$\leq \limsup_{\varepsilon \to 0} \int_{\Omega_\varepsilon^-} |\mu_\varepsilon| dx \leq |\mu|(\Omega),$$

which proves the proposition. ∎

Finally, we prove the following properties of integrals of regularizations of measures.

Proposition 4.1.3. Let $\mu \in \mathcal{M}_{\mathrm{loc}}(\mathbf{R}^n)$. For every $\varepsilon > 0$ let μ_ε be the regularization of μ defined by means of (4.1.1). Then, for every $\varphi \in C_0^0(\mathbf{R}^n)$ it results that

$$\int_{\mathbf{R}^n} \varphi \mu_\varepsilon dx = \int_{\mathbf{R}^n} \varphi_\varepsilon d\mu,$$

and

$$\int_{\mathbf{R}^n} \varphi \frac{\partial^{|\alpha|} \mu_\varepsilon}{\partial x^\alpha} dx = (-1)^{|\alpha|} \int_{\mathbf{R}^n} \frac{\partial^{|\alpha|} \varphi_\varepsilon}{\partial x^\alpha} d\mu \text{ for every } \alpha \in (\mathbf{N} \cup \{0\})^n.$$

Proof. Let φ be as above. Then by Fubini's theorem applied to both the positive and the negative parts of μ, and the symmetry of the mollifier, it follows that

$$\int_{\mathbf{R}^n} \varphi \mu_\varepsilon dx = \frac{1}{\varepsilon^n} \int_{\mathbf{R}^n} \varphi(x) \int_{\mathbf{R}^n} \rho\left(\frac{x-y}{\varepsilon}\right) d\mu(y) dx =$$

$$= \frac{1}{\varepsilon^n} \int_{\mathbf{R}^n} \int_{\mathbf{R}^n} \rho\left(\frac{x-y}{\varepsilon}\right) \varphi(x) dx d\mu(y) = \int_{\mathbf{R}^n} \varphi_\varepsilon d\mu.$$

If now $\alpha \in (\mathbf{N} \cup \{0\})^n$, by Theorem 4.1.1, and again Fubini's theorem and the symmetry of the mollifier, it follows that

$$\int_{\mathbf{R}^n} \varphi \frac{\partial^{|\alpha|} \mu_\varepsilon}{\partial x^\alpha} dx = \frac{1}{\varepsilon^{n+|\alpha|}} \int_{\mathbf{R}^n} \varphi(x) \int_{\mathbf{R}^n} \frac{\partial^{|\alpha|} \rho}{\partial z^\alpha}\left(\frac{x-y}{\varepsilon}\right) d\mu(y) dx =$$

$$= \frac{1}{\varepsilon^{n+|\alpha|}} \int_{\mathbf{R}^n} \int_{\mathbf{R}^n} \frac{\partial^{|\alpha|} \rho}{\partial z^\alpha}\left(\frac{x-y}{\varepsilon}\right) \varphi(x) dx d\mu(y) =$$

$$= (-1)^{|\alpha|} \int_{\mathbf{R}^n} \frac{\partial^{|\alpha|} \varphi_\varepsilon}{\partial x^\alpha} d\mu. \qquad \blacksquare$$

The regularization process provides also approximation in energy, at least for summable functions. However, as we will prove in Chapter 7, such property holds also in the more general context of functions taking values in a Hausdorff locally convex topological vector space.

The main tool needed to get such approximation is Jensen's inequality, whose proof will be given in Chapter 7 in the above mentioned context.

Theorem 4.1.4 (Jensen's Inequality). *Let $(\Omega, \mathcal{E})$ be a measure space, μ be a positive measure defined on $\mathcal{E}$ with $\mu(\Omega) = 1$, and $f \colon \mathbf{R}^m \to [0, +\infty]$ be convex and lower semicontinuous. Then*

$$f\left(\int_\Omega w d\mu\right) \leq \int_\Omega f(w) d\mu \text{ for every } w \in (L^1(\Omega, \mu))^m.$$

We can now prove the approximation in energy result.

Lemma 4.1.5. *Let $\Omega \in \mathcal{A}(\mathbf{R}^n)$, and $f \colon \mathbf{R}^m \to [0, +\infty]$ be convex and lower semicontinuous. Then*

$$\int_{\Omega_\varepsilon^-} f(w_\varepsilon) dx \leq \int_\Omega f(w) dx \text{ for every } w \in (L^1_{\text{loc}}(\Omega))^m, \ \varepsilon > 0.$$

Proof. Let $w \in (L^1_{\text{loc}}(\Omega))^m$, and $\varepsilon > 0$. Let $x \in \Omega_\varepsilon^-$, and let $\mu_{\varepsilon,x}$ be the positive measure defined for every $E \in \mathcal{L}_n(\Omega)$ by $\mu_{\varepsilon,x}(E) = \frac{1}{\varepsilon^n} \int_E \rho(\frac{x-y}{\varepsilon}) dy$. Then, since $\mu_{\varepsilon,x}(\Omega) = 1$, by Jensen's inequality applied to $\mu_{\varepsilon,x}$, and by Theorem 2.3.2 it follows that

$$f(w_\varepsilon(x)) = f\left(\frac{1}{\varepsilon^n} \int_\Omega \rho\left(\frac{x-y}{\varepsilon}\right) w(y) dy\right) = f\left(\int_\Omega w(y) d\mu_{\varepsilon,x}(y)\right) \leq$$

$$\leq \int_\Omega f(w(y)) d\mu_{\varepsilon,x}(y) = \frac{1}{\varepsilon^n} \int_\Omega \rho\left(\frac{x-y}{\varepsilon}\right) f(w(y)) dy = (f(w))_\varepsilon(x)$$

$$\text{for every } x \in \Omega_\varepsilon^-.$$

By integrating the above inequality over Ω_ε^-, and by using Fubini's theorem we conclude that

$$\int_{\Omega_\varepsilon^-} f(w_\varepsilon(x))dx \le \int_{\Omega_\varepsilon^-} \frac{1}{\varepsilon^n} \int_\Omega \rho\left(\frac{x-y}{\varepsilon}\right) f(w(y))dydx =$$

$$= \int_\Omega f(w(y)) \frac{1}{\varepsilon^n} \int_{\Omega_\varepsilon^-} \rho\left(\frac{x-y}{\varepsilon}\right) dxdy \le \int_\Omega f(w(y))dy,$$

which proves the lemma. ∎

Theorem 4.1.6. *Let $\Omega \in \mathcal{A}(\mathbf{R}^n)$, and $f: \mathbf{R}^m \to [0, +\infty]$ be convex and lower semicontinuous. Then the limit $\lim_{\varepsilon \to 0} \int_{\Omega_\varepsilon^-} f(w_\varepsilon)dx$ exists, and*

$$\lim_{\varepsilon \to 0} \int_{\Omega_\varepsilon^-} f(w_\varepsilon)dx = \int_\Omega f(w)dx \text{ for every } w \in (L^1_{\text{loc}}(\Omega))^m.$$

Proof. Let $u \in (L^1_{\text{loc}}(\Omega))^m$, and $A \subset\subset \Omega$. Then (4.1.6), Fatou's lemma, and Lemma 4.1.5 yield

$$\limsup_{\varepsilon \to 0} \int_A f(u)dx \le \liminf_{\varepsilon \to 0} \int_A f(u_\varepsilon)dx \le \liminf_{\varepsilon \to 0} \int_{\Omega_\varepsilon^-} f(u_\varepsilon)dx \le$$

$$\le \limsup_{\varepsilon \to 0} \int_{\Omega_\varepsilon^-} f(u_\varepsilon)dx \le \int_\Omega f(u)dx,$$

from which the proof follows letting A increase to Ω. ∎

§4.2 *BV* **Spaces**

Let $\Omega \in \mathcal{A}(\mathbf{R}^n)$.

For every $\mu \in \mathcal{M}(\Omega)$, and $i \in \{1, \dots, n\}$ we say that the *i-th weak partial derivative* of μ is in $\mathcal{M}(\Omega)$ if there exists $\nu \in \mathcal{M}(\Omega)$ such that

$$\int_\Omega \nabla_i \varphi d\mu = - \int_\Omega \varphi d\nu \text{ for every } \varphi \in C_0^\infty(\Omega).$$

If this is the case, we denote by $D_i\mu$ the i-th weak partial derivative of μ, and define the *weak gradient* of μ as the Borel vector measure $D\mu = (D_1\mu, \dots, D_n\mu)$.

Definition 4.2.1. *Let $\Omega \in \mathcal{A}(\mathbf{R}^n)$. We define the space $BV(\Omega)$ of the functions of bounded variation in Ω as*

$$BV(\Omega) = \{(\mu, \nu) \in \mathcal{M}(\Omega) \times (\mathcal{M}(\Omega))^n : \nu = D\mu\},$$

and the space of the functions of locally bounded variation in Ω as

$$BV_{\mathrm{loc}}(\Omega) = \{(\mu, \nu) \in \mathcal{M}_{\mathrm{loc}}(\Omega) \times (\mathcal{M}_{\mathrm{loc}}(\Omega))^n :$$

$$(\mu, \nu) \in BV(A) \text{ for every } A \subset\subset \Omega\}.$$

According to Definition 4.2.1, given $\Omega \in \mathcal{A}(\mathbf{R}^n)$, $BV(\Omega)$ is a vector subspace of $(\mathcal{M}(\Omega))^{n+1}$. Nevertheless, by using just the definition of weak partial derivative, and the density of $C_0^\infty(\Omega)$ in $C_0^0(\Omega)$ endowed with the $C_{\mathrm{b}}^0(\Omega)$ topology, it is easy to verify that, if $(\mu, D\mu) \in BV(\Omega)$, then $D\mu$ is uniquely determined by μ. Consequently, the application $(\mu, D\mu) \in BV(\Omega) \mapsto \mu \in \mathcal{M}(\Omega)$ turns out to be an injection, that allows $BV(\Omega)$ to be identified with $\{\mu \in \mathcal{M}(\Omega) : (\mu, D\mu) \in BV(\Omega)\}$, and therefore to see it as a space of Borel real measures on Ω.

Actually, even more can be said, and all is based on the following inequality for smooth functions, that we prove for the sake of completeness.

Lemma 4.2.2. *Let $u \in C_0^1(\mathbf{R}^n)$. Then*

$$\int_{\mathbf{R}^n} |T[h]u - u| dx \le |h| \int_{\mathbf{R}^n} |\nabla u| dx \text{ for every } h \in \mathbf{R}^n.$$

Proof. Let h be as above. Then obviously

$$|u(x+h) - u(x)| = \left| \int_0^1 \nabla u(x+th) \cdot h \, dt \right| \le |h| \int_0^1 |\nabla u(x+th)| dt$$

$$\text{for every } x \in \mathbf{R}^n,$$

from which, by using Fubini's theorem, we obtain that

$$\int_{\mathbf{R}^n} |u(x+h) - u(x)| dx \le |h| \int_{\mathbf{R}^n} \int_0^1 |\nabla u(x+th)| dt dx =$$

$$= |h| \int_0^1 \int_{\mathbf{R}^n} |\nabla u(x+th)| dx dt = |h| \int_0^1 \int_{\mathbf{R}^n} |\nabla u(y)| dy dt =$$

$$= |h| \int_{\mathbf{R}^n} |\nabla u(y)| dy,$$

which proves the lemma. ∎

Proposition 4.2.3. *Let $\Omega \in \mathcal{A}(\mathbf{R}^n)$, and $\mu \in BV(\Omega)$. Then μ is absolutely continuous with respect to $\mathcal{L}^n$.*

Proof. By using Corollary 2.4.3 it suffices to prove that the restriction of μ to $\mathcal{B}(K)$ is absolutely continuous with respect to $\mathcal{L}^n$ for every compact subset K of Ω, and it is clear that this holds if the restriction of μ to $\mathcal{B}(B)$ is absolutely continuous with respect to $\mathcal{L}^n$ for every open ball $B \subset\subset \Omega$.

Let $B_r(x_0) \subset\subset \Omega$ be an open ball centred in $x_0 \in \Omega$ and with radius r, and let, for every $\varepsilon > 0$ sufficiently small $\mu_\varepsilon : x \in B_{r-\varepsilon}(x_0) \mapsto \frac{1}{\varepsilon^n} \int_{B_r(x_0)} \rho(\frac{x-y}{\varepsilon}) d\mu(y)$ be the regularization of μ defined in (4.1.1).

Let $\varepsilon > 0$. Then, since for every $x \in B_{r-\varepsilon}(x_0)$ the function $\rho(\frac{x-\cdot}{\varepsilon}) \in C_0^\infty(B_r(x_0))$, by Theorem 4.1.1 it follows that

$$\nabla\mu_\varepsilon(x) = -\frac{1}{\varepsilon^n} \int_{B_r(x_0)} \nabla_y \rho\left(\frac{x-\cdot}{\varepsilon}\right)(y) d\mu(y),$$

from which, since $\mu \in BV(\Omega)$, it results that

$$(4.2.1) \qquad \nabla\mu_\varepsilon(x) = \frac{1}{\varepsilon^n} \int_{B_r(x_0)} \rho\left(\frac{x-y}{\varepsilon}\right)(y) dD\mu(y) = (D\mu)_\varepsilon(x)$$

$$\text{for every } x \in B_{r-\varepsilon}(x_0).$$

Let now $\sigma \in]0, r[$, and $\varphi \in C_0^\infty(B_r(x_0))$ be such that $0 \le \varphi(x) \le 1$ for every $x \in \mathbf{R}^n$, and $\varphi(x) = 1$ for every $x \in B_{r-\sigma}(x_0)$. Then, by (4.2.1), and by Theorem 4.1.1, it thus comes that

$$(4.2.2) \qquad \int_{\mathbf{R}^n} |\varphi\mu_\varepsilon| dx \le \int_{\mathrm{spt}(\varphi)} |\mu_\varepsilon| dx \le \int_{B_{r-\varepsilon}(x_0)} |\mu_\varepsilon| dx \le |\mu|(B_r(x_0))$$

$$\text{for every } \varepsilon \in]0, \mathrm{dist}(\mathrm{spt}(\varphi), \partial B_r(x_0))[,$$

and

$$\int_{\mathbf{R}^n} |\nabla(\varphi\mu_\varepsilon)| dx \le \int_{\mathrm{spt}(\varphi)} |\nabla\mu_\varepsilon| dx + \int_{\mathrm{spt}(\varphi)} |\mu_\varepsilon||\nabla\varphi| dx \le$$

$$\le \int_{B_{r-\varepsilon}(x_0)} |(D\mu)_\varepsilon| dx + \||\nabla\varphi|\|_{C^0(B_r(x_0))} \int_{B_{r-\varepsilon}(x_0)} |\mu_\varepsilon| dx \le$$

$$\le |D\mu|(B_r(x_0)) + \||\nabla\varphi|\|_{C^0(B_r(x_0))}|\mu|(B_r(x_0))$$

$$\text{for every } \varepsilon \in]0, \mathrm{dist}(\mathrm{spt}(\varphi), \partial B_r(x_0))[,$$

from which, together with Lemma 4.2.2, it results that

$$(4.2.3) \qquad \int_{\mathbf{R}^n} |T[h](\varphi\mu_\varepsilon) - \varphi\mu_\varepsilon| dx \le$$

$$\leq |h| \left(|D\mu|(B_r(x_0)) + \|\nabla\varphi\|_{C^0(B_r(x_0))} |\mu|(B_r(x_0)) \right)$$

for every $\varepsilon \in\,]0, \mathrm{dist}(\mathrm{spt}(\varphi), \partial B_r(x_0)[,\ h \in \mathbf{R}^n$.

By (4.2.2) and (4.2.3), Theorem 2.2.8 applies, and the compactness of $\{\varphi\mu_\varepsilon\}_{\varepsilon\in]0,\mathrm{dist}(\mathrm{spt}(\varphi),\partial B_r(x_0)[}$ in $L^1(\mathbf{R}^n)$ follows. Because of this, also $\{\mu_\varepsilon\}_{\varepsilon\in]0,\mathrm{dist}(\mathrm{spt}(\varphi),\partial B_r(x_0)[}$ turns out to be compact in $L^1(B_{r-\sigma}(x_0))$. This, together with the weak*-$(\mathcal{M}(B_{r-\sigma}(x_0)))^m$-convergence of $\{\mu_\varepsilon\mathcal{L}^n\}_{\varepsilon>0}$ to μ given by Theorem 4.1.1, in turn implies that, for every $\sigma \in\,]0, r[$, the restriction of μ to $\mathcal{B}(B_{r-\sigma}(x_0))$ is absolutely continuous with respect to $\mathcal{L}^n$.

Such property provides that the restriction of μ to $\mathcal{B}(B_r(x_0))$ is absolutely continuous with respect to $\mathcal{L}^n$, and hence the proposition. ∎

Let $\Omega \in \mathcal{A}(\mathbf{R}^n)$.

If for every $\mu \in \mathcal{M}(\Omega)$ absolutely continuous with respect to $\mathcal{L}^n$ we identify μ with its Radon-Nikodym derivative, and vice-versa for every $u \in L^1(\Omega)$ we identify u with $u\mathcal{L}^n$, and set, provided $D(u\mathcal{L}^n) \in (\mathcal{M}(\Omega))^n$, $Du = D(u\mathcal{L}^n)$, then by Proposition 4.2.3 we conclude that

$$BV(\Omega) = \{u \in L^1(\Omega) : Du \in (\mathcal{M}(\Omega))^n\},$$

and of course that

$$BV_{\mathrm{loc}}(\Omega) = \{u \in L^1_{\mathrm{loc}}(\Omega) : u \in BV(A) \text{ for every } A \subset\subset \Omega\}.$$

We observe explicitly that, according to the definitions of Chapter 2, if $u \in BV_{\mathrm{loc}}(\Omega)$, then Du turns out to be a Radon measure on Ω.

Since $BV(\Omega)$ is a subset of $(\mathcal{M}(\Omega))^{n+1}$, it naturally inherits its topological structures.

In particular, $BV(\Omega)$ becomes a normed space with the $(\mathcal{M}(\Omega))^{n+1}$-norm. We denote such norm functional as

$$\| \cdot \|_{BV(\Omega)}\colon u \in BV(\Omega) \mapsto \|u\|_{L^1(\Omega)} + |Du|(\Omega),$$

and, as usual, we denote again by $BV(\Omega)$ the strong topology of $BV(\Omega)$.

In addition, $BV(\Omega)$ also inherits the weak*-$(\mathcal{M}(\Omega))^{n+1}$ topology of $(\mathcal{M}(\Omega))^{n+1}$. In this case, the following result holds.

Proposition 4.2.4. *Let $\Omega \in \mathcal{A}(\mathbf{R}^n)$. Then $BV(\Omega)$ is a weak*-$(\mathcal{M}(\Omega))^{n+1}$ closed subspace of $(\mathcal{M}(\Omega))^{n+1}$.*

Proof. Follows immediately once we observe that

$$BV(\Omega) =$$

$$= \cap_{\varphi\in C_0^\infty(\Omega)} \left\{ (\mu, \nu) \in \mathcal{M}(\Omega) \times (\mathcal{M}(\Omega))^n : \int_\Omega \nabla\varphi d\mu + \int_\Omega \varphi d\nu = 0 \right\},$$

and that the sets in the right-hand side of the above equality are weak*-$(\mathcal{M}(\Omega))^{n+1}$ closed since, for every $\varphi \in C_0^\infty(\Omega)$, the functional $(\mu, \nu) \in \mathcal{M}(\Omega) \times (\mathcal{M}(\Omega))^n \mapsto \int_\Omega \nabla\varphi d\mu + \int_\Omega \varphi d\nu$ is weak*-$(\mathcal{M}(\Omega))^{n+1}$ continuous. ∎

Proposition 4.2.4 has some important consequences. First of all, it implies that $BV(\Omega)$ is weak*-$(\mathcal{M}(\Omega))^{n+1}$ sequentially complete, since it is a subspace of $(\mathcal{M}(\Omega))^{n+1}$ closed in weak*-$(\mathcal{M}(\Omega))^{n+1}$, and since $(\mathcal{M}(\Omega))^{n+1}$ is sequentially weak*-$(\mathcal{M}(\Omega))^{n+1}$ complete by the Banach-Steinhaus Theorem. Moreover, since $BV(\Omega)$ is also a $(\mathcal{M}(\Omega))^{n+1}$ closed subspace of $(\mathcal{M}(\Omega))^{n+1}$, it turns out to be a Banach space too.

A sequence in $BV(\Omega)$ that converges in the weak*-$(\mathcal{M}(\Omega))^{n+1}$ topology turns out to be bounded in $BV(\Omega)$. Conversely, the following compactness result holds.

Proposition 4.2.5. *Let $\Omega \in \mathcal{A}(\mathbf{R}^n)$, and $\{u_h\} \subseteq BV(\Omega)$ be $BV(\Omega)$-bounded. Then there exist $\{h_k\} \subseteq \mathbf{N}$ strictly increasing, and $u \in BV(\Omega)$ such that*

$$u_{h_k} \to u \text{ in weak*-}(\mathcal{M}(\Omega))^{n+1}.$$

Proof. Follows from Theorem 2.4.10, and Proposition 4.2.4. ∎

With an abuse of notation, we denote by weak*-$BV(\Omega)$ the topology on $BV(\Omega)$ induced from the one of $\mathcal{M}(\Omega) \times (\text{weak*-}(\mathcal{M}(\Omega))^n)$, namely from the product of the strong topology of $\mathcal{M}(\Omega)$ and of the weak* one of $(\mathcal{M}(\Omega))^n$. In particular, once we see $BV(\Omega)$ as a space of functions, given $\{u_h\} \subseteq BV(\Omega)$ and $u \in BV(\Omega)$, it turns out that $u_h \to u$ in weak*-$BV(\Omega)$ if and only if $u_h \to u$ in $L^1(\Omega)$ and $Du_h \to Du$ in weak*-$(\mathcal{M}(\Omega))^n$.

It is clear that the weak*-$BV(\Omega)$ topology is finer than the weak*-$(\mathcal{M}(\Omega))^{n+1}$ one. Because of this and of the completeness of $\mathcal{M}(\Omega)$, once we endow $BV(\Omega)$ with the weak*-$BV(\Omega)$ topology, $BV(\Omega)$ becomes a sequentially complete Hausdorff locally convex topological vector space. Therefore, if $\{u_h\} \subseteq BV(\Omega)$ is such that $\{u_h\}$ is a Cauchy sequence in $L^1(\Omega)$, and for every $\psi \in (\widehat{C}_0^0(\Omega))^n$, $\{\int_\Omega \psi Du \ _h\}$ is a Cauchy sequence in $\mathbf{R}$, then there exists $u \in BV(\Omega)$ such that $u_h \to u$ in weak*-$BV(\Omega)$, namely such that $u_h \to u$ in $L^1(\Omega)$, and $Du_h \to Du$ in weak*-$(\mathcal{M}(\Omega))^n$. We refer to [ABF, Remark 3.12] for a description of BV spaces as a dual spaces.

Given $\Omega \in \mathcal{A}(\mathbf{R}^n)$, and u in $BV(\Omega)$, Lebesgue Decomposition Theorem yields $Du = (Du)^a + (Du)^s$, where $(Du)^a$ is the absolutely continuous part of Du with respect to Lebesgue measure and $(Du)^s$ is its singular part. For the sake of simplicity, it is standard to set $D^a u = (Du)^a$, $D^s u = (Du)^s$, and denote by $\nabla u = (\nabla_1 u, \ldots, \nabla_n u)$ the Radon-Nikodym derivative of $D^a u$ with respect to $\mathcal{L}^n$, i.e. $\nabla u = \frac{dD^a u}{d\mathcal{L}^n}$. We also denote by $\nabla^s u$ the Radon-Nikodym derivative of $D^s u$ with respect to $|D^s u|$, i.e. $\nabla^s u = \frac{dD^s u}{d|D^s u|}$.

If $u \in BV(\mathbf{R}^n)$, and $x_0 \in \mathbf{R}^n$ it results that $T[x_0]u \in BV(\mathbf{R}^n)$. In fact

$$\int_{\mathbf{R}^n} T[x_0]u\nabla\varphi dx = \int_{\mathbf{R}^n} u\nabla(T[-x_0]\varphi)dx = -\int_{\mathbf{R}^n} T[-x_0]\varphi dDu =$$

$$= -\int_{\mathbf{R}^n} \varphi d(T[x_0]Du) \text{ for every } \varphi \in C_0^\infty(\mathbf{R}^n),$$

from which we also conclude that $D(T[x_0]u) = T[x_0]Du$. Because of this, and by Theorem 2.3.5, we thus deduce that $\nabla T[x_0]u = T[x_0]\nabla u$, and $\nabla^s T[x_0]u = T[x_0]\nabla^s u$.

A quite different way to introduce BV functions is by mean of variations.

Let $\Omega \in \mathcal{A}(\mathbf{R}^n)$. For every $u \in L^1_{loc}(\Omega)$ the symbol $\int_\Omega |Du|$ denotes the variation of u on Ω defined as

$$\int_\Omega |Du| = \sup\left\{ \int_\Omega u\,\mathrm{div}\varphi dx : \varphi \in (C_0^1(\Omega))^n, \ |\varphi| \le 1 \text{ in } \Omega \right\}.$$

For every $u \in L^1_{loc}(\Omega)$, the variation of u on Ω in general belongs to $[0, +\infty]$, and it can actually assume the value $+\infty$. Nevertheless, the following result characterizes, by means of BV spaces, the set where it is finite.

Proposition 4.2.6. *Let $\Omega \in \mathcal{A}(\mathbf{R}^n)$, and $u \in L^1(\Omega)$. Then*

$$\int_\Omega |Du| = \begin{cases} |Du|(\Omega) & \text{if } u \in BV(\Omega) \\ +\infty & \text{if } u \in L^1(\Omega) \setminus BV(\Omega). \end{cases}$$

Proof. If $u \in BV(\Omega)$, then the density of $C_0^1(\Omega)$ in $\widehat{C}_0^0(\Omega)$ yields

$$\int_\Omega |Du| = \sup\left\{ \int_\Omega \varphi \cdot dD_i u : \varphi \in (\widehat{C}_0^0(\Omega))^n, \ |\varphi| \le 1 \text{ in } \Omega \right\} = |Du|(\Omega).$$

If now $u \in L^1(\Omega) \setminus BV(\Omega)$, we assume by contradiction that $\int_\Omega |Du| < +\infty$. Then the functional $\varphi \in (C_0^1(\Omega))^n \mapsto \int_\Omega u\,\mathrm{div}\varphi dx$ turns out to be linear and continuous on $(C_0^1(\Omega))^n$ endowed with the $(C_b^0(\Omega))^n$ topology, and, again by the above density argument, it can be extended to a linear and continuous functional, say L, on $(\widehat{C}_0^0(\Omega))^n$ such that $\|L\| = \int_\Omega |Du|$.

On the other side, the Riesz Representation Theorem provides $\mu \in (\mathcal{M}(\Omega))^n$ for which $L(\varphi) = \int_\Omega \varphi d\mu$ for every $\varphi \in (\widehat{C}_0^0(\Omega))^n$, and $\|L\| = |\mu|(\Omega)$. Consequently, it turns out that

$$\int_\Omega \varphi d\mu = \int_\Omega u\,\mathrm{div}\varphi dx \text{ for every } \varphi \in (C_0^1(\Omega))^n,$$

from which we conclude that $u \in BV(\Omega)$, thus getting a contradiction. Therefore $\int_\Omega |Du| = +\infty$, and the proof follows. ∎

In particular, by Proposition 4.2.6 it follows that, for a given $\Omega \in \mathcal{A}(\mathbf{R}^n)$, $u \in BV(\Omega)$ if and only if $u \in L^1(\Omega)$ and $\int_\Omega |Du| < +\infty$.

Again by Proposition 4.2.6, the following lower semicontinuity result for the variation of L^1 functions holds.

Theorem 4.2.7. Let $\Omega \in \mathcal{A}(\mathbf{R}^n)$. Then $u \in L^1(\Omega) \mapsto \int_\Omega |Du|$ is $L^1(\Omega)$-lower semicontinuous.

Proof. Follows once we observe that the variation functional is the pointwise supremum of the family of $L^1(\Omega)$-lower semicontinuous functionals $u \in L^1(\Omega) \mapsto \int_\Omega u \operatorname{div}\varphi dx$, as φ varies among the elements of $(C_0^1(\Omega))^n$ satisfying $|\varphi| \leq 1$ in Ω. ■

BV spaces may possess genuine discontinuities. For example, if $\Omega \in \mathcal{A}_0$ has Lipschitz boundary, and satisfies $\mathcal{H}^{n-1}(\partial\Omega) < +\infty$, then it is possible to prove that $\chi_\Omega \in BV(\mathbf{R}^n)$, and that $D\chi_\Omega(B) = D^s\chi_\Omega(B) = -\int_{B \cap \partial\Omega} \mathbf{n}_\Omega d\mathcal{H}^{n-1}$ for every $B \in \mathcal{B}(\Omega)$.

Since the gradients of BV functions, in general, need not be absolutely continuous with respect to $\mathcal{L}^n$, we conclude that smooth functions cannot be dense in BV spaces endowed with their norm topology. Nevertheless the following weaker results hold.

Proposition 4.2.8. Let $\Omega \in \mathcal{A}(\mathbf{R}^n)$, and $u \in BV_{\text{loc}}(\Omega)$. For every $\varepsilon > 0$ let u_ε be the regularization of u defined by (4.1.2). Then $u_\varepsilon \to u$ in weak*-$BV(A)$ for every $A \subset\subset \Omega$. Moreover, the limit $\lim_{\varepsilon \to 0} \int_{\Omega_\varepsilon^-} |\nabla u_\varepsilon| dx$ exists, and

$$\lim_{\varepsilon \to 0} \int_{\Omega_\varepsilon^-} |\nabla u_\varepsilon| dx = |Du|(\Omega).$$

Proof. The same argument used in the proof of Proposition 4.2.3 yields $\nabla u_\varepsilon(x) = (Du)_\varepsilon(x)$ for every $x \in \Omega_\varepsilon^-$. Consequently, the part of the proof concerning the convergence of $\{u_\varepsilon\}_{\varepsilon>0}$ follows from Theorem 4.1.1, whilst the remaining one from Proposition 4.1.2. ■

Theorem 4.2.9. Let $\Omega \in \mathcal{A}(\mathbf{R}^n)$, and $u \in BV(\Omega)$. Then there exists $\{u_h\} \subseteq BV(\Omega) \cap C^\infty(\Omega)$ such that $u_h \to u$ in $L^1(\Omega)$, and

$$\lim_{h \to +\infty} \int_\Omega |\nabla u_h| dx = |Du|(\Omega).$$

As already sketched in Lemma 4.2.2, the property of BV functions to posses weak gradients that are measures with finite total variation implies a higher summability property on the functions themselves.

Theorem 4.2.10. Let $\Omega \in \mathcal{A}_0$ have Lipschitz boundary. Then $BV(\Omega)$ continuously embeds in $L^{\frac{n}{n-1}}(\Omega)$, and there exists $C_{n,\Omega} > 0$ such that

$$\|u\|_{L^{\frac{n}{n-1}}(\Omega)} \leq C_{n,\Omega}\|u\|_{BV(\Omega)} \text{ for every } u \in BV(\Omega).$$

The above imbedding theorem is specified by the following compactness result.

Theorem 4.2.11. *Let $\Omega \in \mathcal{A}_0$ have Lipschitz boundary, and let $\{u_h\}$ be bounded in $BV(\Omega)$. Then there exist $\{h_k\} \subseteq \mathbf{N}$ strictly increasing, and $u \in BV(\Omega)$ such that*

$$u_{h_k} \to u \ in \ \cap_{q \in [1, \frac{n}{n-1}[} L^q(\Omega).$$

If Ω has Lipschitz boundary, then it turns out that the functions in $BV(\Omega)$ have traces on $\partial\Omega$ in the sense that for every $u \in BV(\Omega)$ it is possible to define a function on $\partial\Omega$ that can be thought as giving the values of u on $\partial\Omega$. Its properties are summarized by the following trace theorem for BV spaces.

Theorem 4.2.12 (Trace Theorem for BV Functions). *Let $\Omega \in \mathcal{A}_0$ have Lipschitz boundary. Then there exists a surjective bounded linear operator $\gamma_\Omega \colon BV(\Omega) \to L^1(\partial\Omega, \mathcal{H}^{n-1})$ such that*

$$(\gamma_\Omega u)(x) = u(x) \ for \ every \ u \in BV(\Omega) \cap C^0(\overline{\Omega}), \ and \ for \ \mathcal{H}^{n-1}\text{-}a.e. \ x \in \partial\Omega.$$

Moreover

$$\int_\Omega u \operatorname{div}\varphi dx = -\int_\Omega \varphi \cdot dDu + \int_{\partial\Omega} \varphi \gamma_\Omega u \cdot \mathbf{n}_\Omega d\mathcal{H}^{n-1}$$

for every $u \in BV(\Omega)$, $\varphi \in (C^1(\mathbf{R}^n))^n$,

and

$$\lim_{r \to 0^+} \frac{1}{\mathcal{L}^n(\Omega \cap B_r(x))} \int_{\Omega \cap B_r(x)} |u(y) - (\gamma_\Omega u)(x)| dy = 0$$

for every $u \in BV(\Omega)$, and for $\mathcal{H}^{n-1}\text{-}a.e. \ x \in \partial\Omega.$

Given Ω as in the Trace Theorem for BV Functions, the operator γ_Ω is called the *trace operator on* $\partial\Omega$, and, if $u \in BV(\Omega)$, the function $\gamma_\Omega u$ is called the *trace of u on* $\partial\Omega$.

BV functions behave quite nicely with respect to extension processes, as proved by the following result.

Proposition 4.2.13. *Let Ω, $\Omega' \in \mathcal{A}_0$ have Lipschitz boundary be such that $\Omega \subset\subset \Omega'$, $u \in BV(\Omega)$, and $v \in BV(\Omega' \setminus \overline{\Omega})$. Then the function w defined as $w = \begin{cases} u & in \ \Omega \\ v & in \ \Omega' \setminus \overline{\Omega} \end{cases}$ is in $BV(\Omega')$, and*

$$Dw(E) = \int_E (\gamma_\Omega v - \gamma_{\Omega' \setminus \overline{\Omega}} u) \mathbf{n}_\Omega d\mathcal{H}^{n-1} \ for \ every \ E \in \mathcal{B}(\partial\Omega).$$

In particular, by Proposition 4.2.13 it follows that, if $\Omega \in \mathcal{A}_0$ has Lipschitz boundary, and $u \in BV(\Omega)$, then the function w defined as $w = \begin{cases} u & \text{in } \Omega \\ 0 & \text{in } \mathbf{R}^n \setminus \Omega \end{cases}$ is in $BV(\mathbf{R}^n)$, and

$$Dw(E) = -\int_E \gamma_\Omega u \mathbf{n}_\Omega d\mathcal{H}^{n-1} \text{ for every } E \in \mathcal{B}(\partial\Omega).$$

§4.3 Sobolev Spaces

Once defined BV spaces, it is straightforward to define Sobolev spaces as their particular subspaces.

Definition 4.3.1. *Let* $\Omega \in \mathcal{A}(\mathbf{R}^n)$, *and* $p \in [1, +\infty]$. *We define the Sobolev space* $W^{1,p}(\Omega)$ *as*

$$W^{1,p}(\Omega) = \{u \in BV(\Omega) : u \in L^p(\Omega), \ D^s u = 0, \ \nabla u \in (L^p(\Omega))^n\},$$

and

$$W_{\text{loc}}^{1,p}(\Omega) = \{u \in BV_{\text{loc}}(\Omega) : u \in W^{1,p}(A) \text{ for every } A \subset\subset \Omega\}.$$

In other words, given $\Omega \in \mathcal{A}(\mathbf{R}^n)$, and $p \in [1, +\infty]$, $W^{1,p}(\Omega)$ is the set of the functions u in $L^p(\Omega)$ such that the weak gradient of $u\mathcal{L}^n$ is absolutely continuous with respect to $\mathcal{L}^n$, and has its Radon-Nikodym derivative in $(L^p(\Omega))^n$. We call such functions *Sobolev functions*.

The examples of the previous section of functions in $BV(\Omega)$ with singular weak gradient prove that, in general, $BV(\Omega) \neq W^{1,1}(\Omega)$.

Given $\Omega \in \mathcal{A}(\mathbf{R}^n)$, and $p \in [1, +\infty]$, $W^{1,p}(\Omega)$ is a vector subspace of $(L^p(\Omega))^{n+1}$. Moreover, since a $BV(\Omega)$ function u uniquely determines Du, it is immediately verified that a $W^{1,p}(\Omega)$ function u uniquely determines ∇u, and therefore that $W^{1,p}(\Omega)$ can be identified with a subspace of $L^p(\Omega)$.

Since $W^{1,p}(\Omega)$ is a subset of $(L^p(\Omega))^{n+1}$, it naturally inherits its topological structures.

In particular, $W^{1,p}(\Omega)$ becomes a normed space with the $(L^p(\Omega))^{n+1}$-norm. We denote such norm functional as

$$\|\cdot\|_{W^{1,p}(\Omega)} : u \in W^{1,p}(\Omega) \mapsto \|u\|_{L^p(\Omega)} + \|\nabla u\|_{L^p(\Omega)},$$

and, as usual, we denote again by $W^{1,p}(\Omega)$ the strong topology of $W^{1,p}(\Omega)$.

We also denote with $W_{\text{loc}}^{1,p}(\Omega)$ the topology on $W_{\text{loc}}^{1,p}(\Omega)$ generated by the family of seminorms $u \in W_{\text{loc}}^{1,p}(\Omega) \mapsto \|u\|_{W^{1,p}(A)}$ as $A \subset\subset \Omega$, and with

$\cap_{q \in [1,p[} W^{1,q}_{\text{loc}}(\Omega)$ the one on $\cap_{q \in [1,p[} W^{1,q}_{\text{loc}}(\Omega)$ generated by the family of semi-norms $u \in \cap_{q \in [1,p[} W^{1,q}_{\text{loc}}(\Omega) \mapsto \|u\|_{W^{1,q}(A)}$ as $q \in [1,p[$, and $A \subset\subset \Omega$. Once endowed with their respective topologies, $W^{1,p}_{\text{loc}}(\Omega)$, and $\cap_{q \in [1,p[} W^{1,q}_{\text{loc}}(\Omega)$ turn out to be complete metrizable topological vector spaces.

If in addition $\Omega \in \mathcal{A}_0$, we denote again with $\cap_{q \in [1,p[} W^{1,q}(\Omega)$ the topology on $\cap_{q \in [1,p[} W^{1,q}(\Omega)$ generated by the family of seminorms $u \in \cap_{q \in [1,p[} W^{1,q}(\Omega) \mapsto \|u\|_{W^{1,q}(\Omega)}$, as q varies in $[1,p[$. Once endowed with the $\cap_{q \in [1,p[} W^{1,q}(\Omega)$ topology, $\cap_{q \in [1,p[} W^{1,q}(\Omega)$ turns out to be a complete metrizable topological vector space.

If $\Omega \in \mathcal{A}(\mathbf{R}^n)$, and $p \in [1,+\infty[$, $W^{1,p}(\Omega)$ also inherits the weak-$(L^p(\Omega))^{n+1}$ topology of $(L^p(\Omega))^{n+1}$, whilst, if $p = +\infty$, $W^{1,\infty}(\Omega)$ inherits the weak*-$(L^\infty(\Omega))^{n+1}$ one of $(L^\infty(\Omega))^{n+1}$. In all the cases, it is easy to prove the following result.

Proposition 4.3.2. *Let $\Omega \in \mathcal{A}(\mathbf{R}^n)$. Then $W^{1,p}(\Omega)$ is a weak-$(L^p(\Omega))^{n+1}$ closed subspace of $(L^p(\Omega))^{n+1}$, for every $p \in [1,+\infty[$, and $W^{1,\infty}(\Omega)$ is a weak*-$(L^\infty(\Omega))^{n+1}$ closed subspace of $(L^\infty(\Omega))^{n+1}$.*

Proof. Follows immediately once we observe that

$$W^{1,p}(\Omega) =$$

$$= \cap_{\varphi \in C_0^\infty(\Omega)} \left\{ (u,v) \in L^p(\Omega) \times (L^p(\Omega))^n : \int_\Omega u \nabla \varphi dx + \int_\Omega v \varphi dx = 0 \right\},$$

and that the sets in the right-hand side of the above equality are weak-$(L^p(\Omega))^{n+1}$ closed for every $p \in [1,+\infty[$, and weak*-$(L^\infty(\Omega))^{n+1}$ closed if $p = +\infty$ since, for every $\varphi \in C_0^\infty(\Omega)$, the functional $(u,v) \in L^p(\Omega) \times (L^p(\Omega))^n \mapsto \int_\Omega u \nabla \varphi dx + \int_\Omega v \varphi dx$ is weak-$(L^p(\Omega))^{n+1}$ continuous for every $p \in [1,+\infty[$, and weak*-$(L^\infty(\Omega))^{n+1}$ continuous if $p = +\infty$. ∎

As consequence, given $\Omega \in \mathcal{A}(\mathbf{R}^n)$, from Proposition 4.3.2 it follows that, for every $p \in [1,+\infty]$, $W^{1,p}(\Omega)$ is also a $(L^p(\Omega))^{n+1}$ closed subspace of $(L^p(\Omega))^{n+1}$. Consequently it turns out to be a Banach space.

Moreover, being for $p \in]1,+\infty[$ $W^{1,p}(\Omega)$ a closed subspace of the reflexive space $(L^p(\Omega))^{n+1}$, it turns out to be reflexive too. Because of this, from Banach-Steinhaus Theorem and Proposition 4.3.2, it also follows that, for $p \in]1,+\infty]$, $W^{1,p}(\Omega)$ is sequentially complete once we endow it with the weak-$(L^p(\Omega))^{n+1}$ topology if $p \in]1,+\infty[$, or with the weak*-$(L^\infty(\Omega))^{n+1}$ one if $p = +\infty$.

If $p \in [1,+\infty[$, we denote by weak-$W^{1,p}(\Omega)$ the topology on $W^{1,p}(\Omega)$ induced from the weak-$(L^p(\Omega))^{n+1}$ one of $(L^p(\Omega))^{n+1}$. Once endowed with the weak-$W^{1,p}(\Omega)$ topology, $W^{1,p}(\Omega)$ becomes a Hausdorff locally convex topological vector space, sequentially complete if $p \in]1,+\infty[$. With an abuse of notation, we denote by weak*-$W^{1,\infty}(\Omega)$ the topology on $W^{1,\infty}(\Omega)$ induced from the weak*-$(L^\infty(\Omega))^{n+1}$ one of $(L^\infty(\Omega))^{n+1}$. Once endowed with the weak*-$W^{1,\infty}(\Omega)$ topology, $W^{1,\infty}(\Omega)$ becomes a sequentially complete Hausdorff locally convex topological vector space.

It is worth while to remark that, if $p \in [1,+\infty[$, the weak-$W^{1,p}(\Omega)$ topology is just the weak topology of the Banach space $W^{1,p}(\Omega)$, as explained by the following result.

Theorem 4.3.3. Let $\Omega \in \mathcal{A}(\mathbf{R}^n)$, and $p \in [1,+\infty[$. Then, for every $L \in (W^{1,p}(\Omega))'$ there exist $v_0 \in L^{p'}(\Omega)$, $v \in (L^{p'}(\Omega))^n$ such that

$$L(u) = \int_\Omega v_0 u\, dx + \int_\Omega v \nabla u\, dx \text{ for every } u \in W^{1,p}(\Omega).$$

If $p = +\infty$, we denote, with an abuse of notation, by weak*-$W^{1,\infty}(\Omega)$ the topology on $W^{1,\infty}(\Omega)$ induced from the weak*-$(L^\infty(\Omega))^{n+1}$ one of $(L^\infty(\Omega))^{n+1}$.

Therefore, if $p \in [1,+\infty]$, and $\{u_h\} \subseteq W^{1,p}(\Omega)$ is such that for every $v \in L^{p'}(\Omega)$, and $w \in (L^{p'}(\Omega))^n$ the sequences $\{\int_\Omega v u_h dx\}$, and $\{\int_\Omega w \nabla u_h dx\}$ converge, then there exists $u \in W^{1,p}(\Omega)$ such that $u_h \to u$ in weak-$W^{1,p}(\Omega)$ if $p \in [1,+\infty[$, or in weak*-$W^{1,\infty}(\Omega)$ if $p = +\infty$.

Finally, we observe that a sequence in $W^{1,p}(\Omega)$ that converges in the weak-$W^{1,p}(\Omega)$ topology if $p \in [1,+\infty[$, or in the weak*-$W^{1,\infty}(\Omega)$ one if $p = +\infty$, turns out to be bounded in $W^{1,p}(\Omega)$, and deduce from Proposition 4.3.2 the following compactness result.

Proposition 4.3.4. Let $\Omega \in \mathcal{A}(\mathbf{R}^n)$, $p \in]1,+\infty]$, and $\{u_h\} \subseteq W^{1,p}(\Omega)$ be $W^{1,p}(\Omega)$-bounded. Then there exist $\{h_k\} \subseteq \mathbf{N}$ strictly increasing, and $u \in W^{1,p}(\Omega)$ such that

$$u_{h_k} \to u \text{ in weak-}W^{1,p}(\Omega) \text{ if } p \in]1,+\infty[, \text{ in weak*-}W^{1,\infty}(\Omega) \text{ if } p = +\infty.$$

Proof. Follows from Theorem 2.2.6, and Proposition 4.3.2. ■

The following results describe the structure of Sobolev spaces with $p = +\infty$, and the Chain Rule in Sobolev spaces.

Theorem 4.3.5. Let $\Omega \in \mathcal{A}(\mathbf{R}^n)$, then

$$W^{1,\infty}_{\text{loc}}(\Omega) = \{u\colon \Omega \to \mathbf{R} : u \text{ is locally Lipschitz continuous in } \Omega\}.$$

If, in addition, Ω is bounded and has Lipschitz boundary, then

$$W^{1,\infty}(\Omega) = \{u\colon \Omega \to \mathbf{R} : u \text{ is Lipschitz continuous in } \Omega\}.$$

Theorem 4.3.6 (Chain Rule). Let $\Omega \in \mathcal{A}(\mathbf{R}^n)$, $p \in [1,+\infty]$, $f\colon \mathbf{R} \to \mathbf{R}$ be Lipschitz continuous, and $u \in W^{1,p}(\Omega)$. Assume that $f \circ u \in L^p(\Omega)$. Then $f \circ u \in W^{1,p}(\Omega)$, and

$$\nabla(f \circ u)(x) = f'(u(x))\nabla u(x) \text{ for a.e. } x \in \Omega.$$

In particular, from the above result we deduce that, if $u \in W^{1,p}(\Omega)$, then $|u|$, u^+, u^- are in $W^{1,p}(\Omega)$. Moreover, if $\mathcal{L}^n(\Omega) < +\infty$, and $k \in [0, +\infty[$, the truncation $T_k u$ of u at levels k and $-k$ given by $T_k u = u - (u-k)^+ - (u+k)^-$ is in $W^{1,p}(\Omega)$, and

$$\nabla T_k u(x) = \begin{cases} \nabla u(x) & \text{if } -k < u(x) < k \\ 0 & \text{otherwise} \end{cases} \quad \text{for a.e. } x \in \Omega.$$

We now turn our attention to density results for smooth functions in Sobolev spaces.

Theorem 4.3.7. *Let $\Omega \in \mathcal{A}(\mathbf{R}^n)$, and $p \in [1, +\infty[$. Then $C^\infty(\Omega) \cap \{u \in W^{1,p}(\Omega) : \|u\|_{W^{1,p}(\Omega)} < +\infty\}$ is dense in $W^{1,p}(\Omega)$.*

In addition, if Ω is bounded and has Lipschitz boundary, then $C^\infty(\mathbf{R}^n)$ is dense in $W^{1,p}(\Omega)$.

Smooth functions with compact support are not dense, in general, in Sobolev spaces.

Given $\Omega \in \mathcal{A}(\mathbf{R}^n)$ and $p \in [1, +\infty[$, we denote by $W_0^{1,p}(\Omega)$ the closure of $C_0^\infty(\Omega)$ in $W^{1,p}(\Omega)$. If $p = +\infty$, we set $W_0^{1,\infty}(\Omega) = \{u \in W^{1,\infty}(\mathbf{R}^n) : u(x) = 0 \text{ for every } x \in \mathbf{R}^n \setminus \Omega\}$.

It is clear that for every $p \in [1, +\infty]$, once we endow it with the $W^{1,p}(\Omega)$ topology, $W_0^{1,p}(\Omega)$ is a Banach subspace of $W^{1,p}(\Omega)$, in general proper.

We observe that, if $u \in W_0^{1,p}(\Omega)$, then the null extension of u to $\mathbf{R}^n$ given by $x \in \mathbf{R}^n \mapsto \begin{cases} u(x) & \text{if } x \in \Omega \\ 0 & \text{if } x \in \mathbf{R}^n \setminus \Omega \end{cases}$ is in $W^{1,p}(\mathbf{R}^n)$. We will always identify the functions in $W_0^{1,p}(\Omega)$ with their null extensions to $\mathbf{R}^n$.

The above extension result is a trivial case of a more general one holding for Sobolev functions under smoothness assumptions on $\partial \Omega$.

Theorem 4.3.8. *Let $\Omega \in \mathcal{A}_0$ have Lipschitz boundary, and $p \in [1, +\infty]$. Then there exists a bounded linear operator $E \colon W^{1,p}(\Omega) \to W^{1,p}(\mathbf{R}^n)$ such that $Eu = u$ a.e. in Ω for every $u \in W^{1,p}(\Omega)$.*

As for BV functions, also Sobolev ones enjoy higher summability properties, or even smoothness ones, due to the presence of p-summable derivatives.

For every $p \in [1, +\infty]$ we denote, as usual, by p^* the Sobolev conjugate of p defined as

$$p^* = \begin{cases} \frac{np}{n-p} & \text{if } p \in [1, n[\\ +\infty & \text{if } p \in [n, +\infty]. \end{cases}$$

If $\Omega \in \mathcal{A}_0$, and $\alpha \in \,]0, 1[$, we denote by $C^{0,\alpha}(\Omega)$ the *space of the Hölder continuous functions* in Ω defined as

$$C^{0,\alpha}(\Omega) = \left\{ u \in C^0(\overline{\Omega}) : \sup_{\substack{x,y \in \Omega \\ x \neq y}} \frac{|u(x) - u(y)|}{|x-y|^\alpha} < +\infty \right\}.$$

We recall that the functional

$$\| \cdot \|_{C^{0,\alpha}(\Omega)} \colon u \in C^{0,\alpha}(\Omega) \mapsto \max_{\overline{\Omega}} |u| + \sup_{\substack{x,y \in \Omega \\ x \neq y}} \frac{|u(x) - u(y)|}{|x - y|^{\alpha}}$$

is a norm on $C^{0,\alpha}(\Omega)$ that makes it a Banach space.

Theorem 4.3.9 (Sobolev Imbedding Theorem). *Let $\Omega \in \mathcal{A}_0$ have Lipschitz boundary, and $p \in [1, +\infty]$. Then the following facts hold:*
i) if $p \in [1, n[$, $W^{1,p}(\Omega)$ continuously embeds in $L^{p^*}(\Omega)$, and there exists $C_{n,\Omega,p} > 0$ such that

$$\|u\|_{L^{p^*}(\Omega)} \leq C_{n,\Omega,p} \|u\|_{W^{1,p}(\Omega)} \ \text{for every } u \in W^{1,p}(\Omega),$$

ii) if $p = n$, $W^{1,p}(\Omega)$ continuously embeds in $\cap_{q \in [1,+\infty[} L^q(\Omega)$, and for every $q \in [1, +\infty[$ there exists $C_{n,\Omega,q} > 0$ such that

$$\|u\|_{L^q(\Omega)} \leq C_{n,\Omega,q} \|u\|_{W^{1,n}(\Omega)} \ \text{for every } u \in W^{1,n}(\Omega),$$

iii) if $p \in \,]n, +\infty]$, $W^{1,p}(\Omega)$ continuously embeds in $C^{0,1-\frac{n}{p}}(\Omega)$, and there exists $C_{n,\Omega,p} > 0$ such that

$$\|u\|_{C^{0,1-\frac{n}{p}}(\Omega)} \leq C_{n,\Omega,p} \|u\|_{W^{1,p}(\Omega)} \ \text{for every } u \in W^{1,p}(\Omega).$$

If $W^{1,p}(\Omega)$ is replaced by $W_0^{1,p}(\Omega)$, the same conclusions in i), ii), and iii) continue to hold without assuming that Ω has Lipschitz boundary.

Remark 4.3.10. We remark explicitly that in case ii) of the Sobolev Imbedding Theorem the embedding of $W^{1,n}(\Omega)$ in $L^\infty(\Omega)$ does not hold. On the other side, a more shrinking result can be proved in the framework of Orlicz spaces in which it can be proved that, if $\Omega \in \mathcal{A}_0$ has Lipschitz boundary, then there exists $C_{n,\Omega} > 0$ such that

$$\inf\left\{\lambda > 0 : \int_\Omega \left[\exp\left(\frac{|u(x)|}{\lambda}\right)^{\frac{n}{n-1}} - 1\right] dx \leq 1\right\} \leq C_{n,\Omega} \|u\|_{W^{1,n}(\Omega)}$$

for every $u \in W^{1,n}(\Omega)$.

Sobolev Imbedding Theorem can be specified by means of the following compact imbedding result.

Theorem 4.3.11 (Rellich-Kondrachov Compactness Theorem).
Let $\Omega \in \mathcal{A}_0$ have Lipschitz boundary, $p \in [1, +\infty]$, and let $\{u_h\}$ be bounded in $W^{1,p}(\Omega)$. Then there exists $\{h_k\} \subseteq \mathbf{N}$ strictly increasing such that the following facts hold:
i) if $p = 1$, there exists $u \in \cap_{q \in [1,1^*[} L^q(\Omega)$ such that

$$u_{h_k} \to u \text{ in } \cap_{q \in [1,1^*[} L^q(\Omega),$$

ii) if $p \in \,]1, n]$, there exists $u \in W^{1,p}(\Omega)$ such that

$$u_{h_k} \to u \text{ in } \cap_{q \in [1,p^*[} L^q(\Omega),$$

iii) if $p \in \,]n, +\infty]$, there exists $u \in W^{1,p}(\Omega)$ such that

$$u_{h_k} \to u \text{ in } C^{0,\alpha}(\Omega) \text{ for every } \alpha \in \left]0, 1 - \frac{n}{p}\right[\text{ if } p \in \,]n, +\infty[,$$

$$\text{for every } \alpha \in \,]0, 1[\text{ if } p = +\infty.$$

If $\{u_h\} \subseteq W_0^{1,p}(\Omega)$, the same conclusions in i), ii), and iii) continue to hold without assuming that Ω has Lipschitz boundary. In this case it turns out that also the limit points in ii) and iii) are in $W_0^{1,p}(\Omega)$.

Let $\Omega \in \mathcal{A}_0$ have Lipschitz boundary, and $p \in \,]n, +\infty]$. Then the Sobolev Imbedding Theorem ensures that, if $u \in W^{1,p}(\Omega)$, then $u \in C^0(\overline{\Omega})$, and consequently that it makes sense to speak of the values of u on $\partial\Omega$. Actually such values enjoy deeper properties, even if $p \in [1, n]$, as shown by the following result.

Theorem 4.3.12 (Trace Theorem for Sobolev Functions). *Let $\Omega \in \mathcal{A}_0$ have Lipschitz boundary, and $p \in [1, +\infty]$. Then the following facts hold:*
i) if $p \in [1, n[$, there exists a bounded linear operator $\gamma_\Omega \colon W^{1,p}(\Omega) \to L^{\frac{(n-1)p}{n-p}}(\partial\Omega, \mathcal{H}^{n-1})$ such that

$$(4.3.1) \qquad\qquad\qquad (\gamma_\Omega u)(x) = u(x)$$

for every $u \in W^{1,p}(\Omega) \cap C^0(\overline{\Omega})$, and for $\mathcal{H}^{n-1}$-a.e. $x \in \partial\Omega$,

ii) if $p = n$, for every $q \in [1, +\infty[$ there exists a bounded linear operator $\gamma_\Omega \colon W^{1,n}(\Omega) \to L^q(\partial\Omega, \mathcal{H}^{n-1})$ such that (4.3.1) holds,
iii) if $p \in \,]n + \infty]$, there exists a bounded linear operator $\gamma_\Omega \colon W^{1,p}(\Omega) \to C^0(\partial\Omega)$ such that

$$(\gamma_\Omega u)(x) = u(x) \text{ for every } u \in W^{1,p}(\Omega), \text{ and every } x \in \partial\Omega.$$

Moreover, in all the cases,

$$\int_\Omega u\,\mathrm{div}\varphi dx = -\int_\Omega \nabla u \cdot \varphi dx + \int_{\partial\Omega} \varphi\gamma_\Omega u \cdot \mathbf{n}_\Omega d\mathcal{H}^{n-1}$$

for every $u \in W^{1,p}(\Omega)$, $\varphi \in (C^1(\mathbf{R}^n))^n$,

and

$$\lim_{r\to 0^+} \frac{1}{\mathcal{L}^n(\Omega \cap B_r(x))} \int_{\Omega\cap B_r(x)} |u(y) - (\gamma_\Omega u)(x)| dy = 0$$

for every $u \in W^{1,p}(\Omega)$, and for $\mathcal{H}^{n-1}$-a.e. $x \in \partial\Omega$.

Finally, if $p = 1$, γ_Ω is surjective.

Given Ω and p as in the Trace Theorem for Sobolev Functions, the operator γ_Ω is again called the *trace operator on* $\partial\Omega$, and, if $u \in W^{1,p}(\Omega)$, the function $\gamma_\Omega u$ is again called the *trace of* u *on* $\partial\Omega$.

It is not for a case that the trace operator for $W^{1,1}$ functions is denoted with the same symbol used to describe the corresponding operator for BV functions, since the latter extends the first, as it can be easily checked by using Theorems 4.2.12 and 4.3.12.

Proposition 4.3.13. *Let* $\Omega \in \mathcal{A}_0$ *have Lipschitz boundary, and* $p \in [1, +\infty[$. *Then* $\{u \in W^{1,p}(\Omega) : \gamma_\Omega u = 0\} = W_0^{1,p}(\Omega)$.

If $\Omega \in \mathcal{A}(\mathbf{R}^n)$, $p \in [1, +\infty[$, and $\Gamma \subseteq \partial\Omega$ we denote by $W_{0,\Gamma}^{1,p}(\Omega)$ the closure in $W^{1,p}(\Omega)$ of $\{u \in W_{\mathrm{loc}}^{1,p}(\mathbf{R}^n) : u = 0$ a.e. in a neighborhood of $\Gamma\}$. If $p = +\infty$ we set $W_{0,\Gamma}^{1,\infty}(\Omega) = \{u \in W_{\mathrm{loc}}^{1,p}(\mathbf{R}^n) : \gamma_\Omega u = 0$ in $\Gamma\}$.

It is clear that, for every $p \in [1, +\infty]$, $W_{0,\Gamma}^{1,p}(\Omega)$ is a Banach space, and that $W_{0,\partial\Omega}^{1,p}(\Omega) = W_0^{1,p}(\Omega)$.

Proposition 4.3.14. *Let* $\Omega \in \mathcal{A}(\mathbf{R}^n)$. *Then, if* $p \in [1, +\infty[$, $W_{0,\Gamma}^{1,p}(\Omega)$ *is closed once we endow it with the weak-*$W^{1,p}(\Omega)$ *topology, and, when* $p = +\infty$ *and* $\Omega \in \mathcal{A}_0$ *has Lipschitz boundary,* $W_{0,\Gamma}^{1,\infty}(\Omega)$ *is closed in weak*-* $W^{1,\infty}(\Omega)$.

Proof. If $p \in [1, +\infty[$, by Theorem 1.1.2, it follows that $W_{0,\Gamma}^{1,p}(\Omega)$, as a strongly closed convex subspace of $W^{1,p}(\Omega)$, is closed in the weak-$W^{1,p}(\Omega)$ topology.

If $p = +\infty$ and $\Omega \in \mathcal{A}_0$ has Lipschitz boundary, by virtue of the Rellich-Kondrachov Compactness Theorem, it follows that $W_{0,\Gamma}^{1,\infty}(\Omega)$ is sequentially closed in weak*-$W^{1,\infty}(\Omega)$. Therefore, since $(L^1(\Omega))^{n+1}$ is separable, and $W_{0,\Gamma}^{1,\infty}(\Omega)$ is convex, the desired closure follows from Theorem 1.1.4. ∎

If $\Omega \in \mathcal{A}_0$ and has Lipschitz boundary, then $W_{0,\Gamma}^{1,p}(\Omega) \subseteq \{u \in W^{1,p}(\Omega) : \gamma_\Omega u = 0 \ \mathcal{H}^{n-1}$-a.e. in $\Gamma\}$, and the following result holds.

Proposition 4.3.15. *Let $\Omega \in \mathcal{A}_0$ have Lipschitz boundary, $\Gamma \subseteq \partial\Omega$ and $p \in [1, +\infty]$. Then*

$$W^{1,p}_{0,\Gamma}(\Omega) \cap C^0(\overline{\Omega}) = \{u \in W^{1,p}(\Omega) \cap C^0(\overline{\Omega}) : \gamma_\Omega u = 0 \ \mathcal{H}^{n-1}\text{-a.e. in } \Gamma\}.$$

Proof. The proposition is obvious if $p = +\infty$.

If $p \in [1, +\infty[$, it is clear that

$$W^{1,p}_{0,\Gamma}(\Omega) \cap C^0(\overline{\Omega}) \subseteq \{u \in W^{1,p}(\Omega) \cap C^0(\overline{\Omega}) : \gamma_\Omega u = 0 \ \mathcal{H}^{n-1}\text{-a.e. in } \Gamma\}.$$

On the other side, let $u \in W^{1,p}(\Omega) \cap C^0(\overline{\Omega})$ be such that $\gamma_\Omega u = 0$ $\mathcal{H}^{n-1}$-a.e. in Γ. Then, by Theorem 4.3.8, it is not restrictive to assume that $u \in W^{1,p}(\mathbf{R}^n)$.

For $h \in \mathbf{N}$ let $\vartheta_h \colon t \in \mathbf{R} \mapsto \max\{t - \frac{1}{h}, \min\{t + \frac{1}{h}, 0\}\}$. Then, by using the uniform continuity of u in $\partial\Omega$, it follows that $\vartheta_h(u) \in W^{1,p}(\Omega)$, and that $\vartheta_h(u) = 0$ in a neighborhood of Γ in Ω for every $h \in \mathbf{N}$. Moreover

$$\limsup_{h \to +\infty} \|\vartheta_h(u) - u\|_{W^{1,p}(\Omega)} \leq$$

$$\leq \limsup_{h \to +\infty} \left\{\frac{1}{h}\mathcal{L}^n(\Omega) + \int_{\{x \in \Omega : 0 < |u(x)| \leq \frac{1}{h}\}} |\nabla u|^p dx\right\} = 0,$$

from which we conclude that $u \in W^{1,p}_{0,\Gamma}(\Omega) \cap C^0(\overline{\Omega})$.

Because of this, and by the above inclusion, the proof follows. ∎

We now report on another important feature of Sobolev functions: the Poincaré and Poincaré-Wirtinger type inequalities.

Both these results follow from the abstract result below due to N.G. Meyers (cf. [Me], [Z, 4.1.3. Lemma]). To state it precisely, we recall that, if U is a a normed space with norm $\|\cdot\|$, and $P \colon U \to U$ is a bounded linear operator, we denote by $\|P\|_{\mathcal{L}(U,U)}$ the usual *operator norm* of P defined as $\|P\|_{\mathcal{L}(U,U)} = \sup\{\|P(u)\|/\|u\| : u \in U \setminus \{0\}\}$, and say that P is a *projection* if $P(P(u)) = P(u)$ for every $u \in U$.

Proposition 4.3.16. *Let U_0 be a normed space with norm $\|\cdot\|_0$, and let U be a Banach subspace of U_0 with norm $\|\cdot\| = \|\cdot\|_0 + \|\cdot\|_1$ for some seminorm $\|\cdot\|_1$ on U. Assume that the bounded sets in U are precompact in U_0. Then there exists $C \in \,]0, +\infty[$ such that*

$$\|u - P(u)\|_0 \leq C\|P\|_{\mathcal{L}(U,U)}\|u\|_1$$

for every projection P satisfying $P(U) = \{v \in U : \|v\|_1 = 0\}$, and $u \in U$.

A direct application of Proposition 4.3.16, and of Sobolev Imbedding Theorem yields the following general form of the Poincaré and Poincaré-Wirtinger inequalities, where, for every $r \in \mathbf{R}$, we identify r with the function identically equal to r.

Theorem 4.3.17. *Let $\Omega \in \mathcal{A}_0$ be connected and with Lipschitz boundary, $p \in [1, +\infty]$, and let T be a linear continuous functional on $W^{1,p}(\Omega)$ such that $T(1) \neq 0$. Then the following facts hold:*
i) if $p \in [1, n[$, there exists $C_{n,\Omega,p,T} > 0$ such that

$$\left\| u - \frac{T(u)}{T(1)} \right\|_{L^{p^*}(\Omega)} \leq C_{n,\Omega,p,T} \|\|\nabla u\|\|_{L^p(\Omega)} \text{ for every } u \in W^{1,p}(\Omega),$$

ii) if $p = n$, for every $q \in [1, +\infty[$ there exists $C_{n,\Omega,q,T} > 0$ such that

$$\left\| u - \frac{T(u)}{T(1)} \right\|_{L^q(\Omega)} \leq C_{n,\Omega,q,T} \|\|\nabla u\|\|_{L^n(\Omega)} \text{ for every } u \in W^{1,n}(\Omega),$$

iii) if $p \in]n, +\infty]$, there exists $C_{n,\Omega,p,T} > 0$ such that

$$\left\| u - \frac{T(u)}{T(1)} \right\|_{C^{0,1-\frac{n}{p}}(\Omega)} \leq C_{n,\Omega,p,T} \|\|\nabla u\|\|_{L^p(\Omega)} \text{ for every } u \in W^{1,p}(\Omega).$$

Proof. Let $p \in [1, +\infty]$.

First of all, let us observe that the operator

$$P: u \in W^{1,p}(\Omega) \mapsto \frac{T(u)}{T(1)} \in W^{1,p}(\Omega)$$

is a projection, and that, by using the connectedness of Ω, it is easy to prove that

$$P\left(W^{1,p}(\Omega) \right) = \mathbf{R} = \left\{ u \in W^{1,p}(\Omega) : \|\|\nabla u\|\|_{L^p(\Omega)} = 0 \right\}.$$

Moreover, by Rellich-Kondrachov Compactness Theorem, it soon follows that the assumptions of Proposition 4.3.16 are fulfilled with the choices $U_0 = L^p(\Omega)$, $\| \cdot \|_{L^p(\Omega)}$, $U = W^{1,p}(\Omega)$, and $\|\|\nabla \cdot \|\|_{L^p(\Omega)}$.

Because of this, Proposition 4.3.16 applied to the P above yields

$$\left\| u - \frac{T(u)}{T(1)} \right\|_{L^p(\Omega)} \leq C \frac{\|T\|_{W^{1,p}(\Omega)'}}{T(1)} \|\|\nabla u\|\|_{L^p(\Omega)} \text{ for every } u \in W^{1,p}(\Omega),$$

from which, the proof follows by applying the Sobolev Imbedding Theorem. ∎

In particular, the classical Poincaré and Poincaré-Wirtinger inequalities below follow from Theorem 4.3.17.

Theorem 4.3.18. *Let $\Omega \in \mathcal{A}_0$, and $p \in [1, +\infty]$. Then there exists $C_{n,\Omega,p} > 0$ such that*

$$\|u\|_{L^{p^*}(\Omega)} \leq C_{n,\Omega,p} \|\,|\nabla u|\,\|_{L^p(\Omega)} \text{ for every } u \in W_0^{1,p}(\Omega).$$

Moreover, if Ω is also connected and has Lipschitz boundary, and $\Gamma \in \mathcal{B}(\partial\Omega)$ satisfies $\mathcal{H}^{n-1}(\Gamma) > 0$, then there exists $C_{n,\Omega,p,\Gamma} > 0$ such that

$$\|u\|_{L^{p^*}(\Omega)} \leq C_{n,\Omega,p,\Gamma} \|\,|\nabla u|\,\|_{L^p(\Omega)}$$

for every $u \in W^{1,p}(\Omega)$ such that $\gamma_\Omega u = 0 \; \mathcal{H}^{n-1}$-a.e. in Γ.

Theorem 4.3.19. *Let $\Omega \in \mathcal{A}_0$ be connected and with Lipschitz boundary, and $p \in [1, +\infty]$. Then there exists $C_{n,\Omega,p} > 0$ such that*

$$\left\| u - \frac{1}{\mathcal{L}^n(\Omega)} \int_\Omega u \, dx \right\|_{L^{p^*}(\Omega)} \leq C_{n,\Omega,p} \|\,|\nabla u|\,\|_{L^p(\Omega)} \text{ for every } u \in W^{1,p}(\Omega).$$

Finally, we recall the following differentiation result for Sobolev functions.

Theorem 4.3.20. *Let $p \in [1, +\infty]$, and $u \in W^{1,p}_{loc}(\mathbf{R}^n)$. Then the following facts hold:*
i) if $p \in [1, n[$,

$$\lim_{r \to 0^+} \frac{1}{r} \left(\frac{1}{r^n} \int_{Q_r(x)} |u(y) - u(x) - \nabla u(x) \cdot (y - x)|^{p^*} dy \right)^{1/p^*} = 0$$

for a.e. $x \in \mathbf{R}^n$,

ii) if $p = n$,

$$\lim_{r \to 0^+} \frac{1}{r} \left(\frac{1}{r^n} \int_{Q_r(x)} |u(y) - u(x) - \nabla u(x) \cdot (y - x)|^q dy \right)^{1/q} = 0$$

for a.e. $x \in \mathbf{R}^n$, and every $q \in [1, +\infty[$,

iii) if $p \in \,]n, +\infty]$

$$\lim_{y \to x} \frac{|u(y) - u(x) - \nabla u(x) \cdot (y - x)|}{|y - x|} = 0 \text{ for a.e. } x \in \mathbf{R}^n.$$

§4.4 Some Compactness Criteria

In the present section we establish some compactness properties for subsets of *BV* and Sobolev spaces that will also be useful in the sequel.

Proposition 4.4.1. Let $\Omega \in \mathcal{A}_0$ have with Lipschitz boundary, $r \in]1, 1^*[$, and λ, b, $c \in]0, +\infty[$. Then the set

$$\{u \in BV(\Omega) : |Du|(\Omega) + \lambda\|u\|^r_{L^r(\Omega)} - b\|u\|_{L^r(\Omega)} \le c\}$$

is sequentially compact in weak*-$BV(\Omega)$, and in $L^r(\Omega)$.

Proof. It is clear that, if $\{u_h\} \subseteq \{u \in BV(\Omega) : |Du|(\Omega) + \lambda\|u\|^r_{L^r(\Omega)} - b\|u\|_{L^r(\Omega)} \le c\}$, then $\{u_h\}$ must be bounded in $L^r(\Omega)$, otherwise, since $r > 1$, it would result that

$$+\infty = \limsup_{h \to +\infty}\{\lambda\|u_h\|^r_{L^r(\Omega)} - b\|u_h\|_{L^r(\Omega)}\} \le$$

$$\le \limsup_{h \to +\infty}\{|Du_h|(\Omega) + \lambda\|u_h\|^r_{L^r(\Omega)} - b\|u_h\|_{L^r(\Omega)}\} \le c.$$

Because of this, we get that also $\{|Du_h|(\Omega)\}$ is bounded, and therefore that actually $\{u_h\}$ is bounded in $BV(\Omega)$. Consequently, by Proposition 4.2.5, and Theorem 4.2.11, there exists $u \in BV(\Omega)$ such that, up to subsequences, $u_h \to u$ in weak*-$BV(\Omega)$, and in $L^r(\Omega)$.

Finally, by the weak*-$BV(\Omega)$-lower semicontinuity of $v \in BV(\Omega) \mapsto |Dv|(\Omega)$, we conclude that

$$|Du|(\Omega) + \lambda\|u\|^r_{L^r(\Omega)} - b\|u\|_{L^r(\Omega)} \le$$

$$\le \liminf_{h \to +\infty}\left\{|Du_h|(\Omega) + \lambda\|u_h\|^r_{L^r(\Omega)} - b\|u_h\|_{L^r(\Omega)}\right\} \le c,$$

from which the desired compactness follows. ∎

Lemma 4.4.2. Let $p \in]1, +\infty]$, $\Omega \in \mathcal{A}_0$ be connected and with Lipschitz boundary, $\{u_h\} \subseteq W^{1,p}(\Omega)$, and $u \in L^1(\Omega)$. Assume that $u_h \to u$ in $L^1(\Omega)$, and that $\{\nabla u_h\}$ is bounded in $(L^p(\Omega))^n$. Then $u_h \to u$ in weak-$W^{1,p}(\Omega)$ if $p \in]1, +\infty[$, or in weak*-$W^{1,\infty}(\Omega)$ if $p = +\infty$.

Proof. We prove the lemma when $p \in]1, +\infty[$, the proof in the other case being similar.

By Theorem 4.3.19, there exists $C_{n,p,\Omega} > 0$ such that

$$\|u_h\|_{L^p(\Omega)} \le \left\|u_h - \frac{1}{\mathcal{L}^n(\Omega)}\int_\Omega u_h dx\right\|_{L^p(\Omega)} + \frac{1}{(\mathcal{L}^n(\Omega))^{1 - \frac{1}{p}}}\int_\Omega |u_h| dx \le$$

$$\le C_{n,p,\Omega}\left(\|\,|\nabla u_h|\,\|_{L^p(\Omega)} + \|u_h\|_{L^1(\Omega)}\right). \text{ for every } h \in \mathbf{N}.$$

Consequently, $\{u_h\}$ turns out to be bounded in $W^{1,p}(\Omega)$, hence relatively compact in the weak-$W^{1,p}(\Omega)$ topology. Because of this, and by the convergence of $\{u_h\}$ to u in $L^1(\Omega)$, the lemma follows. ∎

Proposition 4.4.3. Let $p \in]1, +\infty]$, $\Omega \in \mathcal{A}_0$ have Lipschitz boundary, $r \in]1, p^*[$, and λ, b, $c \in]0, +\infty[$. Then,
i) if $p \in]1, n]$, the set

$$\left\{ u \in W^{1,p}(\Omega) : \||\nabla u|\|^p_{L^p(\Omega)} + \lambda \|u\|^r_{L^r(\Omega)} - b\|u\|_{W^{1,p}(\Omega)} \leq c \right\}$$

is relatively sequentially compact in weak-$W^{1,p}(\Omega)$, and in $\cap_{s \in [1,p^*[} L^s(\Omega)$,
ii) if $p \in]n, +\infty[$, the same set is relatively sequentially compact in weak-$W^{1,p}(\Omega)$, and in $L^\infty(\Omega)$,
iii) if $p = +\infty$, and $R > 0$, the set

$$\left\{ u \in W^{1,\infty}(\Omega) : \||\nabla u|\|_{L^\infty(\Omega)} \leq R, \ \lambda \|u\|^r_{L^r(\Omega)} - b\|u\|_{W^{1,\infty}(\Omega)} \leq c \right\}$$

is relatively sequentially compact in weak*-$W^{1,\infty}(\Omega)$, and in $L^\infty(\Omega)$.

Proof. We prove the proposition only in case i), the proof of the other ones being analogous.

Let $\Omega' \in \mathcal{A}_0$ be connected and with Lipschitz boundary such that $\Omega \subseteq \Omega'$, and let $E: W^{1,p}(\Omega) \to W^{1,p}(\mathbf{R}^n)$ be the extension operator given by Theorem 4.3.8.

Let $\{u_h\} \subseteq \{u \in W^{1,p}(\Omega) : \||\nabla u|\|^p_{L^p(\Omega)} + \lambda \|u\|^r_{L^r(\Omega)} - b\|u\|_{W^{1,p}(\Omega)} \leq c\}$. Then (for the sake of simplicity we continue to use the same symbols for the constants involved) it turns out that

$$\||\nabla E(u_h)|\|^p_{L^p(\Omega')} + \lambda \|E(u_h)\|^r_{L^r(\Omega')} - b\|E(u_h)\|_{W^{1,p}(\Omega')} \leq c \text{ for every } h \in \mathbf{N}.$$

Moreover, by Sobolev Imbedding Theorem, and again by using the same symbols for the constants, it also follows that

$$(4.4.1) \qquad \||\nabla E(u_h)|\|^p_{L^p(\Omega')} + \lambda \|E(u_h)\|^r_{L^r(\Omega')} -$$

$$-b\||\nabla E(u_h)|\|_{L^p(\Omega')} - b\|E(u_h)\|_{L^r(\Omega')} \leq c \text{ for every } h \in \mathbf{N}.$$

Condition (4.4.1) yields that $\{E(u_h)\}$ is bounded in $L^r(\Omega')$, and that $\{\nabla E(u_h)\}$ is bounded in $(L^p(\Omega'))^n$. In fact, if this does not occur, since $r > 1$ and $p > 1$, as in Proposition 4.4.1 condition (4.4.1) would be contradicted.

Consequently, by the Rellich-Kondrachov Compactness Theorem, it follows that, up to subsequences, there exists $u \in W^{1,\min\{p,r\}}(\Omega')$ such that $E(u_h) \to u$ in $L^{\min\{p,r\}}(\Omega')$, and, by Lemma 4.4.2, that $u \in W^{1,p}(\Omega')$, and $E(u_h) \to u$ in weak-$W^{1,p}(\Omega')$. Because of this, we conclude also that $E(u_h) \to u$ in $\cap_{s \in [1,p^*[} L^s(\Omega')$, from which the desired compactness follows. ∎

Proposition 4.4.4. Let $p \in]1, +\infty]$, $\Omega \in \mathcal{A}_0$, and b, $c \in]0, +\infty[$. Then,
i) if $p \in]1, n]$, the set

$$\left\{ u \in W^{1,p}_0(\Omega) : \||\nabla u|\|^p_{L^p(\Omega)} - b\|u\|_{W^{1,p}(\Omega)} \leq c \right\}$$

is relatively sequentially compact in weak-$W^{1,p}(\Omega)$, and in $\cap_{s\in[1,p^*[}L^s(\Omega)$,
ii) if $p \in]n,+\infty[$, the same set is relatively sequentially compact in weak-$W^{1,p}(\Omega)$, and in $L^\infty(\Omega)$,
iii) if $p = +\infty$, and $R > 0$, the set

$$\{u \in W_0^{1,\infty}(\Omega) : \||\nabla u|\|_{L^\infty(\Omega)} \leq R\}$$

is relatively sequentially compact in weak*-$W^{1,\infty}(\Omega)$, and in $L^\infty(\Omega)$.

Proof. We prove the proposition only in case i), the proof of the other ones being analogous.

Let $\{u_h\} \subseteq W^{1,p}(\Omega)$ be such that $\||\nabla u_h|\|_{L^p(\Omega)}^p - b\|u_h\|_{W^{1,p}(\Omega)} \leq c$ for every $h \in \mathbf{N}$, and let $C_{n,\Omega,p}$ be given by Theorem 4.3.18. Then, since by Theorem 4.3.18

$$\||\nabla u_h|\|_{L^p(\Omega)}^p - (1 + C_{n,\Omega,p})^p b\||\nabla u_h|\|_{L^p(\Omega)} \leq$$

$$\leq \||\nabla u_h|\|_{L^p(\Omega)}^p - b\|u_h\|_{W^{1,p}(\Omega)} \leq c,$$

it follows that $\{|\nabla u_h|\}$ is bounded in $L^p(\Omega)$, and, by Theorem 4.3.18, that $\{u_h\}$ is bounded in $W^{1,p}(\Omega)$. Consequently, by the Rellich-Kondrachov Compactness Theorem, it follows that, up to subsequences, there exists $u \in W_0^{1,p}(\Omega)$ such that $u_h \to u$ in $\cap_{s\in[1,p^*[}L^s(\Omega)$, and in weak-$W^{1,p}(\Omega)$. This completes the proof. ∎

Proposition 4.4.5. *Let $\Omega \in \mathcal{A}_0$ be connected and with Lipschitz boundary, $\Gamma \in \mathcal{B}(\partial\Omega)$ satisfy $\mathcal{H}^{n-1}(\Gamma) > 0$, $\phi: \mathbf{R}^n \to [0,+\infty]$ be Borel with $\lim_{z\to+\infty}\frac{\phi(z)}{|z|} = +\infty$, and $b, c \in]0,+\infty[$. Then the set*

$$\left\{u \in W_{0,\Gamma}^{1,1}(\Omega) : \int_\Omega \phi(\nabla u)dx - b\|u\|_{W^{1,1}(\Omega)} \leq c\right\}$$

is relatively sequentially compact in weak-$W^{1,1}(\Omega)$, and in $L^1(\Omega)$.

Proof. Let $C_{n,\Omega,1,\Gamma}$ be the constant appearing in the second part of Theorem 4.3.18 with $p = 1$, and let $b' > 2b(C_{n,\Omega,1,\Gamma}+1)$. Then the assumptions on ϕ guarantee the existence of $R > 0$ such that $\phi(z) \geq b'|z|$ for every $z \in \mathbf{R}^n$ with $|z| > R$, from which we obtain that

(4.4.2) $\phi(z) \geq b'|z| - Rb'$ for every $z \in \mathbf{R}^n$.

Let $\{u_h\} \subseteq \{u \in W^{1,1}(\Omega) : \int_\Omega \phi(\nabla u)dx - b\|u\|_{W^{1,1}(\Omega)} \leq c\}$. Then (4.4.2) implies that

$$b'\||\nabla u_h|\|_{L^1(\Omega)} - Rb'\mathcal{L}^n(\Omega) - b\|u_h\|_{W^{1,1}(\Omega)} \leq c \text{ for every } h \in \mathbf{N},$$

from which, together with Theorem 4.3.18, we infer that

$$\frac{b'}{2}\|\|\nabla u_h\|\|_{L^1(\Omega)} + \frac{b'}{2C_{n,\Omega,1,\Gamma}}\|u_h\|_{L^1(\Omega)} - Rb'\mathcal{L}^n(\Omega) - b\|u_h\|_{W^{1,1}(\Omega)} \le c$$

$$\text{for every } h \in \mathbf{N},$$

that is

$$\left(\frac{b'}{2} - b\right)\|\|\nabla u_h\|\|_{L^1(\Omega)} + \left(\frac{b'}{2C_{n,\Omega,1,\Gamma}} - b\right)\|u_h\|_{L^1(\Omega)} - Rb'\mathcal{L}^n(\Omega) \le c$$

$$\text{for every } h \in \mathbf{N}.$$

By keeping into account that $b' > 2b(C_{n,\Omega,1,\Gamma}+1)$, the above inequality provides that $\{u_h\}$ is bounded in $W^{1,1}(\Omega)$, from which we conclude that actually $\{\int_\Omega \phi(\nabla u_h)dx\}$ too is bounded. Therefore, by using the Rellich-Kondrachov Compactness Theorem, and the Dunford-Pettis-de la Vallée Poussin Theorem, it follows that, up to subsequences, there exists $u \in W^{1,1}(\Omega)$ such that $u_h \to u$ in $L^1(\Omega)$, and in weak-$W^{1,1}(\Omega)$. This completes the proof. ∎

§4.5 Periodic Sobolev Functions

In this section we make some remarks on periodic Sobolev functions, that are of particular interest in homogenization theory.

Let $p \in [1, +\infty]$, and set

$$W_{\text{per}}^{1,p}(Y) = \{v \in W^{1,p}(Y) : \gamma_Y v \text{ takes the same values}$$

$$\text{on the opposite faces of } Y\}.$$

We call the elements of $W_{\text{per}}^{1,p}(Y)$ *periodic Sobolev functions*.

It is clear that $W_{\text{per}}^{1,p}(Y)$ is vector subspace of $W^{1,p}(Y)$.

Proposition 4.5.1. *Let $p \in [1, +\infty]$. Then $W_{\text{per}}^{1,p}(Y)$ is closed in the weak-$W^{1,p}(Y)$ topology if $p \in [1, +\infty[$, or in the weak*-$W^{1,\infty}(Y)$ one if $p = +\infty$.*

Proof. If $p \in [1, +\infty[$, the proof follows from the convexity of $W_{\text{per}}^{1,p}(Y)$, from its closure in $W^{1,p}(Y)$, and from Theorem 1.1.2.

If $p = +\infty$, $W^{1,\infty}(Y)$ turns out to be sequentially closed in weak*-$W^{1,\infty}(Y)$ by virtue of Rellich-Kondrachov Compactness Theorem, therefore, since $(L^1(Y))^{n+1}$ is separable, and $W_{\text{per}}^{1,\infty}(Y)$ is convex, the desired closure follows from Theorem 1.1.4. ∎

Functions in $W_{\text{per}}^{1,p}(Y)$ can be extended by means of periodic replicas to the whole of $\mathbf{R}^n$, getting periodic functions in $W_{\text{loc}}^{1,p}(\mathbf{R}^n)$.

To see this, for every $m \in \mathbf{N}$, and every function $u \in (L^1(Y))^m$ let us denote by $u^\#$ the function defined a.e. in $\mathbf{R}^n$ as

$$u^\#(x) = u(x - (i_1, \ldots, i_n))$$

$$\text{provided } x \in (i_1, \ldots, i_n) + Y \text{ for } (i_1, \ldots, i_n) \in \mathbf{Z}^n.$$

Then obviously $u^\#$ turns out to be Y-periodic.

Proposition 4.5.2. Let $p \in [1, +\infty]$, and $u \in W^{1,p}_{\text{per}}(Y)$. Then $u^\# \in W^{1,p}_{\text{loc}}(\mathbf{R}^n)$, and $\nabla u^\# = (\nabla u)^\#$.

Proof. To prove the proposition, it suffices to verify that

$$(4.5.1) \qquad \int_{Q_k(0)} u^\# \nabla \varphi \, dx = -\int_{Q_k(0)} \varphi (\nabla u)^\# \, dx$$

for every $k \in \mathbf{N}$, and every $\varphi \in C^\infty_0(Q_k(0))$.

Let k, φ be as in (4.5.1), and set $I_k = \{-k, -k+1, \ldots, k-1\}$. Then by the Trace Theorem for Sobolev Functions we obtain that

$$\int_{Q_k(0)} u^\# \nabla \varphi \, dx = \sum_{i \in I_k^n} \int_{i+Y} u^\# \nabla \varphi \, dx = \sum_{i \in I_k^n} \int_Y u^\#(x+i) \nabla \varphi(x+i) \, dx =$$

$$= \sum_{i \in I_k^n} \int_Y u \nabla (T[i]\varphi) \, dx =$$

$$= -\sum_{i \in I_k^n} \int_Y (T[i]\varphi) \nabla u \, dx + \sum_{i \in I_k^n} \int_{\partial Y} T[i]\varphi \gamma_Y u \mathbf{n}_Y \, d\mathcal{H}^{n-1} =$$

$$= -\sum_{i \in I_k^n} \int_Y \varphi(x+i) \nabla u(x+i) \, dx + \sum_{i \in I_k^n} \int_{\partial Y} T[i]\varphi \gamma_Y u \mathbf{n}_Y \, d\mathcal{H}^{n-1} =$$

$$= -\sum_{i \in I_k^n} \int_{i+Y} \varphi (\nabla u)^\# \, dx + \sum_{i \in I_k^n} \int_{\partial Y} T[i]\varphi \gamma_Y u \mathbf{n}_Y \, d\mathcal{H}^{n-1} =$$

$$= -\int_{Q_k(0)} \varphi (\nabla u)^\# \, dx + \sum_{i \in I_k^n} \int_{\partial Y} T[i]\varphi \gamma_Y u \mathbf{n}_Y \, d\mathcal{H}^{n-1}.$$

To complete the proof, we now observe that, if $S_1, \ldots, S_{2n}$ are the faces of Y so that S_{n+1} is opposite to S_1, S_{n+2} to S_2, and so on, it turns out that

$$\sum_{i \in I_k^n} \int_{\partial Y} T[i]\varphi \gamma_Y u \mathbf{n}_Y \, d\mathcal{H}^{n-1} = 0,$$

since, by using the properties of $\gamma_Y u$, for every $h = 1, \ldots, n$, and $i \in I_k^n$ there exists $j_{h,i} \in I_k^n$ such that

$$\int_{S_h} T[i]\varphi \gamma_Y u \mathbf{n}_Y \, d\mathcal{H}^{n-1} + \int_{S_{n+h}} T[j_{h,i}]\varphi \gamma_Y u \mathbf{n}_Y \, d\mathcal{H}^{n-1} = 0. \qquad \blacksquare$$

By virtue of Proposition 4.5.2, the elements of $W^{1,p}_{\text{per}}(Y)$ can be thought as periodic functions defined a.e. in $\mathbf{R}^n$. In the sequel we will always assume such identification.

Chapter 5

Lower Semicontinuity
and Minimization
of Integral Functionals

In this chapter we introduce the study of some types of integral functionals of the calculus of variations, i.e. those of the kind

$$F(\Omega, u) = \int_\Omega f(x, \nabla u) dx$$

on "regular" functions, that will be our energy functionals.

We prove some lower semicontinuity and minimization properties of certain convex functionals of this kind, when they are defined in BV and Sobolev spaces.

§5.1 Functionals on BV Spaces

Let $\Omega \in \mathcal{A}(\mathbf{R}^n)$, and $f \colon \mathbf{R}^n \to [0, +\infty]$ be convex and lower semicontinuous. Then the study of the lower semicontinuity properties of a functional of the type

$$u \mapsto \int_\Omega f(\nabla u) dx,$$

when settled in the framework of BV spaces, naturally leads to the problem of the "correct" definition of the functional itself, due to the presence of singular parts in the gradients of BV functions that are not taken into account in the above integral.

A possible approach to this problem has been proposed by C. Goffman and J. Serrin in 1964 (cf. [GS], and Theorem 6.3.3 in the next chapter) with the introduction of suitable convex functionals defined on spaces of measures, according to the point of view expressed in Remark 2.3.6.

In this section we consider a convex functional defined on the space of Borel measures and strictly linked to those considered in [GS] (cf. [Bu2] for additional references on the subject), of which we study the lower semicontinuity properties with respect to weak* topology. This approach allows us to deduce a lower semicontinuity result for the corresponding functionals defined in BV spaces.

Let Ω be a Hausdorff locally compact space, μ be a σ-finite Borel positive measure on Ω, and $f: \mathbf{R}^m \to [0, +\infty]$ be convex and lower semicontinuous. Then f^∞ turns out to be well defined, convex and lower semicontinuous. Consequently, we can consider the functional F defined as

$$(5.1.1) \qquad F: \nu \in (\mathcal{M}(\Omega))^m \mapsto \int_\Omega f\left(\frac{d\nu^a}{d\mu}\right) d\mu + \int_\Omega f^\infty\left(\frac{d\nu^s}{d|\nu^s|}\right) d|\nu^s|.$$

First of all, let us observe that

$$F(\nu) = \int_\Omega f\left(\frac{d\nu^a}{d\mu}\right) d\mu$$

whenever $\nu \in (\mathcal{M}(\Omega))^m$ is absolutely continuous with respect to μ,

and that, if $\lim_{z \to \infty} \frac{f(z)}{|z|} = +\infty$, then $f^\infty(0) = 0$, $f^\infty(z) = +\infty$ for every $z \neq 0$, and

$$F(\nu) = +\infty \text{ whenever } \nu \in (\mathcal{M}(\Omega))^m \text{ satisfies } |\nu^s|(\Omega) \neq 0.$$

Let us now prove some preparatory results.

Lemma 5.1.1. *Let $(\Omega, \mathcal{E})$ be a measure space, and μ be a positive measure on $\mathcal{E}$. For every $h \in \mathbf{N}$ let $g_h: \Omega \to [0, +\infty]$ be $\mathcal{E}$-measurable, and set $g: x \in \Omega \mapsto \sup_{h \in \mathbf{N}} g_h(x)$. Then*

$$\int_\Omega g \, d\mu = \sup\left\{\sum_{j \in J} \int_{B_j} g_j \, d\mu : \{B_j\}_{j \in J} \subseteq \mathcal{E} \text{ finite partition of } \Omega\right\}$$

Proof. It is clear that

$$(5.1.2) \qquad \int_\Omega g \, d\mu \geq$$

$$\geq \sup\left\{\sum_{j \in J} \int_{B_j} g_j \, d\mu : \{B_j\}_{j \in J} \subseteq \mathcal{E} \text{ finite partition of } \Omega\right\}.$$

For every $h \in \mathbf{N}$ we set $f_h: x \in \Omega \mapsto \max\{g_1(x), \ldots, g_h(x)\}$. Then it is clear that $\lim_{h \to +\infty} f_h(x) = g(x)$ for μ-a.e. $x \in \Omega$, and, by the Monotone Convergence Theorem, that

$$\int_\Omega g d\mu = \lim_{h \to +\infty} \int_\Omega f_h d\mu.$$

Because of this, once we observe that for every $h \in \mathbf{N}$ a finite partition $\{B_1^h, \ldots, B_{m_h}^h\} \subseteq \mathcal{E}$ of Ω can be found such that $f_h = g_j$ in B_j^h for every $j \in \{B_1^h, \ldots, B_{m_h}^h\}$, we conclude that

(5.1.3)
$$\int_\Omega g d\mu = \lim_{h \to +\infty} \sum_{j=1}^{m_h} \int_{B_j^h} g_j d\mu \le$$

$$\le \sup\left\{ \sum_{j \in J} \int_{B_j} g_j d\mu : \{B_j\}_{j \in J} \subseteq \mathcal{E} \text{ finite partition of } \Omega \right\}.$$

By (5.1.2), and (5.1.3) the lemma follows. ∎

Lemma 5.1.2. *Let Ω be a Hausdorff locally compact space. Then, for every couple of disjoint compact subsets K_1, and K_2 of Ω there exist two open sets A_1, and A_2 having compact closures such that $K_1 \subseteq A_1$, $K_2 \subseteq A_2$, and $\overline{A_1} \cap \overline{A_2} = \emptyset$.*

Proof. Let $x_2 \in K_2$. Then, being Ω Hausdorff and locally compact, for every $x_1 \in K_1$ there exist $I_{x_1} \in \mathcal{N}(x_1)$ having compact closure, and $I_{x_2} \in \mathcal{N}(x_2)$ having compact closure such that $I_{x_1} \cap I_{x_2} = \emptyset$.

It is clear that the family $\{I_{x_1} : x_1 \in K_1\}$ forms a covering of K_1, therefore, by extracting a finite subcovering, we can construct an open set B_1, depending on x_2 and having compact closure, and $J_{x_2} \in \mathcal{N}(x_2)$ having compact closure such that $K_1 \subseteq B_1$, and $B_1 \cap J_{x_2} = \emptyset$.

We now observe that the family $\{J_{x_2} : x_2 \in K_2\}$ forms a covering of K_2, therefore, by extracting a finite subcovering, we can construct two open set A_1 and B_2 with compact closure such that $K_1 \subseteq A_1$, $K_2 \subseteq B_2$ and $A_1 \cap B_2 = \emptyset$. Moreover, it also turns out that $\overline{A_1} \cap K_2 = \emptyset$.

Because of this, and by repeating the same above arguments applied to K_2 and $\overline{A_1}$ in place of K_1 and K_2, we construct an open set A_2 with compact closure such that $K_2 \subseteq A_2$, and $\overline{A_1} \cap \overline{A_2} = \emptyset$. This completes the proof. ∎

We are now in position to prove the lower semicontinuity result.

Theorem 5.1.3. *Let Ω be a Hausdorff locally compact space, μ be a σ-finite Borel positive measure on Ω, and $f: \mathbf{R}^m \to [0, +\infty]$ be convex and* *lower semicontinuous. Let F be defined by (5.1.1). Then F is weak*-$(\mathcal{M}(\Omega))^m$-lower semicontinuous.*

Proof. By using Proposition 1.1.12, we deduce the existence of $\{a_h\} \subseteq \mathbf{R}^m$, and of $\{b_h\} \subseteq \mathbf{R}$ for which, by setting for every $h \in \mathbf{N}$, $f_h : z \in \mathbf{R}^m \mapsto (a_h \cdot z + b_h)^+$, it results

$$(5.1.4) \qquad f(z) = \sup\{f_h(z) : h \in \mathbf{N}\} \text{ for every } z \in \mathbf{R}^m.$$

By Lebesgue Decomposition Theorem, for every $\nu \in (\mathcal{M}(\Omega))^m$ let $N_0 \in \mathcal{B}(\Omega)$ satisfy $\mu(N_0) = 0$, and $|\nu^s|(\Omega \setminus N_0) = 0$. Moreover, by Radon-Nikodym Theorem, let N^a, $N^s \in \mathcal{B}(\Omega)$ with $\mu(N^a) = 0$, and $|\nu^s|(\Omega \setminus N^s) = 0$, be such that $\frac{d\nu^a}{d\mu}(x)$ exists for every $x \in \Omega \setminus N^a$, and $\frac{d\nu^s}{d|\nu^s|}(x)$ exists for every $x \in N^s$. We can clearly assume that $N^s \subseteq N_0 \subseteq N^a$.

For every $h \in \mathbf{N}$, and $\nu \in (\mathcal{M}(\Omega))^m$ let us define

$$(5.1.5) \qquad g_h : x \in (\Omega \setminus N^a) \cup N^s \mapsto \begin{cases} f_h\left(\frac{d\nu^a}{d\mu}(x)\right) & \text{if } x \in \Omega \setminus N^a \\ f_h^\infty\left(\frac{d\nu^s}{d|\nu^s|}(x)\right) & \text{if } x \in N^s, \end{cases}$$

then g_h turns out to be defined $\mu + |\nu^s|$-a.e. in Ω, since

$$(\mu + |\nu^s|)(N^a \setminus N^s) = \mu(N^a \setminus N^s) + |\nu^s|(N^a \setminus N^s) \le \mu(N^a) + |\nu^s|(\Omega \setminus N^s) = 0,$$

and

$$(5.1.6) \qquad \int_\Omega g_h d(\mu + |\nu^s|) =$$

$$= \int_{\Omega \setminus N^a} f_h\left(\frac{d\nu^a}{d\mu}\right) d(\mu + |\nu^s|) + \int_{N^s} f_h^\infty\left(\frac{d\nu^s}{d|\nu^s|}\right) d(\mu + |\nu^s|) =$$

$$= \int_\Omega f_h\left(\frac{d\nu^a}{d\mu}\right) d\mu + \int_\Omega f_h^\infty\left(\frac{d\nu^s}{d|\nu^s|}\right) d|\nu^s|.$$

We now observe that (5.1.4) trivially implies that

$$\sup_{h \in \mathbf{N}} g_h(x) = \begin{cases} f\left(\frac{d\nu^a}{d\mu}(x)\right) & \text{if } x \in \Omega \setminus N^a \\ f^\infty\left(\frac{d\nu^s}{d|\nu^s|}(x)\right) & \text{if } x \in N^s, \end{cases} \qquad \text{for every } x \in (\Omega \setminus N^a) \cup N^s,$$

therefore, by (5.1.6), and Lemma 5.1.1, we conclude that

$$F(\nu) = \sup\left\{\sum_{j \in J} \int_{B_j} f_j\left(\frac{d\nu^a}{d\mu}\right) d\mu + \int_{B_j} f_j^\infty\left(\frac{d\nu^s}{d|\nu^s|}\right) d|\nu^s| : \right.$$

$\{B_j\}_{j \in J} \subseteq \mathcal{B}(\Omega)$ finite partition of $\Omega \Big\}$ for every $\nu \in (\mathcal{M}(\Omega))^m$,

from which, by using the regularity properties of Borel positive measures, we also have that

$$(5.1.7) \qquad F(\nu) = \sup \left\{ \sum_{j \in J} \int_{K_j} f_j \left(\frac{d\nu^a}{d\mu} \right) d\mu + \int_{K_j} f_j^\infty \left(\frac{d\nu^s}{d|\nu^s|} \right) d|\nu^s| : \right.$$

$\{K_j\}_{j \in J}$ finite set of pairwise disjoint compact subsets of $\Omega \Big\}$

for every $\nu \in (\mathcal{M}(\Omega))^m$.

In addition, by (5.1.7), and Lemma 5.1.2, we also obtain that

$$F(\nu) = \sup \left\{ \sum_{j \in J} \int_{A_j} f_j \left(\frac{d\nu^a}{d\mu} \right) d\mu + \int_{A_j} f_j^\infty \left(\frac{d\nu^s}{d|\nu^s|} \right) d|\nu^s| : \right.$$

$\{A_j\}_{j \in J}$ finite set of pairwise disjoint open subsets of $\Omega \Big\}$

for every $\nu \in (\mathcal{M}(\Omega))^m$,

therefore, to prove the theorem, we only have to prove that for every $A \in \mathcal{A}(\Omega)$, and $h \in \mathbf{N}$, the functional $\nu \in (\mathcal{M}(\Omega))^m \mapsto \int_A f_h(\frac{d\nu^a}{d\mu})d\mu + \int_A f_h^\infty(\frac{d\nu^s}{d|\nu^s|})d|\nu^s|$ is weak*-$(\mathcal{M}(\Omega))^m$-lower semicontinuous.

To see this, we first observe that

$$f_h^\infty(z) = (a_h \cdot z)^+ \text{ for every } h \in \mathbf{N}, \ z \in \mathbf{R}^m,$$

and that, by (5.1.5),

$$g_h(x) = \left(\left(a_h \cdot \frac{d\nu^a}{d\mu}(x) + b_h \right) \chi_{\Omega \setminus N^a}(x) + a_h \cdot \frac{d\nu^s}{d|\nu^s|}(x) \chi_{N^s}(x), 0 \right)^+$$

for every $h \in \mathbf{N}$, and $(\mu + |\nu^s|)$-a.e. $x \in \Omega$.

Consequently, by using also (5.1.6) and Corollary 2.4.7, for every $A \in \mathcal{A}(\Omega)$, and $h \in \mathbf{N}$ we have that

$$(5.1.8) \qquad \int_A f_h \left(\frac{d\nu^a}{d\mu} \right) d\mu + \int_A f_h^\infty \left(\frac{d\nu^s}{d|\nu^s|} \right) d|\nu^s| = \int_A g_h d(\mu + |\nu^s|) =$$

$$= \sup \left\{ \int_{A \setminus N^a} \left(a_h \cdot \frac{d\nu^a}{d\mu} + b_h \right) \varphi d(\mu + |\nu^s|) + \int_{A \cap N^s} a_h \cdot \frac{d\nu^s}{d|\nu^s|} \varphi d(\mu + |\nu^s|) : \right.$$

$$\varphi \in C_0^0(A),\ 0 \le \varphi \le 1 \Big\} =$$

$$= \sup \Big\{ a_h \cdot \int_A \varphi d\nu + b_h \int_A \varphi d\mu : \varphi \in C_0^0(A),\ 0 \le \varphi \le 1 \Big\}$$

for every $\nu \in (\mathcal{M}(\Omega))^m$.

Because of this, the proof follows, since the functionals in the right-hand side of (5.1.8) are weak*-$(\mathcal{M}(\Omega))^m$ continuous. ∎

Coming back to the lower semicontinuity problem in BV spaces, Theorem 5.1.3 suggests the introduction, for every $\Omega \in \mathcal{A}(\mathbf{R}^n)$ and $f: \mathbf{R}^n \to [0, +\infty]$, of the functional

(5.1.9) $\qquad G: u \in BV(\Omega) \mapsto \int_\Omega f(\nabla u)dx + \int_\Omega f^\infty(\nabla^s u)d|D^s u|.$

First of all, let us observe that

$$G(u) = \int_\Omega f(\nabla u)dx \text{ whenever } u \in W^{1,1}(\Omega),$$

and that, if $\lim_{z \to \infty} \frac{f(z)}{|z|} = +\infty$, then $f^\infty(0) = 0$, $f^\infty(z) = +\infty$ for every $z \ne 0$, and

$$G(u) = +\infty \text{ whenever } u \in BV(\Omega) \setminus W^{1,1}(\Omega).$$

Then, from Theorem 5.1.3 the following lower semicontinuity result for functionals on BV spaces follows.

Theorem 5.1.4. *Let $f: \mathbf{R}^n \to [0, +\infty]$ be convex and lower semicontinuous, $\Omega \in \mathcal{A}(\mathbf{R}^n)$, and let G be defined by (5.1.9). Then G is weak*-$BV(\Omega)$-lower semicontinuous.*

Proof. Follows from Theorem 5.1.3. ∎

§5.2 Functionals on Sobolev Spaces

In this section we discuss the lower semicontinuity properties, with respect to weak convergence, of integral functionals of the kind

(5.2.1) $\qquad u \to \int_\Omega f(x, \nabla u(x))dx,$

defined in Sobolev spaces, where $\Omega \in \mathcal{A}(\mathbf{R}^n)$, and f is a function defined in $\Omega \times \mathbf{R}^n$.

Integrands in (5.2.1) are obtained through the composition of f with measurable functions, thus getting, in general, non-necessarily measurable integrands. The following result provides conditions ensuring the measurability of such compositions.

Proposition 5.2.1. *Let $\Omega \in \mathcal{L}_n(\mathbf{R}^n)$, and $f: (x, z) \in \Omega \times \mathbf{R}^n \mapsto f(x, z) \in [-\infty, +\infty]$ be $(\mathcal{L}_n(\Omega) \times \mathcal{B}(\mathbf{R}^n))$-measurable. Then, for every $m: \Omega \to \mathbf{R}^n$ measurable, the composition $x \to f(x, m(x))$ is measurable.*

Proof. Let m be as above, and set $\widetilde{m}: x \in \Omega \mapsto (x, m(x)) \in \Omega \times \mathbf{R}^n$.

Then, the measurability of m implies that

$$(5.2.2) \qquad \widetilde{m}^{-1}(A \times B) = A \cap m^{-1}(B) \in \mathcal{L}_n(\Omega)$$

for every $A \in \mathcal{L}_n(\Omega)$, $B \in \mathcal{B}(\mathbf{R}^n)$.

Now, it is easy to verify that the set $\{X \subseteq \Omega \times \mathbf{R}^n : \widetilde{m}^{-1}(X) \in \mathcal{L}_n(\Omega)\}$ is a σ-algebra, therefore (5.2.2) yields

$$\mathcal{L}_n(\Omega) \times \mathcal{B}(\mathbf{R}^n) \subseteq \{X \subseteq \Omega \times \mathbf{R}^n : \widetilde{m}^{-1}(X) \in \mathcal{L}_n(\Omega)\}.$$

Because of this, the measurability of the composition follows since

$$\{x \in \Omega : f(x, m(x)) > \lambda\} = \widetilde{m}^{-1}\left(f^{-1}(]\lambda, +\infty[)\right) \text{ for every } \lambda \in \mathbf{R}. \qquad \blacksquare$$

In particular, given $\Omega \in \mathcal{L}_n(\mathbf{R}^n)$, we point out a class of particularly significant $(\mathcal{L}_n(\Omega) \times \mathcal{B}(\mathbf{R}^n))$-measurable functions: the one of the indicator functions of balls with varying radius. More precisely, if $\varphi: \Omega \to]0, +\infty[$ is $\mathcal{L}_n(\Omega)$-measurable, then $(x, z) \in \Omega \times \mathbf{R}^n \to I_{\overline{B_{\varphi(x)}(0)}}(z)$ turns out to be $(\mathcal{L}_n(\Omega) \times \mathcal{B}(\mathbf{R}^n))$-measurable, since, as λ varies in $\mathbf{R}$, $\{(x, z) \in \Omega \times \mathbf{R}^n : f(x, z) \geq \lambda\}$ can be equal to $\Omega \times \mathbf{R}^n$, or to $\{(x, z) \in \Omega \times \mathbf{R}^n : \varphi(x) < |z|\}$ that is clearly $(\mathcal{L}_n(\Omega) \times \mathcal{B}(\mathbf{R}^n))$-measurable.

We also observe that, if Ω, φ are as above, and $f: \Omega \times \mathbf{R}^n \to]-\infty, +\infty]$ is Borel, then $(x, z) \in \Omega \times \mathbf{R}^n \to f(x, z) + I_{\overline{B_{\varphi(x)}(0)}}(z)$ is $(\mathcal{L}_n(\Omega) \times \mathcal{B}(\mathbf{R}^n))$-measurable.

We can now prove the lower semicontinuity result.

Theorem 5.2.2. *Let $p \in [1, +\infty]$, $\Omega \in \mathcal{A}(\mathbf{R}^n)$, and $f: \Omega \times \mathbf{R}^n \to [0, +\infty]$ be $(\mathcal{L}_n(\Omega) \times \mathcal{B}(\mathbf{R}^n))$-measurable, and such that*

$$f(x, \cdot) \text{ is convex and lower semicontinuous for a.e. } x \in \Omega.$$

Then the functional

$$F: u \in W^{1,p}(\Omega) \mapsto \int_\Omega f(x, \nabla u)dx$$

is sequentially weak-$W^{1,p}(\Omega)$-lower semicontinuous if $p \in [1, +\infty[$, sequentially weak*-$W^{1,\infty}(\Omega)$-lower semicontinuous if $p = +\infty$.

Proof. Let us first assume that $p \in [1, +\infty[$.

First of all, let us observe that F is lower semicontinuous in the strong $W^{1,p}(\Omega)$ topology. In fact, if $\{u_h\} \subseteq W^{1,p}(\Omega)$, $u \in W^{1,p}(\Omega)$ satisfy $u_h \to u$ in $W^{1,p}(\Omega)$, let $\{u_{h_k}\} \subseteq \{u_h\}$ be such that $\nabla u_{h_k} \to \nabla u$ a.e. in Ω, and $\liminf_{k \to +\infty} F(u_{h_k}) = \liminf_{h \to +\infty} F(u_h)$. Then Fatou's lemma yields

$$\int_\Omega f(x, \nabla u)dx \leq \int_\Omega \liminf_{k \to +\infty} f(x, \nabla u_{h_k})dx \leq$$

$$\leq \liminf_{k \to +\infty} \int_\Omega f(x, \nabla u_{h_k})dx = \liminf_{h \to +\infty} \int_\Omega f(x, \nabla u_h)dx,$$

from which the $W^{1,p}(\Omega)$-lower semicontinuity of F follows.

Because of this, and Theorem 1.1.13 the proof follows if $p \in [1, +\infty[$.

If $p = +\infty$, let $\{u_h\} \subseteq W^{1,\infty}(\Omega)$, $u \in W^{1,\infty}(\Omega)$ be such that $u_h \to u$ in weak*-$W^{1,\infty}(\Omega)$, and let $A \in \mathcal{A}_0$ with $A \subset\subset \Omega$. Then, for every $q \in [1, +\infty[$, $u_h \to u$ in weak-$W^{1,q}(A)$, and by the above treated case it follows that

$$\int_A f(x, \nabla u)dx \leq \liminf_{h \to +\infty} \int_A f(x, \nabla u_h)dx \leq \liminf_{h \to +\infty} \int_\Omega f(x, \nabla u_h)dx,$$

from which the proof follows letting A increase to Ω. ∎

§5.3 Minimization of Integral Functionals

In the present section we apply the abstract minimization results of Chapter 3 to the concrete case of the integral functionals considered in the previous sections, and for some Dirichlet and Neumann minimum problems.

Of course, the minimum problems that we consider here, as well as those in the next chapters, have an illustrative value, and the integral functionals to be minimized contain pieces that make them fulfil the necessary coerciveness assumptions.

We start with the case of functionals on BV spaces.

Theorem 5.3.1. *Let $f: \mathbf{R}^n \to [0, +\infty]$ be convex, lower semicontinuous, and satisfying*

(5.3.1) $$|z| \leq f(z) \text{ for every } z \in \mathbf{R}^n.$$

Then, for every $\Omega \in \mathcal{A}_0$ with Lipschitz boundary, $\lambda \in]0, +\infty[$, $r \in]1, 1^*[$, and $\beta \in L^{r'}(\Omega)$ the problem

$$\min \left\{ \int_\Omega f(\nabla u)dx + \int_\Omega f^\infty(\nabla^s u)d|D^s u| + \lambda \int_\Omega |u|^r dx + \int_\Omega \beta u dx : \right.$$

$$\left. u \in BV(\Omega) \right\}$$

has a solution.

Proof. Let Ω, λ, r, β be as above. The proof follows from Theorem 3.1.4, once we prove that the functional

$$F\colon u \in BV(\Omega) \mapsto \int_\Omega f(\nabla u)dx + \int_\Omega f^\infty(\nabla^s u)d|D^s u| + \lambda \int_\Omega |u|^r dx + \int_\Omega \beta u dx$$

is sequentially weak*-$BV(\Omega)$ lower semicontinuous and sequentially coercive in the same topology.

To prove the sequential weak*-$BV(\Omega)$ lower semicontinuity of F let $\{u_h\} \subseteq BV(\Omega)$, $u \in BV(\Omega)$ be such that $u_h \to u$ in weak*-$BV(\Omega)$, and assume for simplicity that $\lim_{h\to+\infty} F(u_h)$ exists. Then $\{u_h\}$ turns out to be bounded in $BV(\Omega)$, and by Theorem 4.2.11, there exist $\{u_{h_k}\} \subseteq \{u_h\}$ and $u \in BV(\Omega)$ such that $u_{h_k} \to u$ in $L^r(\Omega)$.

Because of this, and by Theorem 5.1.4, we conclude that

$$F(u) \le \liminf_{k\to+\infty} F(u_{h_k}) = \liminf_{h\to+\infty} F(u_h),$$

that is the desired lower semicontinuity.

To prove the sequential coerciveness of F in the weak*-$BV(\Omega)$ topology, let us first observe that (5.3.1) implies that

(5.3.2) $$|z| \le f^\infty(z) \text{ for every } z \in \mathbf{R}^n,$$

therefore, by (5.3.1), and (5.3.2) it follows that

$$|Du|(\Omega) + \lambda\|u\|^r_{L^r(\Omega)} - \|\beta\|_{L^{r'}(\Omega)}\|u\|_{L^r(\Omega)} \le F(u) \text{ for every } u \in BV(\Omega),$$

from which we conclude that, for every $c \in \mathbf{R}$, $\{v \in BV(\Omega) : F(v) \le c\} \subseteq \{v \in BV(\Omega) : |Dv|(\Omega) + \lambda\|v\|^r_{L^r(\Omega)} - \|\beta\|_{L^{r'}(\Omega)}\|v\|_{L^r(\Omega)} \le c\}$.

Now Proposition 4.4.1 implies that this last set is compact once we equip it with the weak*-$BV(\Omega)$ topology, so, if $c \in \mathbf{R}$, and $\{u_h\} \subseteq \{v \in BV(\Omega) : F(v) \le c\}$, there exist $\{u_{h_k}\} \subseteq \{u_h\}$ and $u \in \{v \in BV(\Omega) : |Dv|(\Omega) + \lambda\|v\|^r_{L^r(\Omega)} - \|\beta\|_{L^{r'}(\Omega)}\|v\|_{L^r(\Omega)} \le c\}$ such that $u_{h_k} \to u$ in weak*-$BV(\Omega)$. Consequently, by the previously proved lower semicontinuity of F, it follows that

$$F(u) \le \liminf_{k\to+\infty} F(u_{h_k}) \le c,$$

from which we conclude that $u \in \{v \in BV(\Omega) : F(v) \le c\}$, and therefore that F is sequentially coercive in the weak*-$BV(\Omega)$ topology. ∎

Theorem 5.3.2. *Let $p \in]1, +\infty]$, $\Omega \in \mathcal{A}_0$ have Lipschitz boundary, and $f\colon \Omega \times \mathbf{R}^n \to [0, +\infty]$ be $(\mathcal{L}_n(\Omega) \times \mathcal{B}(\mathbf{R}^n))$-measurable, and such that*

$$f(x, \cdot) \text{ is convex and lower semicontinuous for a.e. } x \in \Omega,$$

(5.3.3) $\begin{cases} |z|^p \leq f(x,z) \text{ for a.e. } x \in \Omega \text{ and every } z \in \mathbf{R}^n & \text{if } p \in]1,+\infty[\\ \text{dom} f(x,\cdot) \subseteq B_R(0) \text{ for a.e. } x \in \Omega & \text{if } p = +\infty \end{cases}$

for some $R > 0$. Then, for every $\lambda \in]0,+\infty[$, $r \in [1,p^*[$, and $\beta \in L^{p'}(\Omega)$ the problem

$$\min\left\{ \int_\Omega f(x,\nabla u)dx + \lambda \int_\Omega |u|^r dx + \int_\Omega \beta u dx :\in W^{1,p}(\Omega) \right\}$$

has a solution.

Proof. Let Ω, λ, r, β be as above. As in Theorem 5.3.1, the proof follows from Theorem 3.1.4, once we prove that the functional

$$F: u \in W^{1,p}(\Omega) \mapsto \int_\Omega f(x,\nabla u)dx + \lambda \int_\Omega |u|^r dx + \int_\Omega \beta u dx$$

is sequentially lower semicontinuous and sequentially coercive in the weak-$W^{1,p}(\Omega)$ topology if $p \in]1,+\infty[$, or in the weak*-$W^{1,\infty}(\Omega)$ one if $p = +\infty$.

The proof of the lower semicontinuity of F follows as in the proof of Theorem 5.3.1, and by exploiting Rellich-Kondrachov Compactness Theorem in place of Theorem 4.2.11, and Theorem 5.2.2 in place of Theorem 5.1.4.

To prove the sequential coerciveness properties of F, let us first consider the case when $p \in]1,+\infty[$. Let us first observe that (5.3.3) implies that

$$\||Du|\|^p_{L^p(\Omega)} + \lambda\|u\|^r_{L^r(\Omega)} - \|\beta\|_{L^{p'}(\Omega)}\|u\|_{L^p(\Omega)} \leq F(u)$$

$$\text{for every } u \in W^{1,p}(\Omega),$$

from which we conclude that, for every $c \in \mathbf{R}$, $\{v \in W^{1,p}(\Omega) : F(v) \leq c\} \subseteq \{v \in W^{1,p}(\Omega) : \||Dv|\|^p_{L^p(\Omega)} + \lambda\|v\|^r_{L^r(\Omega)} - \|\beta\|_{L^{p'}(\Omega)}\|v\|_{W^{1,p}(\Omega)} \leq c\}$.

Now Proposition 4.4.3 implies that this last set is relatively sequentially compact once we equip it with the weak-$W^{1,p}(\Omega)$ topology, and the proof completes as in the one of Theorem 5.3.1.

Finally, when $p = +\infty$, (5.3.3) implies that

$$I_{]-R,R[}\left(\||Du|\|_{L^\infty(\Omega)}\right) + \lambda\|u\|^r_{L^r(\Omega)} - \|\beta\|_{L^\infty(\Omega)}\|u\|_{L^1(\Omega)} \leq F(u)$$

$$\text{for every } u \in W^{1,\infty}(\Omega),$$

from which we conclude that, for every $c \in \mathbf{R}$, $\{v \in W^{1,\infty}(\Omega) : F(v) \leq c\} \subseteq \{v \in W^{1,\infty}(\Omega) : \||Dv|\|_{L^\infty(\Omega)} \leq R, \ \lambda\|v\|^r_{L^r(\Omega)} - \mathcal{L}^n(\Omega)\|\beta\|_{L^\infty(\Omega)}\|v\|_{W^{1,\infty}(\Omega)} \leq c\}$.

Now Proposition 4.4.3 implies that this last set is relatively sequentially compact once we equip it with the weak*-$W^{1,\infty}(\Omega)$ topology, and the proof completes as in the one of Theorem 5.3.1. ∎

Theorem 5.3.3. Let p, $\Omega \in \mathcal{A}_0$, and f be as in Theorem 5.3.2. Then, for ever $\beta \in L^{p'}(\Omega)$ the problem

$$\min\left\{ \int_\Omega f(x, \nabla u)dx + \int_\Omega \beta u dx :\in W_0^{1,p}(\Omega)\right\}$$

has a solution.

Proof. Let Ω, β be as above. As in Theorem 5.3.1, the proof follows from Theorem 3.1.4, once we prove that the functional

$$F: u \in W_0^{1,p}(\Omega) \mapsto \int_\Omega f(x, \nabla u)dx + \int_\Omega \beta u dx$$

is sequentially lower semicontinuous and sequentially coercive in the weak-$W^{1,p}(\Omega)$ topology if $p \in \,]1, +\infty[$, or in the weak*-$W^{1,\infty}(\Omega)$ one if $p = +\infty$.

The lower semicontinuity of F follows directly from Theorem 5.2.2, and the continuity in the weak-$W^{1,p}(\Omega)$ topology if $p \in \,]1, +\infty[$, or in the weak*-$W^{1,\infty}(\Omega)$ one if $p = +\infty$ of $u \in W_0^{1,p}(\Omega) \mapsto \int_\Omega \beta u dx$.

To prove the sequential coerciveness properties of F, let us first consider the case when $p \in \,]1, +\infty[$. Let us first observe that (5.3.3) implies that

$$\|Du\|_{L^p(\Omega)}^p - \|\beta\|_{L^{p'}(\Omega)}\|u\|_{L^p(\Omega)} \leq F(u) \text{ for every } u \in W_0^{1,p}(\Omega),$$

from which we conclude that, for every $c \in \mathbf{R}$, $\{v \in W_0^{1,p}(\Omega) : F(v) \leq c\} \subseteq \{v \in W_0^{1,p}(\Omega) : \|Dv\|_{L^p(\Omega)}^p - \|\beta\|_{L^{p'}(\Omega)}\|v\|_{W^{1,p}(\Omega)} \leq c\}$.

Now Proposition 4.4.4 implies that this last set is relatively sequentially compact once we equip it with the weak-$W^{1,p}(\Omega)$ topology, and the proof completes as in the one of Theorem 5.3.1.

Finally, when $p = +\infty$, (5.3.3) implies that

$$I_{]-R,R[}\left(\|Du\|_{L^\infty(\Omega)}\right) - \|\beta\|_{L^{r'}(\Omega)}\|u\|_{L^r(\Omega)} \leq F(u) \text{ for every } u \in W_0^{1,\infty}(\Omega),$$

from which we conclude that, for every $c \in \mathbf{R}$, $\{v \in W_0^{1,\infty}(\Omega) : F(v) \leq c\} \subseteq \{v \in W_0^{1,\infty}(\Omega) : \|Dv\|_{L^\infty(\Omega)} \leq R\}$.

Now Proposition 4.4.4 implies that this last set is relatively sequentially compact once we equip it with the weak*-$W^{1,\infty}(\Omega)$ topology, and the proof completes as in the one of Theorem 5.3.1. ∎

Chapter 6

Classical Results and Mathematical Models Originating Unbounded Functionals

The present chapter constitutes a brief introduction to unique extension, integral representation, relaxation, and homogenization problems by means of a presentation of some well established results in literature dealing with finite valued integral functionals of the calculus of variations. Obviously they are not necessarily the finest ones, but we hope they are significant enough to illustrate the main features of the above problems.

In the next chapters we will start our study on similar problems, but for functionals possibly taking also not finite values.

Finally, we describe the mathematical aspects of some physical models as an introduction to unbounded functionals. We emphasize that the essential difference between this classical theory, and the one that we introduce here and develop in the next chapters, is that in the first one integrands assume only finite values.

§6.1 Classical Unique Extension Results

A classical mathematical problem deals with the extension of a given function to a larger definition set preserving some of its properties. For example, a classical item in this framework is given by Hahn-Banach Theorem.

A similar problem arises for example when X is a dense subset of a topological space (Y, τ), and $f: X \to [-\infty, +\infty]$. In this case, besides the problem of the existence of an extension of f from X to Y preserving certain properties, one may also ask whether such extension can be constructed in an essentially unique way.

The most elementary case occurs when Y is a metric space, and f is uniformly continuous. In this case, f can be extended to Y in a unique way, preserving the uniform continuity modulus. For example, a classical item in this framework is given by the definition of elementary functions.

A finer case occurs when one considers the non-parametric area functional A defined on functions belonging to $C^1(\mathbf{R}^n)$.

Several methods had been developed in order to extend A to all continuous functions. The oldest is due to Lebesgue, and another is due to Caccioppoli.

Roughly speaking, in the first method the extension is given by the semicontinuous envelope of A in the uniform convergence topology. In the second one the extension is given by the semicontinuous envelope of A in the L^1 topology. Then, a well known result (cf. [Mi2]) establishes that the two extensions agree. In this case, it is essentially convexity that is responsible for coincidence.

We refer to these types of problems as to unique extension problems.

§6.2 Classical Integral Representation Results

Integral representation problems appear naturally in many situations, typically in the framework of functional analysis, relaxation, or of Γ-convergence of integrals of the calculus of variations, in which one has an abstract functional defined on some function spaces and verifying suitable assumptions, and has to deduce that it itself actually has an integral form.

For example, Riesz Representation Theorem can be reread in this setting as a result under linearity assumptions.

The situation becomes more involved when the dependence of the functional on the elements of the function spaces turns out to be nonlinear, or through their first or higher order derivatives.

In the framework of relaxation theory for variational integrals on BV spaces, an implicit integral representation problem is studied in [S1], [S2], and finally in [GS] by means of convex functions of measures.

An explicit integral representation theorem is proved in [DG3], where the following result is proved (cf. [DG3, Lemma II]).

We denote by $\mathcal{P}_n$ the class of the finite unions of open intervals of $\mathbf{R}^n$ with endpoints in $\mathbf{Q}^n$.

Theorem 6.2.1. *Let* $F \colon \mathcal{P}_n \times C^1(\mathbf{R}^n) \to [0, +\infty[$ *satisfy for some* $s > 0$

$$\int_A |\nabla u| dx \le F(A, u) \le s \int_A (1 + |u| + |\nabla u|) dx$$

$$\text{for every } (A, u) \in \mathcal{P}_n \times C^1(\mathbf{R}^n),$$

$$|F(A, u') - F(A, u'')| \le s \int_A (|u' - u''| + |\nabla u' - \nabla u''|) dx$$

$$\text{for every } A \in \mathcal{P}_n, \ u', u'' \in C^1(\mathbf{R}^n),$$

$$F(A, u) = F(A', u) - F(A'', u) \text{ for every } A, A', A'' \in \mathcal{P}_n, \ u \in C^1(\mathbf{R}^n)$$

$$\text{such that } A' \cap A'' = \emptyset, \ A' \cup A'' \subseteq A, \ \mathcal{L}^n(A \setminus (A' \cup A'')) = 0.$$

Then there exists $f \colon \mathbf{R}^n \times \mathbf{R} \times \mathbf{R}^n \to [0, +\infty[\ \mathcal{L}_n(\mathbf{R}^n)$-measurable with respect to the first group of variables, and satisfying

$$|z| \le f(x, y, z) \le s(1 + |y| + |z|) \text{ for every } (x, y, z) \in \mathbf{R}^n \times \mathbf{R} \times \mathbf{R}^n,$$

$$|f(x, y', z') - f(x, y'', z'')| \le s(|y' - y''| + |z' - z''|)$$

$$\text{for every } (x, y', z'), (x, y'', z'') \in \mathbf{R}^n \times \mathbf{R} \times \mathbf{R}^n$$

such that

$$F(A, u) = \int_A f(x, u, \nabla u) dx \text{ for every } (A, u) \in \mathcal{P}_n \times C^1(\mathbf{R}^n).$$

The above result has been the starting point of a wide literature on integral representation problems. We refer to [Bu2] and [DM2] for more complete references on the subject, and also for a treatment in more general situations.

For the sake of clearness, we report now an integral representation result due to G. Buttazzo and G. Dal Maso.

Let $\Omega \in \mathcal{A}(\mathbf{R}^n)$, and $f \colon \Omega \times \mathbf{R}^n \to \mathbf{R}$. We recall that f is said to be a *Carathéodory integrand* if $f(\cdot, z)$ is measurable for every $z \in \mathbf{R}^n$, and $f(x, \cdot)$ is continuous for a.e. $x \in \Omega$.

It is well known that, if f is a Carathéodory integrand, and $m \colon \Omega \to \mathbf{R}^n$ is measurable, then the composition $x \in \Omega \mapsto f(x, m(x))$ too is measurable.

Theorem 6.2.2. *Let $\Omega \in \mathcal{A}_0$, $p \in [1, +\infty]$, and $F \colon \mathcal{A}(\Omega) \times W^{1,p}(\Omega) \to \mathbf{R}$. Assume that*
i) $F(A, u) = F(A, v)$ whenever $A \in \mathcal{A}(\Omega)$, $u, v \in W^{1,p}(\Omega)$ satisfy $u = v$ a.e. in A,
ii) for every $u \in W^{1,p}(\Omega)$, $F(\cdot, u)$ is the restriction to $\mathcal{A}(\Omega)$ of a real Borel measure,
iii) if $p \in [1, +\infty[$ there exist $a \in L^1(\Omega)$, and $b \ge 0$ such that

$$|F(A, u)| \le \int_A (a(x) + b|\nabla u|^p) dx \text{ for every } A \in \mathcal{A}(\Omega), \ u \in W^{1,p}(\Omega),$$

iv) if $p = +\infty$ for every $r \ge 0$ there exists $a_r \in L^1(\Omega)$ such that

$$|F(A, u)| \le \int_A a_r(x) dx$$

for every $A \in \mathcal{A}(\Omega)$, $u \in W^{1,\infty}(\Omega)$ with $|\nabla u| \leq r$ a.e. in A,

v) $F(A, u + c) = F(A, u)$ for every $A \in \mathcal{A}(\Omega)$, $u \in W^{1,p}(\Omega)$, $c \in \mathbf{R}$,

vi) for every $A \in \mathcal{A}(\Omega)$, $F(A, \cdot)$ is sequentially weak-$W^{1,p}(\Omega)$-lower semicontinuous if $p \in [1, +\infty[$, weak*-$W^{1,\infty}(\Omega)$-lower semicontinuous if $p = +\infty$.
Then there exists a Carathéodory integrand $f: \Omega \times \mathbf{R}^n \to \mathbf{R}$ such that
i) if $p \in [1, +\infty[$,

$$|f(x, z)| \leq a(x) + b|z|^p \text{ for a.e. } x \in \Omega, \text{ and every } z \in \mathbf{R}^n,$$

ii) if $p = +\infty$, for every $r \geq 0$

$$|f(x, z)| \leq a_r(x) \text{ for a.e. } x \in \Omega, \text{ and every } z \in \mathbf{R}^n \text{ with } |z| \leq r,$$

iii) for a.e. $x \in \Omega$, $f(x, \cdot)$ is convex,
iv) the following integral representation formula holds

$$F(A, u) = \int_A f(x, \nabla u) dx \text{ for every } (A, u) \in \mathcal{A}(\Omega) \times W^{1,p}(\Omega).$$

The following integral representation result holds under a translation invariance property (cf. [DM2, Theorem 23.4]).

Theorem 6.2.3. Let $p \in [1, +\infty[$, and $F: \mathcal{A}_0 \times L^p_{\text{loc}}(\mathbf{R}^n) \to [0, +\infty]$. Assume that F is increasing, convex, $L^p_{\text{loc}}(\mathbf{R}^n)$-lower semicontinuous, and such that
i) $F(A - x_0, T[x_0]u) = F(A, u)$ for every $A \in \mathcal{A}_0$, $u \in L^p_{\text{loc}}(\mathbf{R}^n)$, $x_0 \in \mathbf{R}^n$,
ii) $F(A, u) = F(A, v)$ whenever $A \in \mathcal{A}_0$, $u, v \in L^p_{\text{loc}}(\mathbf{R}^n)$ satisfy $u = v$ a.e. in A,
iii) for every $u \in L^p_{\text{loc}}(\mathbf{R}^n)$, $F(\cdot, u)$ is the restriction to $\mathcal{A}_0$ of a Borel positive measure,
iv) $F(A, u + c) = F(A, u)$ for every $A \in \mathcal{A}_0$, $u \in L^p_{\text{loc}}(\mathbf{R}^n)$, $c \in \mathbf{R}$,
v) there exist $a, b \in \mathbf{R}$ such that

$$F(A, u) \leq \int_A (a + b|\nabla u|^p) dx \text{ for every } (A, u) \in \mathcal{A}_0 \times (W^{1,1}_{\text{loc}}(\mathbf{R}^n) \cap L^p_{\text{loc}}(\mathbf{R}^n)).$$

Then there exists $f: \mathbf{R}^n \to [0, +\infty[$ convex, such that

$$f(z) \leq a + b|z|^p \text{ for every } z \in \mathbf{R}^n,$$

and

$$F(A, u) = \int_A f(\nabla u) dx \text{ for every } A \in \mathcal{A}_0, \ u \in L^p_{\text{loc}}(\mathbf{R}^n) \cap W^{1,1}_{\text{loc}}(A).$$

§6.3 Classical Relaxation Results

The relevance of the relaxed functional of a given function F is linked to the qualitative property, described in Chapter 3, ensuring that the infimum of F is equal to the minimum of its relaxed functional.

In the calculus of variations one is often led to consider minimization problems for a functional defined on a "regular class" of functions, where generally no minimum points exist. So, a relevant strategy of attach consists in the extension of the functional to the whole L^1, by defining it equal to $+\infty$ out of the original definition set in order to preserve infima, and then in the analysis of its relaxed functionals in the L^1 topology, hoping in some compactness property to obtain "relaxed" minimum points.

We now describe two relevant and classical examples, where this approach works, and which inspired it in its full generality. The former is concerned with Dirichlet integral, the latter with the area functional.

Theorem 6.3.1. *For every* $\Omega \in \mathcal{A}_0$ *let*

$$D(\Omega, \cdot): u \in L^1(\Omega) \mapsto \begin{cases} \int_\Omega |\nabla u|^2 dx & \text{if } u \in C^1(\Omega) \\ +\infty & \text{if } u \in L^1(\Omega) \setminus C^1(\Omega). \end{cases}$$

Then, for every $\Omega \in \mathcal{A}_0$, $u \in L^1(\Omega)$ *it results that*

$$\mathrm{sc}^-(L^1(\Omega))D(\Omega, u) = \begin{cases} \int_\Omega |\nabla u|^2 dx & \text{if } u \in W^{1,2}(\Omega) \\ +\infty & \text{if } u \in L^1(\Omega) \setminus W^{1,2}(\Omega). \end{cases}$$

Theorem 6.3.2. *For every* $\Omega \in \mathcal{A}_0$ *let*

$$A(\Omega, \cdot): u \in L^1(\Omega) \mapsto \begin{cases} \int_\Omega \sqrt{1 + |\nabla u|^2} dx & \text{if } u \in C^1(\Omega) \\ +\infty & \text{if } u \in L^1(\Omega) \setminus C^1(\Omega). \end{cases}$$

Then, for every $\Omega \in \mathcal{A}_0$, $u \in L^1(\Omega)$ *it results that*

$$\mathrm{sc}^-(L^1(\Omega))A(\Omega, u) =$$

$$= \begin{cases} \int_\Omega \sqrt{1 + |\nabla u|^2} dx + |D^s u|(\Omega) & \text{if } u \in BV(\Omega) \\ +\infty & \text{if } u \in L^1(\Omega) \setminus BV(\Omega). \end{cases}$$

More generally, the result below holds (cf. [S2], [GS], and also [CEDA2, Proposition 1.7], [CEDA5, Theorem 2.4]).

Theorem 6.3.3. *Let $f : \mathbf{R}^n \to [0, +\infty[$ be convex. For every $\Omega \in \mathcal{A}_0$ let*

$$F(\Omega, \cdot) : u \in L^1(\Omega) \mapsto \begin{cases} \int_\Omega f(\nabla u) dx & \text{if } u \in C^1(\Omega) \\ +\infty & \text{if } u \in L^1(\Omega) \setminus C^1(\Omega). \end{cases}$$

Then, for every $\Omega \in \mathcal{A}_0$, $u \in BV(\Omega)$ it results that

$$\text{sc}^-(L^1(\Omega)) F(\Omega, u) = \int_\Omega f(\nabla u) dx + \int_\Omega f^\infty(\nabla^s u) d|D^s u|.$$

In all the above results, the relaxation problem was settled for convex integral functionals defined on sets of smooth functions. On the contrary, in the refined and well established result below, no convexity condition is assumed.

For every $\Omega \in \mathcal{A}(\mathbf{R}^n)$ let $\overline{\mathcal{F}}$ be defined by

$$\overline{\mathcal{F}} = \{ G : W^{1,p}(\Omega) \to [-\infty, +\infty] :$$

G is sequentially weak-$W^{1,p}(\Omega)$-lower semicontinuous$\}$,

and, for every $F : \mathcal{A}(\Omega) \times W^{1,p}(\Omega) \to [-\infty, +\infty]$, let $\overline{F}$ be given by

$$\overline{F} : (A, u) \in \mathcal{A}(\Omega) \times W^{1,p}(\Omega) \to$$

$$\sup\{ G(u) : G \in \overline{\mathcal{F}}, \ G(v) \leq F(A, v) \text{ for every } v \in W^{1,p}(\Omega) \}.$$

Theorem 6.3.4. *Let $\Omega \in \mathcal{A}_0$, $f : \Omega \times \mathbf{R}^n \to [0, +\infty]$ be Borel, and satisfy*
i) if $p \in [1, +\infty[$ there exist $a \in L^1(\Omega)$ and $b \geq 0$ such that

$$f(x, z) \leq a(x) + b|z|^p \text{ for a.e. } x \in \Omega, \text{ and every } z \in \mathbf{R}^n,$$

ii) if $p = +\infty$ for every $r \geq 0$ there exists $a_r \in L^1(\Omega)$ such that

$$f(x, z) \leq a_r(x) \text{ for a.e. } x \in \Omega, \text{ and every } z \in \mathbf{R}^n \text{ with } |z| \leq r.$$

Let $F : (A, u) \in \mathcal{A}(\Omega) \times W^{1,p}(\Omega) \mapsto \int_A f(x, \nabla u) dx$. Then there exists a Carathéodory integrand $\overline{f} : \Omega \times \mathbf{R}^n \to [0, +\infty[$ such that
i) if $p \in [1, +\infty[$,

$$\overline{f}(x, z) \leq a(x) + b|z|^p \text{ for a.e. } x \in \Omega, \text{ and every } z \in \mathbf{R}^n,$$

ii) if $p = +\infty$, for every $r \geq 0$

$$\overline{f}(x, z) \leq a_r(x) \text{ for a.e. } x \in \Omega, \text{ and every } z \in \mathbf{R}^n \text{ with } |z| \leq r,$$

iii) for a.e. $x \in \Omega$, $\overline{f}(x, \cdot)$ is convex,
iv) the following integral representation formula holds

$$\overline{F}(A, u) = \int_A \overline{f}(x, \nabla u) dx \text{ for every } (A, u) \in \mathcal{A}(\Omega) \times W^{1,p}(\Omega).$$

Moreover, if for a.e. $x \in \Omega$ $f(x, \cdot)$ is upper semicontinuous, then $\overline{f}(x, \cdot) = f^{**}(x, \cdot)$ for a.e. $x \in \Omega$.

Relaxation problems in BV spaces for integral functionals with integrands depending also on the space variable have been treated in [GMS1], and [DM1], also for Dirichlet type variational problems. For example, the following result has been proved in [GMS1].

Theorem 6.3.5. Let $\Omega \in \mathcal{A}_0$ be smooth, $u_0 \in W^{1,1}(\Omega)$, $f: \Omega \times \mathbf{R}^n \to [0, +\infty[$ be continuous, with $f(x, \cdot)$ convex for every $x \in \Omega$, and satisfying for some $M \geq 0$

$$|z| \leq f(x, z) \leq M(1 + b|z|) \text{ for every } x \in \Omega, \text{ and } z \in \mathbf{R}^n.$$

Let

$$F: u \in L^1(\Omega) \mapsto \begin{cases} \int_\Omega f(x, \nabla u) dx & \text{if } u \in u_0 + W_0^{1,1}(\Omega) \\ +\infty & \text{if } u \notin u_0 + W_0^{1,1}(\Omega). \end{cases}$$

Then, for every $u \in L^1(\Omega)$ it results that

$$\text{sc}^-(L^1(\Omega))F(u) = \begin{cases} \int_\Omega f(x, \nabla u) dx + \int_\Omega f^\infty(x, \nabla^s u) d|D^s u| + \\ \quad + \int_{\partial\Omega} f^\infty((u_0 - \gamma_\Omega u)\mathbf{n}_\Omega) d\mathcal{H}^{n-1} & \text{if } u \in BV(\Omega) \\ +\infty & \text{if } u \notin BV(\Omega). \end{cases}$$

§6.4 Classical Homogenization Results

Homogenization theory origins from the double exigency of describing a nonhomogeneous, finely grained material with two or more components mixed in a periodic manner by a homogeneous one, and, vice-versa, of simulating a homogeneous material by a composite one, possibly enjoying a microstructure emphasizing some special features.

In our framework, the simulation is to be intended in the sense that the energy of the homogeneous material is approximated by those of the nonhomogeneous ones for every exterior force.

One of the first significant results mathematically well established is the following one due to E. De Giorgi and S. Spagnolo (cf. [DGS]), and inspired also from conversations with E. Sànchez-Palencia.

Theorem 6.4.1. Let $\{a_{ij}\}$ be a $n \times n$ symmetric matrix of measurable Y-periodic functions on $\mathbf{R}^n$ satisfying for some $0 < \lambda \leq \Lambda < +\infty$

$$\lambda|z|^2 \leq \sum_{i,j=1}^n a_{ij}(x)z_i z_j \leq \Lambda|z|^2 \text{ for a.e. } x \in \mathbf{R}^n, \text{ and every } z \in \mathbf{R}^n.$$

Then, for every $\Omega \in \mathcal{A}_0$, and every $g \in L^2(\Omega)$ the family $\{u_\varepsilon(g)\}_{\varepsilon > 0}$ of the unique solutions of the problems

$$m_\varepsilon(g) = \min\left\{ \int_\Omega \sum_{i,j=1}^n a_{ij}\left(\frac{x}{\varepsilon}\right) \nabla_i u \nabla_j u dx + \int_\Omega g u dx : u \in W_0^{1,2}(\Omega) \right\}$$

converges in $L^2(\Omega)$ as $\varepsilon \to 0^+$ to the unique solution $u(g)$ of the problem

$$m_{\text{hom}}(g) = \min\left\{\int_\Omega \sum_{i,j=1}^n a_{ij}^{\text{hom}} \nabla_i u \nabla_j u\, dx + \int_\Omega gu\, dx : u \in W_0^{1,2}(\Omega)\right\},$$

and $m_\varepsilon(g) \to m_{\text{hom}}(g)$, where

$$\sum_{i,j=1}^n a_{ij}^{\text{hom}} z_i z_j = \min\left\{\int_\Omega \sum_{i,j=1}^n a_{ij}(y) \nabla_i v \nabla_j v\, dx : u \in u_z + W_{\text{per}}^{1,2}(Y)\right\}$$

for every $z \in \mathbf{R}^n$.

A more general result in this setting is the following (cf. [CEDA1]).

Theorem 6.4.2. *Let f satisfy*

$$\begin{cases} f \colon (x,z) \in \mathbf{R}^n \times \mathbf{R}^n \mapsto f(x,z) \in [0,+\infty[\\ f(\cdot, z)\ Y\text{-periodic and in } L^1(Y) \text{ for every } z \in \mathbf{R}^n \\ f(x, \cdot)\ \text{convex for a.e. } x \in \mathbf{R}^n, \end{cases}$$

and

$$|z| \leq f(x,z) \text{ for a.e. } x \in \mathbf{R}^n \text{ and every } z \in \mathbf{R}^n.$$

Then, for every $q \in [1,+\infty]$, $\Omega \in \mathcal{A}_0$ with Lipschitz boundary, $\beta \in L^\infty(\Omega)$, $\lambda > 0$, and $r \in]1,1^*[$ the values

$$i_\varepsilon = \inf\left\{\int_\Omega f\left(\frac{x}{\varepsilon}, \nabla u\right) dx + \int_\Omega \beta u\, dx + \lambda \int_\Omega |u|^r dx : u \in W_0^{1,q}(\Omega)\right\}$$

converge as $\varepsilon \to 0^+$ to

$$m_{\text{hom}} = \min\left\{\int_\Omega f_{\text{hom}}^q(\nabla u) dx + \int_\Omega (f_{\text{hom}}^q)^\infty(\nabla^s u) d|D^s u| + \right.$$

$$\left. + \int_{\partial\Omega} (f_{\text{hom}}^q)^\infty(-\gamma_\Omega u \mathbf{n}_\Omega) d\mathcal{H}^{n-1} + \int_\Omega \beta u\, dx + \lambda \int_\Omega |u|^r dx : u \in BV(\Omega)\right\},$$

where

$$f_{\text{hom}}^q(z) = \inf\left\{\int_Y f(y, z + \nabla v) dy : v \in W_{\text{per}}^{1,q}(Y)\right\} \text{ for every } z \in \mathbf{R}^n.$$

Moreover, if for every $\varepsilon > 0$ $u_\varepsilon \in W_0^{1,q}(\Omega)$ is such that

$$\lim_{\varepsilon \to 0^+}\left\{\int_\Omega f\left(\frac{x}{\varepsilon}, \nabla u_\varepsilon\right) dx + \int_\Omega \beta u_\varepsilon dx + \lambda \int_\Omega |u_\varepsilon|^r dx - i_\varepsilon\right\} = 0,$$

then $\{u_\varepsilon\}_{\varepsilon>0}$ is compact in $L^1(\Omega)$, and its converging subsequences converge to solutions of m_{hom}.

We point out that in the above result the dependence on q can be a true one, as proved in [CEDA1] where an example is proposed in which f_{hom}^q is not constant with respect to q.

At present, literature on homogenization is very large, and offers different approaches to various types of problems. We refer e.g. to [DM2] for a wide bibliography, at least until the first years of the nineties, and to [CD].

We point out that homogenization problems have been the starting point of the development of several analytical methods in Applied Mathematics. The Γ-convergence of E. De Giorgi, the heuristic multiscale method introduced by N.S. Bakhvalov and deeply used and largely diffused by J.L. Lions, and the energy method of L. Tartar, with the contribution of F. Murat, had been developed just to study this kind of problems, at least in the scalar case.

Finally, we remark that, sometimes in the following, when we are looking at properties of mixing materials, we use the term *at mesoscopic level*, and when we speak of properties of the homogenized material, we use the term *at macroscopic level*.

§6.5 Mathematical Aspects of Some Physical Models Originating Unbounded Functionals

Some physical models lead to minimization problems for integral energies of the type $\int_\Omega f(x, \nabla u)dx$ defined on sets of "regular configurations" on the open set Ω, and with densities f possibly taking the value $+\infty$, and satisfying conditions like

$$\begin{cases} f\colon (x, z) \in \Omega \times \mathbf{R}^n \mapsto f(x, z) \in [0, +\infty] \\ f\ (\mathcal{L}_n(\Omega) \times \mathcal{B}(\mathbf{R}^n))\text{-measurable} \\ f(x, \cdot) \text{ convex for a.e. } x \in \Omega. \end{cases}$$

We now recall briefly some examples where the energy densities effectively assume the value $+\infty$.

The first one is concerned with elastic-plastic torsion problems (cf. [DLi], [GL]), where densities f of the following kind have been proposed

$$f(x, z) = |z|^2 + I_{\overline{B_{\varphi(x)}(0)}}(z) \text{ for a.e. } x \in \mathbf{R}^n, \text{ and every } z \in \mathbf{R}^n,$$

with $\varphi\colon \mathbf{R}^n \to {]0, +\infty[}$ measurable, and bounded.

In the electrostatic screening problem (cf. e.g. [RT]), densities f of the type

$$f(x, z) = |z|^2 + I_{\{\zeta \in \mathbf{R}^n : |\zeta| \le \varphi(x)\}}(z) \text{ for a.e. } x \in \mathbf{R}^n, \text{ and every } z \in \mathbf{R}^n,$$

where φ is measurable on $\mathbf{R}^n$ and takes only the values 0 and $+\infty$, have been considered. This case corresponds to the one of a composite material in which perfect conductors are included, where the potential to be determined is subject to be constant.

Finally, in the modelling of rubber-like nonlinear elastomers, the following densities f have been introduced by Treloar (cf. [Tr]) when $n = 1$

$$f(x, z) = \frac{1}{2}G(x)\left(z^2 + \frac{2}{z} - 3\right) \text{ for a.e. } x \in \mathbf{R}, \text{ and every } z \in \mathbf{R},$$

$$f(x, z) = \frac{1}{2}G(x)\left(z - \frac{1}{z}\right)^2 \text{ for a.e. } x \in \mathbf{R}, \text{ and every } z \in \mathbf{R},$$

$$f(x, z) = C_1(x)\left(z^2 + \frac{2}{z} - 3\right) + C_2(x)\left(\frac{1}{z^2} + 2z - 3\right)$$

$$\text{for a.e. } x \in \mathbf{R}, \text{ and every } z \in \mathbf{R},$$

G, C_1, and C_2 being measurable, and bounded from above and below by positive constants.

We point out that in this last case the densities explode near some values, and that also a loss of symmetry occurs.

It is straightaway verified that in all the above examples the densities f are $(\mathcal{L}_n(\Omega) \times \mathcal{B}(\mathbf{R}^n))$-measurable, and convex in the z variable for a.e. x.

Chapter 7

Abstract Regularization and Jensen's Inequality

In the present chapter we exploit the properties of convex functions and of measure spaces to prove a general approximation in energy result of the elements of a subspace of $L^1_{\text{loc}}(\mathbf{R}^n)$ with functions in $C^\infty(\mathbf{R}^n)$, by assuming just convexity hypotheses on the energy functional.

The main tool is the notion of integral of a function with values in a locally convex topological vector space that enables us to prove a general version of Jensen's inequality.

Finally, the approximation result is applied to deduce a lower semicontinuity result, for functionals defined in BV spaces, with respect to a very weak notion of convergence: the one in the sense of distributions.

§7.1 Integral of Functions with Values in Locally Convex Topological Vector Spaces

The approximation result expressed in Theorem 4.1.6 can be extended to much more general situations. To do this, we make use of the notion of integral of functions with values in topological vector spaces given by R.S. Phillips in 1940 (cf. [Ph]).

Definition 7.1.1. *Let $(\Omega, \mathcal{E})$ be a measure space, μ a finite positive measure on $\mathcal{E}$, U a Hausdorff locally convex topological vector space, and $f: \Omega \to U$. We say that f is U-integrable on Ω if for every $S \in \mathcal{E}$, $u(S) \in U$ can be found such that for every $I \in \mathcal{N}(u(S))$ there exist a subdivision $\{B_{S,I,j}\}_{j\in\mathbf{N}} \subseteq \mathcal{E}$ of S into pairwise disjoint sets whose union is S, and $N_{S,I} \subseteq \mathbf{N}$ finite such that, whenever $N \subseteq \mathbf{N}$ is finite and contains $N_{S,I}$, it results*

$$\sum_{j\in N} \mu(B_{S,I,j}) f(x_j) \in I \text{ whenever } x_j \in B_{S,I,j} \text{ for every } j \in N.$$

The vector $u(S)$ is the value of the integral of f on S, and is denoted by $(U)\int_S f d\mu$.

The above defined integral satisfies the main structure properties of the integral of real valued functions. In fact in [Ph] it is proved to be linear, countably additive, and satisfying a suitable absolute continuity property.

Remark 7.1.2. It is clear that, if V is another Hausdorff locally convex topological vector space containing U and having a topology less fine than the one of U, and if $f:\Omega \to U$ is U-integrable on Ω, then f turns out to be also V-integrable on Ω, and

$$(V)\int_S f d\mu = (U)\int_S f d\mu \text{ for every } S \in \mathcal{E}.$$

The results below provides an integrability condition.

Theorem 7.1.3. *Let $\Omega \in \mathcal{A}(\mathbf{R}^n)$, μ be a finite positive measure on $\mathcal{L}_n(\Omega)$, U a sequentially complete Hausdorff locally convex topological vector space, and let $f:\Omega \to U$ be continuous and with compact support. Then f is U-integrable on Ω.*

Proof. Let $\{p_\theta\}_{\theta \in \mathcal{T}}$ be a family of seminorm defining the topology of U.

Let us first observe that, since f is continuous with compact support, f is uniformly continuous in the sense that for every $\theta \in \mathcal{T}$, $\eta > 0$ there exists $\delta_{\theta,\eta} > 0$ such that $p_\theta(f(u) - f(v)) < \eta$ whenever $x, y \in \Omega$ satisfy $|x - y| < \delta_{\theta,\eta}$.

Let $S \in \mathcal{L}_n(\Omega)$. For every $h \in \mathbf{N}$ let $\mathcal{R}_h = \{Q_j^h\}_{j\in\mathbf{N}}$ be a partition of $\mathbf{R}^n$ made up by half open cubes with sidelength $1/h$, and set, for every $j \in \mathbf{N}$, $S_j^h = S \cap Q_j^h$. Then, since $\text{spt}(f)$ is compact, it is not restrictive to assume the existence of $m_h \in \mathbf{N}$, and of a compact set K not depending on h such that $S_j^h \cap \text{spt}(f) \neq \emptyset$ if and only if $j \in \{1, \ldots, m_h\}$, and $\cup_{j=1}^{m_h} S_j^h \subseteq K$.

For every $j \in \{1, \ldots, m_h\}$ we choose $x_j^h \in S_j^h$, and define $u_h = \sum_{j=1}^{m_h} f(x_j^h)\mu(S_j^h)$. Let us prove that $\{u_h\}$ is a Cauchy sequence in U.

To do this, let $\theta \in \mathcal{T}$, $\eta > 0$, and let $\delta_{\theta,\eta}$ be given by the uniform continuity of f. Let $\nu \in \mathbf{N}$ be such that $\frac{1}{\nu} < \frac{\delta_{\theta,\eta}}{2\sqrt{n}}$. Then, for every h, $k > \nu$, it results that

$$(7.1.1) \qquad p_\theta(u_h - u_k) \le p_\theta\left(\sum_{j=1}^{m_h} f(x_j^h)\mu(S_j^h) - \sum_{j=1}^{m_k} f(x_j^k)\mu(S_j^k) \right) =$$

$$= p_\theta\left(\sum_{i=1}^{m_h} f(x_i^h)\sum_{j=1}^{m_k}\mu(S_i^h \cap S_j^k) - \sum_{j=1}^{m_k} f(x_j^k)\sum_{i=1}^{m_h}\mu(S_j^k \cap S_i^h) \right) \le$$

$$\leq \sum_{i=1}^{m_h} \sum_{j=1}^{m_k} \mu(S_i^h \cap S_j^k) p_\theta \left(f(x_i^h) - f(x_j^k) \right).$$

We now observe that if $S_i^h \cap S_j^k \neq \emptyset$, then $|x_i^h - x_j^k| < \delta_{\theta,\eta}$, consequently, by (7.1.1), and the uniform continuity of f, we deduce that

$$p_\theta(u_h - u_k) \leq \sum_{i=1}^{m_h} \sum_{j=1}^{m_k} \mu(S_i^h \cap S_j^k) \eta \leq \mu(K)) \eta \text{ for every } h, \ k > \nu,$$

from which we conclude that $\{u_h\}$ is a Cauchy sequence. Therefore, by the sequential completeness of U, we deduce the existence of $u(S) \in U$ such that $u_h \to u(S)$.

We now need to remark that a priori $u(S)$ depends on the particular choice of the vectors $\{x_j^h\}$. Nevertheless, by using again the uniform continuity of F, it turns out that it does not.

To see this, let, for every $h \in \mathbf{N}$, $u_h^1 = \sum_{j=1}^{m_h} f(x_j^{1,h}) \mu(S_j^h)$, $u_h^2 = \sum_{j=1}^{m_h} f(x_j^{2,h}) \mu(S_j^h)$ be two sequences constructed as above, and relative to two different choices of the vectors $\{x_j^h\}$, and let $u^1(S)$, $u^2(S)$ be their limits. Let $\theta \in \mathcal{T}$, and $\eta > 0$. Then

$$p_\theta \left(u^1(S) - u^2(S) \right) \leq p_\theta \left(u^1(S) - u_h^1 \right) + p_\theta \left(u_h^1 - u_h^2 \right) + p_\theta \left(u_h^2 - u^2(S) \right) \leq$$

$$\leq p_\theta \left(u^1(S) - u_h^1 \right) + \sum_{j=1}^{m_h} \mu(S_j^h) p_\theta \left(f(x_j^{1,h}) - f(x_j^{2,h}) \right) + p_\theta \left(u_h^2 - u^2(S) \right) \leq$$

$$\leq p_\theta \left(u^1(S) - u_h^1 \right) + \mu(K) \eta + p_\theta \left(u_h^2 - u^2(S) \right)$$

for every $h \in \mathbf{N}$ sufficiently large,

from which we conclude that $p_\theta(u^1(S) - u^2(S)) = 0$ for every $\theta \in \mathcal{T}$, and, being U Hausdorff, that $u^1(S) = u^2(S)$.

Because of this, and again the uniform continuity of f, and by using an argument similar to the above one, it is now easy to prove that for every $\theta \in \mathcal{T}$, and $\eta > 0$ it results that

$$\sup \left\{ p_\theta \left(\sum_{j=1}^{m_h} f(x_j^h) \mu(S_j^h) - u(S) \right) : x_j^h \in S_j^h \text{ for every } j \in \{1, \ldots, m_h\} \right\} <$$

$$< \eta \text{ for every } h \in \mathbf{N} \text{ sufficiently large,}$$

that is $u(S) = (U) \int_S f d\mu$. ∎

The above notion of integral behaves nicely with respect to composition with convex functions. In fact, the following Jensen type inequality holds in the framework of locally convex topological vector spaces.

Theorem 7.1.4. Let $(\Omega, \mathcal{E})$ be a measure space, μ a finite positive measure on $\mathcal{E}$ satisfying $\mu(\Omega) = 1$, U a Hausdorff locally convex topological vector space, and let $\Phi \colon U \to [0, +\infty]$ be convex and lower semicontinuous. Then, for every $w \colon \Omega \to U$ U-integrable on Ω it results that

$$(7.1.2) \qquad \Phi\left((U)\int_\Omega w \, d\mu \right) \leq \int_\Omega (\Phi \circ w) \, d\mu.$$

Proof. Let $w \colon \Omega \to U$ be U-integrable on Ω.

We first prove the theorem by assuming in addition that $\Phi(0) < +\infty$.

Let $t < \Phi((U)\int_\Omega w \, d\mu)$. Then, by the lower semicontinuity of Φ, we deduce the existence of $I_t \in \mathcal{N}((U)\int_\Omega w \, d\mu)$ such that

$$t < \Phi(v) \text{ for every } v \in I_t.$$

Consequently, there exist a subdivision $\{B_{\Omega, I_t, j}\}_{j \in \mathbf{N}} \subseteq \mathcal{E}$ of Ω into pairwise disjoint sets whose union is Ω, and $N_{\Omega, I_t} \subseteq \mathbf{N}$ finite such that, whenever $N \subseteq \mathbf{N}$ is finite and contains N_{Ω, I_t}, it results

$$(7.1.3) \qquad t < \Phi\left(\sum_{j \in N} \mu(B_{\Omega, I_t, j}) f(x_j) \right)$$

whenever $x_j \in B_{\Omega, I_t, j}$ for every $j \in N$.

We now take $\varepsilon_t > 0$, and $N_t \subseteq \mathbf{N}$ finite and containing N_{Ω, I_t} such that ε_t vanishes as t approaches $\Phi((U)\int_\Omega w \, d\mu)$, and $\mu(\Omega \setminus \cup_{j \in N_t} B_{\Omega, I_t, j}) \Phi(0) < \varepsilon_t$.

Let us set $A_t = \Omega \setminus \cup_{j \in N_t} B_{\Omega, I_t, j}$. Then $\mu(A_t) + \sum_{j \in N_t} \mu(B_{\Omega, I_t, j}) = 1$, and by (7.1.3) and the convexity of Φ, we obtain that

$$t < \Phi\left(\mu(A_t) 0 + \sum_{j \in N_t} \mu(B_{\Omega, I_t, j}) f(x_j) \right) \leq$$

$$\leq \mu(A_t) \Phi(0) + \sum_{j \in N_t} \mu(B_{\Omega, I_t, j}) \Phi(f(x_j))$$

whenever $x_j \in B_{\Omega, I_t, j}$ for every $j \in N_t$,

and hence that

$$(7.1.4) \qquad t < \varepsilon_t + \sum_{j \in N_t} \mu(B_{\Omega, I_t, j}) \inf_{B_{\Omega, I_t, j}} \Phi \circ f.$$

Now it is clear that the function $\sum_{j \in N_t} \chi_{B_{\Omega, I_t, j}} \inf_{B_{\Omega, I_t, j}} (\Phi \circ f)$ is simple $\mathcal{E}$-measurable, and that $\sum_{j \in N_t} \chi_{B_{\Omega, I_t, j}}(x) \inf_{B_{\Omega, I_t, j}} (\Phi \circ f) \leq (\Phi \circ f)(x)$ for every $x \in \Omega$. Consequently, by (7.1.4), we deduce that

$$t < \varepsilon_t + \int_\Omega (\Phi \circ f) \, d\mu \text{ for every } t < \Phi\left((U)\int_\Omega w \, d\mu \right),$$

that provides (7.1.2) as t increases to $\Phi((U)\int_\Omega w d\mu)$, under the additional assumption $\Phi(0) < +\infty$.

Finally, in the general case, by Theorem 1.1.11 it follows that for every $t < \Phi((U)\int_\Omega w d\mu)$ there exist $L \in U'$ and $c \in \mathbf{R}$ such that

$$t < L\left((U)\int_\Omega w d\mu\right) + c, \quad L(v) + c \le \Phi(v) \text{ for every } v \in U,$$

from which, since Φ is nonnegative, we also obtain that

$$t < (L+c)^+\left((U)\int_\Omega w d\mu\right), \quad (L+c)^+(v) \le \Phi(v) \text{ for every } v \in U.$$

Now it is clear that $(L+c)^+$ is convex and lower semicontinuous, and that $(L+c)^+(0) < +\infty$. Consequently, by the previously treated case, we infer that

$$t < (L+c)^+\left((U)\int_\Omega w d\mu\right) \le \int_\Omega \left((L+c)^+ \circ f\right) d\mu \le \int_\Omega (\Phi \circ f) d\mu,$$

that again provides (7.1.2) as t increases to $\Phi((U)\int_\Omega w d\mu)$, and completes the proof. ∎

Remark 7.1.5. We point out that Jensen's inequality actually provides a characterization of convex lower semicontinuous functions, provided μ is surjective.

To see this, let $(\Omega, \mathcal{E})$, μ, U be as in Theorem 7.1.4, assume that $\mu(\mathcal{E}) = [0, 1]$, and let $\Phi: U \to [0, +\infty]$ be lower semicontinuous and satisfying (7.1.2) whenever $w: \Omega \to U$ is U-integrable on Ω. Then, if w_1, $w_2 \in U$, $t \in [0, 1]$, $E \in \mathcal{E}$ is such that $\mu(E) = t$, and $w: x \in \Omega \mapsto \chi_E(x)w_1 + \chi_{\Omega \setminus E}(x)w_2$, it is easy to verify that w is U-integrable on Ω, that $(U)\int_\Omega w d\mu = tw_1 + (1-t)w_2$, that $\Phi \circ w$ is $\mathcal{E}$-measurable, and that, by (7.1.2),

$$\Phi(tw_1 + (1-t)w_2) = \Phi\left((U)\int_\Omega w d\mu\right) \le \int_\Omega (\Phi \circ w) d\mu = t\Phi(w_1) + (1-t)\Phi(w_2),$$

that is the convexity of Φ.

For what concerns the surjectivity properties of μ, we recall that a μ as above turns out to be surjective if for every $A \in \mathcal{E}$ with $\mu(A) > 0$ there exists $B \in \mathcal{E}$ such that $0 < \mu(B) < \mu(A)$.

§7.2 On the Definition of a Functional on Functions and on Their Equivalence Classes

Throughout the book, and starting in particular from the present chapter, we consider functions and equivalence classes of functions, with respect to identity a.e., that we need to compare. In addition, we also consider different types of functionals defined on such equivalence classes, that occasionally are computed on their elements. To do this properly, it is necessary to make explicitly some simple considerations.

First of all we recall that $L^1_{loc}(\mathbf{R}^n)$ is a space of equivalence classes of functions defined on $\mathbf{R}^n$, being two such functions equivalent if they agree everywhere on $\mathbf{R}^n$ except possibly for a set of Lebesgue zero measure, and that, as usual, its elements are thought as functions defined almost everywhere in $\mathbf{R}^n$. Thus, when considering a subspace W of $L^1_{loc}(\mathbf{R}^n)$, we will think to its elements as to equivalence classes of summable functions on $\mathbf{R}^n$, or to functions defined almost everywhere in $\mathbf{R}^n$. In particular this holds when $W = C^\infty(\mathbf{R}^n)$.

On the other side, $C^\infty(\mathbf{R}^n)$, especially if endowed with the $C^\infty(\mathbf{R}^n)$ topology, is naturally a space of functions defined everywhere in $\mathbf{R}^n$, therefore a way to identify its elements with their equivalence classes, and to introduce the corresponding topology on this set, is needed.

To do this, let us denote, for the moment and for the sake of clearness, by $C^\infty_{fct}(\mathbf{R}^n)$ the set of the C^∞-functions on $\mathbf{R}^n$, and by $C^\infty_{cls}(\mathbf{R}^n)$ the one of the equivalence classes of the elements of $C^\infty_{fct}(\mathbf{R}^n)$. Then it is obvious that for every $\mathbf{u} \in C^\infty_{cls}(\mathbf{R}^n)$ there exists a unique $J\mathbf{u} \in C^\infty_{fct}(\mathbf{R}^n)$ such that $J\mathbf{u} \in \mathbf{u}$.

Because of this, the application $J: \mathbf{u} \in C^\infty_{cls}(\mathbf{R}^n) \mapsto J\mathbf{u} \in C^\infty_{fct}(\mathbf{R}^n)$ turns out to be well defined, linear, and one-to-one. Consequently $\{J^{-1}(A) : A$ open set in $C^\infty(\mathbf{R}^n)\}$ turns out to be a topology on $C^\infty_{cls}(\mathbf{R}^n)$ that makes it a complete metrizable topological vector space, and J an isomorphism between topological vector spaces that allows the identification of classes with each of their elements.

In addition, given $F: C^\infty_{fct}(\mathbf{R}^n) \to [0, +\infty]$, we also identify it with the functional $F_{cls} = F \circ J$ defined on $C^\infty_{cls}(\mathbf{R}^n)$, thus preserving its vectorial and topological properties, and keep to denote F_{cls} by F.

So, given $u \in C^\infty_{fct}(\mathbf{R}^n)$, we allow F to act directly on all the functions in $J^{-1}u$, by defining $F(v) = F(u)$ for every $v \in J^{-1}u$. In this sense, we can say that if $u \in C^\infty_{fct}(\mathbf{R}^n)$ and $v \in L^1_{loc}(\mathbf{R}^n)$ is such that $v = u$ a.e. in $\mathbf{R}^n$, then $F(v) = F(u)$.

It is obvious that now $C^\infty_{fct}(\mathbf{R}^n)$ and $C^\infty_{cls}(\mathbf{R}^n)$ can be identified and denoted by $C^\infty(\mathbf{R}^n)$.

This standard identification procedure is fundamental: it allows to translate problems defined on regular classes of functions into "regular" Lebesgue equivalence classes.

This point of view agrees with the one described in Remark 2.3.6, in which the identification of $C^\infty(\Omega)$ with a space of measures, given by $u \in C^\infty(\Omega) \mapsto u\mathcal{L}^n$, is examined.

We also point out that in some situations such identification procedure is impracticable. For example, the classical total variation functional can produce different values when evaluated on two functions, one of which possibly smooth, differing just in one point.

§7.3 Regularization of Functions in Locally Convex Topological Vector Subspaces of $L^1_{\text{loc}}(\mathbf{R}^n)$

In this section we study the properties of the regularizations of functions in a locally convex topological vector subspace U of $L^1_{\text{loc}}(\mathbf{R}^n)$, by proving approximation via regularizations results analogous to those of §4.1. The main idea to do this is to see the regularization of an element of U as the integral of a particular function taking its values in U, and then apply Jensen's inequality.

Lemma 7.3.1. *Let $u \in L^1_{\text{loc}}(\mathbf{R}^n)$, and ρ be a symmetric mollifier. Then, for every $\varepsilon > 0$ the function $y \in \mathbf{R}^n \mapsto \rho(y)T[\varepsilon y]u \in L^1_{\text{loc}}(\mathbf{R}^n)$ is $L^1_{\text{loc}}(\mathbf{R}^n)$-integrable on $\mathbf{R}^n$, and*

$$\left((L^1_{\text{loc}}(\mathbf{R}^n)) \int_{\mathbf{R}^n} \rho(y)T[\varepsilon y]u\,dy \right)(x) = u_\varepsilon(x) \text{ for a.e. } x \in \mathbf{R}^n,$$

u_ε being the regularization of u defined in (4.1.2).

Proof. First of all, let us observe that $L^1_{\text{loc}}(\mathbf{R}^n)$, with its natural topology, is a Hausdorff locally convex sequentially complete topological vector space, and that, by Theorem 2.2.7, $y \in \mathbf{R}^n \mapsto \rho(y)T[\varepsilon y]u \in L^1_{\text{loc}}(\mathbf{R}^n)$ is continuous, and with compact support. Consequently, by Theorem 7.1.3, $\rho T[\varepsilon \cdot]u$ is $L^1_{\text{loc}}(\mathbf{R}^n)$-integrable on $\mathbf{R}^n$.

Let $\varepsilon > 0$, Q be a half open cube of $\mathbf{R}^n$ with sidelength l satisfying $B_1(0) \subseteq Q$, and let us observe that the proof of Theorem 7.1.3 actually provides an approximating sequence of $(L^1_{\text{loc}}(\mathbf{R}^n)) \int_Q \rho(y)T[\varepsilon y]u\,dy$. In fact, if for every $h \in \mathbf{N}$ we take a partition $\mathcal{R}_h = \{Q^h_j\}_{j \in \{1,\ldots,h^n\}}$ of Q made up by half open cubes with faces parallel to the ones of Q, and sidelength l/h, then in the proof of Theorem 7.1.3 it is proved that

(7.3.1)
$$\lim_{h \to +\infty} \sup \left\{ \int_Q \left| \sum_{j=1}^{h^n} \rho(y^h_j)T[\varepsilon y^h_j]u \mathcal{L}^n(Q^h_j) - \right. \right.$$

$$\left. \left. -(L^1_{\text{loc}}(\mathbf{R}^n)) \int_Q \rho(y)T[\varepsilon y]u\,dy \right|\,dx : y^h_j \in Q^h_j \text{ for every } j \in \{1,\ldots,h^n\} \right\} = 0.$$

Let $\eta > 0$. Then, because of (7.3.1), Theorem 2.2.7, and of the compactness of $\text{spt}(\rho)$, there exists $h \in \mathbf{N}$ such that, if $\mathcal{R}_h = \{Q^h_j\}_{j \in \{1,\ldots,h^n\}}$ is a partition as above, and, for every $j \in \{1,\ldots,h^n\}$, $y^h_j \in Q^h_j$, then

(7.3.2)
$$\int_Q \left| \sum_{j=1}^{h^n} \rho(y^h_j)T[\varepsilon y^h_j]u \mathcal{L}^n(Q^h_j) - (L^1_{\text{loc}}(\mathbf{R}^n)) \int_Q \rho(y)T[\varepsilon y]u\,dy \right|\,dx < \eta,$$

and

(7.3.3) $$\int_Q |\rho(y_1)T[\varepsilon y_1]u - \rho(y_2)T[\varepsilon y_2]u|dx < \eta$$

whenever y_1, $y_2 \in \mathbf{R}^n$ satisfy $|y_1 - y_2| < \dfrac{\sqrt{n}}{h}$.

Then, by (7.3.2), (7.3.3), and Fubini's theorem, once we recall that $u_\varepsilon(x) = \int_Q \rho(y)u(x + \varepsilon y)dy$ for every $x \in \mathbf{R}^n$, and that $\mathrm{spt}(\rho) \subseteq Q$, we have that

$$\int_Q \left|(L^1_{\mathrm{loc}}(\mathbf{R}^n))\int_Q \rho(y)T[\varepsilon y]udy - u_\varepsilon\right|dx \le$$

$$\le \eta + \int_Q \left|\sum_{j=1}^{h^n} \rho(y_j^h)T[\varepsilon y_j^h]u\mathcal{L}^n(Q_j^h) - u_\varepsilon\right|dx =$$

$$= \eta + \int_Q \left|\sum_{j=1}^{h^n} \rho(y_j^h)u(x + \varepsilon y_j^h)\mathcal{L}^n(Q_j^h) - \int_Q \rho(y)u(x + \varepsilon y)dy\right|dx \le$$

$$\le \eta + \sum_{j=1}^{h^n} \int_Q \int_{Q_j^h} |\rho(y_j^h)u(x + \varepsilon y_j^h) - \rho(y)u(x + \varepsilon y)|dydx =$$

$$= \eta + \sum_{j=1}^{h^n} \int_{Q_j^h} \int_Q |\rho(y_j^h)u(x + \varepsilon y_j^h) - \rho(y)u(x + \varepsilon y)|dxdy <$$

$$< \eta + \eta\sum_{j=1}^{h^n} \mathcal{L}^n(Q_j^h) = (1 + \mathcal{L}^n(Q))\eta \text{ for every } \eta > 0,$$

from which, together with the arbitrariness of Q, the lemma follows. ∎

Let now $\mathcal{O} \subseteq \mathcal{A}(\mathbf{R}^n)$, and U be a Hausdorff locally convex topological vector subspace of $L^1_{\mathrm{loc}}(\mathbf{R}^n)$ such that

(7.3.4) $x_0 + \Omega \in \mathcal{O}$ whenever $x_0 \in \mathbf{R}^n$, and $\Omega \in \mathcal{O}$,

(7.3.5) $T[x_0]u \in U$ whenever $x_0 \in \mathbf{R}^n$, and $u \in U$,

(7.3.6)   the topology of U is finer that $L^1_{\mathrm{loc}}(\mathbf{R}^n)$,

(7.3.7)   for every $u \in U$, $y \in \mathbf{R}^n \mapsto T[y]u \in U$ is continuous.

The results below proves that the regularizations of an element of U can be regarded as integrals of a function taking values in U.

Proposition 7.3.2. *Let U be a sequentially complete Hausdorff locally convex topological vector subspace of $L^1_{\text{loc}}(\mathbf{R}^n)$ satisfying (7.3.5)÷(7.3.7). Let ρ be a symmetric mollifier, $u \in U$, and, for every $\varepsilon > 0$, let u_ε be the regularization of u defined by (4.1.2). Then, for every $\varepsilon > 0$, $\rho(\cdot)T[\varepsilon\cdot]u$ is U-integrable on $\mathbf{R}^n$, and*

$$\left((U)\int_{\mathbf{R}^n} \rho(y)T[\varepsilon y]u\,dy \right)(x) = u_\varepsilon(x) \text{ for a.e. } x \text{ in } \mathbf{R}^n.$$

In particular, $u_\varepsilon \in U$ for every $\varepsilon > 0$.

Proof. Let $\varepsilon > 0$.

First of all, let us observe that by Lemma 7.3.1, $\rho(\cdot)T[\varepsilon\cdot]u$ turns out to be $L^1_{\text{loc}}(\mathbf{R}^n)$-integrable on $\mathbf{R}^n$, and that

$$(7.3.8) \qquad \left((L^1_{\text{loc}}(\mathbf{R}^n))\int_{\mathbf{R}^n} \rho(y)T[\varepsilon y]u\,dy \right)(x) = u_\varepsilon(x) \text{ for a.e. } x \text{ in } \mathbf{R}^n.$$

Consequently, by using (7.3.7) it results that $\rho(\cdot)T[\varepsilon\cdot]u$ too is continuous with compact support, and therefore, by Theorem 7.1.3, that $\rho(\cdot)T[\varepsilon\cdot]u$ is also U-integrable on $\mathbf{R}^n$. This, together with (7.3.6) and Remark 7.1.2, implies that

$$(U)\int_{\mathbf{R}^n} \rho(y)T[\varepsilon y]u\,dy = (L^1_{\text{loc}}(\mathbf{R}^n))\int_{\mathbf{R}^n} \rho(y)T[\varepsilon y]u\,dy,$$

from which, making also use of (7.3.8), the first part of the proposition follows.

From what just proved it is now trivial to deduce that $u_\varepsilon \in U$ for every $\varepsilon > 0$. In fact, for every $\varepsilon > 0$, u_ε turns out to agree a.e. with an element of U. ∎

Proposition 7.3.2 allows us to study the behaviour of the regularizations of the elements of U as $\varepsilon \to 0^+$.

Proposition 7.3.3. *Let U be a sequentially complete Hausdorff locally convex topological vector subspace of $L^1_{\text{loc}}(\mathbf{R}^n)$ satisfying (7.3.5)÷(7.3.7). Let $u \in U$, and, for every $\varepsilon > 0$, let u_ε be the regularization of u defined by (4.1.2). Then $\{u_\varepsilon\}_{\varepsilon>0} \subseteq U$, and $u_\varepsilon \to u$ in U as $\varepsilon \to 0^+$.*

Proof. Let ρ be the symmetric mollifier appearing in (4.1.2). Then Proposition 7.3.2, yields that for every $\varepsilon > 0$, $\rho(\cdot)T[\varepsilon\cdot]u$ is U-integrable on $\mathbf{R}^n$, and that $\{u_\varepsilon\}_{\varepsilon>0} \subseteq U$.

Let $\{p_\theta\}_{\theta \in \mathcal{T}}$ be a family of seminorms generating the topology of U, $\theta \in \mathcal{T}$, $\eta > 0$. Then, by (7.3.7), there exists $\varepsilon_{\theta,\eta} > 0$ such that

$$(7.3.9) \qquad \sup\{p_\theta(T[\varepsilon y]u - u) : y \in B_1(0)\} < \eta \text{ for every } \varepsilon \in \,]0, \varepsilon_{\theta,\eta}[.$$

Therefore, by Theorem 7.1.4 applied to $\Phi = p_\theta$, and (7.3.9), we conclude that

$$p_\theta\left((U)\int_{\mathbf{R}^n} \rho(y)T[\varepsilon y]u\,dy - u\right) = p_\theta\left((U)\int_{\mathbf{R}^n} \rho(y)(T[\varepsilon y]u - u)dy\right) \leq$$

$$\leq \int_{\mathbf{R}^n} p_\theta(T[\varepsilon y]u - u)\rho(y)dy < \eta \text{ for every } \varepsilon \in \,]0, \varepsilon_{\theta,\eta}[,$$

that is the convergence in U of $\{(U)\int_{\mathbf{R}^n} \rho(y)T[\varepsilon y]u\,dy\}_{\varepsilon>0}$ to u as ε goes to 0.

Because of this, and by Proposition 7.3.2, the proof follows. ∎

We emphasize that the properties established in Proposition 7.3.3 are somewhat surprising once we observe that no assumption on the existence of smooth functions in U is made.

We conclude this section with the approximation in energy result of an element of U via its regularizations.

Let $\Phi\colon \mathcal{O} \times U \to\,]-\infty, +\infty]$. We say that Φ is *translation invariant* if

$$\Phi(\Omega - x_0, T[x_0]u) = \Phi(\Omega, u) \text{ for every } \Omega \in \mathcal{O}, \ x_0 \in \mathbf{R}^n, \ u \in U.$$

We say that Φ is *convex* if for every $\Omega \in \mathcal{O}$, $\Phi(\Omega, \cdot)$ is convex, and say that Φ is *U-lower semicontinuous* if for every $\Omega \in \mathcal{O}$, $\Phi(\Omega, \cdot)$ is U-lower semicontinuous.

Lemma 7.3.4. *Let $\mathcal{O} \subseteq \mathcal{A}(\mathbf{R}^n)$, U be a sequentially complete Hausdorff locally convex topological vector subspace of $L^1_{\text{loc}}(\mathbf{R}^n)$ satisfying (7.3.4)÷ (7.3.7), and let $\Phi\colon \mathcal{O} \times U \to [0, +\infty]$ be translation invariant, convex, and U-lower semicontinuous. Then $\{u_\varepsilon\}_{\varepsilon>0} \subseteq U$, and*

$$\Phi(A, u_\varepsilon) \leq \Phi_-(\Omega, u)$$

for every $\Omega \in \mathcal{A}(\mathbf{R}^n)$, $A \in \mathcal{O}$ with $A \subset\subset \Omega$, $\varepsilon \in\,]0, \text{dist}(A, \partial\Omega)[, \ u \in U$.

Proof. Proposition 7.3.3 provides that $\{u_\varepsilon\}_{\varepsilon>0} \subseteq U$.

Let Ω, A, ε, u be as above, and let ρ be a symmetric mollifier as in (4.1.2). Then, by Theorem 7.1.4 applied to $\Phi(A, \cdot)$ and $\mu = \rho\mathcal{L}^n$, once we observe that the U-integrability of $\rho T[\varepsilon\cdot]$ on $\mathbf{R}^n$ with respect to Lebesgue measure implies also the U-integrability of $T[\varepsilon\cdot]$ on $\mathbf{R}^n$ with respect to the measure $\rho\mathcal{L}^n$, we deduce that

$$(7.3.10) \qquad \Phi\left(A, (U)\int_{\mathbf{R}^n} \rho(y)T[\varepsilon y]u\,dy\right) \leq \int_{\mathbf{R}^n} \Phi(A, T[\varepsilon y]u)\rho(y)dy.$$

On the other side, being Φ translation invariant, by (7.3.10) it follows that

$$\Phi\left(A, (U)\int_{\mathbf{R}^n} \rho(y)T[\varepsilon y]u\,dy\right) \leq \int_{\text{spt}(\rho)} \Phi\left(A + \varepsilon y, u\right)\rho(y)dy \leq$$

$$\leq \int_{B_1(0)} \Phi_-(\Omega, u)\rho(y)dy = \Phi_-(\Omega, u),$$

from which, together with Proposition 7.3.2, the lemma follows. ∎

Theorem 7.3.5. *Let $\mathcal{O} \subseteq \mathcal{A}(\mathbf{R}^n)$, U be a sequentially complete Hausdorff locally convex topological vector subspace of $L^1_{\mathrm{loc}}(\mathbf{R}^n)$ satisfying (7.3.4)÷ (7.3.7), and let $\Phi \colon \mathcal{O} \times U \to [0, +\infty]$ be translation invariant, convex, and U-lower semicontinuous. Then $\{u_\varepsilon\}_{\varepsilon>0} \subseteq U$, the limit $\lim_{\varepsilon \to 0} \Phi_-(\Omega_\varepsilon^-, u_\varepsilon)$ exists, and*

$$\lim_{\varepsilon \to 0} \Phi_-(\Omega_\varepsilon^-, u_\varepsilon) = \Phi_-(\Omega, u) \text{ for every } \Omega \in \mathcal{A}(\mathbf{R}^n), \ u \in U.$$

Proof. By Lemma 7.3.4 it follows that

(7.3.11) $\Phi_-(\Omega_\varepsilon^-, u_\varepsilon) \leq \Phi_-(\Omega, u)$ for every $\varepsilon > 0$ sufficiently small.

Consequently, fixed $A \in \mathcal{O}$ with $A \subset\subset \Omega$, by the lower semicontinuity of $\Phi(A, \cdot)$, Proposition 7.3.3, and (7.3.11), it results that

$$\Phi(A, u) \leq \liminf_{\varepsilon \to 0} \Phi(A, u_\varepsilon) \leq \liminf_{\varepsilon \to 0} \Phi_-(\Omega_\varepsilon^-, u_\varepsilon) \leq$$

$$\leq \limsup_{\varepsilon \to 0} \Phi_-(\Omega_\varepsilon^-, u_\varepsilon) \leq \Phi_-(\Omega, u),$$

from which the proof follows letting A increase to Ω. ∎

§7.4 Applications to Convex Functionals on BV Spaces

In this section we exploit the abstract approximation by regularizations method developed in this chapter to improve the lower semicontinuity results of Chapter 5 for convex functionals defined in BV spaces. Finally, for the same class of functional, an approximation in energy result via regularizations is established.

As in Chapter 5, we first prove some general results for convex functionals defined on spaces of of measures.

Let Ω be a Hausdorff locally compact space, μ be a σ-finite Borel positive measure on Ω, $f \colon \mathbf{R}^m \to [0, +\infty]$ be convex and lower semicontinuous, and let F be defined by (5.1.1).

We first study the convexity properties of F.

To do this, we first prove a preparatory result.

Proposition 7.4.1. *Let $(\Omega, \mathcal{E})$ be a measure space, λ be a σ-finite positive measure on $\mathcal{E}$, μ be a finite positive measure on $\mathcal{E}$, and $\nu: \mathcal{E} \to \mathbf{R}$ be a $\mathbf{R}^m$-valued vector measure on $\mathcal{E}$. Assume that ν is absolutely continuous with respect to μ, and that μ is absolutely continuous with respect to λ. Moreover, let $g: \mathbf{R}^m \to [0, +\infty]$ be positively 1-homogeneous. Then*

$$\int_\Omega g\left(\frac{d\nu}{d\mu}\right) d\mu = \int_\Omega g\left(\frac{d\nu}{d\lambda}\right) d\lambda.$$

Proof. By Radon-Nikodym Theorem, and Theorem 2.3.2 it follows that

$$\nu(A) = \int_\Omega \frac{d\nu}{d\lambda} d\lambda = \int_\Omega \frac{d\nu}{d\mu}\frac{d\mu}{d\lambda} d\lambda,$$

from which, by using also the uniqueness of the Radon-Nikodym derivative of ν with respect to λ, we conclude that $\frac{d\nu}{d\lambda}(x) = \frac{d\nu}{d\mu}(x)\frac{d\mu}{d\lambda}(x)$ for λ-a.e. $x \in \Omega$.

Because of this, by the homogeneity properties of g, and by Theorem 2.3.2 we thus obtain that

$$\int_\Omega g\left(\frac{d\nu}{d\lambda}\right) d\lambda = \int_\Omega g\left(\frac{d\nu}{d\mu}\frac{d\mu}{d\lambda}\right) d\lambda = \int_\Omega g\left(\frac{d\nu}{d\mu}\right)\frac{d\mu}{d\lambda} d\lambda = \int_\Omega g\left(\frac{d\nu}{d\mu}\right) d\mu,$$

from which the proof follows. ∎

Theorem 7.4.2. *Let Ω be a Hausdorff locally compact space, μ be a σ-finite Borel positive measure on Ω, and $f: \mathbf{R}^m \to [0, +\infty]$ be convex and lower semicontinuous. Let F be defined by (5.1.1). Then F is convex.*

Proof. Let $\nu_1, \nu_2 \in (\mathcal{M}(\Omega))^m$, $t \in [0, 1]$. Then the uniqueness of the Lebesgue decomposition of $t\nu_1 + (1 - t)\nu_2$ with respect to μ, it follows that $(t\nu_1 + (1 - t)\nu_2)^a = t\nu_1^a + (1 - t)\nu_2^a$, and $(t\nu_1 + (1 - t)\nu_2)^s = t\nu_1^s + (1 - t)\nu_2^s$. Consequently, by the uniqueness of the Radon-Nikodym derivative of $(t\nu_1 + (1 - t)\nu_2)^a$ with respect to μ, we conclude that $\frac{d(t\nu_1 + (1-t)\nu_2)^a}{d\mu} = t\frac{d\nu_1^a}{d\mu} + (1 - t)\frac{d\nu_2^a}{d\mu}$.

Because of this, and by the convexity of f, we infer that

$$(7.4.1) \qquad \int_\Omega f\left(\frac{d(t\nu_1 + (1 - t)\nu_2)^a}{d\mu}\right) d\mu =$$

$$= \int_\Omega f\left(t\frac{d\nu_1^a}{d\mu} + (1 - t)\frac{d\nu_2^a}{d\mu}\right) d\mu \le t\int_\Omega f\left(\frac{d\nu_1^a}{d\mu}\right) d\mu + (1 - t)\int_\Omega f\left(\frac{d\nu_2^a}{d\mu}\right) d\mu.$$

In order to treat the singular part of F, we observe that $|(t\nu_1 + (1 - t)\nu_2)^s|$ is clearly absolutely continuous with respect to $|\nu_1| + |\nu_2|$, and that, again by the uniqueness of the Radon-Nikodym derivative of $(t\nu_1 + (1 -$

$t)\nu_2)^{\mathrm{s}}$ with respect to $|\nu_1| + |\nu_2|$, $\frac{d(t\nu_1 + (1-t)\nu_2)^{\mathrm{s}}}{d(|\nu_1| + |\nu_2|)}(x) = t\frac{d\nu_1^{\mathrm{s}}}{d(|\nu_1| + |\nu_2|)}(x) + (1 - t)\frac{d(t\nu_1 + (1-t)\nu_2)^{\mathrm{s}}}{d(|\nu_1| + |\nu_2|)}(x)$ for $(|\nu_1| + |\nu_2|)$-a.e. $x \in \Omega$. Therefore, by a double application of Proposition 7.4.1, and the convexity of f^∞, we obtain that

$$(7.4.2) \qquad \int_\Omega f^\infty \left(\frac{d(t\nu_1 + (1-t)\nu_2)^{\mathrm{s}}}{d|(t\nu_1 + (1-t)\nu_2)^{\mathrm{s}}|} \right) d|(t\nu_1 + (1-t)\nu_2)^{\mathrm{s}}| =$$

$$= \int_\Omega f^\infty \left(\frac{d(t\nu_1 + (1-t)\nu_2)^{\mathrm{s}}}{d(|\nu_1| + |\nu_2|)} \right) d(|\nu_1| + |\nu_2|) \leq$$

$$\leq t \int_\Omega f^\infty \left(\frac{d\nu_1^{\mathrm{s}}}{d(|\nu_1| + |\nu_2|)} \right) d(|\nu_1| + |\nu_2|) +$$

$$+ (1 - t) \int_\Omega f^\infty \left(\frac{d\nu_2^{\mathrm{s}}}{d(|\nu_1| + |\nu_2|)} \right) d(|\nu_1| + |\nu_2|) =$$

$$= t \int_\Omega f^\infty \left(\frac{d\nu_1^{\mathrm{s}}}{d|\nu_1^{\mathrm{s}}|} \right) d|\nu_1^{\mathrm{s}}| + (1 - t) \int_\Omega f^\infty \left(\frac{d\nu_2^{\mathrm{s}}}{d|\nu_2^{\mathrm{s}}|} \right) d|\nu_2^{\mathrm{s}}|.$$

By (7.4.1), and (7.4.2) the convexity of F follows. ∎

Finally, we prove a translation invariance property of F when $\Omega \in \mathcal{B}(\mathbf{R}^n)$, and $\mu = \mathcal{L}^n$.

Theorem 7.4.3. Let $f : \mathbf{R}^m \to [0, +\infty]$ be convex and lower semicontinuous. Then

$$\int_{\Omega - x_0} f \left(\frac{d(T[x_0]\nu)^{\mathrm{a}}}{d\mathcal{L}^n} \right) dx + \int_{\Omega - x_0} f^\infty \left(\frac{d(T[x_0]\nu)^{\mathrm{s}}}{d|(T[x_0]\nu)^{\mathrm{s}}|} \right) d|(T[x_0]\nu)^{\mathrm{s}}| =$$

$$= \int_\Omega f \left(\frac{d\nu^{\mathrm{a}}}{d\mathcal{L}^n} \right) dx + \int_\Omega f^\infty \left(\frac{d\nu^{\mathrm{s}}}{d|\nu^{\mathrm{s}}|} \right) d|\nu^{\mathrm{s}}|$$

for every $\Omega \in \mathcal{B}(\mathbf{R}^n)$, $\nu \in (\mathcal{M}(\Omega))^m$, $x_0 \in \mathbf{R}^n$.

Proof. Let Ω, ν, x_0 be as above. Then, because of the uniqueness of the Lebesgue decomposition of ν, it follows that $(T[x_0]\nu)^{\mathrm{a}} = T[x_0]\nu^{\mathrm{a}}$, $(T[x_0]\nu)^{\mathrm{s}} = T[x_0]\nu^{\mathrm{s}}$, and, consequently, that $|(T[x_0]\nu)^{\mathrm{s}}| = T[x_0]|\nu^{\mathrm{s}}|$. Hence, by Theorem 2.3.5, we infer that

$$\frac{d(T[x_0]\nu)^{\mathrm{a}}}{d\mathcal{L}^n}(x) = \lim_{r \to 0} \frac{(T[x_0]\nu)^{\mathrm{a}}(Q_r(x))}{r^n} = \lim_{r \to 0} \frac{\nu^{\mathrm{a}}(Q_r(x_0 + x))}{r^n} =$$

$$= \frac{d\nu^{\mathrm{a}}}{d\mathcal{L}^n}(x_0 + x) = T[x_0]\frac{d\nu^{\mathrm{a}}}{d\mathcal{L}^n}(x) \text{ for } \mathcal{L}^n\text{-a.e. } x \in \Omega,$$

and

$$\frac{d(T[x_0]\nu)^{\mathrm{s}}}{d|(T[x_0]\nu)^{\mathrm{s}}|}(x) = \lim_{r \to 0} \frac{(T[x_0]\nu)^{\mathrm{s}}(Q_r(x))}{|(T[x_0]\nu)^{\mathrm{s}}|(Q_r(x)} = \lim_{r \to 0} \frac{\nu^{\mathrm{s}}(Q_r(x_0 + x))}{|\nu^{\mathrm{s}}|(Q_r(x_0 + x))} =$$

$$= \frac{d\nu^{\mathrm{s}}}{d|\nu^{\mathrm{s}}|}(x_0 + x) = T[x_0]\frac{d\nu^{\mathrm{s}}}{d|\nu^{\mathrm{s}}|}(x) \text{ for } |\nu^{\mathrm{s}}|\text{-a.e. } x \in \Omega.$$

Because of this, and by (2.1.5), we therefore conclude that

$$\int_{\Omega-x_0} f\left(\frac{d(T[x_0]\nu)^{\mathrm{a}}}{d\mathcal{L}^n}\right) dx + \int_{\Omega-x_0} f^\infty\left(\frac{d(T[x_0]\nu)^{\mathrm{s}}}{d|(T[x_0]\nu)^{\mathrm{s}}|}\right) d|(T[x_0]\nu)^{\mathrm{s}}| =$$

$$= \int_{\Omega-x_0} f\left(T[x_0]\frac{d\nu^{\mathrm{a}}}{d\mathcal{L}^n}\right) dx + \int_{\Omega-x_0} f^\infty\left(T[x_0]\frac{d\nu^{\mathrm{s}}}{d|\nu^{\mathrm{s}}|}\right) dT[x_0]|\nu^{\mathrm{s}}| =$$

$$= \int_\Omega f\left(\frac{d\nu^{\mathrm{a}}}{d\mathcal{L}^n}\right) dx + \int_\Omega f^\infty\left(\frac{d\nu^{\mathrm{s}}}{d|\nu^{\mathrm{s}}|}\right) d|\nu^{\mathrm{s}}|,$$

which proves the theorem. ∎

We now come to integrals functionals defined on BV spaces. To prove the announced lower semicontinuity property, we need to establish the following approximation from below in energy result.

Lemma 7.4.4. *Let* $f: \mathbf{R}^n \to [0, +\infty]$ *be convex and lower semicontinuous. Then*

$$\int_{\Omega_\varepsilon^-} f(\nabla u_\varepsilon) dx \le \int_\Omega f(\nabla u) dx + \int_\Omega f^\infty(\nabla^{\mathrm{s}} u) d|D^{\mathrm{s}} u|$$

for every $\Omega \in \mathcal{A}(\mathbf{R}^n)$, $u \in BV_{\mathrm{loc}}(\Omega)$, *and* $\varepsilon > 0$.

Proof. By Theorem 7.4.3, and the properties of the translated of $BV(\mathbf{R}^n)$ functions, we obtain that the functional

$$G: (A, u) \in \mathcal{A}(\mathbf{R}^n) \times BV(\mathbf{R}^n) \mapsto \int_A f(\nabla u) dx + \int_A f^\infty(\nabla^{\mathrm{s}} u) d|D^{\mathrm{s}} u|$$

is translation invariant.

Moreover, by Theorem 7.4.2, and Theorem 5.1.4, G turns out to be also convex, and weak*-$BV(\mathbf{R}^n)$-lower semicontinuous.

Because of such properties, Lemma 7.3.4 with $\mathcal{O} = \mathcal{A}(\mathbf{R}^n)$, $U = BV(\mathbf{R}^n)$ endowed with the weak*-$BV(\mathbf{R}^n)$ topology, and $\Phi = G$ applies since $BV(\mathbf{R}^n)$ endowed with the weak*-$BV(\mathbf{R}^n)$ topology is sequentially complete. We thus obtain the lemma when $u \in BV(\mathbf{R}^n)$.

If now $u \in BV_{\mathrm{loc}}(\Omega)$, for every $h \in \mathbf{N}$ let $A_h \in \mathcal{A}_0$ have Lipschitz boundary, and satisfy $A_h \subset\subset A_{h+1} \subset\subset \Omega$, $\cup_{h=1}^{+\infty} A_h = \Omega$, and let v_h be the zero extensions of u out of A_h. Then, $v_h \in BV(\mathbf{R}^n)$, $Dv_h = Du$ in A_h, and consequently $\nabla v_h = \nabla u$ $\mathcal{L}^n$-a.e. in A_h and $D^{\mathrm{s}} v_h = D^{\mathrm{s}} u$ in A_h for every $h \in \mathbf{N}$.

Because of this, and by the previously treated case, we infer that

$$\int_{(A_h)_\varepsilon^-} f(\nabla u_\varepsilon) dx = \int_{(A_h)_\varepsilon^-} f(\nabla(v_h)_\varepsilon) dx \le$$

$$\leq \int_{A_h} f(\nabla v_h)dx + \int_{A_h} f^\infty(\nabla^s v_h)d|D^s v_h| =$$

$$= \int_{A_h} f(\nabla u)dx + \int_{A_h} f^\infty(\nabla^s u)d|D^s u| \leq$$

$$\leq \int_\Omega f(\nabla u)dx + \int_\Omega f^\infty(\nabla^s u)d|D^s u| \text{ for every } h \in \mathbf{N}, \ \varepsilon > 0,$$

from which the lemma follows as h diverges. ∎

Let $\Omega \in \mathcal{A}(\mathbf{R}^n)$, and let $\{\mu_h\} \subseteq \mathcal{M}_{\mathrm{loc}}(\Omega)$, $\mu \in \mathcal{M}_{\mathrm{loc}}(\Omega)$. We recall that $\{\mu_h\}$ converges to μ in the sense of distributions in Ω if

$$\int_\Omega \varphi d\mu_h \to \int_\Omega \varphi d\mu \text{ for every } \varphi \in C_0^\infty(\Omega).$$

If $\{u_h\} \subseteq L_{\mathrm{loc}}^1(\Omega)$, and $u \in L_{\mathrm{loc}}^1(\Omega)$, we say that $\{u_h\}$ converges to u in the sense of distributions in Ω if $u_h\mathcal{L}^n \to u\mathcal{L}^n$ in the sense of distributions in Ω.

The result below shows that converging sequences of Radon measures improve their convergence after a regularization process.

Proposition 7.4.5. Let $\Omega \in \mathcal{A}(\mathbf{R}^n)$, and $\{\mu_h\} \subseteq \mathcal{M}_{\mathrm{loc}}(\Omega)$, $\mu \in \mathcal{M}_{\mathrm{loc}}(\Omega)$ be such that $\mu_h \to \mu$ in the sense of distributions in Ω. Then, for every $\varepsilon > 0$, $\mu_{h,\varepsilon} \to \mu_\varepsilon$ in $C^\infty(\Omega_\varepsilon^-)$.

Proof. Let $\varepsilon > 0$.

We first treat the case in which $\{\mu_h\} \subseteq \mathcal{M}(\Omega)$ and $\mu \in \mathcal{M}(\Omega)$.

For every $h \in \mathbf{N}$ let us define $\tilde{\mu}_h$ and $\tilde{\mu}$ by

$$\tilde{\mu}_h \colon E \in \mathcal{B}(\mathbf{R}^n) \mapsto \mu_h(E \cap \Omega), \quad \tilde{\mu} \colon E \in \mathcal{B}(\mathbf{R}^n) \mapsto \mu(E \cap \Omega).$$

Then clearly $\{\tilde{\mu}_h\} \subseteq \mathcal{M}(\mathbf{R}^n)$, $\tilde{\mu} \in \mathcal{M}(\mathbf{R}^n)$.

Let $A \in \mathcal{A}_0$ have Lipschitz boundary be such that $\overline{A} \subseteq \Omega_\varepsilon^-$. Then, once we observe that the null extension of $\varphi \in \widehat{C}_0^0(A)$ to $\mathbf{R}^n$ is actually in $C_0^0(\mathbf{R}^n)$, Proposition 4.1.3 yields that

$$(7.4.3) \quad \int_A \frac{\partial^{|\alpha|} \tilde{\mu}_{h,\varepsilon}}{\partial x^\alpha}\varphi dx = \int_{\mathbf{R}^n} \frac{\partial^{|\alpha|} \tilde{\mu}_{h,\varepsilon}}{\partial x^\alpha}\varphi dx = (-1)^{|\alpha|} \int_{\mathbf{R}^n} \frac{\partial^{|\alpha|} \varphi_\varepsilon}{\partial x^\alpha}d\tilde{\mu}_h \to$$

$$\to (-1)^{|\alpha|} \int_{\mathbf{R}^n} \frac{\partial^{|\alpha|} \varphi_\varepsilon}{\partial x^\alpha}d\tilde{\mu} = \int_{\mathbf{R}^n} \frac{\partial^{|\alpha|} \tilde{\mu}_\varepsilon}{\partial x^\alpha}\varphi dx = \int_A \frac{\partial^{|\alpha|} \tilde{\mu}_\varepsilon}{\partial x^\alpha}\varphi dx$$

for every $\varphi \in \widehat{C}_0^0(A)$, and every $\alpha \in (\mathbf{N} \cup \{0\})^n$.

Condition (7.4.3) actually guarantees that $\frac{\partial^{|\alpha|} \tilde{\mu}_{h,\varepsilon}}{\partial x^\alpha}\mathcal{L}^n \to \frac{\partial^{|\alpha|} \tilde{\mu}_\varepsilon}{\partial x^\alpha}\mathcal{L}^n$ in weak*-$\mathcal{M}(A)$ for every $\alpha \in (\mathbf{N} \cup \{0\})^n$, from which we conclude that for every $\alpha \in (\mathbf{N} \cup \{0\})^n$, $\{\|\frac{\partial^{|\alpha|} \tilde{\mu}_{h,\varepsilon}}{\partial x^\alpha}\|_{L^1(A)}\}_{h \in \mathbf{N}}$ is bounded.

We now observe that for every $\alpha \in (\mathbf{N} \cup \{0\})^n$, and $h \in \mathbf{N}$, $\frac{\partial^{|\alpha|} \tilde{\mu}_{h,\varepsilon}}{\partial x^\alpha} \in W^{1,1}(A)$, consequently an iterated use of the Rellich-Kondrachov Compactness Theorem provides that $\tilde{\mu}_{h,\varepsilon} \to \tilde{\mu}_\varepsilon$ in $C^m(\overline{A})$ for every $m \in \mathbf{N} \cup \{0\}$.

Because of this, the proof follows, once we observe that $\tilde{\mu}_{h,\varepsilon}(x) = \mu_{h,\varepsilon}(x)$, and $\tilde{\mu}_\varepsilon(x) = \mu_\varepsilon(x)$ for every $h \in \mathbf{N}$, and $x \in \Omega_\varepsilon^-$.

Finally, if $\{\mu_h\} \subseteq \mathcal{M}_{\mathrm{loc}}(\Omega)$ and $\mu \in \mathcal{M}_{\mathrm{loc}}(\Omega)$, we take $B \in \mathcal{A}_0$ with $B \subset\subset \Omega$, and define for every $h \in \mathbf{N}$, $\overline{\mu}_h$ and $\overline{\mu}$ by

$$\overline{\mu}_h \colon E \in \mathcal{B}(B) \mapsto \mu_h(E \cap B), \quad \overline{\mu} \colon E \in \mathcal{B}(B) \mapsto \mu(E \cap B).$$

Then clearly $\{\overline{\mu}_h\} \subseteq \mathcal{M}(B)$, $\overline{\mu} \in \mathcal{M}(B)$, and by the above considered case we conclude that $\overline{\mu}_{h,\varepsilon} \to \overline{\mu}_\varepsilon$ in $C^\infty(B_\varepsilon^-)$.

Because of this, the proposition also in this case follows, once we observe that for every compact set $K \subseteq \Omega_\varepsilon^-$ there exists $B \in \mathcal{A}_0$ with $B \subset\subset \Omega$ such that $K \subseteq B_\varepsilon^-$, and that $\overline{\mu}_{h,\varepsilon}(x) = \mu_{h,\varepsilon}(x)$ and $\overline{\mu}_\varepsilon(x) = \mu_{h,\varepsilon}(x)$ for every $h \in \mathbf{N}$ and every $x \in B_\varepsilon^-$. ∎

Theorem 7.4.6. *Let* $f \colon \mathbf{R}^n \to [0, +\infty]$ *be convex and lower semicontinuous. Then*

$$\int_\Omega f(\nabla u) dx + \int_\Omega f^\infty(\nabla^s u) d|D^s u| \le$$

$$\le \liminf_{h \to +\infty} \int_\Omega f(\nabla u_h) dx + \int_\Omega f^\infty(\nabla^s u_h) d|D^s u_h|$$

whenever $\Omega \in \mathcal{A}(\mathbf{R}^n)$, $\{u_h\} \subseteq BV_{\mathrm{loc}}(\Omega)$, $u \in BV_{\mathrm{loc}}(\Omega)$

are such that $u_h \to u$ *in the sense of distributions in* Ω.

Proof. Let Ω, $\{u_h\}$, u be as above, $A \in \mathcal{A}_0$ with $A \subset\subset \Omega$, and $\varepsilon \in \,]0, \mathrm{dist}(A, \partial\Omega)[$.

For every $h \in \mathbf{N}$ let $u_{h,\varepsilon}$ be the regularization of u_h. Then, by Lemma 7.4.4, we get that

$$(7.4.4) \qquad \int_A f(\nabla u_{h,\varepsilon}) dx \le \int_\Omega f(\nabla u_h) dx + \int_\Omega f^\infty(\nabla^s u_h) d|D^s u_h|$$

for every $h \in \mathbf{N}$,

whilst, by Proposition 7.4.5, Fatou's lemma, and (7.4.4) we deduce that

$$(7.4.5) \qquad \int_A f(\nabla u_\varepsilon) dx \le \liminf_{h \to +\infty} \int_A f(\nabla u_{h,\varepsilon}) dx \le$$

$$\le \liminf_{h \to +\infty} \left\{ \int_\Omega f(\nabla u_h) dx + \int_\Omega f^\infty(\nabla^s u_h) d|D^s u_h| \right\}$$

for every $\varepsilon \in \,]0, \mathrm{dist}(A, \partial\Omega)[$.

Finally, from Proposition 4.2.8, Theorem 5.1.4, and (7.4.5), we conclude that

$$\int_A f(\nabla u)dx + \int_A f^\infty(\nabla^s u)d|D^s u| \leq \liminf_{\varepsilon \to 0^+} \int_A f(\nabla u_\varepsilon)dx \leq$$

$$\leq \liminf_{h \to +\infty} \left\{ \int_\Omega f(\nabla u_h)dx + \int_\Omega f^\infty(\nabla^s u_h)d|D^s u_h| \right\},$$

from which the proof follows letting A increase to Ω. ∎

By the above results we deduce an approximation in energy result for BV functions.

Proposition 7.4.7. Let $f: \mathbf{R}^n \to [0, +\infty]$ be convex and lower semicontinuous. Then for every $\Omega \in \mathcal{A}(\mathbf{R}^n)$, and $u \in BV_{\text{loc}}(\Omega)$ the limit $\lim_{\varepsilon \to 0^+} \int_{\Omega_\varepsilon^-} f(\nabla u_\varepsilon)dx$ exists, and

$$\lim_{\varepsilon \to 0^+} \int_{\Omega_\varepsilon^-} f(\nabla u_\varepsilon)dx = \int_\Omega f(\nabla u)dx + \int_\Omega f^\infty(\nabla^s u)d|D^s u|.$$

Proof. Let Ω, u be as above, and $A \in \mathcal{A}(\Omega)$ with $A \subset\subset \Omega$. Then Theorem 7.4.6 and Lemma 7.4.4 yield that

$$\int_A f(\nabla u)dx + \int_A f^\infty(\nabla^s u)d|D^s u| \leq \liminf_{\varepsilon \to 0^+} \int_A f(\nabla u_\varepsilon)dx \leq$$

$$\leq \limsup_{\varepsilon \to 0^+} \int_{\Omega_\varepsilon^-} f(\nabla u_\varepsilon)dx \leq \int_\Omega f(\nabla u)dx + \int_\Omega f^\infty(\nabla^s u)d|D^s u|,$$

from which the proof follows letting A increase to Ω. ∎

Chapter 8

Unique Extension Results

In this chapter we begin a systematic treatment of unbounded functionals. In particular, we deal here with unique extension problems.

Starting from the well celebrated example of H.A. Schwarz (in 1880) and G. Peano (in 1882), the problem of the definition of the concept of area of a surface and of the study of its properties, both in the parametric and non-parametric cases, and possibly also in the noncontinuous framework, interested many important mathematicians.

The researches developed produced a great amount of fruitful ideas and techniques. We refer to the book of Cesari (cf. [Cs1]) for a survey and a bibliography up to 1956, and to [DGCP], [F], [GMS2], [Gu], [MaM], [M] and to the references quoted therein.

To analyse the problem, various kinds of approaches were proposed, among which also some of axiomatic type in which conditions on an abstract functional, defined on sets of "generalized surfaces" and furnishing the value of the area on the smooth ones, were proposed in order to uniquely identify the area one. These last approaches were essentially based on the topological (e.g. lower semicontinuity) and the measure theoretic properties of the area functional.

In this chapter, we want to make some remarks in order to obtain uniqueness of the extension for classes of functionals, including the area one, in an axiomatic context. So, having in mind the non-parametric area case, we enlarge the classical point of view by keeping into account also a vectorial property of the area functional: the convexity.

Then, we consider an abstract functional, say F, given on a collection of elementary smooth functions and open sets, and taking values in $[0, +\infty]$, and propose sets of conditions fulfilled by F that select classes of functionals, defined on spaces of less smooth functions and open sets, in which F possesses a unique extension. This (unique) extension turns out to be strongly linked to the relaxed functional of F in the L^1 topology introduced, in the case of integral functionals, in [S1] and [S2], and represented in [GS].

The result is obtained under inner regularity, translation invariance, and lower semicontinuity assumptions on F, besides convexity. We emphasize that such notions are classical in the framework of area definition. Indeed the notion of inner regularity is of measure theoretic nature, the one of translation invariance is of geometric type (cf. [Fr], [Le], [J]), and the one of lower semicontinuity is classical and well recognized when dealing with extension procedures (cf. [Fr]). We also point out that the notion of convexity is linked to energy and statistics type considerations: in fact the convexity property that we will exploit is essentially the feature of a functional to take values on averages of configurations smaller than the corresponding average of the ones on the single configurations (Jensen's inequality).

Similar unique extension results have been treated in [DM, Chapter 23], but in the more restrictive framework of integral representation theory, and essentially in the finite valued case.

The results obtained are then applied to the problem of the unique extension of certain integral functionals of the calculus of variations, similarly to what has already been done for the area functional.

The results of the present chapter form the basis of the relaxation approach to variational problems when no a priori singularities on the admissible configurations are allowed, approach that we follow in the present volume.

Nevertheless, it must be pointed out that such approach is not the only possible one, and actually one may expect to obtain, in general, different results, as exposed in the last section of the chapter.

§8.1 Unique Extension Results for Inner Regular Functionals

In this section we deal with unique extension results under inner regularity assumptions on the functionals taken into account.

To do this, it is worth while to recall that, by Proposition 7.3.2, for every sequentially complete Hausdorff locally convex topological vector space U satisfying $(7.3.5) \div (7.3.7)$ it turns out that

$$C^\infty(\mathbf{R}^n) \cap U \neq \emptyset.$$

In the following we will take $\mathcal{E}_0 \subseteq \mathcal{A}_0$ satisfying

(8.1.1) $x_0 + \Omega \in \mathcal{E}_0$ whenever $x_0 \in \mathbf{R}^n$, $\Omega \in \mathcal{E}_0$.

Proposition 8.1.1. *Let $\mathcal{E}_0 \subseteq \mathcal{A}_0$ satisfy (8.1.1), U be a sequentially complete Hausdorff locally convex topological vector space satisfying $(7.3.5) \div$*

(7.3.7), and G, $H\colon \mathcal{E}_0 \times U \to [0, +\infty]$. *Assume that H is translation invariant and convex, that G and H are U-lower semicontinuous, and that*

$$(8.1.2) \qquad G(\Omega, u) \leq H(\Omega, u) \text{ for every } (\Omega, u) \in \mathcal{E}_0 \times (C^\infty(\mathbf{R}^n) \cap U).$$

Then

$$G_{\mathcal{E}_0-}(\Omega, u) \leq H_{\mathcal{E}_0-}(\Omega, u) \text{ for every } (\Omega, u) \in \mathcal{A}_0 \times U.$$

Proof. The proposition is clearly true if $\{A \in \mathcal{E}_0 : A \subset\subset \Omega\} = \emptyset$.

Otherwise, let $(\Omega, u) \in \mathcal{A}_0 \times U$. Then by (8.1.2), and Lemma 7.3.4 applied with $\mathcal{O} = \mathcal{E}_0$, $\Phi = H$, we get

$$G(A, u_\varepsilon) \leq H(A, u_\varepsilon) \leq H_{\mathcal{E}_0-}(\Omega, u)$$

for every $A \in \mathcal{E}_0$ with $A \subset\subset \Omega$, $\varepsilon \in \,]0, \mathrm{dist}(A, \partial\Omega)[$,

from which, together with the U-lower semicontinuity of G, and Proposition 7.3.3, the proof follows. ∎

Then the unique extension result is the following.

Theorem 8.1.2. *Let $\mathcal{E} \subseteq \mathcal{A}_0$, $\mathcal{E}_0 \subseteq \mathcal{E}$ be dense with respect to $\mathcal{E}$, and satisfying (8.1.1). Let U be a sequentially complete Hausdorff locally convex topological vector space satisfying (7.3.5)÷(7.3.7), and G, $H\colon \mathcal{E} \times U \to [0, +\infty]$. Assume that G and H are inner regular, that their restrictions to $\mathcal{E}_0 \times U$ are translation invariant, convex, U-lower semicontinuous, and that*

$$G(\Omega, u) = H(\Omega, u) \text{ for every } (\Omega, u) \in \mathcal{E}_0 \times (C^\infty(\mathbf{R}^n) \cap U).$$

Then

$$G(\Omega, u) = H(\Omega, u) \text{ for every } (\Omega, u) \in \mathcal{E} \times U.$$

Proof. By a double application of Proposition 8.1.1 to the restrictions of G and H to $\mathcal{E}_0 \times U$, we infer that

$$(8.1.3) \qquad G_{\mathcal{E}_0-}(\Omega, u) = H_{\mathcal{E}_0-}(\Omega, u) \text{ for every } (\Omega, u) \in \mathcal{A}_0 \times U.$$

On the other hand, by ii) of Proposition 2.6.9 we immediately deduce that

$$G_{\mathcal{E}-}(\Omega, u) = G_{\mathcal{E}_0-}(\Omega, u), \ H_{\mathcal{E}-}(\Omega, u) = H_{\mathcal{E}_0-}(\Omega, u)$$

for every $(\Omega, u) \in \mathcal{E} \times U,$

from which, together with the inner regularity of G and H, and (8.1.3), the proof follows. ∎

We point out that Theorem 8.1.2 fails if the convexity assumptions are dropped, as shown in the example below.

As usual, for every $x \in \mathbf{R}$, we denote by δ_x the Dirac measure defined for every $B \in \mathcal{B}(\Omega)$ by $\delta_x(B) = 1$ if $x \in B$, and $\delta_x(B) = 0$ if $x \notin B$. Moreover, we denote by $\#$ the counting measure defined on $\emptyset$ as $\#(\emptyset) = 0$, and on $B \in \mathcal{B}(\Omega)$ as the cardinality of B.

For every open subset Ω of $\mathbf{R}$ we set denote by $BV^{\#}(\Omega)$ the set of the functions $u \in BV(\Omega)$ such that $D^s u = \sum_{h=1}^{+\infty} c_h \delta_{x_h}$ for some $\{c_h\} \subseteq \mathbf{R}$ satisfying $\sum_{h=1}^{+\infty} |c_h| < +\infty$, and $\{x_h\} \subseteq \Omega$. For every $u \in BV^{\#}(\Omega)$ we set $S_u = \cup_{h=1}^{+\infty} \{x_h\}$.

Example 8.1.3. Let $n = 1$, $\mathcal{E}_0 = \mathcal{E} = \mathcal{A}_0$, $U = BV(\mathbf{R})$ endowed with the weak*-$BV(\mathbf{R})$ topology,

$$G : (\Omega, u) \in \mathcal{A}_0 \times BV(\mathbf{R}) \mapsto \begin{cases} \int_{\Omega} |\nabla u|^2 dx & \text{if } u \in W^{1,2}(\Omega) \\ +\infty & \text{if } u \notin W^{1,2}(\Omega), \end{cases}$$

and

$$H : (\Omega, u) \in \mathcal{A}_0 \times BV(\mathbf{R}) \mapsto \begin{cases} \int_{\Omega} |\nabla u|^2 dx + \#(S_u) & \text{if } u \in BV^{\#}(\Omega) \\ +\infty & \text{if } u \notin BV^{\#}(\Omega). \end{cases}$$

With such choices, G and H are inner regular, translation invariant, and

$$G(\Omega, u) = H(\Omega, u) \text{ for every } (\Omega, u) \in \mathcal{A}_0 \times (C^{\infty}(\mathbf{R}) \cap BV(\mathbf{R})).$$

Moreover, the weak*-$BV(\mathbf{R})$-lower semicontinuity of G follows from Theorems 6.3.1 and 4.2.11, and the one of H from [BoB, Remark 3.5].

This notwithstanding, G and H are different, since H is not convex.

§8.2 Existence and Uniqueness Results

In the present section we discuss the problem of the existence of the extension.

Lemma 8.2.1. *Let $\mathcal{E}_0 \subseteq \mathcal{A}_0$ satisfy (8.1.1), and $F : \mathcal{E}_0 \times C^{\infty}(\mathbf{R}^n) \to [0, +\infty]$ be translation invariant, convex, and $C^{\infty}(\mathbf{R}^n)$-lower semicontinuous. Then $F_{\mathcal{E}_0-}$ is $L^1_{\text{loc}}(\mathbf{R}^n)$-lower semicontinuous.*

Proof. Let $\Omega \in \mathcal{A}_0$, $u \in C^{\infty}(\mathbf{R}^n)$, $\{u_h\} \subseteq C^{\infty}(\mathbf{R}^n)$ be such that $u_h \to u$ in $L^1_{\text{loc}}(\mathbf{R}^n)$.

It is clear that, if $\{A \in \mathcal{E}_0 : A \subset\subset \Omega\} = \emptyset$, then

$$F_{\mathcal{E}_0-}\Omega, u) = 0 = \liminf_{h \to +\infty} F_{\mathcal{E}_0-}(\Omega, u_h).$$

Otherwise, for every $h \in \mathbf{N}$, $\varepsilon > 0$, let $u_{h,\varepsilon}$ be the regularization of u_h defined by means of (4.1.1).

Let $A \in \mathcal{E}_0$ be such that $A \subset\subset \Omega$, and $\varepsilon \in]0, \operatorname{dist}(A, \partial\Omega)[$. Then Proposition 7.4.5 provides that $u_{h,\varepsilon} \to u_\varepsilon$ in $C^\infty(\mathbf{R}^n)$. By Lemma 7.3.4 applied with $\mathcal{O} = \mathcal{E}_0$, $U = C^\infty(\mathbf{R}^n)$, $\Phi = F$, and by the $C^\infty(\mathbf{R}^n)$-lower semicontinuity of F, we get

$$(8.2.1) \qquad F(A, u_\varepsilon) \le \liminf_{h\to+\infty} F(A, u_{h,\varepsilon}) \le \liminf_{h\to+\infty} F_{\mathcal{E}_0-}(\Omega, u_h).$$

By (8.2.1), and again the $C^\infty(\mathbf{R}^n)$-lower semicontinuity of F, once we observe that $u_\varepsilon \to u$ in $C^\infty(\mathbf{R}^n)$, we conclude that

$$F(A, u) \le \liminf_{\varepsilon\to0+} F(A, u_\varepsilon) \le \liminf_{h\to+\infty} F_{\mathcal{E}_0-}(\Omega, u_h)$$

for every $A \in \mathcal{E}_0$ with $A \subset\subset \Omega$,

from which the lemma follows. ∎

For every $\mathcal{E}_0 \subseteq \mathcal{A}_0$, $\Phi: \mathcal{E}_0 \times C^\infty(\mathbf{R}^n) \to [0, +\infty]$, and $\Omega \in \mathcal{E}_0$ for the sake of simplicity we denote in this chapter by $\overline{\Phi}(\Omega, \cdot)$ the $L^1_{\mathrm{loc}}(\mathbf{R}^n)$-lower semicontinuous envelope of

$$u \in L^1_{\mathrm{loc}}(\mathbf{R}^n) \mapsto \begin{cases} \Phi(\Omega, u) & \text{if } u \in C^\infty(\mathbf{R}^n) \\ +\infty & \text{if } u \in L^1_{\mathrm{loc}}(\mathbf{R}^n) \setminus C^\infty(\mathbf{R}^n), \end{cases}$$

i.e.

$$\overline{\Phi}: (\Omega, u) \in \mathcal{E}_0 \times L^1_{\mathrm{loc}}(\mathbf{R}^n) \mapsto$$

$$\inf \left\{ \liminf_{h\to+\infty} \Phi(\Omega, u_h) : \{u_h\} \subseteq C^\infty(\mathbf{R}^n),\ u_h \to u \text{ in } L^1_{\mathrm{loc}}(\mathbf{R}^n) \right\}.$$

Then $\overline{\Phi}: (\Omega, \cdot)$ is $L^1_{\mathrm{loc}}(\mathbf{R}^n)$-lower semicontinuous, and it turns out to be the greatest $L^1_{\mathrm{loc}}(\mathbf{R}^n)$-lower semicontinuous functional on $L^1_{\mathrm{loc}}(\mathbf{R}^n)$ less than or equal to $\Phi(\Omega, \cdot)$ on $C^\infty(\mathbf{R}^n)$.

Proposition 8.2.2. Let $\mathcal{E}_0 \subseteq \mathcal{A}_0$ satisfy (8.1.1), and $F: \mathcal{E}_0 \times C^\infty(\mathbf{R}^n) \to [0, +\infty]$. Assume that F is inner regular, translation invariant, convex, and $C^\infty(\mathbf{R}^n)$-lower semicontinuous. Then $(\overline{F_{\mathcal{E}_0-}})_{\mathcal{A}_0-}$ is translation invariant, convex, and agrees with F on $\mathcal{E}_0 \times C^\infty(\mathbf{R}^n)$. For every topological vector space $U \subseteq L^1_{\mathrm{loc}}(\mathbf{R}^n)$ satisfying (7.3.6), its restriction to $\mathcal{A}_0 \times U$ is U-lower semicontinuous, and for every $\mathcal{E} \subseteq \mathcal{A}_0$ perfect with respect to $\mathcal{A}_0$, its restriction to $\mathcal{E} \times L^1_{\mathrm{loc}}(\mathbf{R}^n)$ is inner regular.

Proof. It is easy to verify that $(\overline{F_{\mathcal{E}_0-}})_{\mathcal{A}_0-}$ is translation invariant and convex.

Moreover, by Lemma 8.2.1, we have that

$$F_{\mathcal{E}_0-}(\Omega, u) = \overline{F_{\mathcal{E}_0-}}(\Omega, u) \text{ for every } (\Omega, u) \in \mathcal{A}_0 \times C^\infty(\mathbf{R}^n),$$

from which, together with the remark that $\mathcal{A}_0$ is perfect with respect to $\mathcal{E}_0$, i) of Proposition 2.6.9, and the inner regularity of F, we deduce the identity of $(\overline{F_{\mathcal{E}_0-}})_{\mathcal{A}_0-}$ with F on $\mathcal{E}_0 \times C^\infty(\mathbf{R}^n)$.

Let now U be as above. Then, by using also (7.3.6), it is easy to deduce that the restriction of $(\overline{F_{\mathcal{E}_0-}})_{\mathcal{A}_0-}$ to $\mathcal{A}_0 \times U$ is U-lower semicontinuous.

Finally, given $\mathcal{E}$ as above, i) of Proposition 2.6.9 yields

$$\left(\left(\overline{F_{\mathcal{E}_0-}} \right)_{\mathcal{A}_0-} \right)_{\mathcal{E}-} (\Omega, u) = \left(\overline{F_{\mathcal{E}_0-}} \right)_{\mathcal{A}_0-} (\Omega, u) \text{ for every } (\Omega, u) \in \mathcal{E} \times L^1_{\mathrm{loc}}(\mathbf{R}^n),$$

from which the inner regularity of the restriction of $(\overline{F_{\mathcal{E}_0-}})_{\mathcal{A}_0-}$ to $\mathcal{E} \times L^1_{\mathrm{loc}}(\mathbf{R}^n)$ follows. ∎

We can now prove the existence and uniqueness result.

To do this, we take $\mathcal{E} \subseteq \mathcal{A}_0$ satisfying

$$(8.2.2) \qquad\qquad x_0 + \Omega \in \mathcal{E} \text{ whenever } x_0 \in \mathbf{R}^n, \ \Omega \in \mathcal{E}.$$

Theorem 8.2.3. *Let $\mathcal{E}_0 \subseteq \mathcal{A}_0$ satisfy (8.1.1), and $F: \mathcal{E}_0 \times C^\infty(\mathbf{R}^n) \to [0, +\infty]$. Assume that F is inner regular, translation invariant, convex, and $C^\infty(\mathbf{R}^n)$-lower semicontinuous. Then, for every $\mathcal{E} \subseteq \mathcal{A}_0$ perfect with respect to $\mathcal{A}_0$, having $\mathcal{E}_0$ as a dense subset, and satisfying (8.2.2), and for every sequentially complete Hausdorff locally convex topological vector space U satisfying (7.3.5)÷(7.3.7), the restriction of $(\overline{F_{\mathcal{E}_0-}})_{\mathcal{A}_0-}$ to $\mathcal{E} \times U$ is the only inner regular translation invariant convex U-lower semicontinuous functional from $\mathcal{E} \times U$ to $[0, +\infty]$ that agrees with F on $\mathcal{E}_0 \times (C^\infty(\mathbf{R}^n) \cap U)$.*

Proof. Let $\mathcal{E}$, U be as above. Then by (8.2.2), and Proposition 8.2.2 it follows that the restriction of $(\overline{F_{\mathcal{E}_0-}})_{\mathcal{A}_0-}$ to $\mathcal{E} \times U$ is an inner regular translation invariant convex U-lower semicontinuous functional from $\mathcal{E} \times U$ to $[0, +\infty]$ that agrees with F on $\mathcal{E}_0 \times (C^\infty(\mathbf{R}^n) \cap U)$. By Theorem 8.1.2, it is the only one with such properties. ∎

§8.3 Unique Extension Results for Measure Like Functionals

In Theorem 8.2.3 a central role is played by inner regularity assumptions. In the theorems below we propose some results in the same order of ideas of Theorem 8.2.3, but under groups of assumptions implying inner regularity conditions, and determining again closed classes of functionals in which the extension processes can be carried out.

Definition 8.3.1. *Let $\mathcal{O} \subseteq \mathcal{A}(\mathbf{R}^n)$, U be a set, and $\Phi: \mathcal{O} \times U \to [0, +\infty]$. We say that Φ is*
i) boundary superadditive for every $u \in U$, so is $\Phi(\cdot, u)$,
ii) boundary subadditive if for every $u \in U$, so is $\Phi(\cdot, u)$,

iii) a Borel positive measure if for every $u \in U$, $\Phi(\cdot, u)$ is the restriction to $\mathcal{O}$ of a Borel positive measure.

Proposition 8.3.2. *Let $\mathcal{E}_0 \subseteq \mathcal{A}_0$ satisfy (8.1.1), and (2.6.1) with $\mathcal{O} = \mathcal{E}_0$, and let $F: \mathcal{E}_0 \times C^\infty(\mathbf{R}^n) \to [0, +\infty]$. Assume that F is increasing, translation invariant, convex, $C^\infty(\mathbf{R}^n)$-lower semicontinuous, boundary superadditive, boundary subadditive, and satisfying the following conditions:*
i) for every $(\Omega, u) \in \mathcal{E}_0 \times C^\infty(\mathbf{R}^n)$ such that $F(\Omega, u) < +\infty$, F is vanishing along the sequences in $\mathcal{E}_0$ that are well decreasing to the empty set with respect to Ω,
ii) for every $(\Omega, u) \in \mathcal{E}_0 \times C^\infty(\mathbf{R}^n)$ such that $F(\Omega, u) = +\infty$, F is diverging along the sequences in $\mathcal{E}_0$ that are well increasing to Ω.
Then, for every $\mathcal{E} \subseteq \mathcal{A}_0$ perfect with respect to $\mathcal{A}_0$, having $\mathcal{E}_0$ as a dense subset, and satisfying (8.2.2) and (2.6.1) with $\mathcal{O} = \mathcal{E}$, and for every sequentially complete Hausdorff locally convex topological vector space U satisfying $(7.3.5) \div (7.3.7)$, the restriction of $(\overline{F_{\mathcal{E}_0-}})_{\mathcal{A}_0-}$ to $\mathcal{E} \times U$ is the only functional from $\mathcal{E} \times U$ to $[0, +\infty]$ that
i) is equal to F on $\mathcal{E}_0 \times (C^\infty(\mathbf{R}^n) \cap U)$,
ii) is increasing, translation invariant, convex, U-lower semicontinuous, boundary superadditive, boundary subadditive,
iii) vanishes along the sequences in $\mathcal{E}$ that are well decreasing to the empty set with respect to Ω, for every $(\Omega, u) \in \mathcal{E} \times U$ where it is finite,
iv) diverges along the sequences in $\mathcal{E}$ that are well increasing to Ω, for every $(\Omega, u) \in \mathcal{E} \times U$ where it is not finite.

Proof. Let $\mathcal{E}$, U be as above.

It is clear that $\mathcal{E}_0$ too is perfect with respect to $\mathcal{A}_0$, therefore by Proposition 2.6.10, the inner regularity of F follows.

Because of this, and of the assumptions on F, Theorem 8.2.3 applies and we conclude that the restriction of $(\overline{F_{\mathcal{E}_0-}})_{\mathcal{A}_0-}$ to $\mathcal{E} \times U$ is the only inner regular translation invariant convex U-lower semicontinuous functional from $\mathcal{E} \times U$ to $[0, +\infty]$ that is equal to F on $\mathcal{E}_0 \times (C^\infty(\mathbf{R}^n) \cap U)$.

We now prove some additional properties of $(\overline{F_{\mathcal{E}_0-}})_{\mathcal{A}_0-}$.

It is obvious that $(\overline{F_{\mathcal{E}_0-}})_{\mathcal{A}_0-}$ is increasing.

Let us prove that the restriction of $(\overline{F_{\mathcal{E}_0-}})_{\mathcal{A}_0-}$ to $\mathcal{E} \times U$ is boundary superadditive.

Let Ω, A, $B \in \mathcal{E}$, with $A \subset\subset B \subset\subset \Omega$, $u \in U$, and by using the properties of $\mathcal{E}_0$ and $\mathcal{E}$, let Ω', $B' \in \mathcal{E}_0$, be such that $B \subset\subset B' \subset\subset \Omega' \subset\subset \Omega$. Then by i) of Proposition 2.6.9, Lemma 7.3.4 applied with $\mathcal{O} = \mathcal{E}$ and $\Phi = (\overline{F_{\mathcal{E}_0-}})_{\mathcal{A}_0-}$ restricted to $\mathcal{E} \times U$, by the properties of $(\overline{F_{\mathcal{E}_0-}})_{\mathcal{A}_0-}$, the inner regularity and the boundary superadditivity of F, and by (2.6.1) with $\mathcal{O} = \mathcal{E}_0$ we get that

$$(8.3.1) \qquad (\overline{F_{\mathcal{E}_0-}})_{\mathcal{A}_0-}(\Omega, u) = \left((\overline{F_{\mathcal{E}_0-}})_{\mathcal{A}_0-}\right)_{\mathcal{E}-}(\Omega, u) \geq$$

$$\geq \left(\overline{F_{\mathcal{E}_0-}}\right)_{\mathcal{A}_0-} (\Omega', u_\varepsilon) = F_{\mathcal{E}_0-}(\Omega', u_\varepsilon) = F(\Omega', u_\varepsilon) \geq$$

$$\geq F(A, u_\varepsilon) + F(\Omega' \setminus \overline{B'}, u_\varepsilon) \geq F_{\mathcal{E}_0-}(A, u_\varepsilon) + F_{\mathcal{E}_0-}(\Omega' \setminus \overline{B'}, u_\varepsilon)$$

for every $\varepsilon > 0$ sufficiently small.

Then, by (8.3.1), and Proposition 7.3.3 we conclude that

$$\left(\overline{F_{\mathcal{E}_0-}}\right)_{\mathcal{A}_0-} (\Omega, u) \geq \left(\overline{F_{\mathcal{E}_0-}}\right)_{\mathcal{A}_0-} (A, u) + \left(\overline{F_{\mathcal{E}_0-}}\right)_{\mathcal{A}_0-} (\Omega' \setminus \overline{B'}, u)$$

for every Ω', $B' \in \mathcal{E}_0$ with $B \subset\subset B' \subset\subset \Omega' \subset\subset \Omega$,

from which, together with the density of $\mathcal{E}_0$ with respect to $\mathcal{E}$, the boundary superadditivity of $(\overline{F_{\mathcal{E}_0-}})_{\mathcal{A}_0-}$ follows as Ω' increases to Ω and B' decreases to B.

Let us now prove that the restriction of $(\overline{F_{\mathcal{E}_0-}})_{\mathcal{A}_0-}$ to $\mathcal{E} \times U$ is boundary subadditive.

Let Ω, A, $B \in \mathcal{E}$, with $A \subset\subset B \subset\subset \Omega$, $u \in U$, and by the density of $\mathcal{E}_0$ with respect to $\mathcal{E}$, let Ω', A', $B' \in \mathcal{E}_0$, be such that $A \subset\subset A' \subset\subset B' \subset\subset B \subset\subset \Omega' \subset\subset \Omega$. Then, by the same arguments used above, we get that

$$(8.3.2) \qquad \left(\overline{F_{\mathcal{E}_0-}}\right)_{\mathcal{A}_0-} (B, u) = \left(\left(\overline{F_{\mathcal{E}_0-}}\right)_{\mathcal{A}_0-}\right)_{\mathcal{E}-} (B, u) \geq$$

$$\geq \left(\overline{F_{\mathcal{E}_0-}}\right)_{\mathcal{A}_0-} (B', u_\varepsilon) = F(B', u_\varepsilon)$$

for every $\varepsilon > 0$ sufficiently small.

Analogously, by (2.6.1) with $\mathcal{O} = \mathcal{E}_0$ we also deduce that

$$(8.3.3) \qquad \left(\overline{F_{\mathcal{E}_0-}}\right)_{\mathcal{A}_0-} (\Omega \setminus \overline{A}, u) \geq F(\Omega' \setminus \overline{A'}, u_\varepsilon)$$

for every $\varepsilon > 0$ sufficiently small.

Therefore by (8.3.2), (8.3.3), and the boundary subadditivity of F we conclude that

$$F(\Omega', u_\varepsilon) \leq F(B', u_\varepsilon) + F(\Omega' \setminus \overline{A'}, u_\varepsilon) \leq$$

$$\leq \left(\overline{F_{\mathcal{E}_0-}}\right)_{\mathcal{A}_0-} (B, u) + \left(\overline{F_{\mathcal{E}_0-}}\right)_{\mathcal{A}_0-} (\Omega \setminus \overline{A}, u)$$

for every $\Omega' \in \mathcal{E}_0$ with $\Omega' \subset\subset \Omega$, $\varepsilon > 0$ sufficiently small,

from which, together with Proposition 7.3.3, we obtain as ε decreases to 0 that

$$(8.3.4) \quad \left(\overline{F_{\mathcal{E}_0-}}\right)_{\mathcal{A}_0-} (\Omega', u) \leq \left(\overline{F_{\mathcal{E}_0-}}\right)_{\mathcal{A}_0-} (B, u) + \left(\overline{F_{\mathcal{E}_0-}}\right)_{\mathcal{A}_0-} (\Omega \setminus \overline{A}, u)$$

for every $\Omega' \in \mathcal{E}_0$ with $\Omega' \subset\subset \Omega$.

By (8.3.4), and the density of $\mathcal{E}_0$ with respect to $\mathcal{E}$ the boundary sub-additivity $(\overline{F_{\mathcal{E}_0-}})_{\mathcal{A}_0-}$ follows as Ω' increases to Ω.

Finally, by Proposition 2.6.10, the vanishing of $(\overline{F_{\mathcal{E}_0-}})_{\mathcal{A}_0-}$ along the sequences in $\mathcal{E}$ that are well decreasing to the empty set with respect to Ω for every $(\Omega, u) \in \mathcal{E} \times U$ for which $(\overline{F_{\mathcal{E}_0-}})_{\mathcal{A}_0-}(\Omega, u) < +\infty$, and the diverging of $(\overline{F_{\mathcal{E}_0-}})_{\mathcal{A}_0-}$ along the sequences in $\mathcal{E}$ that are well increasing to Ω for every $(\Omega, u) \in \mathcal{E} \times U$ for which $(\overline{F_{\mathcal{E}_0-}})_{\mathcal{A}_0-}(\Omega, u) = +\infty$ too follow.

In conclusion, since by Proposition 2.6.10 every functional satisfying i)÷iv) is inner regular, also the uniqueness part of the proposition follows. ∎

By Proposition 8.3.2 we deduce the following result.

Proposition 8.3.3. *Let $\mathcal{E}_0 \subseteq \mathcal{A}_0$ be dense in $\mathcal{A}_0$, satisfy (8.1.1) and (2.6.1) with $\mathcal{O} = \mathcal{E}_0$, and let $F: \mathcal{E}_0 \times C^\infty(\mathbf{R}^n) \to [0, +\infty]$. Assume that F is translation invariant, convex, $C^\infty(\mathbf{R}^n)$-lower semicontinuous, and a Borel positive measure. Then, for every $\mathcal{E} \subseteq \mathcal{A}_0$ with $\mathcal{E}_0 \subseteq \mathcal{E}$, and satisfying (8.2.2) and (2.6.1) with $\mathcal{O} = \mathcal{E}$, and for every sequentially complete Hausdorff locally convex topological vector space U satisfying (7.3.5)÷(7.3.7), the restriction of $(\overline{F_{\mathcal{E}_0-}})_{\mathcal{A}_0-}$ to $\mathcal{E} \times U$ is the only translation invariant convex U-lower semicontinuous functional from $\mathcal{E} \times U$ to $[0, +\infty]$ that is equal to F on $\mathcal{E}_0 \times (C^\infty(\mathbf{R}^n) \cap U)$, and is a Borel positive measure.*

Proof. We first observe that every translation invariant convex U-lower semicontinuous functional from $\mathcal{E} \times U$ to $[0, +\infty]$ equal to F on $\mathcal{E}_0 \times (C^\infty(\mathbf{R}^n) \cap U)$, and that is a Borel positive measure, actually fulfils also conditions i)÷iv) of Proposition 8.3.2. Then the result follows from Proposition 8.3.2, once we prove that $(\overline{F_{\mathcal{E}_0-}})_{\mathcal{A}_0-}$ is a Borel positive measure.

To do this, we prove that the conditions of Theorem 2.6.12 with $\mathcal{O} = \mathcal{E}$ are fulfilled.

Let us start with the superadditivity condition.

Let $u \in U$, $\Omega, \Omega_1, \Omega_2 \in \mathcal{E}$ with $\Omega_1 \cup \Omega_2 \subseteq \Omega$ and $\Omega_1 \cap \Omega_2 = \emptyset$, and let Ω_1', $\Omega_2' \in \mathcal{E}_0$ be such that $\Omega_1' \subset\subset \Omega_1$, $\Omega_2' \subset\subset \Omega_2$. By using the properties of $\mathcal{E}_0$ and $\mathcal{E}$, let $\Omega' \in \mathcal{E}_0$ satisfying $\Omega' \subset\subset \Omega$, $\Omega_1' \subset\subset \Omega' \cap \Omega_1$, and $\Omega_2' \subset\subset \Omega' \cap \Omega_2$. Then by Lemma 7.3.4 applied with $\mathcal{O} = \mathcal{E}$, and $\Phi = (\overline{F_{\mathcal{E}_0-}})_{\mathcal{A}_0-}$, the inner regularity of $(\overline{F_{\mathcal{E}_0-}})_{\mathcal{A}_0-}$, the properties of $(\overline{F_{\mathcal{E}_0-}})_{\mathcal{A}_0-}$, the measure theoretic properties of F, and (2.6.1) we get that

$$(8.3.5) \qquad (\overline{F_{\mathcal{E}_0-}})_{\mathcal{A}_0-}(\Omega, u) \geq (\overline{F_{\mathcal{E}_0-}})_{\mathcal{A}_0-}(\Omega', u_\varepsilon) = F(\Omega', u_\varepsilon) \geq$$

$$\geq F(\Omega_1', u_\varepsilon) + F(\Omega_2', u_\varepsilon) \geq F_{\mathcal{E}_0-}(\Omega_1', u_\varepsilon) + F_{\mathcal{E}_0-}(\Omega_2', u_\varepsilon)$$

for every $\varepsilon > 0$ sufficiently small.

By (8.3.5), and Proposition 7.3.3 we conclude that

$$(\overline{F_{\mathcal{E}_0-}})_{\mathcal{A}_0-}(\Omega, u) \geq (\overline{F_{\mathcal{E}_0-}})_{\mathcal{A}_0-}(\Omega_1', u) + (\overline{F_{\mathcal{E}_0-}})_{\mathcal{A}_0-}(\Omega_2', u)$$

for every Ω'_1, $\Omega'_2 \in \mathcal{E}_0$ with $\Omega'_1 \subset\subset \Omega_1$, $\Omega'_2 \subset\subset \Omega_2$,

from which, using again the properties of $\mathcal{E}_0$ and $\mathcal{E}$, and Proposition 2.6.9, it follows that

$$\left(\overline{F_{\mathcal{E}_0-}}\right)_{\mathcal{A}_0-}(\Omega, u) \geq \left(\overline{F_{\mathcal{E}_0-}}\right)_{\mathcal{A}_0-}(\Omega_1, u) + \left(\overline{F_{\mathcal{E}_0-}}\right)_{\mathcal{A}_0-}(\Omega_2, u)$$

for every Ω, Ω_1, $\Omega_2 \in \mathcal{E}$ with $\Omega_1 \cup \Omega_2 \subseteq \Omega$ and $\Omega_1 \cap \Omega_2 = \emptyset$, $u \in U$.

We now prove the subadditivity condition.

Let $u \in U$, Ω, Ω_1, $\Omega_2 \in \mathcal{E}$ with $\Omega \subseteq \Omega_1 \cup \Omega_2$, and let $\Omega' \in \mathcal{E}_0$ be such that $\Omega' \subset\subset \Omega$. By the properties of $\mathcal{E}_0$ and $\mathcal{E}$, let Ω'_1, $\Omega'_2 \in \mathcal{E}_0$ with $\Omega'_1 \subset\subset \Omega_1$, $\Omega'_2 \subset\subset \Omega_2$, and $\Omega' \subseteq \Omega'_1 \cup \Omega'_1$. Then by Lemma 7.3.4 applied with $\mathcal{O} = \mathcal{E}$, and $\Phi = \left(\overline{F_{\mathcal{E}_0-}}\right)_{\mathcal{A}_0-}$, the inner regularity of $\left(\overline{F_{\mathcal{E}_0-}}\right)_{\mathcal{A}_0-}$, the properties of $\left(\overline{F_{\mathcal{E}_0-}}\right)_{\mathcal{A}_0-}$, the measure theoretic properties of F, and (2.6.1) we get that

$$(8.3.6) \qquad \left(\overline{F_{\mathcal{E}_0-}}\right)_{\mathcal{A}_0-}(\Omega_1, u) + \left(\overline{F_{\mathcal{E}_0-}}\right)_{\mathcal{A}_0-}(\Omega_2, u) \geq$$

$$\geq \left(\overline{F_{\mathcal{E}_0-}}\right)_{\mathcal{A}_0-}(\Omega'_1, u_\varepsilon) + \left(\overline{F_{\mathcal{E}_0-}}\right)_{\mathcal{A}_0-}(\Omega'_2, u_\varepsilon) = F(\Omega'_1, u_\varepsilon) + F(\Omega'_2, u_\varepsilon) \geq$$

$$\geq F(\Omega', u_\varepsilon) \geq F_{\mathcal{E}_0-}(\Omega', u_\varepsilon) \text{ for every } \varepsilon > 0 \text{ sufficiently small.}$$

By (8.3.6), and Proposition 7.3.3 we conclude that

$$\left(\overline{F_{\mathcal{E}_0-}}\right)_{\mathcal{A}_0-}(\Omega_1, u) + \left(\overline{F_{\mathcal{E}_0-}}\right)_{\mathcal{A}_0-}(\Omega_2, u) \geq \left(\overline{F_{\mathcal{E}_0-}}\right)_{\mathcal{A}_0-}(\Omega', u)$$

for every $\Omega' \in \mathcal{E}_0$ with $\Omega' \subset\subset \Omega$,

from which it follows that

$$\left(\overline{F_{\mathcal{E}_0-}}\right)_{\mathcal{A}_0-}(\Omega, u) \leq \left(\overline{F_{\mathcal{E}_0-}}\right)_{\mathcal{A}_0-}(\Omega_1, u) + \left(\overline{F_{\mathcal{E}_0-}}\right)_{\mathcal{A}_0-}(\Omega_2, u)$$

for every Ω, Ω_1, $\Omega_2 \in \mathcal{E}$ with $\Omega \subseteq \Omega_1 \cup \Omega_2$, $u \in U$.

By the above conditions, and the inner regularity of $\left(\overline{F_{\mathcal{E}_0-}}\right)_{\mathcal{A}_0-}$ the proof follows by using Theorem 2.6.12. ∎

§8.4 Some Applications

In the present section we apply the results of the previous ones to some integral functionals of the calculus of variations.

Proposition 8.4.1. *Let $\mathcal{E}_0 \subseteq \mathcal{A}_0$ satisfy (8.1.1), $k \in \mathbf{N}$, $f: \mathbf{R} \times \mathbf{R}^n \times \mathbf{R}^{n^2} \times \ldots \times \mathbf{R}^{n^k} \to [0, +\infty]$ be convex and lower semicontinuous, and $F: (\Omega, u) \in \mathcal{E}_0 \times C^\infty(\mathbf{R}^n) \mapsto \int_\Omega f(u, \nabla u, \nabla^2 u, \ldots, \nabla^k u) dx$. Then, for every $\mathcal{E} \subseteq \mathcal{A}_0$ perfect with respect to $\mathcal{A}_0$, having $\mathcal{E}_0$ as a dense subset, and satisfying (8.2.2), and for every sequentially complete Hausdorff locally convex topological vector space U satisfying (7.3.5)÷(7.3.7), the restriction of $(\overline{F_{\mathcal{E}_0-}})_{\mathcal{A}_0-}$ to $\mathcal{E} \times U$ is the only inner regular (respectively measure, provided (2.6.1) with $\mathcal{O} = \mathcal{E}_0$ and $\mathcal{O} = \mathcal{E}$ is fulfilled) translation invariant convex U-lower semicontinuous functional from $\mathcal{E} \times U$ to $[0, +\infty]$ that agrees with F on $\mathcal{E}_0 \times (C^\infty(\mathbf{R}^n) \cap U)$.*

Proof. Follows trivially from Theorem 8.2.3 (respectively from Proposition 8.3.3). ∎

Proposition 8.4.2. *Let $\mathcal{E}_0 \subseteq \mathcal{A}_0$ satisfy (8.1.1), $f: \mathbf{R}^n \to [0, +\infty]$ be convex and lower semicontinuous, and let*

$$F: (\Omega, u) \in \mathcal{E}_0 \times C^\infty(\mathbf{R}^n) \mapsto \int_\Omega f(\nabla u) dx.$$

Then, for every $\mathcal{E} \subseteq \mathcal{A}_0$ perfect with respect to $\mathcal{A}_0$, having $\mathcal{E}_0$ as a dense subset and satisfying (8.2.2), the functional

$$\widetilde{F}: (\Omega, u) \in \mathcal{E} \times BV(\mathbf{R}^n) \mapsto \int_\Omega f(\nabla u) dx + \int_\Omega f^\infty(\nabla^s u) d|D^s u|$$

is the only inner regular (respectively measure, provided (2.6.1) with $\mathcal{O} = \mathcal{E}_0$ and $\mathcal{O} = \mathcal{E}$ are fulfilled) translation invariant convex $L^1_{\mathrm{loc}}(\mathbf{R}^n)$-lower semicontinuous functional from $\mathcal{E} \times BV(\mathbf{R}^n)$ to $[0, +\infty]$ equal to F on $\mathcal{E}_0 \times (C^\infty(\mathbf{R}^n) \cap BV(\mathbf{R}^n))$.

If, in addition, f satisfies

$$(8.4.1) \qquad\qquad |z| \leq f(z) \text{ for every } z \in \mathbf{R}^n,$$

then, for every $\mathcal{E} \subseteq \mathcal{A}_0$ perfect with respect to $\mathcal{A}_0$, having $\mathcal{E}_0$ as a dense subset, and satisfying (8.2.2), the functional

$$\widehat{F}: (\Omega, u) \in \mathcal{E} \times L^1_{\mathrm{loc}}(\mathbf{R}^n) \mapsto$$

$$\begin{cases} \int_\Omega f(\nabla u) dx + \int_\Omega f^\infty(\nabla^s u) d|D^s u| & \text{if } u \in BV(\Omega) \\ +\infty & \text{if } u \in L^1_{\mathrm{loc}}(\mathbf{R}^n) \setminus BV(\Omega) \end{cases}$$

is the only inner regular (respectively measure, provided (2.6.1) with $\mathcal{O} = \mathcal{E}_0$ and $\mathcal{O} = \mathcal{E}$ are fulfilled) translation invariant convex $L^1_{\mathrm{loc}}(\mathbf{R}^n)$-lower semicontinuous functional from $\mathcal{E} \times L^1_{\mathrm{loc}}(\mathbf{R}^n)$ to $[0, +\infty]$ equal to F on $\mathcal{E}_0 \times C^\infty(\mathbf{R}^n)$.

Proof. We prove only the part of the proposition under inner regularity assumptions, the other one being analogous.

We start with the part relative to $\widetilde{F}$.

In this case we observe that the properties of $\mathcal{E}$, Theorem 7.4.3, the properties of the translated of $BV(\mathbf{R}^n)$ functions, Theorem 7.4.2, and Theorem 5.1.4 provide that $\widetilde{F}$ is inner regular, translation invariant, convex, and weak*-$BV(\mathbf{R}^n)$-lower semicontinuous. Consequently, Theorem 8.2.3 applies with $U = BV(\mathbf{R}^n)$ equipped with the weak*-$BV(\mathbf{R}^n)$ topology, and we conclude that $\widetilde{F}$ is the only inner regular translation invariant convex weak*-$BV(\mathbf{R}^n)$-lower semicontinuous functional from $\mathcal{E} \times BV(\mathbf{R}^n)$ to $[0, +\infty]$ equal to F on $\mathcal{E}_0 \times (C^\infty(\mathbf{R}^n) \cap BV(\mathbf{R}^n))$.

We now observe that, because of Theorem 7.4.6, $\widetilde{F}$ is actually $L^1(\mathbf{R}^n)$-lower semicontinuous. This implies the proposition for $\widetilde{F}$, once we observe that every $L^1(\mathbf{R}^n)$-lower semicontinuous functional on $\mathcal{E} \times BV(\mathbf{R}^n)$ is also weak*-$BV(\mathbf{R}^n)$-lower semicontinuous.

We now treat the part relative to $\widehat{F}$.

In this case the proof follows from Theorem 8.2.3, once we prove that

$$(8.4.2) \qquad \left(\overline{F_{\mathcal{E}_0-}}\right)_{\mathcal{A}_0-}(\Omega, u) = \widehat{F}(\Omega, u) \text{ for every } (\Omega, u) \in \mathcal{E} \times L^1_{\mathrm{loc}}(\mathbf{R}^n).$$

To do this let us first prove that $\widehat{F}$ is $L^1_{\mathrm{loc}}(\mathbf{R}^n)$-lower semicontinuous.

Let $(\Omega, u) \in \mathcal{E} \times L^1_{\mathrm{loc}}(\mathbf{R}^n)$, let $\{u_h\} \subseteq L^1_{\mathrm{loc}}(\mathbf{R}^n)$ be such that $u_h \to u$ in $L^1_{\mathrm{loc}}(\mathbf{R}^n)$, and let us assume that the limit $\lim_{h \to +\infty} \widehat{F}(\Omega, u_h)$ exists and is finite. Because of this, we infer that $u_h \in BV(\Omega)$ for every $h \in \mathbf{N}$ and by using (8.4.1) and Proposition 4.2.5, that $u \in BV(\Omega)$.

The proof of the $L^1_{\mathrm{loc}}(\mathbf{R}^n)$-lower semicontinuity of $\widehat{F}$ is thus reduced to the one of the $L^1_{\mathrm{loc}}(\mathbf{R}^n)$-lower semicontinuity of its restriction to $\mathcal{E} \times BV(\Omega)$, and this holds by Theorem 7.4.6.

The $L^1_{\mathrm{loc}}(\mathbf{R}^n)$-lower semicontinuity of $\widehat{F}$ implies that

$$\widehat{F}(A, u) \leq \overline{F_{\mathcal{E}_0-}}(B, u) \leq \left(\overline{F_{\mathcal{E}_0-}}\right)_{\mathcal{A}_0-}(\Omega, u)$$

for every Ω, $B \in \mathcal{A}_0$, $A \in \mathcal{E}_0$ with $A \subset\subset B \subset\subset \Omega$, $u \in L^1_{\mathrm{loc}}(\mathbf{R}^n)$,

from which, being $\mathcal{E}_0$ dense with respect to $\mathcal{E}$, we conclude that

$$(8.4.3) \qquad \widehat{F}(\Omega, u) = \widehat{F}_{\mathcal{E}_0-}(\Omega, u) \leq \left(\overline{F_{\mathcal{E}_0-}}\right)_{\mathcal{A}_0-}(\Omega, u)$$

for every $(\Omega, u) \in \mathcal{E} \times L^1_{\mathrm{loc}}(\mathbf{R}^n)$.

Conversely, we observe that Theorem 7.4.3 and the properties of the translated of $BV(\mathbf{R}^n)$ functions yield the translation invariance of $\widehat{F}$, and that Theorem 7.4.2 provides its convexity. Then, by Lemma 7.3.4 applied with $\mathcal{O} = \mathcal{E}_0$, $U = L^1_{\mathrm{loc}}(\mathbf{R}^n)$, and $\Phi = \widehat{F}$ we get that

$$\widehat{F}(\Omega, u) \geq \widehat{F}_{\mathcal{E}_0-}(\Omega, u) \geq \widehat{F}(A, u_\varepsilon) = F(A, u_\varepsilon) = F_{\mathcal{E}_0-}(A, u_\varepsilon)$$

for every $\Omega \in \mathcal{E}$, $A \in \mathcal{E}_0$ with $A \subset\subset \Omega$, $\varepsilon \in \,]0, \mathrm{dist}(A, \partial\Omega[$, $u \in L^1_{\mathrm{loc}}(\mathbf{R}^n)$, from which it follows that

$$(8.4.4) \qquad \widehat{F}(\Omega, u) \geq \overline{F_{\mathcal{E}_0-}}(A, u)$$

for every $\Omega \in \mathcal{E}$, $A \in \mathcal{E}_0$ with $A \subset\subset \Omega$, $u \in L^1_{\mathrm{loc}}(\mathbf{R}^n)$.

By (8.4.4), and ii) of Proposition 2.6.9 we get that

$$\widehat{F}(\Omega, u) \geq \left(\overline{F_{\mathcal{E}_0-}}\right)_{\mathcal{E}_0-}(\Omega, u) = \left(\overline{F_{\mathcal{E}_0-}}\right)_{\mathcal{E}-}(\Omega, u)$$

for every $(\Omega, u) \in \mathcal{E} \times L^1_{\mathrm{loc}}(\mathbf{R}^n)$,

from which, being $\mathcal{E}$ perfect with respect to $\mathcal{A}_0$, we conclude that

$$(8.4.5) \qquad \widehat{F}(\Omega, u) \geq \left(\overline{F_{\mathcal{E}_0-}}\right)_{\mathcal{A}_0-}(\Omega, u) \text{ for every } (\Omega, u) \in \mathcal{E} \times L^1_{\mathrm{loc}}(\mathbf{R}^n).$$

By (8.4.3) and (8.4.5), equality (8.4.2) follows. This completes the proof of the proposition. ∎

Corollary 8.4.3. *Let $\mathcal{E}_0 \subseteq \mathcal{A}_0$ satisfy (8.1.1), and*

$$A \colon (\Omega, u) \in \mathcal{E}_0 \times C^\infty(\mathbf{R}^n) \mapsto \int_\Omega \sqrt{1 + |\nabla u|^2} dx.$$

Then, for every $\mathcal{E} \subseteq \mathcal{A}_0$ perfect with respect to $\mathcal{A}_0$, having $\mathcal{E}_0$ as a dense subset, and satisfying (8.2.2) the functional

$$\widehat{A} \colon (\Omega, u) \in \mathcal{E} \times L^1_{\mathrm{loc}}(\mathbf{R}^n) \mapsto$$

$$\begin{cases} \int_\Omega \sqrt{1 + |\nabla u|^2} dx + |D^s u|(\Omega) & \text{if } u \in BV(\Omega) \\ +\infty & \text{if } u \in L^1_{\mathrm{loc}}(\mathbf{R}^n) \setminus BV(\Omega) \end{cases}$$

is the only inner regular (respectively measure, provided (2.6.1) with $\mathcal{O} = \mathcal{E}_0$ and $\mathcal{O} = \mathcal{E}$ are fulfilled) translation invariant convex $L^1_{\mathrm{loc}}(\mathbf{R}^n)$-lower semicontinuous functional from $\mathcal{E} \times L^1_{\mathrm{loc}}(\mathbf{R}^n)$ to $[0, +\infty]$ equal to A on $\mathcal{E}_0 \times C^\infty(\mathbf{R}^n)$.

Proof. Follows from Proposition 8.4.2. ∎

§8.5 A Note on Lavrentiev Phenomenon

In this section we make some simple remark to emphasize the connections of Lavrentiev phenomenon with the unique extension processes studied in this chapter.

In [La], in connection with Tonelli's partial regularity theorem for the minimizers of one dimensional Dirichlet minimum problems, (cf. [To]), M. Lavrentiev observed the occurrence of the surprising feature of some Dirichlet minimum problems for integral functionals to depend critically on slight variations of the set of admissible functions. He produced an example of a rather elaborated one dimensional integral functional whose minimum on Sobolev classes is strictly smaller than the infimum on sets of smooth functions.

It is to be emphasized that this feature is surprising since the proposed functional enjoyed some convexity and weak lower semicontinuity properties, and smooth functions are dense in Sobolev spaces.

Starting from Lavrentiev's work, many papers have been devoted to the study of the phenomenon (cf. for example [M2], [HM], [BM2], [An], [Cs2], [CPSC]), and, in some recent papers (cf. [BuM1], [DA1]) an abstract interpretation of Lavrentiev phenomenon by means of relaxation has been proposed.

Given a topological space (U, τ), a τ-dense subset V of U, and a τ-lower semicontinuous functional $F \colon U \to \,] - \infty, +\infty]$, the τ-lower semicontinuous envelope $\overline{F_V}$ of

$$F_V \colon u \in U \mapsto \begin{cases} F(u) & \text{if } u \in V \\ +\infty & \text{otherwise} \end{cases}$$

has been considered, and it has been observed that, since $\inf\{F(u) : u \in V\} = \inf\{\overline{F_V}(u) : u \in U\}$, the nonoccurrence of the Lavrentiev phenomenon for F, U and V, i.e. the equality $\inf\{F(u) : u \in U\} = \inf\{F(u) : u \in V\}$, can be deduced by the equality $\overline{F_V} = F$. In this framework the occurrence of the Lavrentiev phenomenon for various classes of minimum problems has been studied in many papers also for multiple integrals of the calculus of variations defined in Sobolev and BV spaces, (cf. for example [AM], [ASC], [BB], [BuM2], [CEDA2], [DA3], [DAT2], [Z2], and the survey paper [BuB]).

In particular, in [DA3] the quadratic form

$$q \colon (x, z) \in (\mathbf{R}^n \setminus \{0\}) \times \mathbf{R}^n \to \lambda |z|^2 + \frac{1}{|x|^{n-1}} \left| \frac{x}{|x|} \cdot z \right|^2$$

$(n \geq 3)$ has been proposed so that, for a suitable choice of λ, the functional

$$F \colon u \in W^{1,1}(B_1(0)) \mapsto \int_{B_1(0)} q(x, \nabla u) dx$$

is $L^1(B_1(0))$-lower semicontinuous, but

$$F(u^*) < \overline{F_{C^1(\mathbf{R}^n)}}(u^*),$$

where $u^*(x) = \frac{x_1}{|x|}$.

This example provides an example in which a convex quadratic functional, namely the restriction of F to $C^1(\mathbf{R}^n)$, possesses two different $L^1(B_1(0))$-lower semicontinuous convex extensions to $W^{1,1}(B_1(0))$.

Chapter 9

Integral Representation
for Unbounded Functionals

In the present chapter we give some characterizations of the unbounded
functionals F, depending on an open set Ω and a function u in Sobolev or
BV spaces, that can be represented in an integral form of the kind

$$F(\Omega, u) = \int_\Omega f(\nabla u)dx,$$

when u is a Sobolev function, or

$$F(\Omega, u) = \int_\Omega f(\nabla u)dx + \int_\Omega f^\infty(\nabla^s u)d|D^s u|,$$

when u is a BV one, for some f taking values in $[0, +\infty]$.

§9.1 Representation on Linear Functions

In the present section and in the next one we prove some integral representation results for an abstract functional F depending on a bounded open
set Ω, and u in $C^1(\mathbf{R}^n)$. We start treating the case when u is a linear
function.

Let us consider a functional

(9.1.1) $\qquad F \colon (\Omega, u) \in \mathcal{A}_0 \times W_{\text{loc}}^{1,\infty}(\mathbf{R}^n) \mapsto F(\Omega, u) \in [0, +\infty]$

satisfying

(9.1.2) $\quad F(x_0 + \Omega, u_z) = F(\Omega, u_z)$ for every $\Omega \in \mathcal{A}_0$, $z \in \mathbf{R}^n$, $x_0 \in \mathbf{R}^n$,

(9.1.3)   for every $z \in \mathbf{R}^n$ $F(\cdot, u_z)$ is increasing,

(9.1.4)   for every $z \in \mathbf{R}^n$ $F(\cdot, u_z)$ is weakly superadditive,

(9.1.5)   for every $z \in \mathbf{R}^n$ $F(\cdot, u_z)$ is weakly subadditive,

and introduce the function f_F defined by

(9.1.6) $f_F \colon z \in \mathbf{R}^n \mapsto F(Y, u_z) \in [0, +\infty]$.

Proposition 9.1.1. *Let F be as in (9.1.1) satisfying (9.1.2)÷(9.1.5), and let f_F be given by (9.1.6). Then*

(9.1.7) $F(\Omega, u_z) \le \mathcal{L}^n(\overline{\Omega}) f_F(z)$ *for every* $\Omega \in \mathcal{A}_0$, $z \in \mathbf{R}^n$,

(9.1.8) $\mathcal{L}^n(\Omega) f_F(z) \le F(\Omega, u_z)$ *for every* $\Omega \in \mathcal{A}_0$, $z \in \mathbf{R}^n$.

　　Proof. Let z be in $\mathbf{R}^n$.

　　If $F(\emptyset, u_z) = 0$ let $F^e(\cdot, u_z) \colon A \in \mathcal{A}(\mathbf{R}^n) \mapsto \sup\{F(B, u_z) : B \in \mathcal{A}_0,\ B \subseteq A\}$, then by (9.1.2)÷(9.1.5) it follows that $F^e(\cdot, u_z)$ extends $F(\cdot, u_z)$, is increasing, weakly superadditive, weakly subadditive, and translation invariant. By Proposition 2.6.15 applied with $\alpha = F^e(\cdot, u_z)$, the proof follows.

　　If $F(\emptyset, u_z) \ne 0$, by (9.1.4), and (9.1.5) it must necessarily result $F(\emptyset, u_z) = +\infty$ from which, together with (9.1.3), (9.1.7) and (9.1.8) follow. ∎

§9.2 Representation on Continuously Differentiable Functions

Let F be as in (9.1.1), and f_F be given by (9.1.6).

　　In order to extend the results of §9.1 to C^1 functions, we assume that F satisfies also the following conditions (recall that for every $u = \sum_{j=1}^{m}(u_{z_j} + s_j)\chi_{P_j} \in PA(\mathbf{R}^n)$ we have set $B_u = \cup_{j=1}^{m}(\overline{P_j} \setminus \text{int}(P_j))$)

(9.2.1) $F(\Omega, u) \le \limsup_{h \to +\infty} F(\Omega \setminus B_{u_h}, u_h)$ for every $\Omega \in \mathcal{A}_0$, $u \in C^1(\mathbf{R}^n)$,

$$\{u_h\} \subseteq PA(\mathbf{R}^n) \text{ with } u_h \to u \text{ in } W^{1,\infty}(\Omega),$$

that looks to be a coupling between lower semicontinuity and control hypotheses, and

(9.2.2)   the restriction of $F(Y, \cdot)$ to $\{u_z : z \in \mathbf{R}^n\}$ is convex.

Remark 9.2.1. Let F be as in (9.1.1), and let f_F be given by (9.1.6). Then it is clear that (9.2.2) implies the convexity of f_F.

Moreover (9.2.1), applied for every $\{z_h\} \subseteq \mathbf{R}^n$ and $z \in \mathbf{R}^n$ with $z_h \to z$, to $\Omega = Y$, $u = u_z$, and $u_h = u_{z_h}$ for every $h \in \mathbf{N}$, implies the lower semicontinuity of f_F, and in particular that, for every $u \in W^{1,1}_{\text{loc}}(\mathbf{R}^n)$, the function $x \in \mathbf{R}^n \mapsto f_F(\nabla u(x))$ is measurable.

Condition (9.2.1) is implied by the following assumptions

$$(9.2.3) \qquad F(\Omega, u) \leq F(\Omega \setminus B_u, u) \text{ for every } (\Omega, u) \in \mathcal{A}_0 \times PA(\mathbf{R}^n),$$

and

$$(9.2.4) \qquad \text{for every } \Omega \in \mathcal{A}_0, \ F(\Omega, \cdot) \text{ is } W^{1,\infty}(\Omega)\text{-lower semicontinuous.}$$

Proposition 9.2.2. *Let F be as in (9.1.1) satisfying (9.2.3) and (9.2.4). Then (9.2.1) holds.*

Proof. Let u, $\{u_h\}$ be as in (9.2.1). Then by (9.2.4), and (9.2.3) we have

$$F(\Omega, u) \leq \liminf_{h \to +\infty} F(\Omega, u_h) \leq \limsup_{h \to +\infty} F(\Omega \setminus B_{u_h}, u_h),$$

that is (9.2.1). ∎

We now assume that if (9.2.1), (9.2.2), the invariance and measure theoretic assumptions below

$$(9.2.5) \qquad F(\Omega, u_z + c) = F(\Omega, u_z) \text{ for every } \Omega \in \mathcal{A}_0, \ z \in \mathbf{R}^n, \ c \in \mathbf{R},$$

$$(9.2.6) \qquad \text{for every } u \in C^1(\mathbf{R}^n), \ F(\cdot, u) \text{ is increasing,}$$

$$(9.2.7) \qquad \text{for every } u \in W^{1,\infty}_{\text{loc}}(\mathbf{R}^n), \ F(\cdot, u) \text{ is superadditive,}$$

$$(9.2.8) \qquad \text{for every } u \in W^{1,\infty}_{\text{loc}}(\mathbf{R}^n), \ F(\cdot, u) \text{ is subadditive,}$$

together with

$$(9.2.9) \qquad F(\Omega - x_0, T[x_0]u_z) = F(\Omega, u_z)$$

$$\text{for every } \Omega \in \mathcal{A}_0, \ z \in \mathbf{R}^n, \ x_0 \in \mathbf{R}^n,$$

$$(9.2.10) \quad \limsup_{r \to 0^+} \frac{1}{r^n} F(Q_r(x_0), u) \geq F(Q_1(x_0), u(x_0) + \nabla u(x_0) \cdot (\cdot - x_0))$$

for every $u \in C^1(\mathbf{R}^n)$, x_0 a.e. in $\mathbf{R}^n$,

$$(9.2.11) \qquad F(\Omega, u) \leq F(\Omega, u_z) \text{ whenever } \Omega \in \mathcal{A}_0,$$

$$z \in \mathbf{R}^n, \ u \in PA(\mathbf{R}^n) \text{ with } u(x) = u_z(x) \text{ for every } x \in \Omega$$

hold, then f_F is convex and lower semicontinuous, and

$$(9.2.12) \qquad F_-(\Omega, u) = \int_\Omega f_F(\nabla u) dx \text{ for every } (\Omega, u) \in \mathcal{A}_0 \times C^1(\mathbf{R}^n).$$

We also prove that if we replace conditions (9.2.1) and (9.2.2) with the following

$$(9.2.13) \quad F(\Omega, u) \leq \limsup_{h \to +\infty} F(\Omega \setminus B_{u_h}, u_h) \text{ for every } \Omega \in \mathcal{A}_0, \ u \in C^1(\mathbf{R}^n),$$

$$\{u_h\} \subseteq PA(\mathbf{R}^n) \text{ with } u_h \to u \text{ in weak*-}W^{1,\infty}(\Omega),$$

then the same conclusions on f_F and (9.2.12) continue to hold.

Lemma 9.2.3. *Let F be as in (9.1.1) satisfying (9.1.3), (9.1.4), (9.2.1), (9.2.2), (9.2.5), (9.2.8), (9.2.9), (9.2.11), and let f_F be given by (9.1.6). Then*

$$(9.2.14) \qquad F(Q, u) \leq \int_Q f_F(\nabla u) dx \text{ for every cube } Q, \ u \in C^1(\mathbf{R}^n).$$

Proof. Let Q, u be as in (9.2.14) with $\int_Q f_F(\nabla u) dx < +\infty$. Then, by (9.2.2) and Remark 9.2.1, $\mathrm{dom} f_F$ turns out to be nonempty and convex. Moreover it is not restrictive to assume that

$$(9.2.15) \qquad\qquad 0 \in \mathrm{ri}(\mathrm{dom} f_F).$$

Let $k \ (\leq n)$ be the dimension of $\mathrm{aff}(\mathrm{dom} f_F)$. If $k < n$ let us denote by 0_k, respectively by 0_{n-k}, the origin of $\mathbf{R}^k$, respectively of $\mathbf{R}^{n-k}$.

Let $R \colon \mathbf{R}^n \to \mathbf{R}^n$ be the identity transformation if $k = n$, an orthogonal linear transformation such that

$$(9.2.16) \qquad\qquad R(\mathrm{aff}(\mathrm{dom} f_F)) = \mathbf{R}^k \times \{0_{n-k}\}$$

if $k < n$ and call again with R the $n \times n$ matrix associated to the transformation.

Let us define u' by

$$(9.2.17) \qquad\qquad u' \colon y \in \mathbf{R}^n \mapsto u(R^{-1}y),$$

then, since $R^{-1} = R^T$ (the transpose of R), we have

$$(9.2.18) \quad \nabla_y u'(y) = \nabla_x u(R^{-1}y)R^{-1} = \nabla_x u(R^{-1}y)R^T = (R\nabla_x(R^{-1}y)^T)^T$$

$$\text{for every } y \in \mathbf{R}^n.$$

Let us fix now $\overline{x} \in Q$. Then, since $\nabla u(x) \in \mathrm{dom} f_F$ for every $x \in Q$, by (9.2.16), (9.2.18) and the convexity of Q, we deduce that u' in (9.2.17) effectively depends only on $(y_1, \ldots, y_k)$ when $(y_1, \ldots, y_n)$ varies in $R(Q)$, $R(Q)$ being a cube centred in $\overline{y} = R\overline{x}$.

Because of these considerations, if Pr_k is the projection operator from $\mathbf{R}^n$ to $\mathbf{R}^k$ given by $\mathrm{Pr}_k \colon (y_1, \ldots, y_n) \in \mathbf{R}^n \mapsto (y_1, \ldots, y_k)$, we can define $\hat{u}$ by

$$(9.2.19) \quad \hat{u} \colon (y_1, \ldots, y_k) \in \mathbf{R}^k \mapsto \begin{cases} u(y_1, \ldots, y_n) & \text{if } k = n \\ u'(y_1, \ldots, y_k, \overline{y}_{k+1}, \ldots, \overline{y}_n) & \text{if } k < n, \end{cases}$$

then by (9.2.19), (9.2.18), and (9.2.16) we get

$$(9.2.20) \qquad \nabla \hat{u}(y_1, \ldots, y_k) \in \mathrm{Pr}_k(R(\mathrm{dom} f_F))$$

$$\text{for every } (y_1, \ldots, y_k) \in \mathrm{Pr}_k(R(Q)).$$

We also observe that by (9.2.15), and (9.2.16) we have

$$(9.2.21) \qquad 0_k \in \mathrm{ri}(\mathrm{Pr}_k(R(\mathrm{dom} f_F))).$$

Let $\{\hat{u}_i\}$ be a sequence in $PA(\mathbf{R}^k)$ given by Theorem 0.8 such that

$$(9.2.22) \qquad \begin{cases} \hat{u}_i \to \hat{u} \text{ uniformly in } \overline{\mathrm{Pr}_k(R(Q))}, \\ \nabla \hat{u}_i \to \nabla \hat{u} \text{ in } (L^\infty(\mathrm{Pr}_k(R(Q))))^n, \end{cases}$$

and let $\{t_h\} \subseteq \,]0, 1[$ with $t_h \to 1$. Then by (9.2.20), (9.2.21), and the convexity of $\mathrm{Pr}_k(R(\mathrm{dom} f_F))$ we get the existence of $\delta_h > 0$ such that

$$(9.2.23) \qquad \mathrm{dist}(t_h \nabla \hat{u}(y_1, \ldots, y_k), \mathbf{R}^k \setminus \mathrm{Pr}_k(R(\mathrm{dom} f_F))) > 2\delta_h$$

$$\text{for every } (y_1, \ldots, y_k) \in \mathrm{Pr}_k(R(Q)),$$

hence, by (9.2.22), and (9.2.23), we deduce that

$$(9.2.24) \qquad \mathrm{dist}(t_h \nabla \hat{u}_i(y_1, \ldots, y_k), \mathbf{R}^k \setminus \mathrm{Pr}_k(R(\mathrm{dom} f_F))) > \delta_h$$

for a.e. $(y_1, \ldots, y_k) \in \mathrm{Pr}_k(R(Q))$, every $h \in \mathbf{N}$, and $i \in \mathbf{N}$ large enough.

By using the functions $\hat{u}_i$ we can define u'_i and u_i as

$$(9.2.25) \qquad \begin{cases} u'_i \colon (y_1, \ldots, y_n) \in R(Q) \mapsto \hat{u}_i(y_1, \ldots, y_k) \\ u_i \colon (x_1, \ldots, x_n) \in Q \mapsto u'_i(R(x_1, \ldots, x_n)), \end{cases}$$

then by (9.2.25), (9.2.22), and (9.2.24) it turns out that the functions u_i are in $PA(\mathbf{R}^n)$, that

$$(9.2.26) \qquad\qquad u_i \to u \text{ in } W^{1,\infty}(Q),$$

and that

$$(9.2.27) \qquad t_h \nabla u_i(x) \in \mathrm{dom} f_F, \ \mathrm{dist}(t_h \nabla u_i(x), \mathrm{rb}(\mathrm{dom} f_F)) > \delta_h$$

for a.e. $x \in Q$, every $h \in \mathbf{N}$, and $i \in \mathbf{N}$ large enough.

By (9.2.26) and (9.2.27), once we recall that f_F, being convex, is locally Lipschitz on $\mathrm{ri}(\mathrm{dom} f_F)$, we obtain that

$$(9.2.28) \qquad \lim_{i \to +\infty} \int_Q f_F(t_h \nabla u_i) dx = \int_Q f_F(t_h \nabla u) dx \text{ for every } h \in \mathbf{N},$$

hence, by (9.2.28), we can construct a subsequence $\{u_{i_h}\}$ of $\{u_i\}$ satisfying

$$(9.2.29) \qquad \int_Q f_F(t_h \nabla u_{i_h}) dx \le \int_Q f_F(t_h \nabla u) dx + \frac{1}{h} \text{ for every } h \in \mathbf{N}.$$

Now let us observe that, setting for every $h \in \mathbf{N}$, $u_{i_h} = \sum_{j=1}^{m_h}(u_{z_j^h} + s_j^h)\chi_{P_j^h}$, by (9.1.7) of Proposition 9.1.1, (9.2.5), and (9.2.11) we have that

$$(9.2.30) \qquad \int_Q f_F(t_h \nabla u_{i_h}) dx = \sum_{j=1}^{m_h} \mathcal{L}^n(Q \cap P_j^h) f_F(t_h z_j^h) \ge$$

$$\ge \sum_{j=1}^{m_h} F(Q \cap \mathrm{int}(P_j^h), t_h(u_{z_j^h} + s_j^h)) = \sum_{j=1}^{m_h} F(Q \cap \mathrm{int}(P_j^h), t_h u_{i_h})$$

for every $h \in \mathbf{N}$,

therefore by (9.2.26), (9.2.1), (9.2.8), and (9.2.30) we get that

$$(9.2.31) \qquad F(Q, u) \le \limsup_{h \to +\infty} F(Q \setminus B_{t_h u_{i_h}}, t_h u_{i_h}) =$$

$$= \limsup_{h \to +\infty} F(Q \cap \cup_{j=1}^{m_h} \mathrm{int}(P_j^h), t_h u_{i_h}) \le \limsup_{h \to +\infty} \sum_{j=1}^{m} F(Q \cap \mathrm{int}(P_j^h), t_h u_{i_h}) \le$$

$$\le \limsup_{h \to +\infty} \int_Q f_F(t_h \nabla u_{i_h}) dx.$$

Finally, by (9.2.31), (9.2.29), (9.2.2), and Remark 9.2.1 we conclude that

$$F(Q, u) \le \limsup_{h \to +\infty} \int_Q f_F(t_h \nabla u) dx \le$$

$$\leq \limsup_{h \to +\infty} \left\{ t_h \int_Q f_F(\nabla u) dx + (1 - t_h) \mathcal{L}^n(Q) f_F(0) \right\},$$

that, together with (9.2.15), yields (9.2.14). ∎

Lemma 9.2.4. *Let F be as in (9.1.1) satisfying (9.1.4), (9.2.1), (9.2.2), (9.2.5), (9.2.6), (9.2.8), (9.2.9), (9.2.11), and let f_F be given by (9.1.6). Then*

$$(9.2.32) \qquad F_-(\Omega, u) \leq \int_\Omega f_F(\nabla u) dx \text{ for every } (\Omega, u) \in \mathcal{A}_0 \times C^1(\mathbf{R}^n).$$

Proof. Let (Ω, u) be as in (9.2.32). We can clearly assume that $\int_\Omega f_F(\nabla u) dx < +\infty$.

Let $\Omega' \in \mathcal{A}(\Omega)$ with $\Omega' \subset\subset \Omega$, $\varepsilon > 0$, $Q_{(1)}, \ldots, Q_{(m)}$ be cubes with $Q_{(j)} \subset\subset \Omega$ for every $j \in \{1, \ldots, m\}$, $\Omega' \subset\subset \cup_{j=1}^m Q_{(j)}$, and

$$(9.2.33) \qquad \sum_{j=1}^m \int_{Q_{(j)}} f_F(\nabla u) dx \leq \int_\Omega f_F(\nabla u) dx + \varepsilon.$$

Then by (9.2.6), (9.2.8), Lemma 9.2.3, and (9.2.33) we get that

$$(9.2.34) \qquad F(\Omega', u) \leq F\left(\cup_{j=1}^m Q_{(j)}, u\right) \leq \sum_{j=1}^m F(Q_{(j)}, u) =$$

$$= \sum_{j=1}^m \int_{Q_{(j)}} f_F(\nabla u) dx \leq \int_\Omega f_F(\nabla u) dx + \varepsilon$$

for every $\Omega' \in \mathcal{A}(\Omega)$ with $\Omega' \subset\subset \Omega$.

As Ω' increases to Ω, and ε decreases to 0, inequality (9.2.32) follows from (9.2.34). ∎

We can now prove the integral representation result.

Theorem 9.2.5. *Let F be as in (9.1.1) satisfying (9.2.1), (9.2.2), (9.2.5)÷ (9.2.11), and let f_F be given by (9.1.6). Then f_F is convex and lower semicontinuous, and*

$$(9.2.35) \qquad F_-(\Omega, u) = \int_\Omega f_F(\nabla u) dx \text{ for every } (\Omega, u) \in \mathcal{A}_0 \times C^1(\mathbf{R}^n).$$

Proof. The properties of f_F come from (9.2.2), (9.2.1), and Remark 9.2.1, whilst equality (9.2.35) from Lemma 9.2.4, (9.2.5), (9.2.9), Proposition 2.6.13 applied with $\alpha = F(\cdot, u)$ and (9.2.10). ∎

From Theorem 9.2.5 we trivially deduce the following corollary.

Corollary 9.2.6. *Let F be a convex functional as in (9.1.1) satisfying (9.2.1), (9.2.5)÷(9.2.11), and let f_F be given by (9.1.6). Then f_F is convex and lower semicontinuous, and (9.2.35) holds.*

We now prove that Theorem 9.2.5 still holds if we replace conditions (9.2.1) and (9.2.2) with (9.2.13).

Lemma 9.2.7. *Let F be as in (9.1.1) satisfying (9.1.3)÷(9.2.5), (9.2.8), (9.2.9), (9.2.11), and (9.2.13). Then (9.2.2) holds.*

Proof. The proof follows the outlines of [BDM2, Lemma 1.5].

Let $z_1, z_2 \in \mathbf{R}^n$ with $z_1 \neq z_2$, $t \in [0,1]$ and set $z_0 = \frac{z_2-z_1}{|z_2-z_1|}$. For every $h \in \mathbf{N}$ and $j \in \mathbf{Z}$, set

$$Q_{h,j}^1 = \left\{ x \in \mathbf{R}^n : \frac{j-1}{h} \leq z_0 \cdot x < \frac{j-1+t}{h} \right\},$$

$$Q_{h,j}^2 = \left\{ x \in \mathbf{R}^n : \frac{j-1+t}{h} \leq z_0 \cdot x < \frac{j}{h} \right\},$$

$$Q_h^- = \{ x \in \mathbf{R}^n : z_0 \cdot x < -h \}, \quad Q_h^+ = \left\{ x \in \mathbf{R}^n : \frac{h^2-1}{h} \leq z_0 \cdot x \right\},$$

$$Q_h^1 = \cup_{j=-h^2+1}^{h^2-1} Q_{h,j}^1, \quad Q_h^2 = \cup_{j=-h^2+1}^{h^2-1} Q_{h,j}^2$$

and observe that

(9.2.36) $\chi_{Q_h^1} \to t$, $\chi_{Q_h^2} \to 1 - t$ in weak*-$W^{1,\infty}(Y)$ as $h \to +\infty$.

For every $h \in \mathbf{N}$, $j \in \mathbf{Z}$ let us set

$$c_{h,j}^1 = \frac{(j-1)(1-t)}{h}|z_2 - z_1|, \quad c_{h,j}^2 = -\frac{jt}{h}|z_2 - z_1|,$$

and define u_h by

$$u_h : x \in \mathbf{R}^n \mapsto \begin{cases} z_2 \cdot x + ht|z_2 - z_1| & \text{if } x \in Q_h^- \\ z_1 \cdot x + c_{h,j}^1 & \text{if } x \in \cup_{j=-h^2+1}^{h^2-1} Q_{h,j}^1 \\ z_2 \cdot x + c_{h,j}^2 & \text{if } x \in \cup_{j=-h^2+1}^{h^2-1} Q_{h,j}^2 \\ z_1 \cdot x + \frac{(h^2-1)(1-t)}{h}|z_2 - z_1| & \text{if } x \in Q_h^+. \end{cases}$$

Then u_h turns out to be in $PA(\mathbf{R}^n)$, and by (9.2.36) we deduce that

(9.2.37) $u_h \to u_{tz_1+(1-t)z_2}$ in weak*-$W^{1,\infty}(Y)$.

By (9.2.37), (9.2.13), (9.2.8), (9.2.5), (9.2.11), Proposition 9.1.1, and (9.2.36) we obtain

(9.2.38) $F(Y, u_{tz_1+(1-t)z_2}) \leq$

$$\leq \limsup_{h \to +\infty} F\left(Y \cap \left(\left(\cup_{j=-h^2+1}^{h^2-1}(\mathrm{int}(Q_{h,j}^1) \cup \mathrm{int}(Q_{h,j}^2))\right) \cup Q_h^- \cup \mathrm{int}(Q_h^+)\right), u_h\right) \leq$$

$$\leq \limsup_{h \to +\infty} \left\{ \sum_{j=-h^2+1}^{h^2-1} \{F(Y \cap \mathrm{int}(Q_{h,j}^1), u_{z_1}) + F(Y \cap \mathrm{int}(Q_{h,j}^2), u_{z_2})\} + \right.$$

$$\left. + F(Y \cap \mathrm{int}(Q_h^+), u_{z_1}) + F(Y \cap Q_h^-, u_{z_2}) \right\} \leq$$

$$\leq \limsup_{h \to +\infty} \{\mathcal{L}^n(Y \cap Q_h^1)F(Y, u_{z_1}) + \mathcal{L}^n(Y \cap Q_h^2)F(Y, u_{z_2})\} =$$

$$= tF(Y, u_{z_1}) + (1-t)F(Y, u_{z_2}).$$

By (9.2.38) condition (9.2.2) follows. ∎

Theorem 9.2.8. *Let F be as in (9.1.1) satisfying (9.2.5)÷(9.2.11) and (9.2.13), and let f_F be given by (9.1.6). Then f_F is convex and lower semicontinuous, and (9.2.35) holds.*

    Proof. Follows by Lemma 9.2.7 and Theorem 9.2.5. ∎

§9.3 Representation on Sobolev Spaces

Let $p \in [1, +\infty]$.

    In the present section we prove, under various sets of assumptions, some characterizations of the functionals F depending on a bounded open set Ω, and u in $W_{\mathrm{loc}}^{1,p}(\mathbf{R}^n)$ that can be represented as

$$(9.3.1) \qquad F(\Omega, u) = \int_\Omega f_F(\nabla u)dx \text{ for every } (\Omega, u) \in \mathcal{A}_0 \times W_{\mathrm{loc}}^{1,p}(\mathbf{R}^n),$$

f_F being a convex lower semicontinuous function from $\mathbf{R}^n$ to $[0, +\infty]$.

    Let us consider a functional

$$(9.3.2) \qquad F: (\Omega, u) \in \mathcal{A}_0 \times W_{\mathrm{loc}}^{1,p}(\mathbf{R}^n) \mapsto F(\Omega, u) \in [0, +\infty],$$

define f_F by (9.1.6) and, as a first case, let us introduce the following assumptions

$$(9.3.3) \qquad \text{for every } u \in W_{\mathrm{loc}}^{1,p}(\mathbf{R}^n), \ F(\cdot, u) \text{ is increasing,}$$

$$(9.3.4) \qquad \text{for every } u \in W_{\mathrm{loc}}^{1,p}(\mathbf{R}^n), \ F(\cdot, u) \text{ is weakly superadditive,}$$

$$(9.3.5) \qquad \text{for every } u \in W_{\mathrm{loc}}^{1,p}(\mathbf{R}^n), \ F(\cdot, u) \text{ is weakly subadditive,}$$

(9.3.6) $\displaystyle\limsup_{r\to 0^+}\frac{1}{r^n}F(Q_r(x_0),u)\geq F(Q_1(x_0),u(x_0)+\nabla u(x_0)\cdot(\cdot-x_0))$

for every $u\in W^{1,p}_{\text{loc}}(\mathbf{R}^n)$, x_0 a.e. in $\mathbf{R}^n$,

(9.3.7)   for every $\Omega\in\mathcal{A}_0$, $F(\Omega,\cdot)$ is

$W^{1,p}(\Omega)$-lower semicontinuous if $p\in[1,+\infty[$,

$\bigcap_{q\in[1,+\infty[}W^{1,q}(\Omega)$-lower semicontinuous if $p=+\infty$,

(9.3.8)   for every $u\in W^{1,p}_{\text{loc}}(\mathbf{R}^n)$, $F(\cdot,u)$ is inner regular.

Lemma 9.3.1. *Let F be as in (9.3.2) with $p=+\infty$. Assume that for every $\Omega\in\mathcal{A}_0$, $F(\Omega,\cdot)$ is $W^{1,\infty}(\Omega)$-lower semicontinuous. Then*

(9.3.9) $F(\Omega,u)=F(\Omega,v)$

for every $\Omega\in\mathcal{A}_0$, $u,\,v\in W^{1,\infty}_{\text{loc}}(\mathbf{R}^n)$ with $u=v$ a.e. in Ω.

Proof. If u, v are as in (9.3.9), by defining for every $h\in\mathbf{N}$, $u_h=u$, we have that $u_h\to v$ in $W^{1,\infty}(\Omega)$ and by the $W^{1,\infty}(\Omega)$-lower semicontinuity of $F(\Omega,\cdot)$, that

(9.3.10) $\displaystyle F(\Omega,v)\leq\liminf_{h\to+\infty}F(\Omega,u_h)=F(\Omega,u).$

By (9.3.10) and its analogous obtained by interchanging the roles of u and v condition (9.3.9) follows. ∎

Theorem 9.3.2. *Let $p\in[1,+\infty]$. Let F be as in (9.3.2) satisfying (9.2.5), (9.2.9), (9.3.3)÷(9.3.6), (9.2.1), (9.2.2), (9.3.7), (9.3.8), and let f_F be given by (9.1.6). Then f_F is convex and lower semicontinuous, and (9.3.1) holds.*
  Conversely, given $f:\mathbf{R}^n\to[0,+\infty]$ convex and lower semicontinuous, and defined F by (9.3.1) with $f_F=f$, it turns out that conditions (9.2.5), (9.2.9), (9.3.3)÷(9.3.6), (9.2.1), (9.2.2), (9.3.7), (9.3.8) are satisfied by F.

Proof. First of all, we prove that the assumptions of Theorem 9.2.5 are fulfilled by the restriction of F to $\mathcal{A}_0\times W^{1,\infty}_{\text{loc}}(\mathbf{R}^n)$.
  In fact, besides (9.2.1), (9.2.2), (9.2.5), and (9.2.9), condition (9.2.6) is implied by (9.3.3), condition (9.2.10) by (9.3.6), and conditions (9.2.7) and (9.2.8) by (9.3.4), (9.3.5), (9.3.8), and Proposition 2.6.8. Moreover, (9.2.11) follows from (9.3.7), and Lemma 9.3.1.
  Consequently, by Theorem 9.2.5 we infer that f_F is convex and lower semicontinuous, and that

(9.3.11) $\displaystyle F_-(\Omega,u)=\int_\Omega f_F(\nabla u)dx$ for every $(\Omega,u)\in\mathcal{A}_0\times C^1(\mathbf{R}^n)$.

At this point we observe that by (9.3.11) the assumptions of Proposition 8.1.1 are fulfilled with $\mathcal{E}_0 = \mathcal{A}_0$, $U = W^{1,p}_{\text{loc}}(\mathbf{R}^n)$ endowed with the $W^{1,p}_{\text{loc}}(\mathbf{R}^n)$ topology if $p \in [1,+\infty[$, or with the $\cap_{q \in]1,+\infty[} W^{1,q}_{\text{loc}}(\mathbf{R}^n)$ one if $p = +\infty$, $G = F_-$, and $H \colon (\Omega, u) \in \mathcal{A}_0 \times W^{1,p}_{\text{loc}}(\mathbf{R}^n) \mapsto \int_\Omega f_F(\nabla u)dx$.

In fact, it is clear that H is translation invariant, convex, and, by Fatou's lemma, also $W^{1,p}_{\text{loc}}(\mathbf{R}^n)$ lower semicontinuous if $p \in [1,+\infty[$, or $\cap_{q \in]1,+\infty[} W^{1,q}_{\text{loc}}(\mathbf{R}^n)$ lower semicontinuous if $p = +\infty$. Moreover, by (9.3.7), G too enjoys the same semicontinuity properties.

By Proposition 8.1.1 and Proposition 2.6.4 we thus get that

$$(9.3.12) \quad F_-(\Omega, u) \leq \int_\Omega f_F(\nabla u)dx \text{ for every } (\Omega, u) \in \mathcal{A}_0 \times W^{1,p}_{\text{loc}}(\mathbf{R}^n).$$

We also note that by (9.3.3), (9.3.4), and (9.3.5) the assumptions of Proposition 2.6.13 are satisfied with $\alpha = F(\cdot, u)$, for fixed $u \in W^{1,p}_{\text{loc}}(\mathbf{R}^n)$. Therefore, by Proposition 2.6.13 and (9.3.6) we conclude that

$$(9.3.13) \quad F_-(\Omega, u) \geq \int_\Omega f_F(\nabla u)dx \text{ for every } (\Omega, u) \in \mathcal{A}_0 \times W^{1,p}_{\text{loc}}(\mathbf{R}^n).$$

Finally by (9.3.12), (9.3.13) and (9.3.8) equality (9.3.1) follows.

The second part of the theorem follows from a direct verification, and by using also Fatou's lemma. ∎

As corollaries, we deduce from Theorem 9.3.2 the results below.

Theorem 9.3.3. *Let* $p \in [1,+\infty]$. *Let* F *be as in (9.3.2) satisfying (9.2.5), (9.2.9), (9.3.3)÷(9.3.6), (9.2.13), (9.3.7), (9.3.8), and let* f_F *be given by (9.1.6). Then* f_F *is convex and lower semicontinuous, and (9.3.1) holds.*

Conversely, given $f \colon \mathbf{R}^n \to [0,+\infty]$ *convex and lower semicontinuous, and defined* F *by (9.3.1) with* $f_F = f$, *it turns out that conditions (9.2.5), (9.2.9), (9.3.3)÷(9.3.6), (9.2.13), (9.3.7), (9.3.8) are satisfied by* F.

Proof. Follows the same outlines of the one of Theorem 9.3.2, using Theorem 9.2.8 in place of Theorem 9.2.5. ∎

Theorem 9.3.4. *Let* $p \in [1,+\infty]$. *Let* F *be as in (9.3.2) satisfying (9.2.5), (9.2.9), (9.3.3)÷(9.3.6), (9.2.2), (9.3.7), (9.3.8), (9.2.3), and let* f_F *be given by (9.1.6). Then* f_F *is convex and lower semicontinuous, and (9.3.1) holds.*

Conversely, given $f \colon \mathbf{R}^n \to [0,+\infty]$ *convex and lower semicontinuous, and defined* F *by (9.3.1) with* $f_F = f$, *it turns out that conditions (9.2.5), (9.2.9), (9.3.3)÷(9.3.6), (9.2.2), (9.3.7), (9.3.8), (9.2.3) are satisfied by* F.

Proof. Follows from Theorem 9.3.2, once we observe that (9.2.3), and (9.3.7) imply (9.2.1). ∎

Again from Theorem 9.3.2 we infer the following result under the assumption that

(9.3.14)   for every $\Omega \in \mathcal{A}_0$, $F(\Omega, \cdot)$ is

weak-$W^{1,p}(\Omega)$-lower semicontinuous if $p \in [1, +\infty[$,

weak*-$W^{1,\infty}(\Omega)$-lower semicontinuous if $p = +\infty$.

Theorem 9.3.5. *Let $p \in [1, +\infty]$. Let F be as in (9.3.2) satisfying (9.2.5), (9.2.9), (9.3.3)÷(9.3.6), (9.3.8), (9.3.14), (9.2.3), and let f_F be given by (9.1.6). Then f_F is convex and lower semicontinuous, and (9.3.1) holds.*
Conversely, given $f : \mathbf{R}^n \to [0, +\infty]$ convex and lower semicontinuous, and defined F by (9.3.1) with $f_F = f$, it turns out that conditions (9.2.5), (9.2.9), (9.3.3)÷(9.3.6), (9.3.8), (9.3.14), (9.2.3) are satisfied by F.

Proof. Follows from Theorem 9.3.3, once we observe that (9.3.14) imply (9.3.7), and that (9.2.3) and (9.3.14) imply (9.2.13). ∎

In order to prove additional new characterizations, we now introduce the following conditions

(9.3.15)   there exist $z_0 \in \mathrm{dom} f_F$, $r_0 > 0$,

and a Radon positive measure μ on $\mathbf{R}^n$ such that

$$F(\Omega, u) \leq \mu(\Omega) \text{ whenever } \Omega \in \mathcal{A}_0, \ u \in PA(\mathbf{R}^n)$$

with $\nabla u(x) \in \mathrm{dom} f_F$ for a.e. $x \in \Omega$ and $\|u - u_{z_0}\|_{W^{1,\infty}(\Omega)} < r_0$,

(9.3.16) for every $\Omega \in \mathcal{A}_0$ the restriction of $F(\Omega, \cdot)$ to $PA(\mathbf{R}^n)$ is convex,

(9.3.17)   for every $\Omega \in \mathcal{A}_0$ the restriction of

$F(\Omega, \cdot)$ to $W^{1,\infty}_{\mathrm{loc}}(\mathbf{R}^n)$ is $W^{1,\infty}(\Omega)$-lower semicontinuous.

Proposition 9.3.6. *Let F be as in (9.3.2) satisfying (9.2.5), (9.1.2), (9.1.3), (9.3.15), (9.2.7), (9.2.8), (9.3.16), (9.3.17). Then (9.2.1) holds.*

Proof. Let Ω, u, $\{u_h\}$ with $u_h = \sum_{j=1}^{m_h} (u_{z_j^h} + s_j^h) \chi_{P_j^h}$ for every $h \in \mathbf{N}$, be as in (9.2.1), f_F be given by (9.1.6), and observe that (9.3.16) obviously implies the convexity of f_F.

Let us first prove that (9.2.1) holds if there exists $\overline{x} \in \Omega$ such that $\nabla u(\overline{x}) \notin \mathrm{dom} f_F$.

In this case, by taking into account the continuity of ∇u, there exist a neighborhood I of $\overline{x}$ in Ω and $r > 0$ such that $\mathrm{dist}(\nabla u(x), \mathrm{dom} f_F) > r$ for every $x \in I$. Therefore, for every $h \in \mathbf{N}$ large enough, we have that $\nabla u_h(x) \notin \mathrm{dom} f_F$ for a.e. $x \in I$ and hence that there exists $j_h \in \{1, \ldots, m_h\}$ with $\Omega \cap \mathrm{int}(P_{j_h}^h) \neq \emptyset$ and $z_{j_h}^h \notin \mathrm{dom} f_F$.

Because of this, and (9.1.8) of Proposition 9.1.1 applied to the restriction of F to $\mathcal{A}_0 \times W_{\mathrm{loc}}^{1,\infty}(\mathbf{R}^n)$, it then results that $F(\Omega \cap \mathrm{int}(P_{j_h}^h), u_{z_{j_h}^h}) = +\infty$, and hence, by (9.2.7), (9.2.5), and Lemma 9.3.1, that

$$\limsup_{h \to +\infty} F(\Omega \cap \cup_{j=1}^{m_h} \mathrm{int}(P_j^h), u_h) \geq \liminf_{h \to +\infty} \sum_{j=1}^{m_h} F(\Omega \cap \mathrm{int}(P_j^h), u_h) =$$

$$= \liminf_{h \to +\infty} \sum_{j=1}^{m_h} F(\Omega \cap \mathrm{int}(P_j^h), u_{z_j^h} + s_j^h) \geq \liminf_{h \to +\infty} F(\Omega \cap \mathrm{int}(P_{j_h}^h), u_{z_{j_h}^h}) = +\infty,$$

from which condition (9.2.1) trivially follows.

Let us now prove that (9.2.1) holds if $\nabla u(x) \in \overline{\mathrm{dom} f_F}$ for every $x \in \Omega$.

To do this we first observe that it suffices to consider the case in which z_0 in (9.3.15) is equal to 0, and hence $\mathrm{aff}(\mathrm{dom} f_F)$ is a vector subspace of $\mathbf{R}^n$, being possible to reduce the general case to this one by considering the functional $F(\cdot, u_{z_0} + \cdot)$. Moreover, again by the same argument, the convexity of f_F, and possibly taking r_0 in (9.3.15) sufficiently small, it is not restrictive to assume that

(9.3.18) $\qquad 0 \in \mathrm{ri}(\mathrm{dom} f_F), \quad \mathrm{dist}(0, \mathrm{rb}(\mathrm{dom} f_F)) < r_0.$

As usual, it is not restrictive to assume that the limit $\lim_{h \to +\infty} F(\Omega \setminus B_{u_h}, u_h)$ exists and is finite, so that it results

(9.3.19) $\qquad \nabla u_h \in \mathrm{dom} f_F$ for a.e. $x \in \Omega$ and every $h \in \mathbf{N}.$

Let $t \in {]0,1[}$, and recall that f_F, being convex, is continuous in $\mathrm{ri}(\mathrm{dom} f_F)$. Therefore, by (9.3.18), the convexity of $\mathrm{dom} f_F$, and our assumptions on u we deduce the existence of a neighborhood A_t of $\overline{\Omega}$, and $M_t > 0$ such that

(9.3.20) $\qquad f_F(t(2-t)\nabla u(x)) \leq M_t$ for every $x \in A_t.$

Let μ, r_0 be given by (9.3.15), $\overline{x}_1, \ldots, \overline{x}_{2^n}$ be the vertices of $Q_2(0)$, and, for every $r > 0$, let us take a sequence $\{Q_r(x_i^r)\}_{i \in \mathbf{N}}$ of pairwise disjoint cubes such that $\mathcal{L}^n(\mathbf{R}^n \setminus \cup_{j=1}^\infty Q_r(x_i^r)) = 0.$

For every $h \in \mathbf{N}$, we take $r_h \in {]0, 1/h[}$, and observe that it is not restrictive to assume that $Q_{r_h}(x_i^{r_h}) \cap \overline{\Omega} \cap B_{u_h} \neq \emptyset$ if and only if $i \in \{1, \ldots, n_h\}$. Moreover, we set for every $i \in \mathbf{N}$, $k \in \{1, \ldots, 2^n\}$, $y_i^{h,k} = x_i^{r_h} + \frac{r_h}{3}\overline{x}_k$, and $z_i^{h,k} = \nabla u(y_i^{h,k})$. Then, by taking into account the continuity of ∇u, $\{r_h\}$ can be chosen so that

(9.3.21) $\qquad\qquad\qquad \lim_{h \to +\infty} n_h r_h^n = 0,$

(9.3.22) $\qquad \overline{\Omega} \cap B_{u_h} \subseteq \cup_{i=1}^{n_h} \cup_{k=1}^{2^n} Q_{r_h}(y_i^{h,k})$ for every $h \in \mathbf{N}$,

(9.3.23) $\qquad \left\| u - T[-y_i^{h,k}] u_{z_i^{h,k}} - u(y_i^{h,k}) \right\|_{W^{1,\infty}(Q_{r_h}(y_i^{h,k}))} < \dfrac{(1-t)r_0}{3t^2(2-t)}$

$$\text{for every } h \in \mathbf{N}, \ i \in \{1, \dots, n_h\}, \ k \in \{1, \dots, 2^n\},$$

(9.3.24) $\displaystyle \sum_{i=1}^{n_h} \sum_{k=1}^{2^n} \mu\left(\Omega \cap Q_{r_h}(y_i^{h,k})\right) = \sum_{k=1}^{2^n} \sum_{i=1}^{n_h} \mu\left(\Omega \cap Q_{r_h}\left(x_i^h + \dfrac{r_h}{3}\overline{x}_k\right)\right) \leq$

$$\leq \sum_{k=1}^{2^n} \mu\left(\Omega \cap \cup_{i=1}^{n_h} Q_{r_h}\left(x_i^h + \dfrac{r_h}{3}\overline{x}_k\right)\right) \leq 2^n \mu(\Omega)$$

$$\text{for every } h \in \mathbf{N} \text{ sufficiently large.}$$

Then, by (9.3.22) and (9.2.8) we have that

(9.3.25) $\qquad\qquad\qquad\qquad\qquad F(\Omega, tu_h) \leq$

$$\leq F(\Omega \setminus B_{u_h}, tu_h) + F\left(\Omega \cap \left(\cup_{i=1}^{n_h} \cup_{k=1}^{2^n} Q_{r_h}(y_i^{h,k})\right), tu_h\right) \leq$$

$$\leq F(\Omega \setminus B_{u_h}, tu_h) + \sum_{i=1}^{n_h} \sum_{k=1}^{2^n} F\left(\Omega \cap Q_{r_h}(y_i^{h,k}), tu_h\right) \text{ for every } h \in \mathbf{N}.$$

Let us fix now $h \in \mathbf{N}$, $i \in 1, \dots, n_h$, and $k \in 1, \dots, 2^n$. Then by (9.3.16), and (9.1.7) of Proposition 9.1.1 applied to the restriction of F to $\mathcal{A}_0 \times W_{\text{loc}}^{1,\infty}(\mathbf{R}^n)$, we obtain

(9.3.26) $\qquad\qquad\qquad F(\Omega \cap Q_{r_h}(y_i^{h,k}), tu_h) =$

$$= F\left(\Omega \cap Q_{r_h}(y_i^{h,k}), t^2(2-t)\left(T[-y_i^{h,k}]u_{z_i^{h,k}} + u(y_i^{h,k})\right) + \right.$$

$$\left. + (1-t)\dfrac{tu_h - t^2(2-t)(T[-y_i^{h,k}]u_{z_i^{h,k}} + u(y_i^{h,k}))}{1-t}\right) \leq$$

$$\leq tF\left(\Omega \cap Q_{r_h}(y_i^{h,k}), t(2-t)\left(T[-y_i^{h,k}]u_{z_i^{h,k}} + u(y_i^{h,k})\right)\right) +$$

$$+ (1-t)F\left(\Omega \cap Q_{r_h}(y_i^{h,k}), \dfrac{tu_h - t^2(2-t)\left(T[-y_i^{h,k}]u_{z_i^{h,k}} + u(y_i^{h,k})\right)}{1-t}\right) \leq$$

$$\leq tr_h^n f_F(t(2-t)z_i^{h,k}) +$$

$$+(1-t)F\left(\Omega\cap Q_{r_h}(y_i^{h,k}),\frac{tu_h-t^2(2-t)\left(T[-y_i^{h,k}]u_{z_i^{h,k}}+u(y_i^{h,k})\right)}{1-t}\right).$$

In order to treat the last term in (9.3.26), we observe that by (9.3.23) we have

$$(9.3.27)\quad \left\|\frac{tu_h-t^2(2-t)\left(T[-y_i^{h,k}]u_{z_i^{h,k}}+u(y_i^{h,k})\right)}{1-t}\right\|_{W^{1,\infty}(\Omega\cap Q_{r_h}(y_i^{h,k}))}\leq$$

$$\leq\frac{t}{1-t}(\|u_h-u\|_{W^{1,\infty}(\Omega)}+(1-t)^2\|u\|_{W^{1,\infty}(\Omega)}+$$

$$+t(2-t)\left\|u-\left(T[-y_i^{h,k}]u_{z_i^{h,k}}+u(y_i^{h,k})\right)\right\|_{W^{1,\infty}(Q_{r_h}(y_i^{h,k}))}\leq$$

$$\leq\frac{t}{1-t}\|u_h-u\|_{W^{1,\infty}(\Omega)}+(1-t)^2\|u\|_{W^{1,\infty}(\Omega)}+\frac{r_0}{3}<r_0$$

provided h is large enough.

Therefore, by (9.3.19), our assumptions on u, (9.3.27), and (9.3.18), we infer that $\frac{t}{1-t}(\nabla u_h-t(2-t)z_i^{h,k})\in \mathrm{ri}(\mathrm{dom}f_F)$ a.e. in $\Omega\cap Q_{r_h}(y_i^{h,k})$, for every $h\in\mathbf{N}$ large enough. Consequently, by (9.3.27) and (9.3.15) (recall that $z_0=0$), we obtain that

$$(9.3.28)\quad F\left(\Omega\cap Q_{r_h}(y_i^{h,k}),\frac{tu_h-t^2(2-t)(T[-y_i^{h,k}]u_{z_i^{h,k}}+u(y_i^{h,k}))}{1-t}\right)\leq$$

$$\leq\mu(\Omega\cap Q_{r_h}(y_i^{h,k}))$$

for every $h\in\mathbf{N}$ large enough, $i\in\{1,\ldots,n_h\}$, $k\in\{1,\ldots,2^n\}$.

In conclusion, by (9.3.25), (9.3.26), (9.3.20), (9.3.28), (9.3.24), (9.3.16), and (9.1.7) of Proposition 9.1.1 applied to the restriction of F to $\mathcal{A}_0\times W_{\mathrm{loc}}^{1,\infty}(\mathbf{R}^n)$, we obtain

$$(9.3.29)\qquad\qquad F(\Omega,tu_h)\leq$$

$$\leq F(\Omega\setminus B_{u_h},tu_h)+2^nn_hr_h^nM_t+(1-t)\sum_{k=1}^{2^n}\sum_{i=1}^{n_h}\mu(\Omega\cap Q_{r_h}(y_i^{h,k}))\leq$$

$$\leq F(\Omega\setminus B_{u_h},u_h)+(1-t)F(\Omega,0)+2^nn_hr_h^nM_t+(1-t)2^n\mu(\Omega)\leq$$

$$\leq F(\Omega\setminus B_{u_h},u_h)+(1-t)\mathcal{L}^n(\overline{\Omega})f_F(0)+2^nn_hr_h^nM_t+(1-t)2^n\mu(\Omega)$$

for every $h\in\mathbf{N}$ large enough.

Therefore, by using (9.3.17) and (9.3.21), we infer by (9.3.29) as h diverges that

$$(9.3.30) \qquad\qquad F(\Omega, tu) \leq \liminf_{h \to +\infty} F(\Omega, tu_h) \leq$$

$$\leq \limsup_{h \to +\infty} F(\Omega \setminus B_{u_h}, u_h) + (1 - t)\mathcal{L}^n(\overline{\Omega})f_F(0) + (1 - t)2^n \mu(\Omega)$$

for every t sufficiently close to 1.

As t increases to 1, condition (9.2.1) follows from (9.3.30), (9.3.17), and (9.3.18). ∎

By using the above result, we are able to prove the following characterizations.

Theorem 9.3.7. *Let $p \in [1, +\infty]$. Let F be as in (9.3.2) satisfying (9.2.5), (9.2.9), (9.3.3)÷(9.3.8), (9.3.15), (9.3.16), and let f_F be given by (9.1.6). Then f_F is convex and lower semicontinuous, and (9.3.1) holds.*

Conversely, given $f : \mathbf{R}^n \to [0, +\infty]$ convex and lower semicontinuous, and defined F by (9.3.1) with $f_F = f$, it turns out that conditions (9.2.5), (9.2.9), (9.3.3)÷(9.3.8), (9.3.15), (9.3.16) are satisfied by F.

Proof. The proof follows from Theorem 9.3.2, once we verify that assumptions (9.2.1) and (9.2.2) are fulfilled.

Assumption (9.2.2) is trivially implied by (9.3.16), whilst (9.2.1) comes from Proposition 9.3.6. Therefore, to complete the proof, we only have to verify that the assumptions of Proposition 9.3.6 are fulfilled.

To do this we observe that (9.1.2) follows from (9.2.5) and (9.2.9), (9.1.3) from (9.3.3), (9.2.7) and (9.2.8) from (9.3.4), (9.3.5), (9.3.8) and Proposition 2.6.8, and finally (9.3.17) from (9.3.7).

Because of this, Proposition 9.3.6 applies, and the proof follows. ∎

Theorem 9.3.8. *Let $p \in]1, +\infty]$. Let F be as in (9.3.2), and f_F be given by (9.1.6). Assume that (9.2.5), (9.3.7), (9.3.8), (9.3.15) hold, and that*

$$(9.3.31) \qquad\qquad F \text{ is translation invariant,}$$

$$(9.3.32) \qquad \text{for every } u \in W^{1,\infty}_{\text{loc}}(\mathbf{R}^n),\ F(\cdot, u) \text{ is increasing,}$$

weakly subadditive, and weakly superadditive,

$$(9.3.33) \quad \limsup_{r \to 0^+} \frac{1}{r^n} F(Q_r(x_0), u) \geq F(Q_1(x_0), u(x_0) + Du(x_0) \cdot (\cdot - x_0))$$

for every $u \in W^{1,\infty}_{\text{loc}}(\mathbf{R}^n)$, and a.e. $x_0 \in \mathbf{R}^n$,

(9.3.34)   for every $\Omega \in \mathcal{A}_0$, $F(\Omega, \cdot)$ is convex.

Then f_F is convex and lower semicontinuous, and (9.3.1) holds.

Conversely, given $f: \mathbf{R}^n \to [0, +\infty]$ convex and lower semicontinuous, and defined F by (9.3.1) with $f_F = f$, it turns out that conditions (9.2.5), (9.3.7), (9.3.8), (9.3.15), (9.3.31)÷(9.3.34) are satisfied by F.

Proof. It is easy to verify that the assumptions of Theorem 9.3.7 with $p = +\infty$ are fulfilled. Consequently, we get that f_F is convex and lower semicontinuous, and that

$$(9.3.35) \quad F(\Omega, u) = \int_\Omega f_F(Du)dx \text{ for every } (\Omega, u) \in \mathcal{A}_0 \times W_{\text{loc}}^{1,\infty}(\mathbf{R}^n).$$

If $p < +\infty$, the assumptions of Theorem 8.1.2 with $\mathcal{E}_0 = \mathcal{E} = \mathcal{A}_0$, $U = W_{\text{loc}}^{1,p}(\mathbf{R}^n)$ endowed with the $U = W_{\text{loc}}^{1,p}(\mathbf{R}^n)$ topology, $G = F_-$, and $H: (\Omega, u) \in \mathcal{A}_0 \times W_{\text{loc}}^{1,p}(\mathbf{R}^n) \mapsto \int_\Omega f_F(\nabla u)dx$ are trivially fulfilled by using (9.3.35), Proposition 2.6.4, and Fatou's lemma.

Therefore, by Theorem 8.1.2 and (9.3.8), the proof follows. ∎

§9.4 Representation on BV Spaces

In the present section we want to prove some characterization results in the same order of ideas of the ones of §9.3, but for functionals F defined on $\mathcal{A}_0 \times BV_{\text{loc}}(\mathbf{R}^n)$. We look for necessary and sufficient conditions to impose on F so that it can be expressed by means of an integral of the calculus of variations of the same kind of those considered in §9.3.

As already observed in the previous chapters, the natural extension to BV spaces of an integral of the type $\int_\Omega f_F(\nabla u)dx$ is given by the Goffman-Serrin formula, hence we look for an integral representation result of the type

$$(9.4.1) \qquad F(\Omega, u) = \int_\Omega f_F(\nabla u)dx + \int_\Omega f_F^\infty(\nabla^s u)d|D^s u|$$

for every $(\Omega, u) \in \mathcal{A}_0 \times BV_{\text{loc}}(\mathbf{R}^n)$,

f_F being a convex and lower semicontinuous function from $\mathbf{R}^n$ to $[0, +\infty]$, and f_F^∞ its recession function.

Let

$$(9.4.2) \qquad F: (\Omega, u) \in \mathcal{A}_0 \times BV_{\text{loc}}(\mathbf{R}^n) \mapsto F(\Omega, u) \in [0, +\infty],$$

and define f_F by (9.1.6).

Having in mind the results of the previous section, we introduce the following conditions

(9.4.3)   for every $u \in BV_{\mathrm{loc}}(\mathbf{R}^n)$, $F(\cdot, u)$ is increasing,

(9.4.4)   for every $u \in W_{\mathrm{loc}}^{1,\infty}(\mathbf{R}^n)$, $F(\cdot, u)$ is weakly superadditive,

(9.4.5)   for every $u \in W_{\mathrm{loc}}^{1,\infty}(\mathbf{R}^n)$, $F(\cdot, u)$ is weakly subadditive,

(9.4.6)   for every $\Omega \in \mathcal{A}_0$,

$$F(\Omega, \cdot) \text{ is weak*-}BV(\Omega)\text{-lower semicontinuous,}$$

(9.4.7)   for every $u \in BV_{\mathrm{loc}}(\mathbf{R}^n)$, $F(\cdot, u)$ is inner regular.

Let us observe that, in spite of the results of §9.3, in general conditions like those assumed in the representation results of the previous section written, if necessary, with $BV_{\mathrm{loc}}(\mathbf{R}^n)$ in place of $W_{\mathrm{loc}}^{1,p}(\mathbf{R}^n)$ are not sufficient in order to characterize the functionals that can be represented as in (9.4.1), as it is shown in the example below. Roughly speaking this is due to the fact that, in general, functionals on $BV_{\mathrm{loc}}(\mathbf{R}^n)$ need not be determined by their values on smooth functions, contrarily to what happens in the case of functionals on Sobolev spaces.

Example 9.4.1. Let $n = 1$, U, G and H be as in Example 8.1.3, and set $F = H$. Then F satisfies conditions (9.2.5), (9.2.9), (9.3.3)÷(9.3.5), (9.2.1), (9.2.2), (9.3.8), (9.2.13), (9.2.3), (9.3.15), (9.3.16), (9.3.31), (9.3.33), (9.4.3), (9.4.6), and (9.4.7) written, where necessary, with $BV_{\mathrm{loc}}(\mathbf{R})$ in place of $W_{\mathrm{loc}}^{1,p}(\mathbf{R})$.

This notwithstanding, F cannot be represented for any bounded interval Ω, as in (9.4.1) for some f_F, otherwise it would be $f_F(z) = |z|^2$ for every $z \in \mathbf{R}$, and therefore F would agree with G on $BV_{\mathrm{loc}}(\mathbf{R})$.

Theorem 9.4.2. *Let F be as in (9.4.2) satisfying (9.2.5), (9.3.31), (9.4.3)÷(9.4.5), (9.2.10), (9.2.1), (9.4.6), (9.3.34), (9.4.7), and let f_F be given by (9.1.6). Then f_F is convex and lower semicontinuous, and (9.4.1) holds.*

Conversely, given $f : \mathbf{R}^n \to [0, +\infty]$ convex and lower semicontinuous, and defined F by (9.4.1) with $f_F = f$, it turns out that conditions (9.2.5), (9.3.31), (9.4.3)÷(9.4.5), (9.2.10), (9.2.1), (9.4.6), (9.3.34), (9.4.7) are satisfied by F.

Proof. By (9.2.1), (9.2.5), (9.4.3)÷(9.4.5), (9.3.31), (9.2.10), (9.2.1), (9.4.6), and Lemma 9.3.1 it follows that the restriction of F to $\mathcal{A}_0 \times$

$W_{\text{loc}}^{1,\infty}(\mathbf{R}^n)$ fulfils the assumptions of Corollary 9.2.6, from which we infer that f_F is convex and lower semicontinuous, and that

(9.4.8) $F_-(\Omega, u) = \displaystyle\int_\Omega f_F(\nabla u)dx$ for every $(\Omega, u) \in \mathcal{A}_0 \times C^1(\mathbf{R}^n)$.

At this point we observe that by (9.4.7), (9.3.34), (9.4.6), and (9.4.8) the assumptions of Theorem 8.1.2 with $\mathcal{E} = \mathcal{E}_0 = \mathcal{A}_0$, $U = BV(\mathbf{R}^n)$ equipped with the weak*-$BV(\mathbf{R}^n)$ topology, $G = F_-$, and $H\colon u \in BV(\mathbf{R}^n)$ $\mapsto \int_\Omega f_F(\nabla u)dx + \int_\Omega f_F^\infty(\nabla^s u)d|D^s u|$ are satisfied. By Theorem 8.1.2, and again (9.4.7) we thus obtain that

$$F(\Omega, u) = \int_\Omega f_F(\nabla u)dx + \int_\Omega f_F^\infty(\nabla^s u)d|D^s u|$$

for every $(\Omega, u) \in \mathcal{A}_0 \times BV(\mathbf{R}^n)$.

In conclusion, let $(\Omega, u) \in \mathcal{A}_0 \times BV_{\text{loc}}(\mathbf{R}^n)$, and let B be an open ball with $\Omega \subset\subset B$. Then $\chi_B u \in BV(\mathbf{R}^n)$ and by using (9.4.6) and an argument similar to the one exploited in the proof of Lemma 9.3.1, we obtain that

$$F(\Omega, u) = F(\Omega, \chi_B u) =$$

$$= \int_\Omega f_F(\nabla(\chi_B u))dx + \int_\Omega f_F^\infty(\nabla^s(\chi_B u))d|D^s(\chi_B u)| =$$

$$= \int_\Omega f_F(\nabla u)dx + \int_\Omega f_F^\infty(\nabla^s u)d|D^s u|,$$

from which the first part of the theorem follows.

The second part follows from a direct verification, and by using also Theorem 5.1.4. ∎

Theorem 9.4.3. *Let F be as in (9.4.2) satisfying (9.2.5), (9.3.31), (9.4.3)÷ (9.4.5), (9.2.10), (9.4.6), (9.3.34), (9.2.3), (9.4.7), and let f_F be given by (9.1.6). Then f_F is convex and lower semicontinuous, and (9.4.1) holds.*

Conversely, given $f\colon \mathbf{R}^n \to [0, +\infty]$ convex and lower semicontinuous, and defined F by (9.4.1) with $f_F = f$, it turns out that conditions (9.2.5), (9.3.31), (9.4.3)÷(9.4.5), (9.2.10), (9.4.6), (9.3.34), (9.2.3), (9.4.7) are satisfied by F.

Proof. Follows by Proposition 9.2.2 and Theorem 9.4.2. ∎

Theorem 9.4.4. *Let F be as in (9.4.2) satisfying (9.2.5), (9.3.31), (9.4.3)÷ (9.4.5), (9.2.10), (9.3.15), (9.4.6), (9.3.34), (9.4.7), and let f_F be given by (9.1.6). Then f_F is convex and lower semicontinuous, and (9.4.1) holds.*

Conversely, given $f\colon \mathbf{R}^n \to [0, +\infty]$ convex and lower semicontinuous, and defined F by (9.4.1) with $f_F = f$, it turns out that conditions (9.2.5),

(9.3.31), (9.4.3)÷(9.4.5), (9.2.10), (9.3.15), (9.4.6), (9.3.34), (9.4.7) are satisfied by F.

Proof. By (9.4.4), (9.4.5), (9.4.7) and Proposition 2.6.8 it follows that for every $u \in W^{1,\infty}_{\text{loc}}(\mathbf{R}^n)$, $F(\cdot, u)$ is superadditive and subadditive, from which, together with (9.2.5), (9.3.31), (9.4.3), (9.3.34), and (9.4.6), we conclude that the assumptions of Proposition 9.3.6 are fulfilled.

In conclusion, by Proposition 9.3.6, (9.2.5), (9.3.31), (9.4.3)÷(9.4.5), (9.2.10), (9.4.6), (9.3.34), and (9.4.7), we obtain that the assumptions of Theorem 9.4.2 hold, and the proof follows from Theorem 9.4.2. ∎

Chapter 10

Relaxation
of Unbounded Functionals

In this chapter we study some relaxation problems for certain classes of unbounded variational integral functionals, in the framework of both Sobolev and BV spaces, and in the topological setting of L^1 spaces. We prove that the corresponding relaxed functionals too are integral functionals of the same type.

The results are then applied to relaxation in presence of various types of boundary data.

Problems of this type are treated in [ET, Chapter X], and in [MS2], but limitedly to some specific cases.

Finally, in this chapter we exploit the study on the possible compositions of lower semicontinuous envelope and convex envelope operators carried out in Chapter 1.

§10.1 Notations and Elementary Properties of Relaxed Functionals in the Neumann Case

Let

$$(10.1.1) \qquad f \colon z \in \mathbf{R}^n \mapsto f(z) \in [0, +\infty]$$

be Borel, and let, for every $\Omega \in \mathcal{A}_0$, $F(\Omega, \cdot)$ be given by

$$(10.1.2) \quad F(\Omega, \cdot) \colon u \in L^1(\Omega) \mapsto \begin{cases} \int_\Omega f(\nabla u)dx & \text{if } u \in W^{1,\infty}_{\text{loc}}(\mathbf{R}^n) \\ +\infty & \text{if } u \in L^1(\Omega) \setminus W^{1,\infty}_{\text{loc}}(\mathbf{R}^n). \end{cases}$$

In this section, for $\Omega \in \mathcal{A}_0$, we start the study of $\text{sc}^-(L^1(\Omega))F(\Omega, \cdot)$, namely of the relaxed functional in the $L^1(\Omega)$ topology of $F(\Omega, \cdot)$. For the sake of simplicity, given $\Omega \in \mathcal{A}_0$, we set

$$(10.1.3) \qquad \overline{F}(\Omega, \cdot) \colon u \in L^1(\Omega) \mapsto \text{sc}^-(L^1(\Omega))F(\Omega, u),$$

and recall that, by Proposition 3.5.3, it results that (as usual here and in the sequel we assume that $\min \emptyset = +\infty$)

$$\overline{F}(\Omega, \cdot) \colon u \in L^1(\Omega) \longmapsto$$

$$\min \left\{ \liminf_{h \to +\infty} \int_\Omega f(\nabla u_h) dx : \{u_h\} \subseteq W^{1,\infty}_{\text{loc}}(\mathbf{R}^n), \ u_h \to u \text{ in } L^1(\Omega) \right\}.$$

It is obvious that

(10.1.4)   for every $\Omega \in \mathcal{A}_0$, $\overline{F}(\Omega, \cdot)$ is $L^1(\Omega)$-lower semicontinuous,

and that

$$\overline{F}(\Omega, u) = \min \left\{ \liminf_{h \to +\infty} \int_\Omega f(\nabla u_h) dx : \{u_h\} \subseteq W^{1,\infty}_{\text{loc}}(\mathbf{R}^n), \right.$$

$$\left. \text{for every } h \in \mathbf{N} \ \nabla u_h(x) \in \text{dom} f \text{ for a.e. } x \in \Omega, \ u_h \to u \text{ in } L^1(\Omega) \right\}$$

$$\text{for every } \Omega \in \mathcal{A}_0, \ u \in L^1(\Omega).$$

It is easy to see that $\overline{F}$ satisfies the following properties

(10.1.5) $\overline{F}(\Omega, u + c) = \overline{F}(\Omega, u)$ for every $\Omega \in \mathcal{A}_0$, $u \in L^1(\Omega)$, $c \in \mathbf{R}$,

(10.1.6) $\overline{F}(\Omega - x_0, T[x_0]u) = \overline{F}(\Omega, u)$

$$\text{for every } \Omega \in \mathcal{A}_0, \ u \in L^1(\Omega), \ x_0 \in \mathbf{R}^n,$$

(10.1.7) $\overline{F}(\Omega, O_t u) = \dfrac{1}{t^n} \overline{F}(t\Omega, u)$ for every $\Omega \in \mathcal{A}_0$, $t > 0$, $u \in L^1(\Omega)$,

and

(10.1.8) $\overline{F}(\Omega_2, u) \leq \overline{F}(\Omega_1, u)$

whenever Ω_1, $\Omega_2 \in \mathcal{A}_0$ satisfy $\Omega_1 \subseteq \Omega_2$, $\mathcal{L}^n(\Omega_2 \setminus \Omega_1) = 0$, $u \in L^1(\Omega_2)$.

Moreover we also have that

(10.1.9) $\overline{F}(\Omega_1, u) \leq \overline{F}(\Omega_2, u)$

whenever Ω_1, $\Omega_2 \in \mathcal{A}_0$ satisfy $\Omega_1 \subseteq \Omega_2$, $u \in L^1(\Omega_2)$,

(10.1.10) $\overline{F}(\Omega_1, u) + \overline{F}(\Omega_2, u) \leq \overline{F}(\Omega_1 \cup \Omega_2, u)$

whenever Ω_1, $\Omega_2 \in \mathcal{A}_0$ are disjoint, $u \in L^1(\Omega_1 \cup \Omega_2)$.

In order to prove additional measure theoretic properties of $\overline{F}$, we need to assume further conditions on f. More precisely that

(10.1.11) $\text{dom} f$ is convex,

(10.1.12) f is locally bounded in $\text{ri}(\text{dom} f)$,

i.e. for every compact subset K of $\text{ri}(\text{dom} f)$ there exists $M_K > 0$ such that $\sup_{z \in K} f(z) \le M_K$, and that

(10.1.13)   for every bounded subset L of $\text{dom} f$ there exists

$z_L \in \text{ri}(\text{dom} f)$ such that the function $t \in [0,1] \mapsto f((1-t)z_L + tz)$

is upper semicontinuous at $t = 1$ uniformly as z varies in L,

i.e. for every $\varepsilon > 0$ there exists $t_\varepsilon < 1$ such that $f((1-t)z_L + tz) \le f(z) + \varepsilon$ for every $t \in]t_\varepsilon, 1]$, and $z \in L$.

Remark 10.1.1. Assumption (10.1.13) looks like a sort of uniform radial upper semicontinuity on bounded subsets of $\text{dom} f$. Nevertheless it does not imply in general (10.1.12) (think for example to the case in which $n = 2$, $f(z_1, z_2) = \frac{|z_2|}{|z_1|}$ if $|z_1|^2 + |z_2|^2 \le 1$ and $z_1 z_2 \ne 0$, $f(z_1, z_2) = 0$ if $|z_1|^2 + |z_2|^2 \le 1$ and $z_1 z_2 = 0$, $f(z_1, z_2) = +\infty$ otherwise in $\mathbf{R}^2$, and $z_L = (0,0)$ independently of L). It is fulfilled if f is finite and continuous in $\mathbf{R}^n$, or if there exists $z_0 \in \text{ri}(\text{dom} f)$ such that the function $t \in [0,1] \mapsto f((1-t)z_0 + tz)$ is increasing for every z in $\text{dom} f$.

Lemma 10.1.2. *Let f be a Borel function as in (10.1.1) satisfying (10.1.11), and $\overline{F}$ be given by (10.1.3). Let $A \subset \mathcal{A}_0$, and $u \in W^{1,1}(A)$ be such that $\overline{F}(A, u) < +\infty$. Then*

(10.1.14) $\nabla u(x) \in \overline{\text{dom} f}$ *for a.e. $x \in A$.*

Proof. Since $\overline{F}(A, u) < +\infty$, there exists $\{u_h\} \subseteq W^{1,\infty}_{\text{loc}}(\mathbf{R}^n)$ such that $u_h \to u$ in $L^1(A)$ and

(10.1.15)   for every $h \in \mathbf{N}$, $\nabla u_h(x) \in \text{dom} f$ for a.e. $x \in A$.

We now observe that, being by (10.1.11) $\overline{\text{dom} f}$ closed and convex, there exist two families $\{\alpha_\theta\}_{\theta \in T} \subseteq \mathbf{R}^n$, and $\{\beta_\theta\}_{\theta \in T} \subseteq \mathbf{R}$ such that $z \in \overline{\text{dom} f}$ if and only if $\alpha_\theta \cdot z + \beta_\theta \ge 0$ for every $\theta \in T$. Therefore, by (10.1.15) we obtain that

(10.1.16) $\alpha_\theta \cdot \int_A \varphi \nabla u_h dx + \beta_\theta \ge 0$

for every $h \in \mathbf{N}$, $\theta \in \mathcal{T}$, and every $\varphi \in C_0^1(A)$ with $\varphi \geq 0$, $\int_A \varphi dx = 1$.

By (10.1.16), taking the limit as h diverges, we deduce that

$$\int_A \varphi \nabla u dx \in \overline{\mathrm{dom} f} \text{ for every } \varphi \in C_0^1(A) \text{ with } \varphi \geq 0, \int_A \varphi dx = 1,$$

from which (10.1.14) follows. ∎

§10.2 Relaxation of Neumann Problems: the Case of Bounded Effective Domain with Nonempty Interior

Let f be a Borel function as in (10.1.1), F be defined by (10.1.2), and $\overline{F}$ by (10.1.3).

The integral representation result for $\overline{F}$ will be proved in some steps, in the first one, that is treated in the present section, we assume that

(10.2.1) $\mathrm{dom} f$ is bounded,

and that

(10.2.2) $\mathrm{int}(\mathrm{dom} f) \neq \emptyset$.

It is clear that, by (10.2.1) it results

(10.2.3) $\overline{F}(\Omega, u) = \inf \left\{ \liminf_{h \to +\infty} \int_\Omega f(\nabla u_h) dx : \{u_h\} \subseteq W_{\mathrm{loc}}^{1,\infty}(\mathbf{R}^n), \text{ for} \right.$

$\left. \text{every } h \in \mathbf{N} \ \nabla u_h(x) \in \mathrm{dom} f \text{ for a.e. } x \in \Omega, \ u_h \to u \text{ in weak*-}W^{1,\infty}(\Omega) \right\}$

for every $\Omega \in \mathcal{A}_0$, $u \in L^1(\Omega)$.

Lemma 10.2.1. *Let f be a Borel function as in (10.1.1) satisfying (10.1.11) $\div$(10.1.13), (10.2.1), (10.2.2), and let $\overline{F}$ be given by (10.1.3). Then*

(10.2.4) $\overline{F}_-(\Omega_1 \cup \Omega_2, u) \leq \overline{F}_-(\Omega_1, u) + \overline{F}_-(\Omega_2, u)$

whenever Ω_1, $\Omega_2 \in \mathcal{A}_0$, $u \in L^1(\Omega_1 \cup \Omega_2)$.

Proof. Let us preliminarily observe that, by (10.2.1), we can take $L = \mathrm{dom} f$ in (10.1.13), and that it is not restrictive to assume that $z_{\mathrm{dom} f} = 0$, otherwise we just have to consider the function $f(z_{\mathrm{dom} f} + \cdot)$. In particular this, together with (10.2.2), yields

(10.2.5) $0 \in \mathrm{int}(\mathrm{dom} f)$.

Let now Ω_1, Ω_2, u be as in (10.2.4), let us fix $A \in \mathcal{A}_0$ with $A \subset\subset \Omega$, and observe that there exist $A_1 \subset\subset \Omega_1$, $A_2 \subset\subset \Omega_2$ such that $A \subset\subset A_1 \cup A_2$. Because of this, in order to prove (10.2.4), it suffices to show that

$$(10.2.6) \qquad \overline{F}(A, u) \leq \overline{F}(A_1, u) + \overline{F}(A_2, u)$$

whenever A, A_1, $A_2 \in \mathcal{A}_0$ satisfy $A_1 \subset\subset \Omega_1$, $A_2 \subset\subset \Omega_2$, $A \subset\subset A_1 \cup A_2$.

To do this, we can obviously assume that the right-hand side of (10.2.6) is finite so that, by (10.2.1) and (10.2.3), for $i = 1$, 2 there exists $\{u_h^i\} \subseteq W_{loc}^{1,\infty}(\mathbf{R}^n)$ such that $u_h^i \to u$ in weak*-$W^{1,\infty}(A_i)$, $\nabla u_h^i(x) \in \mathrm{dom} f$ for a.e. $x \in A_i$ and every $h \in \mathbf{N}$, and

$$(10.2.7) \qquad \overline{F}(A_i, u) = \lim_{h \to +\infty} \int_{A_i} f(\nabla u_h^i) dx.$$

Let $B_1 \in \mathcal{A}_0$ with $B_1 \subset\subset A_1$ such that $A \subset\subset B_1 \cup A_2$, let $\varphi \in C_0^1(A_1)$ satisfying

$$(10.2.8) \quad 0 \leq \varphi \leq 1 \text{ in } \mathbf{R}^n, \quad \varphi = 1 \text{ in } B_1, \quad \|\|\nabla\varphi\|\|_{L^\infty(\mathbf{R}^n)} \leq \frac{2}{\mathrm{dist}(B_1, \partial A_1)},$$

and set, for every $h \in \mathbf{N}$, $w_h = \varphi u_h^1 + (1 - \varphi)u_h^2$. Then $w_h \to u$ in weak*-$W^{1,\infty}(A)$, and by (10.2.8), we have that

$$(10.2.9) \qquad \overline{F}(A, tu) \leq \liminf_{h \to +\infty} \int_A f(t\nabla w_h) dx \leq$$

$$\leq \limsup_{h \to +\infty} \int_{A \cap B_1} f(t\nabla u_h^1) dx + \limsup_{h \to +\infty} \int_{A_2} f(t\nabla u_h^2) dx +$$

$$+ \limsup_{h \to +\infty} \int_{A \cap (A_1 \backslash B_1)} f(t\nabla w_h) dx \text{ for every } t \in [0, 1].$$

Let us fix now $t \in [0, 1[$. Then, since for every $h \in \mathbf{N}$ $\nabla w_h = \varphi \nabla u_h^1 + (1 - \varphi)\nabla u_h^2 + (u_h^1 - u_h^2)\nabla\varphi$, and $\nabla u_h^i(x) \in \mathrm{dom} f$ for $i = 1$, 2 and a.e. $x \in A_i$, by (10.1.11) it results that for every $h \in \mathbf{N}$, $t\varphi(x)\nabla u_h^1(x) + t(1 - \varphi(x))\nabla u_h^2(x) \in t\mathrm{dom} f$ for a.e. $x \in A$.

Because of this, once we recall that, by (10.2.5) and (10.1.11), $\overline{t\mathrm{dom} f} \subseteq \mathrm{int}(\mathrm{dom} f)$, and that $u_h^i \to u$ in $L^\infty(A)$ for $i = 1$, 2, we obtain that there exist a compact subset K_t of $\mathrm{int}(\mathrm{dom} f)$ (depending only on t), and $h_{t,A_1,B_1} \in \mathbf{N}$ (depending on t, A_1, and B_1) such that for every $h \geq h_{t,A_1,B_1}$, $t\nabla w_h(x) \in K_t$ for a.e. $x \in A$.

This, together with (10.1.12), yields that

$$(10.2.10) \qquad \text{there exist } M_t > 0, \text{ and } h_{t,A_1,B_1} \in \mathbf{N} \text{ such that}$$

for every $h \geq h_{t,A_1,B_1}$ $f(t\nabla w_h(x)) \leq M_t$ for a.e. $x \in A$.

We now fix $\varepsilon > 0$. Then, by (10.1.13) we obtain the existence of $t_\varepsilon \in [0, 1[$ such that

(10.2.11) $$\int_{A_1} f(t\nabla u_h^1)dx \leq \int_{A_1} f(\nabla u_h^1)dx + \varepsilon \mathcal{L}^n(A_1),$$

$$\int_{A_2} f(t\nabla u_h^2)dx \leq \int_{A_2} f(\nabla u_h^2)dx + \varepsilon \mathcal{L}^n(A_2)$$

for every $t \in]t_\varepsilon, 1[$, $h \in \mathbf{N}$,

hence by (10.2.9)$\div$(10.2.11), and (10.2.7) we deduce that

(10.2.12) $\overline{F}(A, tu) \leq \limsup\limits_{h\to+\infty} \int_{A_1} f(\nabla u_h^1)dx + \limsup\limits_{h\to+\infty} \int_{A_2} f(\nabla u_h^2)dx+$

$$+\varepsilon(\mathcal{L}^n(A_1) + \mathcal{L}^n(A_2)) + M_t\mathcal{L}^n(A \cap (A_1 \setminus B_1)) \leq$$

$$\leq \overline{F}(A_1, u) + \overline{F}(A_2, u) + \varepsilon(\mathcal{L}^n(A_1) + \mathcal{L}^n(A_2)) + M_t\mathcal{L}^n(A \cap (A_1 \setminus B_1))$$

for every $t \in]t_\varepsilon, 1[$.

As B_1 increases to A_1, and then t tends to 1^-, we deduce from (10.1.4), and (10.2.12) that

$$\overline{F}(A, u) \leq \liminf\limits_{t\to 1^-} \overline{F}(A, tu) \leq \overline{F}(A_1, u) + \overline{F}(A_2, u) + \varepsilon(\mathcal{L}^n(A_1) + \mathcal{L}^n(A_2)),$$

from which inequality (10.2.6) follows as ε tends to zero. ∎

Lemma 10.2.2. *Let f be a Borel function as in (10.1.1) satisfying (10.1.11) $\div$(10.1.13), (10.2.1), (10.2.2), and let $\overline{F}$ be given by (10.1.3). Then*

(10.2.13) $\overline{F}_-(\Omega, u) = \overline{F}(\Omega, u)$ *for every $\Omega \in \mathcal{A}_0$, $u \in W_{loc}^{1,\infty}(\mathbf{R}^n)$.*

Proof. Let Ω, u be as in (10.2.13). Then, since $\overline{F}(\cdot, u)$ is increasing in Ω, we immediately have that

(10.2.14) $\overline{F}_-(\Omega, u) \leq \overline{F}(\Omega, u)$.

In order to prove the reverse inequality in (10.2.14), we can obviously assume that $\overline{F}_-(\Omega, u) < +\infty$, so that $\overline{F}(A, u) < +\infty$ for every $A \in \mathcal{A}_0$ with $A \subset\subset \Omega$, and, by Lemma 10.1.2, that

(10.2.15) $\nabla u(x) \in \overline{\mathrm{dom}f}$ for a.e. $x \in \Omega$.

Let now A, $B \in \mathcal{A}_0$ with $A \subset\subset B \subset\subset \Omega$. Then, by (10.2.1) and (10.2.3) there exists $\{u_h\} \subseteq W_{\text{loc}}^{1,\infty}(\mathbf{R}^n)$ such that $u_h \to u$ in weak*-$W^{1,\infty}(B)$, and

$$\overline{F}(B, u) = \lim_{h \to +\infty} \int_B f(\nabla u_h)dx.$$

Let $\varphi \in C_0^1(B)$ be such that

$$(10.2.16) \quad 0 \le \varphi \le 1 \text{ in } \mathbf{R}^n, \ \varphi = 1 \text{ in } A, \ \|\nabla\varphi\|_{L^\infty(\mathbf{R}^n)} \le \frac{2}{\text{dist}(A, \partial B)},$$

and define for every $h \in \mathbf{N}$, $w_h = \varphi u_h + (1 - \varphi)u$. Then obviously $w_h \in W_{\text{loc}}^{1,\infty}(\mathbf{R}^n)$ for every $h \in \mathbf{N}$, and $w_h \to u$ in weak*-$W^{1,\infty}(\Omega)$.

By (10.2.1), assuming as in Lemma 10.2.1 that $z_{\text{dom}f}$ in (10.1.13) relatively to $L = \text{dom}f$ is equal to 0 (and thus getting (10.2.5)), and by using (10.1.11)÷(10.1.13), (10.2.5), (10.2.15), (10.2.16), and an argument similar to the one employed to get (10.2.10), we obtain that

(10.2.17) for every $t \in [0, 1[$ there exist $M_t > 0$ and $h_{t,B,A} \in \mathbf{N}$ such that

for every $h \ge h_{t,B,A}$ $f(t\nabla w_h(x)) + f(t\nabla u(x)) \le M_t$ for a.e. $x \in \Omega$,

and that for fixed $\varepsilon > 0$ there exists $t_\varepsilon \in]0, 1[$ such that

$$(10.2.18) \qquad \int_B f(t\nabla u_h)dx \le \int_B f(\nabla u_h)dx + \varepsilon\mathcal{L}^n(B)$$

for every $t \in]t_\varepsilon, 1[$, $h \in \mathbf{N}$.

By (10.2.16)÷(10.2.18) we conclude that

$$(10.2.19) \qquad \overline{F}(\Omega, tu) \le \liminf_{h \to +\infty} \int_\Omega f(t\nabla w_h)dx \le$$

$$\le \liminf_{h \to +\infty} \int_B f(t\nabla u_h)dx + \limsup_{h \to +\infty} \int_{B\setminus A} f(t\nabla w_h)dx + \int_{\Omega\setminus B} f(t\nabla u)dx \le$$

$$\le \limsup_{h \to +\infty} \int_B f(\nabla u_h)dx + \varepsilon\mathcal{L}^n(B) + M_t\mathcal{L}^n(\Omega \setminus A) \le$$

$$\le \overline{F}_-(\Omega, u) + \varepsilon\mathcal{L}^n(\Omega) + M_t\mathcal{L}^n(\Omega \setminus A) \text{ for every } t \in]t_\varepsilon, 1[.$$

As A increases to Ω, and then t tends to 1^-, we deduce from (10.1.4), and (10.2.19) that

$$(10.2.20) \qquad \overline{F}(\Omega, u) \le \liminf_{t \to 1^-} \overline{F}(\Omega, tu) \le \overline{F}_-(\Omega, u) + \varepsilon\mathcal{L}^n(\Omega),$$

hence, as ε tends to zero, by (10.2.20), and (10.2.14) equality (10.2.13) follows. ∎

Lemma 10.2.3. *Let f be a Borel function as in (10.1.1), and let $\overline{F}$ be given by (10.1.3). Then*

$$\limsup_{r \to 0^+} \frac{1}{r^n} \overline{F}(Q_r(x_0), u) \geq \overline{F}(Q_1(0), \nabla u(x_0) \cdot (\cdot))$$

for every $u \in W^{1,1}_{\mathrm{loc}}(\mathbf{R}^n)$, x_0 a.e. in $\mathbf{R}^n$.

Proof. Let $u \in W^{1,1}_{\mathrm{loc}}(\mathbf{R}^n)$. Then

$$\int_{Q_1(0)} |O_r T[x_0](u - u(x_0))(x) - \nabla u(x_0) \cdot x| dx =$$

$$= \frac{1}{r^{n+1}} \int_{Q_r(0)} |u(x_0 + y) - u(x_0) - \nabla u(x_0) \cdot y| dy$$

for every $x_0 \in \mathbf{R}^n$, $r > 0$.

Consequently, by Theorem 4.3.20, we have that

$$(10.2.21) \qquad \lim_{r \to 0^+} \int_{Q_1(0)} |O_r T[x_0](u - u(x_0))(x) - \nabla u(x_0) \cdot x| dx = 0$$

for a.e. $x_0 \in \mathbf{R}^n$.

We now observe that, by Lebesgue Differentiation Theorem,

$$(10.2.22) \qquad \lim_{r \to 0^+} \int_{Q_1(0)} |\nabla(O_r T[x_0](u - u(x_0))) - \nabla u(x_0)| dx = 0$$

for a.e. $x_0 \in \mathbf{R}^n$,

therefore, by (10.2.21) and (10.2.22), we get that

$$(10.2.23) \quad O_r T[x_0](u - u(x_0)) \to \nabla u(x_0) \cdot (\cdot) \text{ in } W^{1,1}(Q_1(0)) \text{ as } r \to 0^+$$

for a.e. $x_0 \in \mathbf{R}^n$.

By (10.2.23), (10.1.4), (10.1.7), and (10.1.5) we thus obtain that

$$\overline{F}(Q_1(0), \nabla u(x_0) \cdot (\cdot)) \leq$$

$$\leq \liminf_{r \to 0^+} \overline{F}(Q_1(0), O_r T[x_0](u - u(x_0))) = \limsup_{r \to 0^+} \frac{1}{r^n} \overline{F}(Q_r(x_0), u),$$

which proves the lemma. ∎

We are now in a position to prove a first integral representation result for $\overline{F}$.

Theorem 10.2.4. *Let f be a Borel function as in (10.1.1) satisfying (10.1.11)÷(10.1.13), (10.2.1), (10.2.2), and let $\overline{F}$ be given by (10.1.3). Then there exists $\phi_f : \mathbf{R}^n \to [0, +\infty]$ convex and lower semicontinuous such that*

$$\overline{F}(\Omega, u) = \int_\Omega \phi_f(\nabla u)dx \text{ for every } \Omega \in \mathcal{A}_0, \ u \in W^{1,\infty}_{\text{loc}}(\mathbf{R}^n).$$

Proof. By (10.1.5), (10.1.6), (10.1.9), (10.1.10), Lemma 10.2.1, Lemma 10.2.3, Lemma 10.2.2, (10.1.8), and (10.1.4) we get that the assumptions of Theorem 9.3.5 with $p = +\infty$ are fulfilled by the restrictions to $W^{1,\infty}_{\text{loc}}(\mathbf{R}^n)$ of the functionals $\overline{F}(\Omega, \cdot)$, $\Omega \in \mathcal{A}_0$. Thus the proof follows from Theorem 9.3.4. ∎

In the following result we specify the function ϕ_f in Theorem 10.2.4.

Proposition 10.2.5. *Let f be a Borel function as in (10.1.1) satisfying (10.1.11)÷(10.1.13), (10.2.1), (10.2.2), and let ϕ_f the one appearing in Theorem 10.2.4. Then $\phi_f = f^{**}$.*

Proof. Since $f \geq f^{**}$ we immediately deduce from Theorem 10.2.4, by the convexity and the lower semicontinuity of f^{**}, and by Theorem 7.4.6 that $\phi_f \geq f^{**}$.

On the other side it is clear that $\phi_f \leq f$. Therefore, by using the properties of ϕ_f, and (1.3.3), we obtain that $\phi_f \leq f^{**}$, and the proof. ∎

§10.3 Relaxation of Neumann Problems: the Case of Bounded Effective Domain with Empty Interior

We now want to consider the case in which assumption (10.2.2) is dropped.

For every $k \in \{1, \ldots, n\}$, we denote by 0_k the origin of $\mathbf{R}^k$, and, for every open set A of $\mathbf{R}^k$ and u in $L^1(A)$, by $\tilde{u}$ the function on $A \times \mathbf{R}^{n-k}$ defined by $\tilde{u} \colon x = (x_1, \ldots, x_n) \in A \times \mathbf{R}^{n-k} \mapsto u(x_1, \ldots, x_k)$.

Lemma 10.3.1. *Let f be a Borel function as in (10.1.1) satisfying (10.1.11) ÷(10.1.13), (10.2.1), and let $\overline{F}$ be given by (10.1.3). Assume that*

(10.3.1) $\qquad \text{aff}(\text{dom}f) = \mathbf{R}^k \times 0_{n-k} \text{ for some } k \in \{1, \ldots, n-1\}.$

Then there exists $f_p : \mathbf{R}^k \to [0, +\infty]$ convex and lower semicontinuous such that

(10.3.2) $\qquad \overline{F}(A \times I, \tilde{u}) = \mathcal{L}^{n-k}(I) \int_A f_p(\nabla u)dy$

whenever A is a bounded open set of $\mathbf{R}^k$,

I is a connected bounded open set of $\mathbf{R}^{n-k}$, $u \in W^{1,\infty}_{\text{loc}}(\mathbf{R}^k)$.

Proof. Let
$$g \colon (z_1, \ldots, z_k) \in \mathbf{R}^k \mapsto f(z_1, \ldots, z_k, 0_{n-k}),$$
define for every bounded open set A of $\mathbf{R}^k$, the functionals

$$G(A, \cdot) \colon u \in L^1(A) \mapsto \begin{cases} \int_A g(\nabla u)dy & \text{if } u \in W^{1,\infty}_{\mathrm{loc}}(\mathbf{R}^k) \\ +\infty & \text{if } u \in L^1(A) \setminus W^{1,\infty}_{\mathrm{loc}}(\mathbf{R}^k), \end{cases}$$

and
$$\overline{G}(A, \cdot) \colon u \in L^1(A) \mapsto \mathrm{sc}^-(L^1(A))G(A, u),$$

and observe that obviously

$$(10.3.3) \qquad \overline{G}(A, u) = \min\left\{ \liminf_{h \to +\infty} \int_A g(\nabla u_h)dy : \{u_h\} \subseteq W^{1,\infty}_{\mathrm{loc}}(\mathbf{R}^k), \right.$$

$$\left. \text{for every } h \in \mathbf{N} \ \nabla u_h(y) \in \mathrm{dom}g \text{ for a.e. } y \in A, \ u_h \to u \text{ in } L^1(A)\right\}$$

for every bounded open set A of $\mathbf{R}^k$, $u \in L^1(A)$.

The function g satisfies all the assumptions of Theorem 10.2.4 with $n = k$. Consequently, by Theorem 10.2.4 we deduce the existence of $g_p \colon \mathbf{R}^k \to [0, +\infty]$ convex and lower semicontinuous such that

$$(10.3.4) \qquad \overline{G}(A, u) = \int_A f_p(\nabla u)dy$$

for every bounded open set A of $\mathbf{R}^k$, $u \in W^{1,\infty}_{\mathrm{loc}}(\mathbf{R}^k)$.

Let now A, I, u be as in (10.3.2). Let us prove that

$$(10.3.5) \qquad \overline{F}(A \times I, \tilde{u}) \leq \mathcal{L}^{n-k}(I) \int_A f_p(\nabla u)dy.$$

To do this we can assume that the right-hand side of (10.3.5) is finite so that, by (10.3.3) and (10.3.4), there exists $\{u_h\} \subseteq W^{1,\infty}_{\mathrm{loc}}(\mathbf{R}^k)$ such that for every $h \in \mathbf{N} \ \nabla u_h(y) \in \mathrm{dom}f_p$ for a.e. $y \in A$, $u_h \to u$ in $L^1(A)$, and

$$(10.3.6) \qquad \int_A f_p(\nabla u)dy = \liminf_{h \to +\infty} \int_A f(\nabla_1 u_h, \ldots, \nabla_k u_h, 0_{n-k})dy.$$

Then obviously $\tilde{u}_h \to \tilde{u}$ in $L^1(A \times I)$, for every $h \in \mathbf{N} \ \nabla \tilde{u}_h(x) \in \mathrm{dom}f$ for a.e. $x \in A \times I$, and by (10.3.6), it turns out that

$$\overline{F}(A \times I, \tilde{u}) \leq \liminf_{h \to +\infty} \int_{A \times I} f(\nabla \tilde{u}_h)dx =$$

$$= \liminf_{h \to +\infty} \mathcal{L}^{n-k}(I) \int_A f(\nabla_1 u_h, \dots, \nabla_k u_h, 0_{n-k}) dy = \mathcal{L}^{n-k}(I) \int_A f_p(\nabla u) dy,$$

that is (10.3.5).

* In order to prove the opposite inequality to (10.3.5), we assume that $\overline{F}(A \times I, \tilde{u}) < +\infty$ so that there exists $\{v_h\} \subseteq W^{1,\infty}_{loc}(\mathbf{R}^n)$ such that for every $h \in \mathbf{N}$ $\nabla v_h(x) \in \mathrm{dom}f$ for a.e. $x \in A \times I$, $v_h \to \tilde{u}$ in $L^1(A \times I)$, and

$$(10.3.7) \qquad +\infty > \overline{F}(A \times I, \tilde{u}) = \lim_{h \to +\infty} \int_{A \times I} f(\nabla v_h) dx.$$

Then, by (10.3.7) and (10.3.1) we have that for every $h \in \mathbf{N}$ $\nabla_{k+1} v_h = \dots = \nabla_n v_h = 0$ a.e. in $A \times I$ from which, by taking into account the connectedness of I, we infer that v_h depends effectively only on its first k variables in $A \times I$ for every $h \in \mathbf{N}$. Because of this, we can assume that for every $h \in \mathbf{N}$ there exists $w_h \in W^{1,\infty}_{loc}(\mathbf{R}^k)$ such that $v_h = \tilde{w}_h$. Then $w_h \to u$ in $L^1(A)$, and by (10.3.7) and (10.3.4), we have that

$$(10.3.8) \qquad \overline{F}(A \times I, \tilde{u}) = \lim_{h \to +\infty} \int_{A \times I} f(\nabla_1 w_h, \dots, \nabla_k w_h, 0_{n-k}) dx =$$

$$= \mathcal{L}^{n-k}(I) \lim_{h \to +\infty} \int_A g(\nabla w_h) dy \geq \mathcal{L}^{n-k}(I) \overline{G}(A, u) =$$

$$= \mathcal{L}^{n-k}(I) \int_A f_p(\nabla u) dy.$$

By (10.3.5) and (10.3.8) equality (10.3.2) follows. ∎

In order to extend (10.3.2) to a wider class of open sets, we need to prove the following subadditivity result.

Lemma 10.3.2. *Let f be a Borel function as in (10.1.1) satisfying (10.1.11) $\div$(10.1.13), (10.2.1), (10.3.1), and let $\overline{F}$ be given by (10.1.3). Then*

$$(10.3.9) \qquad \overline{F}\left(\cup_{i=1}^m (A_i \times I_i), \tilde{u}\right) \leq \sum_{i=1}^m \overline{F}(A_i \times I_i, \tilde{u})$$

whenever $A_1, \dots, A_m$ are pairwise disjoint bounded open subsets of $\mathbf{R}^k$, $I_1, \dots, I_m$ are connected bounded open subsets of $\mathbf{R}^{n-k}$, $u \in W^{1,\infty}_{loc}(\mathbf{R}^k)$.

Proof. Let $A_1, \dots, A_m, I_1, \dots, I_m, u$ be as in (10.3.9). It is obvious that we can assume the right-hand side of (10.3.9) to be finite, so that, by Lemma 10.1.2, we get that

$$(10.3.10) \qquad \nabla \tilde{u}(x) \in \overline{\mathrm{dom}f} \text{ for a.e. } x \in \cup_{i=1}^m (A_i \times I_i).$$

Moreover, by (10.2.1), it is not restrictive to assume that $z_{\mathrm{dom}f}$ in (10.1.13) is equal to the origin of $\mathbf{R}^n$, thus getting

$$(10.3.11) \qquad\qquad 0_n \in \mathrm{ri}(\mathrm{dom}f).$$

By the finiteness of $\sum_{i=1}^m \overline{F}(A_i \times I_i, \tilde{u})$, (10.2.1), and (10.2.3) for every $i \in \{1, \ldots, m\}$ we deduce the existence of $\{u_h^i\} \subseteq W_{\mathrm{loc}}^{1,\infty}(\mathbf{R}^n)$ such that for every $h \in \mathbf{N}$ $\nabla u_h^i(x) \in \mathrm{dom}f$ for a.e. $x \in A_i \times I_i$, $u_h^i \to \tilde{u}$ in weak*-$W^{1,\infty}(A_i \times I_i)$ as h diverges, and

$$(10.3.12) \qquad \overline{F}(A_i \times I_i, \tilde{u}) = \lim_{h \to +\infty} \int_{A_i \times I_i} f(\nabla u_h^i)dx$$

$$\text{for every } i \in \{1, \ldots, m\}.$$

For every $i \in \{1, \ldots, m\}$, by (10.3.1) and the connectedness of I_i, we obtain that for every $h \in \mathbf{N}$ the functions u_h^i depend effectively only on their first k variables in $A_i \times I_i$. Because of this, from now onwards we will think of them as elements of $W_{\mathrm{loc}}^{1,\infty}(\mathbf{R}^k)$.

For every $i \in \{1, \ldots, m\}$ let B_i be an open subset of $\mathbf{R}^k$ with $B_i \subset\subset A_i$, and let $\varphi_i \in C_0^1(A_i)$ satisfying

$$(10.3.13) \qquad \begin{cases} 0 \le \varphi_i \le 1 \text{ in } \mathbf{R}^k, \ \varphi_i = 1 \text{ in } B_i, \\ \|\nabla\varphi_i\|_{L^\infty(\mathbf{R}^k)} \le \frac{2}{\mathrm{dist}(B_i, \partial A_i)}. \end{cases}$$

For every $h \in \mathbf{N}$ we set $w_h = \sum_{i=1}^m \varphi_i u_h^i + (1 - \sum_{i=1}^m \varphi_i)u$. Then $w_h \to u$ in weak*-$W^{1,\infty}(\cup_{i=1}^m A_i)$, and $\tilde{w}_h \to \tilde{u}$ in weak*-$W^{1,\infty}(\cup_{i=1}^m (A_i \times I_i))$.

Let us now observe that, being $A_1, \ldots, A_m$ pairwise disjoint, it turns out that the values $\varphi_1(y), \ldots, \varphi_m(y)$ are all equal to zero except at most for one as y varies in $\cup_{i=1}^m A_i$, hence we have that

$$\nabla\tilde{w}_h = \sum_{i=1}^m \tilde{\varphi}_i \nabla\tilde{u}_h^i + \left(1 - \sum_{i=1}^m \tilde{\varphi}_i\right)\nabla\tilde{u} + \sum_{i=1}^m \left(\tilde{u}_h^i - \tilde{u}\right)\nabla\tilde{\varphi}_i.$$

Moreover, once we recall that $u_h^i \to u$ in $L^\infty(\mathrm{spt}(\varphi_i))$ for every $i \in \{1, \ldots, m\}$, by arguing as in the proof of Lemma 10.2.1, we get by (10.1.11), (10.3.11), (10.3.10), and (10.3.13) that

(10.3.14) for every $t \in [0, 1[$ there exist a compact subset K_t of $\mathrm{ri}(\mathrm{dom}f)$

$$\text{and } h_t \in \mathbf{N} \text{ such that for every } h \ge h_t$$

$$t\nabla\tilde{w}_h(x) \in K_t \text{ for a.e. } x \in \cup_{i=1}^m(A_i \times I_i).$$

By (10.3.14), being $A_1, \ldots, A_m$ pairwise disjoint, we conclude that

$$(10.3.15) \qquad \overline{F}\left(\cup_{i=1}^m(A_i \times I_i), t\tilde{u}\right) \le \liminf_{h \to +\infty} \int_{\cup_{i=1}^m(A_i \times I_i)} f(t\nabla\tilde{w}_h)dx \le$$

$$\leq \sum_{i=1}^{m} \limsup_{h \to +\infty} \int_{A_i \times I_i} f(t\nabla \tilde{w}_h) dx \leq$$

$$\leq \sum_{i=1}^{m} \limsup_{h \to +\infty} \int_{A_i \times I_i} f(t\nabla \tilde{u}_h^i) dx + \sum_{i=1}^{m} \limsup_{h \to +\infty} \int_{(A_i \setminus B_i) \times I_i} f(t\nabla \tilde{w}_h) dx.$$

Let us now fix $\varepsilon > 0$. Then by (10.1.13) we obtain $t_\varepsilon \in]0, 1[$ such that

$$(10.3.16) \qquad \int_{A_i \times I_i} f(t\nabla \tilde{u}_h^i) dx \leq \int_{A_i \times I_i} f(\nabla \tilde{u}_h^i) dx + \varepsilon \mathcal{L}^k(A_i) \mathcal{L}^{n-k}(I_i)$$

for every $i \in \{1, \ldots, m\}$, $h \in \mathbf{N}$,

and, by (10.3.14) and (10.1.12), that

$$(10.3.17) \qquad \text{for every } t \in]0, 1[\text{ there exists } M_t > 0 \text{ such that}$$

for every $h \geq h_t$ $f(t\nabla \tilde{w}_h(x)) \leq M_t$ for a.e. $x \in \cup_{i=1}^{m}(A_i \times I_i)$.

By (10.3.15)÷(10.3.17), and (10.3.12) we conclude that

$$(10.3.18) \qquad \overline{F}(\cup_{i=1}^{m}(A_i \times I_i), t\tilde{u}) \leq$$

$$\leq \sum_{i=1}^{m} \overline{F}(A_i \times I_i, \tilde{u}) + \varepsilon \sum_{i=1}^{m} \mathcal{L}^k(A_i) \mathcal{L}^{n-k}(I_i) + M_t \sum_{i=1}^{m} \mathcal{L}^k(A_i \setminus B_i) \mathcal{L}^{n-k}(I_i).$$

Letting first B_i increase to A_i for every $i \in \{1, \ldots, m\}$, then t tend to 1^-, and finally ε go to 0^+, we obtain (10.3.9) by (10.3.18), and (10.1.4). ∎

We can now prove the representation result for $\overline{F}$ under assumption (10.2.1).

Theorem 10.3.3. *Let f be a Borel function as in (10.1.1) satisfying (10.1.11)÷(10.1.13), (10.2.1), and let $\overline{F}$ be given by (10.1.3). Then there exists $\phi_f : \mathbf{R}^n \to [0, +\infty]$ convex and lower semicontinuous such that*

$$(10.3.19) \qquad \overline{F}(\Omega, u) = \int_{\Omega} \phi_f(\nabla u) dx$$

for every $\Omega \in \mathcal{A}_0$ convex, $u \in W_{loc}^{1,\infty}(\mathbf{R}^n)$.

Proof. Let us assume for a moment that (10.3.1) holds.

Let Ω, u be as in (10.3.19), and assume that $\overline{F}(\Omega, u) < +\infty$. Then, by Lemma 10.1.2, we get that $\nabla u(x) \in \overline{\text{dom} f}$ for a.e. $x \in \Omega$, and therefore, by taking into account (10.3.1) and the convexity of Ω, that u depends only on its first k variables in Ω. Let $v \in W_{loc}^{1,\infty}(\mathbf{R}^k)$ be such that $u = \tilde{v}$ in Ω. Then it is clear that

$$(10.3.20) \qquad \overline{F}(\Omega, u) = \overline{F}(\Omega, \tilde{v}).$$

For every $\nu \in \mathbf{N}$ let $\mathcal{R}_\nu$ be a partition of $\mathbf{R}^n$, up to a set of zero measure, made up by open cubes $A_i \times I_j$ ($i, j \in \mathbf{N}$) with faces parallel to the coordinate planes, where for every $i, j \in \mathbf{N}$, A_i is an open cube of $\mathbf{R}^k$ and I_j is an open cube of $\mathbf{R}^{n-k}$, and let $S^\nu = \{(i,j) \in \mathbf{N} \times \mathbf{N} : A_i \times I_j \subset\subset \Omega\}$.

Let us fix $\nu \in \mathbf{N}$. By (10.3.20), (10.1.9), (10.1.10), and Lemma 10.3.1 we deduce the existence of $f_p : \mathbf{R}^k \to [0, +\infty]$ convex and lower semicontinuous such that

$$(10.3.21) \qquad \overline{F}(\Omega, u) \geq \overline{F}\left(\cup_{(i,j)\in S^\nu}(A_i \times I_j), \tilde{v}\right) \geq$$

$$\geq \sum_{(i,j)\in S^\nu} \overline{F}(A_i \times I_j, \tilde{v}) = \sum_{(i,j)\in S^\nu} \mathcal{L}^{n-k}(I_j) \int_{A_i} f_p(\nabla v)dy.$$

At this point, if we define ϕ_f by

$$(10.3.22) \quad \phi_f : (z_1, \ldots, z_n) \in \mathbf{R}^n \mapsto \begin{cases} f_p(z_1, \ldots, z_k) & \text{if } z_{k+1} = \ldots = z_n = 0 \\ +\infty & \text{otherwise,} \end{cases}$$

ϕ_f turns out to be convex and lower semicontinuous. Moreover by (10.3.21) we obtain that

$$(10.3.23) \qquad\qquad \overline{F}(\Omega, u) \geq$$

$$\geq \sum_{(i,j)\in S^\nu} \int_{A_i \times I_j} \phi_f(\nabla u)dx = \int_{\cup_{(i,j)\in S^\nu}(A_i \times I_j)} \phi_f(\nabla u)dx.$$

As ν diverges we deduce from (10.3.23) that

$$(10.3.24) \qquad\qquad \overline{F}(\Omega, u) \geq \int_\Omega \phi_f(\nabla u)dx$$

for every $\Omega \in \mathcal{A}_0$ convex, $u \in W^{1,\infty}_{\text{loc}}(\mathbf{R}^n)$.

In order to prove the reverse inequality in (10.3.24), again when (10.3.1) holds, let f_p be given by Lemma 10.3.1, ϕ_f by (10.3.22), and Ω, u as in (10.3.19).

We can clearly assume that $\int_\Omega \phi_f(\nabla u)dx < +\infty$. Because of this, and by the convexity of Ω, we get that u depends effectively only on its first k variables in Ω and, as before, let $v \in W^{1,\infty}_{\text{loc}}(\mathbf{R}^k)$ be such that $u = \tilde{v}$ in Ω. Moreover, for every $\nu \in \mathbf{N}$, let $\mathcal{R}_\nu$, and S^ν be as above.

Let us fix $\nu \in \mathbf{N}$. For every $i \in \mathbf{N}$ let us define $S_i^\nu = \{j \in \mathbf{N} : (i,j) \in S^\nu\}$, and assume, for the sake of simplicity, that $S_i^\nu \neq \emptyset$ if and only if $i \in \{1, \ldots, m_\nu\}$.

For every $i \in \{1, \ldots, m_\nu\}$ set $C_i = \text{int}(\cup_{j \in S_i^\nu} \overline{I_j})$. Then, by using the convexity of Ω, it turns out that C_i is connected, and $\cup_{i=1}^{m_\nu}(A_i \times C_i) \subset\subset \Omega$. Moreover, by (10.3.22), Lemma 10.3.1, and Lemma 10.3.2 we have that

$$(10.3.25) \qquad \int_\Omega \phi_f(\nabla u)dx = \int_\Omega f_p(\nabla v)dx \geq \int_{\cup_{i=1}^{m_\nu}(A_i \times C_i)} f_p(\nabla v)dx =$$

$$= \sum_{i=1}^{m_\nu} \mathcal{L}^{n-k}(C_i) \int_{A_i} f_p(\nabla v)dy = \sum_{i=1}^{m_\nu} \overline{F}(A_i \times C_i, \tilde{v}) = \overline{F}\left(\cup_{i=1}^{m_\nu}(A_i \times C_i), \tilde{v}\right).$$

Let us now set $\Omega_\nu = \text{int}(\overline{\cup_{i=1}^{m_\nu}(A_i \times C_i)})$. Then, by (10.3.25), and (10.1.8), we deduce that

$$(10.3.26) \qquad \int_\Omega \phi_f(\nabla u)dx \geq \overline{F}(\Omega_\nu, \tilde{v}) = \overline{F}(\Omega_\nu, u),$$

therefore, as ν diverges, we obtain by (10.3.26) that

$$(10.3.27) \qquad \int_\Omega \phi_f(\nabla u)dx \geq \overline{F}_-(\Omega, u)$$

for every $\Omega \in \mathcal{A}_0$ convex, $u \in W_{\text{loc}}^{1,\infty}(\mathbf{R}^n)$.

Finally by (10.1.9), (10.1.4), (10.1.6), and (10.1.7) it follows that the assumptions of Proposition 2.7.4 with $\mathcal{O} = \mathcal{A}_0$, $U = W_{\text{loc}}^{1,\infty}(\mathbf{R}^n)$, and $\Phi = \overline{F}$ are fulfilled. Consequently, by Proposition 2.7.4, (2.5.4), and (10.3.27) we infer that

$$(10.3.28) \qquad \int_\Omega \phi_f(\nabla u)dx \geq \overline{F}(\Omega, u)$$

for every $\Omega \in \mathcal{A}_0$ convex, $u \in W_{\text{loc}}^{1,\infty}(\mathbf{R}^n)$.

By (10.3.28), and (10.3.24) we get (10.3.19) under assumption (10.3.1). We now consider the general case, when (10.3.1) is not assumed.

If $\text{aff}(\text{dom} f) = \mathbf{R}^n$, the proof follows from Theorem 10.2.4, hence we can assume that the dimension k of $\text{aff}(\text{dom} f)$ is strictly smaller than n.

If $k = 0$, $\text{dom} f$ consists of a single point and (10.3.19) follows trivially, hence we can assume that $k \in \{1, \ldots, n-1\}$.

Let $A \colon \mathbf{R}^n \to \mathbf{R}^n$ be an affine transformation such that, denoting by M_A the matrix associated to the linear part of A, $\det M_A = 1$, and $A(\text{aff}(\text{dom} f)) = \mathbf{R}^k \times \{0_{n-k}\}$. Let us set

$$f_A \colon (z_1, \ldots, z_n) \in \mathbf{R}^n \mapsto f(A^{-1}(z_1, \ldots, z_n)).$$

Then f_A satisfies (10.1.11)÷(10.1.13), and $\text{aff}(\text{dom} f_A) = \mathbf{R}^k \times \{0_{n-k}\}$.

Let $\overline{F_A}$ be the functional defined by (10.1.3) with $f = f_A$. Let us observe that, for every $\Omega \in \mathcal{A}_0$ convex, the set $A(\Omega)$ is again convex, bounded, and open.

By the particular case considered above we get $\phi_{f_A} : \mathbf{R}^n \to [0, +\infty]$ convex and lower semicontinuous such that

$$(10.3.29) \qquad \overline{F_A}(A^{-1}(\Omega), u^A) = \int_{A^{-1}(\Omega)} \phi_{f_A}(\nabla u^A) dy$$

for every $\Omega \in \mathcal{A}_0$ convex, $u \in W^{1,\infty}_{\text{loc}}(\mathbf{R}^n)$,

u^A being defined by $u^A : y \in \mathbf{R}^n \mapsto u(A(y))$.

Let us observe now that

$$(10.3.30) \quad \overline{F_A}(A^{-1}(\Omega), u^A) = \overline{F}(\Omega, u) \text{ for every } \Omega \in \mathcal{A}_0, \ u \in W^{1,\infty}_{\text{loc}}(\mathbf{R}^n),$$

and define ϕ_f by $\phi_f : z \in \mathbf{R}^n \mapsto \phi_{f_A}(A(z))$. Then obviously $\phi_{f_A}(z) = \phi_f(A^{-1}(z))$ for every $z \in \mathbf{R}^n$, and by (10.3.30) and (10.3.29), we get that

$$\overline{F}(\Omega, u) = \overline{F_A}(A^{-1}(\Omega), u^A) = \int_{A^{-1}(\Omega)} \phi_{f_A}(\nabla_y u^A(y)) dy =$$

$$= \int_{A^{-1}(\Omega)} \phi_f(A^{-1}(\nabla_y u^A(y))) dy = \int_{A^{-1}(\Omega)} \phi_f((\nabla_x u)(A(y))) dy =$$

$$= \int_{\Omega} \phi_f(\nabla_x u) dx \text{ for every } \Omega \in \mathcal{A}_0, \ u \in W^{1,\infty}_{\text{loc}}(\mathbf{R}^n),$$

which proves the theorem. ∎

In the following result we specify the function ϕ_f in (10.3.19).

Proposition 10.3.4. *Let f be a Borel function as in (10.1.1) satisfying (10.1.11)÷(10.1.13), (10.2.1), and ϕ_f the one appearing in Theorem 10.3.3. Then $\phi_f = f^{**}$.*

Proof. Similar to the one of Proposition 10.2.5, but by using Theorem 10.3.3 in place of Theorem 10.2.4. ∎

§10.4 Relaxation of Neumann Problems: a First Result without Boundedness Assumptions on the Effective Domain

Let f be a Borel function as in (10.1.1), F be defined by (10.1.2), and $\overline{F}$ by (10.1.3).

The present section yields some preliminaries to the integral representation result for $\overline{F}$ when assumption (10.2.1) is dropped. This is done by studying the integral representation properties, for every $\Omega \in \mathcal{A}_0$, of the sequential lower value of the restriction of $F(\Omega, \cdot)$ to $W^{1,\infty}(\Omega)$ defined by

$$(10.4.1) \qquad F^{(\infty)}(\Omega, \cdot): u \in W^{1,\infty}(\Omega) \mapsto \inf \left\{ \liminf_{h \to +\infty} \int_\Omega f(\nabla u_h) dx : \right.$$

$$\left. \{u_h\} \subseteq W^{1,\infty}_{\text{loc}}(\mathbf{R}^n), \; u_h \to u \text{ in weak*-}W^{1,\infty}(\Omega) \right\}.$$

As already observed in Chapter 3, in general, for a given $\Omega \in \mathcal{A}_0$, $F^{(\infty)}(\Omega, \cdot)$ needs not be sequentially weak*-$W^{1,\infty}(\Omega)$-lower semicontinuous, and

$$(10.4.2) \qquad \overline{F}(\Omega, u) \le F^{(\infty)}(\Omega, u) \text{ for every } \Omega \in \mathcal{A}_0, \; u \in W^{1,\infty}(\Omega).$$

Theorem 10.4.1. Let f be a Borel function as in (10.1.1) satisfying (10.1.11)÷(10.1.13), and let $F^{(\infty)}$ be defined by (10.4.1). Then there exists $\phi_f: \mathbf{R}^n \to [0, +\infty]$ convex and Borel such that

$$(10.4.3) \qquad F^{(\infty)}(\Omega, u) \ge \int_\Omega \phi_f(\nabla u) dx$$

for every $\Omega \in \mathcal{A}_0$ convex, $u \in W^{1,\infty}(\Omega)$,

$$(10.4.4) \qquad F^{(\infty)}(\Omega, u) = \int_\Omega \phi_f(\nabla u) dx$$

for every $\Omega \in \mathcal{A}_0$ convex, $u \in W^{1,\infty}(\Omega)$ such that $F^{(\infty)}(\Omega, u) < +\infty$.

If in addition $\text{int}(\text{dom} f) \ne \emptyset$, then

$$(10.4.5) \quad F^{(\infty)}(\Omega, u) \ge \int_\Omega \phi_f(\nabla u) dx \text{ for every } \Omega \in \mathcal{A}_0, \; u \in W^{1,\infty}_{\text{loc}}(\mathbf{R}^n),$$

$$(10.4.6) \qquad F^{(\infty)}(\Omega, u) = \int_\Omega \phi_f(\nabla u) dx$$

for every $\Omega \in \mathcal{A}_0$, $u \in W^{1,\infty}_{\text{loc}}(\mathbf{R}^n)$ such that $F^{(\infty)}(\Omega, u) < +\infty$.

Proof. Let us prove (10.4.3).

For every $m \in \mathbf{N}$ set $f_m = f + I_{Q_m(0)}$, and define for every $\Omega \in \mathcal{A}_0$, $\overline{F}_m(\Omega, \cdot)$ as in (10.1.3) with f_m in place of f.

It is clear that the sequence $\{f_m\}$ is decreasing, hence for every $\Omega \in \mathcal{A}_0$, and u in $L^1(\Omega)$ so is also $\overline{F}_m(\Omega, u)$. Moreover we also have that

$$(10.4.7) \quad F^{(\infty)}(\Omega, u) = \inf_{m \in \mathbf{N}} \overline{F}_m(\Omega, u) \text{ for every } \Omega \in \mathcal{A}_0, \; u \in W^{1,\infty}(\Omega).$$

For fixed $m \in \mathbf{N}$, f_m satisfies the assumptions of Theorem 10.3.3. Consequently there exists $\phi_{f_m} \colon \mathbf{R}^n \to [0, +\infty]$ convex and lower semicontinuous such that

$$(10.4.8) \qquad \overline{F_m}(\Omega, u) = \int_\Omega \phi_{f_m}(\nabla u) dx$$

for every $\Omega \in \mathcal{A}_0$ convex, $u \in W^{1,\infty}_{\mathrm{loc}}(\mathbf{R}^n)$, $m \in \mathbf{N}$.

Since for every $\Omega \in \mathcal{A}_0$ and u in $L^1(\Omega)$, $\overline{F_m}(\Omega, u)$ is decreasing, it results that for every $z \in \mathbf{R}^n$ the sequence $\{\phi_{f_m}(z)\}$ too satisfies the same property. Therefore if we define ϕ_f by

$$(10.4.9) \qquad \phi_f \colon z \in \mathbf{R}^n \mapsto \inf_{m \in \mathbf{N}} \phi_{f_m}(z),$$

we get that ϕ_f is convex and Borel and, by (10.4.7) and (10.4.8), that

$$(10.4.10) \qquad F^{(\infty)}(\Omega, u) = \inf_{m \in \mathbf{N}} \overline{F_m}(\Omega, u) = \inf_{m \in \mathbf{N}} \int_\Omega \phi_{f_m}(\nabla u) dx \geq$$

$$\geq \int_\Omega \phi_f(\nabla u) dx \text{ for every } \Omega \in \mathcal{A}_0 \text{ convex}, \ u \in W^{1,\infty}_{\mathrm{loc}}(\mathbf{R}^n),$$

that is (10.4.3) once we recall that, being Ω convex, every element of $W^{1,\infty}(\Omega)$ can be extended to an element of $W^{1,\infty}_{\mathrm{loc}}(\mathbf{R}^n)$.

In order to prove (10.4.4) let us observe that $\phi_f(z) = \lim_{m \to +\infty} \phi_{f_m}(z)$ for every $z \in \mathbf{R}^n$, and that, if $\Omega \in \mathcal{A}_0$ is convex, $u \in W^{1,\infty}_{\mathrm{loc}}(\mathbf{R}^n)$, and $F^{(\infty)}(\Omega, u) < +\infty$, then (10.4.10) yields $\int_\Omega \phi_{f_{m_0}}(\nabla u) dx < +\infty$ for some $m_0 \in \mathbf{N}$. Consequently, by (10.4.7), (10.4.8), and Lebesgue Dominated Convergence Theorem, we conclude that

$$(10.4.11) \qquad F^{(\infty)}(\Omega, u) = \lim_{m \to +\infty} \int_\Omega \phi_{f_m}(\nabla u) dx = \int_\Omega \phi_f(\nabla u) dx$$

for every $\Omega \in \mathcal{A}_0$ convex, $u \in W^{1,\infty}_{\mathrm{loc}}(\mathbf{R}^n)$ such that $F^{(\infty)}(\Omega, u) < +\infty$.

By (10.4.11) equality (10.4.4) follows once we recall that, being Ω convex, every element of $W^{1,\infty}(\Omega)$ can be thought as an element of $W^{1,\infty}_{\mathrm{loc}}(\mathbf{R}^n)$.

Finally, the proofs of (10.4.5) and (10.4.6) follow exactly as above, but by using Theorem 10.2.4 in place of Theorem 10.3.3. ∎

Remark 10.4.2. We point out that, by (10.4.11), and Proposition 10.3.4, under the assumptions of Theorem 10.4.1 the following representation formula for $F^{(\infty)}$ hold

$$F^{(\infty)}(\Omega, u) = \lim_{m \to +\infty} \int_\Omega (f + I_{Q_m(0)})^{**}(\nabla u) dx =$$

$$= \inf_{m \in \mathbf{N}} \int_\Omega (f + I_{Q_m(0)})^{**}(\nabla u) dx \text{ for every } \Omega \in \mathcal{A}_0 \text{ convex}, \ u \in W^{1,\infty}(\Omega),$$

or for every $\Omega \in \mathcal{A}_0$, $u \in W^{1,\infty}_{\mathrm{loc}}(\mathbf{R}^n)$ if $\mathrm{int}(\mathrm{dom} f) \neq \emptyset$.

In the following result we describe the function ϕ_f in Theorem 10.4.1.

Proposition 10.4.3. *Let f be a Borel function as in (10.1.1) satisfying (10.1.11)÷(10.1.13), and ϕ_f be the one appearing in Theorem 10.4.1. Then $\phi_f = \mathrm{co}(\mathrm{sc}^- f)$.*

Proof. Follows from (10.4.9), (10.4.8), and Proposition 1.4.4. ∎

§10.5 Relaxation of Neumann Problems: Relaxation in BV Spaces

Let f be a Borel function as in (10.1.1), F be defined by (10.1.2), and $\overline{F}$ by (10.1.3).

In the present section we prove the representation result for $\overline{F}$ on BV spaces.

Lemma 10.5.1. *Let f be a Borel function as in (10.1.1) satisfying (10.1.11) ÷(10.1.13), and let $\overline{F}$ be given by (10.1.3). Then there exists $\phi_f \colon \mathbf{R}^n \to [0, +\infty]$ convex and lower semicontinuous such that*

$$(10.5.1) \qquad \overline{F}_-(\Omega, u) = \int_\Omega \phi_f(\nabla u)dx + \int_\Omega \phi_f^\infty(\nabla^s u)d|D^s u|$$

for every $\Omega \in \mathcal{A}_0$ convex, $u \in BV(\Omega)$.

If in addition $\mathrm{int}(\mathrm{dom} f) \neq \emptyset$, then

$$(10.5.2) \qquad \overline{F}_-(\Omega, u) = \int_\Omega \phi_f(\nabla u)dx + \int_\Omega \phi_f^\infty(\nabla^s u)d|D^s u|$$

for every $\Omega \in \mathcal{A}_0$, $u \in BV(\Omega)$.

Proof. Let us prove (10.5.1).

For every $\Omega \in \mathcal{A}_0$ let $F^{(\infty)}(\Omega, \cdot)$ be given by (10.4.1), and let ϕ be the convex Borel function given by Theorem 10.4.1. Let us set $\phi_f = (\phi + I_{\mathrm{dom} f})^{**}$. Then it is clear that ϕ_f is convex, lower semicontinuous, and that, since obviously $\phi \leq f$, $\phi_f \leq \phi + I_{\mathrm{dom} f} \leq f$.

Because of this, and of Theorem 7.4.6 we get that

$$(10.5.3) \qquad \overline{F}_-(\Omega, u) \geq \int_\Omega \phi_f(\nabla u)dx + \int_\Omega \phi_f^\infty(\nabla^s u)d|D^s u|$$

for every $\Omega \in \mathcal{A}_0$, $u \in BV(\Omega)$.

In order to prove the reverse inequality in (10.5.3), let us first observe that $I_{\mathrm{dom} f} \leq \phi + I_{\mathrm{dom} f} \leq f$, from which we conclude that $\mathrm{dom}(\phi + I_{\mathrm{dom} f}) = \mathrm{dom} f$ and, together with (10.1.11), the convexity of ϕ, and Proposition 1.3.2, that it results

$$(10.5.4) \qquad \mathrm{ri}(\mathrm{dom} \phi_f) = \mathrm{ri}(\mathrm{dom}(\phi + I_{\mathrm{dom} f})) = \mathrm{ri}(\mathrm{dom} f),$$

(10.5.5) $\phi_f(z) = \phi(z) + I_{\mathrm{dom}f}(z) = \phi(z)$ for every $z \in \mathrm{ri}(\mathrm{dom}f)$.

Let Ω be as in (10.5.1), and assume for the moment that $u \in C^\infty(\mathbf{R}^n)$. Let $z_1 \in \mathrm{ri}(\mathrm{dom}f)$, $t \in [0,1[$, and observe that we can assume $\int_\Omega f(\nabla u)dx < +\infty$ so that $\nabla u(x) \in \mathrm{dom}f$ for every $x \in \Omega$, and there exists a compact subset K_t of $\mathrm{ri}(\mathrm{dom}f)$ such that

$$t\nabla u(x) + (1-t)z_1 \in K_t \text{ for every } x \in \Omega.$$

By (10.5.4) it follows that $K_t \subseteq \mathrm{ri}(\mathrm{dom}\phi_f)$, and hence, by using also (10.1.12), that

$$F^{(\infty)}(\Omega, tu + (1-t)u_{z_1}) \leq \int_\Omega f(t\nabla u + (1-t)z_1)dx < +\infty.$$

This, together with (10.1.9), (10.4.2), Theorem 10.4.1, (10.5.5), and the convexity of ϕ_f implies that

(10.5.6) $\overline{F}_-(\Omega, tu + (1-t)u_{z_1}) \leq F^{(\infty)}(\Omega, tu + (1-t)u_{z_1}) =$

$$= \int_\Omega \phi(t\nabla u + (1-t)z_1)dx = \int_\Omega \phi_f(t\nabla u + (1-t)z_1)dx \leq$$

$$\leq \int_\Omega \phi_f(\nabla u)dx + (1-t)\phi_f(z_1)\mathcal{L}^n(\Omega).$$

Hence, as t increases to 1, we obtain by (10.5.6) and (10.1.4) that

(10.5.7) $\overline{F}_-(\Omega, u) \leq \int_\Omega \phi_f(\nabla u)dx$

for every $\Omega \in \mathcal{A}_0$ convex, $u \in C^\infty(\mathbf{R}^n)$.

We now observe that the assumptions of Proposition 8.1.1 are fulfilled with $\mathcal{E}_0$ equal to the family of the convex bounded open subsets of $\mathbf{R}^n$, $U = BV(\mathbf{R}^n)$ equipped with its weak*-$BV(\mathbf{R}^n)$ topology, G equal to the restriction of $\overline{F}_-$ to $\mathcal{E}_0 \times BV(\mathbf{R}^n)$, and $H:(\Omega, u) \in \mathcal{E}_0 \times BV(\mathbf{R}^n) \mapsto \int_\Omega \phi_f(\nabla u)dx + \int_\Omega \phi_f^\infty(\nabla^s u)d|D^s u|$. In fact an argument similar to the one proposed in the proof of Lemma 7.4.4 yields that H is translation invariant and convex, and by Theorem 5.1.4, it turns out to be weak*-$BV(\mathbf{R}^n)$-lower semicontinuous. Moreover so is also $\overline{F}$, and (10.5.7) holds.

By Proposition 8.1.1 we thus obtain that

$$(\overline{F}_-)_{\mathcal{E}_0-}(\Omega, u) \leq$$

$$\leq \sup\left\{\int_A \phi_f(\nabla u)dx + \int_A \phi_f^\infty(\nabla^s u)d|D^s u| : A \in \mathcal{E}_0, \ A \subset\subset \Omega\right\}$$

$$\text{for every } \Omega \in \mathcal{A}_0, \ u \in BV(\mathbf{R}^n),$$

from which, once we observe that $\mathcal{E}_0$ is perfect in $\mathcal{A}_0$, by using Proposition 2.6.9, Proposition 2.6.4, and an argument similar to the one exploited in the proof of Theorem 9.4.2, we conclude that

$$(10.5.8) \qquad \overline{F}_-(\Omega, u) \leq \int_\Omega \phi_f(\nabla u)dx + \int_\Omega \phi_f^\infty(\nabla^s u)d|D^s u|$$

$$\text{for every } \Omega \in \mathcal{A}_0 \text{ convex. } u \in BV(\Omega).$$

By (10.5.8) and (10.5.3), equality (10.5.1) follows.

The proof of (10.5.2) follows exactly as above with the only difference that in this case (10.5.7) holds for every bounded open set, and by taking $\mathcal{E}_0 = \mathcal{A}_0$ in the application of Proposition 8.1.1. ∎

Theorem 10.5.2. *Let f be a Borel function as in (10.1.1) satisfying (10.1.11)÷(10.1.13), and let $\overline{F}$ be given by (10.1.3). Then there exists $\phi_f: \mathbf{R}^n \to [0, +\infty]$ convex and lower semicontinuous such that*

$$(10.5.9) \qquad \overline{F}(\Omega, u) = \int_\Omega \phi_f(\nabla u)dx + \int_\Omega \phi_f^\infty(\nabla^s u)d|D^s u|$$

for every $\Omega \in \mathcal{A}_0$ convex, $u \in BV(\Omega)$.

Proof. Let ϕ_f be given by Lemma 10.5.1. Then by (10.1.9), (10.1.4), (10.1.6), and (10.1.7) Proposition 2.7.4 applies with $U = BV_{\text{loc}}(\mathbf{R}^n)$, $\Phi = \overline{F}$. Because of this, (2.5.4), and Lemma 10.5.1 we conclude that (10.5.9) holds. ∎

In the following proposition we identify the function ϕ_f in Theorem 10.5.2.

Proposition 10.5.3. *Let f be a Borel function as in (10.1.1) satisfying (10.1.11)÷(10.1.13), and let ϕ_f be the one given by Theorem 10.5.2. Then $\phi_f = f^{**}$.*

Proof. By (1.3.3), and Proposition 1.4.1 we have

$$(10.5.10) \quad f^{**} = (f^{**})^{**} \leq (\text{co}(\text{sc}^- f) + I_{\text{dom}f})^{**} \leq (f + I_{\text{dom}f})^{**} = f^{**},$$

therefore, by the definition of f in Lemma 10.5.1, Proposition 10.4.3, and (10.5.10) the proof follows. ∎

By the above results we deduce the following corollaries.

Corollary 10.5.4. *Let f as in (10.1.1) be convex and lower semicontinuous, and let $\overline{F}$ be given by (10.1.3). Then*

$$\overline{F}(\Omega, u) = \int_\Omega f(\nabla u)dx + \int_\Omega f^\infty(\nabla^s u)d|D^s u|$$

for every $\Omega \in \mathcal{A}_0$ convex, $u \in BV(\Omega)$.

Proof. Let us prove that the assumptions of Theorem 10.5.2 are fulfilled.

To do this, by using the convexity of f, we only have to verify that (10.1.13) is fulfilled. But this holds since the convexity of f yields the following estimate

$$f((1-t)z_L + tz) - f(z) \le (1-t)f(z_L) + tf(z) - f(z) \le (1-t)f(z_L)$$

for every bounded subset L of $\mathrm{dom}f$, $z_L \in \mathrm{int}(\mathrm{dom}f)$, $z \in L$, $t \in [0,1]$,

from which (10.1.13) follows.

The proof now follows from Theorem 10.5.2. ∎

Corollary 10.5.5. *Let $g: \mathbf{R}^n \to [0, +\infty[$ be continuous, and $C \subseteq \mathbf{R}^n$ be convex. Then*

$$\inf \left\{ \liminf_{h \to +\infty} \int_\Omega g(\nabla u_h)dx : \{u_h\} \subseteq W^{1,\infty}_{\mathrm{loc}}(\mathbf{R}^n), \right.$$

for every $h \in \mathbf{N}$ $\nabla u_h(x) \in C$ for a.e. $x \in \Omega$, $u_h \to u$ in $L^1(\Omega) \Big\} =$

$$= \int_\Omega (g + I_C)^{**}(\nabla u)dx + \int_\Omega ((g + I_C)^{**})^\infty(\nabla^s u)d|D^s u|$$

for every $\Omega \in \mathcal{A}_0$ convex, $u \in BV(\Omega)$.

Proof. Follows from Theorem 10.5.2, and Proposition 10.5.3 applied with $f = g + I_C$, once we observe that $g + I_C$ satisfies conditions (10.1.11)÷(10.1.13), the last two being fulfilled by exploiting the uniform continuity of g on the bounded subsets of $\mathbf{R}^n$. ∎

§10.6 Notations and Elementary Properties of Relaxed Functionals in the Dirichlet Case

In the present section we want to deduce analogous representation results, on Sobolev and BV spaces, for relaxed functionals of integrals of the calculus of variations of the type of those considered in this chapter, but relatively to the case in which boundary data, possibly nonhomogeneous, are taken into account.

We point out that such deduction is not a direct consequence of the results of the previous sections. Therefore, given a Borel function f as in (10.1.1), and a boundary datum $u_0 \in W_{\text{loc}}^{1,1}(\mathbf{R}^n)$, we set for every $\Omega \in \mathcal{A}_0$

$$(10.6.1) \qquad F_0(u_0, \Omega, \cdot) \colon u \in L^1(\Omega) \mapsto$$

$$\begin{cases} \int_\Omega f(\nabla u_h)dx & \text{if } u \in u_0 + W_0^{1,\infty}(\Omega) \\ +\infty & \text{if } u \in L^1(\Omega) \setminus (u_0 + W_0^{1,\infty}(\Omega)), \end{cases}$$

and prove some integral representation results, when $\Omega \in \mathcal{A}_0$, for the relaxed functional $\text{sc}^-(L^1(\Omega))F_0(u_0, \Omega, \cdot)$ of $F_0(u_0, \Omega, \cdot)$.

To carry out this program the properties of u_0 will play a crucial role, and the results will rely deeply on whether $\text{dom} f$ has interior points or not.

As in the previous sections, given $\Omega \in \mathcal{A}_0$, we set for the sake of simplicity

$$(10.6.2) \qquad \overline{F_0}(u_0, \Omega, \cdot) \colon u \in L^1(\Omega) \mapsto \text{sc}^-(L^1(\Omega))F_0(u_0, \Omega, u),$$

and recall that, by Proposition 3.5.3, it results that

$$\overline{F_0}(u_0, \Omega, \cdot) \colon u \in L^1(\Omega) \mapsto \min \left\{ \liminf_{h \to +\infty} \int_\Omega f(\nabla u_h)dx : \right.$$

$$\{u_h\} \subseteq u_0 + W_0^{1,\infty}(\Omega), \text{ for every } h \in \mathbf{N} \ \nabla u_h(x) \in \text{dom} f \text{ for a.e. } x \in \Omega,$$

$$\left. u_h \to u \text{ in } L^1(\Omega) \right\} \text{ for every } \Omega \in \mathcal{A}_0, \ u \in L^1(\Omega).$$

It is obvious that

$$(10.6.3) \qquad \text{for every } u_0 \in W_{\text{loc}}^{1,1}(\mathbf{R}^n) \text{ and every } \Omega \in \mathcal{A}_0,$$

$$\overline{F_0}(u_0, \Omega, \cdot) \text{ is } L^1(\Omega)\text{-lower semicontinuous}.$$

Proposition 10.6.1. Let f be a Borel function as in (10.1.1), let $u_0 \in W_{\text{loc}}^{1,1}(\mathbf{R}^n)$, and let $\overline{F_0}(u_0, \cdot, \cdot)$ be given by (10.6.2). Then

$$\overline{F_0}(u_0, \Omega', u) \leq \overline{F_0}(u_0, \Omega, u) + \int_{\Omega' \setminus \Omega} f(\nabla u_0)dx$$

whenever $\Omega, \ \Omega' \in \mathcal{A}_0$ satisfy $\Omega \subseteq \Omega'$, $u \in L^1(\Omega)$ with $u = u_0$ a.e. in $\Omega' \setminus \Omega$.

Proof. Let $\Omega, \ \Omega', \ u$ be as above. Clearly we can assume that $\overline{F_0}(u_0, \Omega, u) + \int_{\Omega' \setminus \Omega} f(\nabla u_0)dx < +\infty$. Then there exists $\{u_h\} \subseteq u_0 + W_0^{1,\infty}(\Omega)$ such that $u_h \to u$ in $L^1(\Omega)$, and

$$(10.6.4) \qquad \overline{F_0}(u_0, \Omega, u) = \liminf_{h \to +\infty} \int_\Omega f(\nabla u_h)dx.$$

It is obvious that, for every $h \in \mathbf{N}$, u_h can be thought as an element of $u_0 + W_0^{1,\infty}(\Omega')$ once we extend it by u_0 out of Ω. Therefore, $u_h \to u$ in $L^1(\Omega')$, and by (10.6.4) it follows that

$$\overline{F_0}(u_0, \Omega', u) \leq \liminf_{h \to +\infty} \int_{\Omega'} f(\nabla u_h) dx = \overline{F_0}(u_0, \Omega, u) + \int_{\Omega' \setminus \Omega} f(\nabla u_0) dx,$$

which proves the proposition. ∎

In order to represent the functional $\overline{F_0}(u_0, \Omega, \cdot)$ for bounded open sets Ω and boundary values u_0, suitable compatibility conditions on Ω and u are needed, depending on whether int$(\text{dom} f)$ is empty or not. This leads to the introduction, for every $\Omega \in \mathcal{A}_0$, of the following classes of admissible boundary data

$$(10.6.5) \quad T_0(f, \Omega) = \left\{ w \in W_{\text{loc}}^{1,\infty}(\mathbf{R}^n) : \nabla w(x) \in \text{dom} f \text{ for a.e. } x \in \Omega \right\},$$

and, if int$(\text{dom} f) \neq \emptyset$, by

$$(10.6.6) \quad T_1(f, \Omega) = \{ w \in W_{\text{loc}}^{1,\infty}(\mathbf{R}^n) : \text{ there exists a compact set}$$

$$K_w \subseteq \text{int}(\text{dom} f) \text{ such that } \nabla w(x) \in K_w \text{ for a.e. } x \in \Omega \}.$$

We observe explicitly that

$$(10.6.7) \qquad \int_{\Omega} f(\nabla u_0) dx < +\infty$$

whenever $\Omega \in \mathcal{A}_0$, $u_0 \in T_1(f, \Omega)$, and provided (10.1.12) holds.

§10.7 Relaxation of Dirichlet Problems

We start by treating the case in which

$$(10.7.1) \qquad\qquad \text{int}(\text{dom} f) \neq \emptyset.$$

Lemma 10.7.1. *Let f be a Borel function as in (10.1.1) satisfying (10.1.11) $\div$(10.1.13), (10.7.1), let $\overline{F_0}$ be given by (10.6.2), and $F^{(\infty)}$ by (10.4.1). Then*

$$\overline{F_0}(u_0, \Omega, u) \leq F^{(\infty)}(\Omega, u)$$

for every $\Omega \in \mathcal{A}_0$, $u_0 \in W_{\text{loc}}^{1,\infty}(\mathbf{R}^n)$, and $u \in u_0 + W_0^{1,\infty}(\Omega)$ such that $\nabla u(x)$ belongs to a compact subset of int$(\text{dom} f)$ for a.e. $x \in \Omega$.

Proof. Let Ω, u be as above, let $K \subseteq \text{int}(\text{dom} f)$ be compact and such that $\nabla u(x) \in K$ for a.e. $x \in \Omega$, and assume that $F^{(\infty)}(\Omega, u) < \infty$.

Let $\varepsilon > 0$, and let $\{u_h\} \subseteq W_{\text{loc}}^{1,\infty}(\mathbf{R}^n)$ be such that $u_h \to u$ in weak*-$W^{1,\infty}(\Omega)$, and

$$(10.7.2) \qquad F^{(\infty)}(\Omega, u) + \varepsilon \geq \lim_{h \to +\infty} \int_\Omega f(\nabla u_h) dx.$$

Moreover, since $F^{(\infty)}(\Omega, u) < +\infty$, let L_ε be a bounded subset of $\text{dom} f$ such that $\nabla u_h(x) \in L_\varepsilon \cap \text{dom} f$ for a.e. $x \in \Omega$ and definitively in h, and let z_{L_ε} be given by (10.1.13). Clearly by (10.7.1) we have that

$$(10.7.3) \qquad\qquad z_{L_\varepsilon} \in \text{int}(\text{dom} f).$$

Let now A, B, $B' \in \mathcal{A}_0$ with $A \subset\subset B$, $B \subset\subset B'$, $B' \subset\subset \Omega$, let $\varphi \in C_0^1(B)$ satisfying

$$(10.7.4) \quad 0 \leq \varphi \leq 1 \text{ in } \mathbf{R}^n, \ \varphi = 1 \text{ in } A, \ \|\nabla \varphi\|_{L^\infty(\mathbf{R}^n)} \leq \frac{2}{\text{dist}(\partial B, A)},$$

$t \in \,]0, 1[$, and $\gamma_t \in C^1(\mathbf{R}^n)$ such that

$$(10.7.5) \qquad t \leq \gamma_t \leq 1 \text{ in } \mathbf{R}^n, \ \gamma_t = t \text{ in } B, \ \gamma_t = 1 \text{ in } \mathbf{R}^n \setminus B',$$

$$\|\nabla \gamma_t\|_{L^\infty(\mathbf{R}^n)} \leq \frac{2(1-t)}{\text{dist}(\partial B', B)}.$$

For every $h \in \mathbf{N}$, set $w_h^t = \gamma_t(\varphi u_h + (1-\varphi)u) + (1-\gamma_t)u_{z_{L_\varepsilon}}$. Then $w_h^t \in u_0 + W_0^{1,\infty}(\Omega)$, and $w_h^t \to \gamma_t u + (1-\gamma_t)u_{z_{L_\varepsilon}}$ in weak*-$W^{1,\infty}(\Omega)$. By (10.7.4) and (10.7.5) we have

$$(10.7.6) \qquad \overline{F_0}(u_0, \Omega, \gamma_t u + (1-\gamma_t)u_{z_{L_\varepsilon}}) \leq \liminf_{h \to +\infty} \int_\Omega f(\nabla w_h^t) dx \leq$$

$$\leq \limsup_{h \to +\infty} \int_\Omega f(t\nabla u_h + (1-t)z_{L_\varepsilon}) dx + \limsup_{h \to +\infty} \int_{B \setminus A} f(\nabla w_h^t) dx +$$

$$+ \limsup_{h \to +\infty} \int_{B' \setminus B} f(\nabla w_h^t) dx + \limsup_{h \to +\infty} \int_{\Omega \setminus B} f(\nabla u) dx.$$

Let us observe now that $\nabla w_h^t = t\varphi \nabla u_h + t(1-\varphi)\nabla u + (1-t)z_{L_\varepsilon} + t(u_h - u)\nabla \varphi$ a.e. in $B \setminus A$, therefore, by (10.7.4) and (10.1.11), we get that $t\varphi(x)\nabla u_h(x) + t(1-\varphi(x))\nabla u(x) + (1-t)z_{L_\varepsilon} \in (1-t)z_{L_\varepsilon} + t\text{dom} f$ for a.e. $x \in B \setminus A$, and h large enough. Moreover, since by (10.7.3) and (10.1.11), $(1-t)z_{L_\varepsilon} + t\text{dom} f \subseteq \text{int}(\text{dom} f)$, by the convergence in $L^\infty(B \setminus A)$ of $\{u_h\}$

to u we deduce the existence of a compact subset K_t of int(dom$f$), and of $h_{t,A,B} \in \mathbf{N}$ such that

$$\nabla w_h^t(x) \in K_t \text{ for a.e. } x \in B \setminus A, \text{ and every } h \geq h_{t,A,B}.$$

Let us also observe that $\nabla w_h^t = \gamma_t \nabla u + (1 - \gamma_t) z_{L_\varepsilon} + u \nabla \gamma_t$ a.e. in $B' \setminus B$, therefore, since $\nabla u(x)$ belongs to a compact subset of int(dom$f$) for a.e. $x \in B' \setminus B$, by (10.7.5) and (10.7.3) we deduce the existence of a compact subset H of int(dom$f$) and of $t_{B,B'} \in \,]0,1[$ such that

$$\nabla w_h^t(x) \in H \text{ for a.e. } x \in B' \setminus B \text{ and every } h \in \mathbf{N}, \text{ provided } t \in \,]t_{B,B'},1[\, .$$

Because of this and (10.1.12) we obtain that

(10.7.7)   there exists $M_t > 0$ such that

$$f(\nabla w_h^t(x)) \leq M_t \text{ for a.e. } x \in B \setminus A, \text{ and every } h \geq h_{t,A,B},$$

and

(10.7.8)   there exists $M > 0$ such that $f(\nabla w_h^t(x)) \leq M$

for a.e. $x \in B' \setminus B$, and every $h \in \mathbf{N}$ provided $t \in \,]t_{B,B'},1[\, .$

In addition, by (10.1.13), there exists $t_\varepsilon \in [0,1[$ such that

$$(10.7.9) \qquad \int_\Omega f(t\nabla u_h + (1-t)z_{L_\varepsilon})dx \leq \int_\Omega f(\nabla u_h)dx + \varepsilon \mathcal{L}^n(\Omega)$$

for every $t \in \,]t_\varepsilon, 1[$, h large enough.

Therefore, by (10.7.6)$\div$(10.7.9), and (10.7.2), we conclude that

$$(10.7.10) \qquad \overline{F_0}(u_0, \Omega, \gamma_t u + (1-\gamma_t)u_{z_{L_\varepsilon}}) \leq$$

$$\leq F^{(\infty)}(\Omega, u) + \varepsilon \mathcal{L}^n(\Omega) + \varepsilon + M_t \mathcal{L}^n(B \setminus A) + M\mathcal{L}^n(\Omega \setminus B) + \sup_{z \in K} f(z)\mathcal{L}^n(\Omega \setminus B).$$

By (10.7.10) and (10.6.3), since $\gamma_t u + (1 - \gamma_t)u_{z_{L_\varepsilon}} \to u$ in $L^1(\Omega)$ as $t \to 1^-$, we deduce the lemma letting first A increase to B, then t increase to 1, B increase to Ω, and finally ε decrease to 0. $\blacksquare$

Theorem 10.7.2. *Let f be a Borel function as in (10.1.1) satisfying (10.1.11)$\div$(10.1.13), (10.7.1), let $\overline{F_0}$ be given by (10.6.2), and $T_1(f,\cdot)$ by (10.6.6). Then*

$$\overline{F_0}(u_0, \Omega, u) = \int_\Omega f^{**}(\nabla u)dx$$

for every $\Omega \in \mathcal{A}_0$, $u_0 \in T_1(f, \Omega)$, $u \in u_0 + W_0^{1,\infty}(\Omega)$.

Proof. Let Ω, u_0, u be as above. Let us first prove that

$$(10.7.11) \qquad \overline{F_0}(u_0, \Omega, u) \leq \int_\Omega f^{**}(\nabla u)dx.$$

To do this, we can assume that $\int_\Omega f^{**}(\nabla u)dx < +\infty$, and that $0 \in \mathrm{int}(\mathrm{dom}f)$, so that $\nabla u(x) \in \mathrm{dom}f^{**}$ for a.e. $x \in \Omega$. Let $K \subseteq \mathrm{int}(\mathrm{dom}f)$ be compact and such that $\nabla u_0(x) \in K$ for a.e. $x \in \Omega$, and take $t \in [0,1[$. Then $tu + (1-t)u_0 \in u_0 + W_0^{1,\infty}(\Omega)$, and $t\nabla u(x) + (1-t)\nabla u_0(x) \in t\mathrm{dom}f^{**} + (1-t)K$ for a.e. $x \in \Omega$.

We now recall that if A, $B \subseteq \mathbf{R}^n$, and if B is relatively compact, then $\overline{A + B} = \overline{A} + \overline{B}$. This, together with (1.3.6) of Proposition 1.3.2, and (10.1.11), implies that

$$\overline{t\mathrm{dom}f^{**} + (1-t)K} = t\overline{\mathrm{dom}f^{**}} + (1-t)K = t\overline{\mathrm{dom}f} + (1-t)K,$$

therefore, since by Proposition 1.1.5 $t\overline{\mathrm{dom}f} + (1-t)z \subseteq \mathrm{ri}(\mathrm{dom}f)$ for every $z \in K$, we conclude that

$$\overline{t\mathrm{dom}f^{**} + (1-t)K} \subseteq \mathrm{ri}(\mathrm{dom}f).$$

In conclusion, we have that $t\nabla u(x) + (1-t)\nabla u_0(x)$ belongs to a compact subset of $\mathrm{ri}(\mathrm{dom}f)$ for a.e. $x \in \Omega$, from which, together with (10.1.12), we deduce that

$$F^{(\infty)}(\Omega, tu + (1-t)u_0) \leq \int_\Omega f(t\nabla u + (1-t)\nabla u_0)dx < +\infty.$$

Because of this, Lemma 10.7.1, Theorem 10.4.1, Proposition 10.4.3, Proposition 1.3.2, (1.4.2) and (1.4.3) of Proposition 1.4.1, and the convexity of f^{**}, we thus obtain that

$$\overline{F_0}(u_0, \Omega, tu + (1-t)u_0) \leq F^{(\infty)}(\Omega, tu + (1-t)u_0) =$$

$$= \int_\Omega (\mathrm{co}(\mathrm{sc}^- f))(t\nabla u + (1-t)\nabla u_0)dx = \int_\Omega f^{**}(t\nabla u + (1-t)\nabla u_0)dx \leq$$

$$\leq t\int_\Omega f^{**}(\nabla u)dx + (1-t)\int_\Omega f^{**}(\nabla u_0)dx \text{ for every } t \in [0,1[,$$

from which, together with (10.6.3), and (10.6.7), we deduce (10.7.11) taking the limit as t increases to 1^-.

In conclusion, by (10.7.11) and Theorem 7.4.6 applied to f^{**}, the proof follows. ∎

In order to extend Theorem 10.7.2 to wider classes of functions, we need the following approximation lemma.

Lemma 10.7.3. *Let $\phi\colon \mathbf{R}^n \to [0, \infty]$ be convex and lower semicontinuous with int(domϕ) $\neq \emptyset$, let $\Omega \in \mathcal{A}_0$, $T_1(\phi, \Omega)$ be given by (10.6.6), and $u_0 \in T_1(\phi, \Omega)$. Then for every $u \in BV(\Omega)$ with $\mathrm{spt}(u - u_0) \subseteq \Omega$ there exists $\{u_h\} \subseteq u_0 + W_0^{1,\infty}(\Omega)$ such that $u_h \to u$ in $L^1(\Omega)$, and*

$$\limsup_{h \to +\infty} \int_\Omega \phi(\nabla u_h)dx \leq \int_\Omega \phi(\nabla u)dx + \int_\Omega \phi^\infty(\nabla^s u)d|D^s u|.$$

Proof. Let us preliminarily observe that, as usual, we can assume that

$$(10.7.12) \qquad\qquad 0 \in \mathrm{int}(\mathrm{dom}\phi).$$

Moreover, since $u_0 \in T_1(\phi, \Omega)$, let $K \subseteq \mathrm{int}(\mathrm{dom}\phi)$ be compact, and such that $\nabla u_0(x) \in K$ for a.e. $x \in \Omega$.

Let $u \in BV(\Omega)$ with $\mathrm{spt}(u - u_0) \subseteq \Omega$, $\Omega' \in \mathcal{A}_0$ with $\Omega' \subset\subset \Omega$, and $\mathrm{spt}(u - u_0) \subseteq \Omega'$. Let $\varepsilon \in\ ]0, \mathrm{dist}(\Omega', \partial\Omega)[$, and u_ε be the regularization of u given by (4.1.2). Then Lemma 7.4.4 yields

$$(10.7.13) \qquad \int_{\Omega'} \phi(\nabla u_\varepsilon)dx \leq \int_\Omega \phi(\nabla u)dx + \int_\Omega \phi^\infty(\nabla^s u)d|D^s u|.$$

Let now $A, B \in \mathcal{A}_0$ with $\mathrm{spt}(u - u_0) \subseteq A$, $A \subset\subset B$, $B \subset\subset \Omega'$, $t \in [0, 1[$, and $\varphi \in C_0^1(B)$ with

$$0 \leq \varphi \leq 1 \text{ in } \mathbf{R}^n, \ \ \varphi = 1 \text{ in } A, \ \ \||\nabla\varphi|\|_{L^\infty(\mathbf{R}^n)} \leq \frac{2}{\mathrm{dist}(\partial B, A)}.$$

Let us set $w_\varepsilon^t = t^2(2 - t)[\varphi u_\varepsilon + (1 - \varphi)u_0] + (1 - t)(1 + t - t^2)u_0$. Then $w_\varepsilon^t \in u_0 + W_0^{1,\infty}(\Omega)$, and $\mathrm{spt}(w_\varepsilon^t - u_0) \subseteq \Omega'$.

By the convexity of ϕ we have

$$(10.7.14) \qquad\qquad \int_{\Omega'} \phi(\nabla w_\varepsilon^t)dx =$$

$$= \int_A \phi(t^2(2 - t)\nabla u_\varepsilon + (1 - t)(1 + t - t^2)\nabla u_0)dx +$$

$$+ \int_{B \setminus A} \phi(t[t(2 - t)\varphi\nabla u_\varepsilon + t(2 - t)(1 - \varphi)\nabla u_0 + t(2 - t)(u_\varepsilon - u_0)\nabla\varphi] +$$

$$+ (1 - t)(1 + t - t^2)\nabla u_0)dx + \int_{\Omega' \setminus B} \phi(\nabla u_0)dx \leq$$

$$\leq t^2(2 - t)\int_A \phi(\nabla u_\varepsilon)dx + (1 - t)(1 + t - t^2)\int_A \phi(\nabla u_0)dx +$$

$$+ t\int_{B \setminus A} \phi(t(2 - t)\varphi\nabla u_\varepsilon + t(2 - t)(1 - \varphi)\nabla u_0 + t(2 - t)(u_\varepsilon - u_0)\nabla\varphi)dx +$$

$$+(1-t)\int_{B\setminus A}\phi((1+t-t^2)\nabla u_0)dx + \int_{\Omega'\setminus B}\phi(\nabla u_0)dx \le$$

$$\le \int_A \phi(\nabla u_\varepsilon)dx + (1-t)(1+t-t^2)\int_A \phi(\nabla u_0)dx+$$

$$+t^2(2-t)\int_{B\setminus A}\phi(\varphi\nabla u_\varepsilon + (1-\varphi)\nabla u_0)dx+$$

$$+[1-t(2-t)]\int_{B\setminus A}\phi\left(\frac{t(2-t)}{1-t(2-t)}(u_\varepsilon - u_0)\nabla\varphi\right)dx+$$

$$+(1-t)\int_{B\setminus A}\phi((1+t-t^2)\nabla u_0)dx + \int_{\Omega'\setminus B}\phi(\nabla u_0)dx \le$$

$$\le \int_A \phi(\nabla u_\varepsilon)dx + (1-t)(1+t-t^2)\int_A \phi(\nabla u_0)dx + \int_{B\setminus A}\phi(\nabla u_\varepsilon)dx+$$

$$+\int_{B\setminus A}\phi(\nabla u_0)dx + [1-t(2-t)]\int_{B\setminus A}\phi\left(\frac{t(2-t)}{1-t(2-t)}(u_\varepsilon - u_0)\nabla\varphi\right)dx+$$

$$+(1-t)\int_{B\setminus A}\phi((1+t-t^2)\nabla u_0)dx + \int_{\Omega'\setminus B}\phi(\nabla u_0)dx.$$

We now observe that $u_\varepsilon \to u_0$ in $L^\infty(B\setminus A)$ from which, together with (10.7.12), and the local boundedness of ϕ in $\mathrm{int}(\mathrm{dom}\phi)$, we get that

$$(10.7.15)\quad \limsup_{\varepsilon\to 0^+}\int_{B\setminus A}\phi\left(\frac{t(2-t)}{1-t(2-t)}(u_\varepsilon - u_0)\nabla\varphi\right)dx = \phi(0)\mathcal{L}^n(B\setminus A)$$

for every $t \in [0,1[$.

Therefore, by (10.7.14), (10.7.13), and (10.7.15), we conclude that

$$(10.7.16)\qquad\qquad \limsup_{\varepsilon\to 0^+}\int_\Omega \phi(\nabla w_\varepsilon^t)dx =$$

$$= \limsup_{\varepsilon\to 0^+}\int_{\Omega'}\phi(\nabla w_\varepsilon^t)dx + \int_{\Omega\setminus\Omega'}\phi(\nabla u_0)dx \le$$

$$\le \int_\Omega \phi(\nabla u)dx + \int_\Omega \phi^\infty(\nabla^s u)d|D^s u| + (1-t)(1+t-t^2)\int_A \phi(\nabla u_0)dx+$$

$$+\int_{\Omega\setminus A}\phi(\nabla u_0)dx + [1-t(2-t)]\phi(0)\mathcal{L}^n(B\setminus A)+$$

$$+(1-t)\int_{B\setminus A}\phi((1+t-t^2)\nabla u_0)dx \text{ for every } t \in [0,1[.$$

At this point we observe that, for t is sufficiently close to 1, $(1+t-t^2)K$ too is a compact subset of int(domϕ), therefore, by (10.7.16), and again the local boundedness of ϕ in int(domϕ), we deduce that

$$(10.7.17) \qquad \limsup_{t \to 1^-} \limsup_{\varepsilon \to 0^+} \int_\Omega \phi(\nabla w_\varepsilon^t)dx \leq$$

$$\leq \int_\Omega \phi(\nabla u)dx + \int_\Omega \phi^\infty(\nabla^s u)d|D^s u| + \int_{\Omega \setminus A} \phi(\nabla u_0)dx.$$

In conclusion, once we observe that, for fixed $t \in [0,1[$, we have that $w_\varepsilon^t \to t^2(2-t)[\varphi u + (1-\varphi)u_0] + (1-t)(1+t-t^2)u_0 = t^2(2-t)u + (1-t)(1+t-t^2)u_0$ in $L^1(\Omega)$ as $e \to 0^+$, and that $t^2(2-t)u + (1-t)(1+t-t^2)u_0 \to u$ in $L^1(\Omega)$ as $t \to 1^-$, by (10.7.17) the proof follows letting also A increase to Ω. ∎

We are now in a position to prove the representation results for $\overline{F_0}$ under assumption (10.7.1). Let us start with a case concerning continuous Sobolev functions.

Theorem 10.7.4. *Let f be a Borel function as in (10.1.1) satisfying (10.1.11)$\div$(10.1.13), (10.7.1), let $\overline{F_0}$ be given by (10.6.2), and $T_1(f,\cdot)$ by (10.6.6),. Then*

$$\overline{F_0}(u_0, \Omega, u) = \int_\Omega f^{**}(\nabla u)dx$$

for every $\Omega \in \mathcal{A}_0$, $u_0 \in T_1(f, \Omega)$, $u \in (u_0 + W_0^{1,1}(\Omega)) \cap C^0(\overline{\Omega})$.

Proof. Let Ω, u_0, u be as above. Let us first prove that

$$(10.7.18) \qquad \overline{F_0}(u_0, \Omega, u) \leq \int_\Omega f^{**}(\nabla u)dx.$$

To do this, let us assume that $\int_\Omega f^{**}(\nabla u)dx < +\infty$, and observe that, by (10.6.7), we have that $\int_\Omega f^{**}(\nabla u_0)dx \leq \int_\Omega f(\nabla u_0)dx < +\infty$.

Let $\sigma > 0$, $\vartheta_\sigma \in W^{1,\infty}(\mathbf{R})$ be given by

$$\vartheta_\sigma: t \in \mathbf{R} \mapsto \max\{\min\{t + \sigma, 0\}, t - \sigma\},$$

and set $v_\sigma = u_0 + \vartheta_\sigma(u - u_0)$. Then, being u continuous in $\mathbf{R}^n$, it turns out that $v_\sigma \in W^{1,1}(\Omega)$, and spt$(v_\sigma - u_0) \subseteq \Omega$.

By Lemma 10.7.3 applied with $\phi = f^{**}$, let $\{u_h\} \subseteq u_0 + W_0^{1,\infty}(\Omega)$ with $u_h \to v_\sigma$ in $L^1(\Omega)$, and

$$(10.7.19) \qquad \limsup_{h \to +\infty} \int_\Omega f^{**}(\nabla u_h)dx \leq \int_\Omega f^{**}(\nabla v_\sigma)dx.$$

Then, by (10.6.3), Theorem 10.7.2, (10.7.19), and the convexity of f^{**} we obtain that

$$(10.7.20) \qquad \overline{F_0}(u_0, \Omega, v_\sigma) \le \liminf_{h \to +\infty} \overline{F_0}(u_0, \Omega, u_h) =$$

$$= \liminf_{h \to +\infty} \int_\Omega f^{**}(\nabla u_h)dx \le \int_\Omega f^{**}(\nabla v_\sigma)dx \le$$

$$\le \int_\Omega \vartheta'_\sigma(u - u_0)f^{**}(\nabla u)dx + \int_\Omega (1 - \vartheta'_\sigma(u - u_0))f^{**}(\nabla u_0)dx =$$

$$= \int_{\{x \in \Omega: |u(x) - u_0(x)| > \sigma\}} f^{**}(\nabla u)dx + \int_{\{x \in \Omega: |u(x) - u_0(x)| \le \sigma\}} f^{**}(\nabla u_0)dx.$$

We now observe that $v_\sigma \to u$ in $L^1(\Omega)$, and that, being $\int_\Omega f^{**}(\nabla u)dx$ and $\int_\Omega f^{**}(\nabla u_0)dx$ finite, it results that

$$(10.7.21) \qquad \lim_{\sigma \to 0^+} \int_{\{x \in \Omega: |u(x) - u_0(x)| > \sigma\}} f^{**}(\nabla u)dx =$$

$$= \int_{\{x \in \Omega: |u(x) - u_0(x)| > 0\}} f^{**}(\nabla u)dx,$$

$$\lim_{\sigma \to 0^+} \int_{\{x \in \Omega: |u(x) - u_0(x)| \le \sigma\}} f^{**}(\nabla u)dx = \int_{\{x \in \Omega: u(x) = u_0(x)\}} f^{**}(\nabla u_0)dx.$$

Hence, once we recall that $\nabla u_0 = \nabla u$ a.e. in $\{x \in \Omega : u(x) = u_0(x)\}$, by (10.6.3), (10.7.20), and (10.7.21) we deduce as $\sigma \to 0^+$ that

$$\overline{F_0}(u_0, \Omega, u) \le \liminf_{\sigma \to 0^+} \overline{F_0}(u_0, \Omega, v_\sigma) \le$$

$$\le \int_{\{x \in \Omega: |u(x) - u_0(x)| > 0\}} f^{**}(\nabla u)dx + \int_{\{x \in \Omega: u(x) = u_0(x)\}} f^{**}(\nabla u_0)dx =$$

$$= \int_{\{x \in \Omega: |u(x) - u_0(x)| > 0\}} f^{**}(\nabla u)dx + \int_{\{x \in \Omega: u(x) = u_0(x)\}} f^{**}(\nabla u)dx =$$

$$= \int_\Omega f^{**}(\nabla u)dx,$$

that is (10.7.18).

Finally, by (10.7.18), and Theorem 7.4.6 applied to f^{**}, the proof follows. ■

In order to prove the representation result for $\overline{F_0}$ on BV spaces, we need the following lemma.

Lemma 10.7.5. *Let f be a Borel function as in (10.1.1), $u_0 \in W^{1,\infty}_{\mathrm{loc}}(\mathbf{R}^n)$, $x_0 \in \mathbf{R}^n$ such that $T[x_0]u_0 - u_0(x_0)$ is positively 1-homogeneous, and let $\overline{F_0}$ be given by (10.6.2). Then*

$$\overline{F_0}(u_0, \Omega, u) \leq \liminf_{t \to 1^+} \overline{F_0}(u_0, x_0 + t(\Omega - x_0), u) \text{ for every } \Omega \in \mathcal{A}_0, u \in L^1(\Omega).$$

Proof. Let Ω, u be as above, let us take $t > 1$, and assume that $\overline{F_0}(u_0, x_0 + t(\Omega - x_0), u) < \infty$, so that there exists $\{u_h\} \subseteq u_0 + W^{1,\infty}_0(x_0 + t(\Omega - x_0))$, with $u_h \to u$ in $L^1(x_0 + t(\Omega - x_0))$, and

$$(10.7.22) \qquad \overline{F_0}(u_0, x_0 + t(\Omega - x_0), u) \geq \liminf_{h \to +\infty} \int_{x_0 + t(\Omega - x_0)} f(\nabla u_h) dy.$$

For every $h \in \mathbf{N}$ we set $v_h = u_0(x_0) + T[-x_0]O_t T[x_0](u_h - u_0(x_0))$. Then $v_h \in u_0(x_0) + T[-x_0]O_t T[x_0](u_0 - u_0(x_0)) + W^{1,\infty}_0(\Omega)$, and hence, by the 1-homogeneity of $T[x_0]u_0 - u_0(x_0)$, $v_h \in u_0 + W^{1,\infty}_0(\Omega)$. Moreover $v_h \to u_0(x_0) + T[-x_0]O_t T[x_0](u - u_0(x_0))$ in $L^1(\Omega)$, and

$$\int_\Omega f(\nabla v_h) dx = \frac{1}{t^n} \int_{x_0 + t(\Omega - x_0)} f(\nabla u_h) dy.$$

This, together with (10.7.22), yields

$$(10.7.23) \qquad \overline{F_0}(u_0, x_0 + t(\Omega - x_0), u) \geq t^n \liminf_{h \to +\infty} \int_\Omega f(\nabla v_h) dx \geq$$

$$\geq t^n \overline{F_0}(u_0, \Omega, u_0(x_0) + T[-x_0]O_t T[x_0](u - u_0(x_0))).$$

In conclusion by (10.7.23), the fact that $u_0(x_0) + T[-x_0]O_t T[x_0](u - u_0(x_0)) \to u$ in $L^1(\Omega)$ as $t \to 1^+$, and (10.6.3), we obtain the lemma. ∎

Theorem 10.7.6. *Let f be a Borel function as in (10.1.1) satisfying (10.1.11)÷(10.1.13), (10.7.1), let $\overline{F_0}$ be given by (10.6.2), and $T_1(f, \cdot)$ by (10.6.6). Then*

$$\overline{F_0}(u_0, \Omega, u) =$$

$$= \int_\Omega f^{**}(\nabla u) dx + \int_\Omega (f^{**})^\infty (\nabla^s u) d|D^s u| + \int_{\partial \Omega} (f^{**})^\infty ((u_0 - u)\mathbf{n}_\Omega) d\mathcal{H}^{n-1}$$

for every $\Omega \in \mathcal{A}_0$ convex, $u_0 \in T_1(f, \Omega)$ for which there exists $x_0 \in \Omega$ such that $T[x_0]u_0 - u_0(x_0)$ is positively 1-homogeneous, and $u \in BV(\Omega)$.

Proof. Let Ω, u_0, u be as above.

Being $u_0 \in T_1(f, \Omega)$, let $K \subseteq \mathrm{int}(\mathrm{dom} f)$ be compact and such that $\nabla u_0(x) \in K$ for a.e. $x \in \Omega$. Let us observe that, by the 1-homogeneity of $T[x_0]u_0 - u_0(x_0)$, $\nabla u_0(x) \in K$ for a.e. $x \in \mathbf{R}^n$.

We can clearly assume that $0 \in \text{int}(\text{dom} f)$, and $x_0 = 0$. Consequently $u_0 - u_0(0)$ turns out to be positively 1-homogeneous.

Let us first prove that

$$(10.7.24) \qquad \overline{F_0}(u_0, \Omega, u) \leq$$

$$\leq \int_\Omega f^{**}(\nabla u)dx + \int_\Omega (f^{**})^\infty (\nabla^s u)d|D^s u| + \int_{\partial\Omega} (f^{**})^\infty((u_0 - u)\mathbf{n}_\Omega)d\mathcal{H}^{n-1}.$$

To do this we first define $\hat{u}$ as the extension of u to the whole $\mathbf{R}^n$ obtained by defining $\hat{u} = u_0$ in $\mathbf{R}^n \setminus \Omega$, and take $t > 1$. Then the convexity of Ω yields that $\Omega \subset\subset t\Omega$ and that $\text{spt}(\hat{u} - u_0) \subseteq t\Omega$. Moreover, being Ω convex, and hence with Lipschitz boundary, we also have that $\hat{u} \in BV(t\Omega)$.

Let $\{u_h\} \subseteq u_0 + W_0^{1,\infty}(t\Omega)$ be given by Lemma 10.7.3 applied with $\phi = f^{**}$. Then $u_h \to \hat{u}$ in $L^1(t\Omega)$ and by (10.6.3) and Theorem 10.7.2, we obtain that

$$(10.7.25) \qquad \overline{F_0}(u_0, t\Omega, \hat{u}) \leq \liminf_{h \to +\infty} \overline{F_0}(u_0, t\Omega, u_h) =$$

$$= \liminf_{h \to +\infty} \int_{t\Omega} f^{**}(\nabla u_h)dx \leq \int_{t\Omega} f^{**}(\nabla \hat{u})dx + \int_{t\Omega} (f^{**})^\infty (\nabla^s \hat{u})d|D^s \hat{u}| \leq$$

$$\leq \int_\Omega f^{**}(\nabla u)dx + \int_{t\Omega \setminus \Omega} f(\nabla u_0)dx +$$

$$+ \int_\Omega (f^{**})^\infty (\nabla^s u)d|D^s u| + \int_{\partial\Omega} (f^{**})^\infty (\nabla^s \hat{u})d|D^s \hat{u}|.$$

At this point, once we recall that $D\hat{u} = (u_0 - u)\mathbf{n}_\Omega \mathcal{H}^{n-1}$ on $\partial\Omega$, by (10.7.25), (10.6.7), and the 1-homogeneity of $(f^{**})^\infty$, we infer that

$$(10.7.26) \qquad \limsup_{t \to 1^+} \overline{F_0}(u_0, t\Omega, \hat{u}) \leq$$

$$\leq \int_\Omega f^{**}(\nabla u)dx + \int_\Omega (f^{**})^\infty (\nabla^s u)d|D^s u| + \int_{\partial\Omega} (f^{**})^\infty((u_0 - u)\mathbf{n}_\Omega)d\mathcal{H}^{n-1}.$$

Therefore, by Lemma 10.7.5 and (10.7.26), since $\overline{F_0}(u_0, \Omega, \hat{u}) = \overline{F_0}(u_0, \Omega, u)$, inequality (10.7.24) follows.

We now prove the reverse inequality in (10.7.24).

Let $t > 1$. Then by Proposition 10.6.1, (10.6.7), and Theorem 7.4.6 we infer that

$$(10.7.27) \qquad \overline{F_0}(u_0, \Omega, u) \geq \overline{F_0}(u_0, t\Omega, \hat{u}) - \int_{t\Omega \setminus \Omega} f(\nabla u_0)dx \geq$$

$$\geq \int_{t\Omega} f^{**}(\nabla \hat{u})dx + \int_{t\Omega} (f^{**})^\infty (\nabla^s \hat{u})d|D^s \hat{u}| - \int_{t\Omega \setminus \Omega} f(\nabla u_0)dx =$$

$$= \int_\Omega f^{**}(\nabla u)dx + \int_{t\Omega\setminus\Omega} f^{**}(\nabla u_0)dx + \int_\Omega (f^{**})^\infty(\nabla^s u)d|D^s u|+$$

$$+ \int_{\partial\Omega} (f^{**})^\infty((u_0 - u)\mathbf{n}_\Omega)d\mathcal{H}^{n-1} - \int_{t\Omega\setminus\Omega} f(\nabla u_0)dx,$$

therefore as $t \to 1^+$, since $\nabla u_0(x) \in K$ for a.e. $x \in \mathbf{R}^n \setminus \Omega$, we deduce from (10.6.7) and (10.7.27) that

$$(10.7.28) \qquad\qquad \overline{F_0}(u_0, \Omega, u) \geq$$

$$\geq \int_\Omega f^{**}(\nabla u)dx + \int_\Omega (f^{**})^\infty(\nabla^s u)d|D^s u| + \int_{\partial\Omega} (f^{**})^\infty((u_0 - u)\mathbf{n}_\Omega)d\mathcal{H}^{n-1}.$$

By (10.7.24) and (10.7.28) the proof follows. ∎

By Theorem 10.7.6 we deduce the following corollaries.

Corollary 10.7.7. *Let f as in (10.1.1) be convex, lower semicontinuous, and satisfying (10.7.1), let $\overline{F_0}$ be given by (10.6.2), and $T_1(f, \cdot)$ by (10.6.6). Then*

$$\overline{F_0}(u_0, \Omega, u) =$$

$$= \int_\Omega f(\nabla u)dx + \int_\Omega f^\infty(\nabla^s u)d|D^s u| + \int_{\partial\Omega} f^\infty((u_0 - u)\mathbf{n}_\Omega)d\mathcal{H}^{n-1}$$

for every $\Omega \in \mathcal{A}_0$ convex, $u_0 \in T_1(f, \Omega)$ for which there exists $x_0 \in \Omega$ such that $T[x_0]u_0 - u_0(x_0)$ is positively 1-homogeneous, and $u \in BV(\Omega)$.

Proof. As in the proof of Corollary 10.5.4, the assumptions on f imply that $(10.1.11)\div(10.1.13)$, and (10.7.1) are fulfilled.

Therefore the proof follows from Theorem 10.7.6. ∎

Corollary 10.7.8. *Let $g: \mathbf{R}^n \to [0, +\infty[$ be continuous, and $C \subseteq \mathbf{R}^n$ be convex and with $\text{int}(C) \neq \emptyset$. Let $\overline{F_0}$ be given by (10.6.2) with $f = g + I_C$, and $T_1(I_C, \cdot)$ by (10.6.6). Then*

$$\overline{F_0}(u_0, \Omega, u) = \int_\Omega (g + I_C)^{**}(\nabla u)dx + \int_\Omega ((g + I_C)^{**})^\infty(\nabla^s u)d|D^s u|+$$

$$+ \int_{\partial\Omega} ((g + I_C)^{**})^\infty((u_0 - u)\mathbf{n}_\Omega)d\mathcal{H}^{n-1}$$

for every $\Omega \in \mathcal{A}_0$ convex, $u_0 \in T_1(I_C, \Omega)$ for which there exists $x_0 \in \Omega$ such that $T[x_0]u_0 - u_0(x_0)$ is positively 1-homogeneous, and $u \in BV(\Omega)$.

Proof. Follows from Theorem 10.7.6 applied with $f = g + I_C$, once we observe that $g + I_C$ satisfies conditions $(10.1.11)\div(10.1.13)$, the last two being fulfilled by exploiting the uniform continuity of g on the bounded subsets of $\mathbf{R}^n$. ∎

We now treat the case when

$$(10.7.29) \qquad\qquad \text{int}(\text{dom} f) = \emptyset,$$

in which the situation is much simpler than the one described under assumption (10.7.1).

Lemma 10.7.9. *Let f be a Borel function as in (10.1.1) satisfying (10.7.29), let $\Omega \in \mathcal{A}_0$, $T_0(f, \Omega)$ be given by (10.6.5), $u_0 \in T_0(f, \Omega)$, and $u \in u_0 + W_0^{1,\infty}(\Omega)$ with $\int_\Omega f(\nabla u)dx < \infty$. Then $u = u_0$.*

Proof. By (10.7.29), let us first prove the lemma by assuming that

$$(10.7.30) \qquad \text{aff}(\text{dom} f) = \mathbf{R}^k \times \{0_{n-k}\} \text{ for some } k \in \{1, \dots, n-1\},$$

$$\text{or } \text{aff}(\text{dom} f) = \{0\}.$$

If $\int_\Omega f(\nabla u)dx < \infty$, then

$$\nabla u(x) \in \text{dom} f \text{ for a.e. } x \in \Omega,$$

therefore, since $u_0 \in T_0(f, \Omega)$, by (10.7.30) we infer that $u - u_0 \in W_0^{1,\infty}(\Omega)$, and that $\nabla_{k+1}(u-u_0) = 0, \dots, \nabla_n(u-u_0) = 0$ a.e. in Ω, or that $\nabla(u-u_0) = 0$ a.e. in Ω. Because of this, we get that $u = u_0$.

In order to treat the case when (10.7.30) is dropped, let us denote by $k \in \{0, 1, \dots, n-1\}$ the dimension of aff(dom f), and let $A: y \in \mathbf{R}^n \mapsto M_A y + b \in \mathbf{R}^n$ be an affine transformation such that $\det M_A = 1$, and $A(\text{aff}(\text{dom} f)) = \mathbf{R}^k \times \{0_{n-k}\}$ if $k > 0$, or $A(\text{aff}(\text{dom} f)) = \{0\}$ if $k = 0$.

Let us set $f_A: z \in \mathbf{R}^n \mapsto f(A^{-1}(z))$, $u_0^A: y \in \mathbf{R}^n \mapsto u_0(A(y)) + b \cdot y$, and $u^A: y \in \mathbf{R}^n \mapsto u(A(y)) + b \cdot y$. Then f_A is a Borel function satisfying (10.7.30) with f_A in place of f, $u_0^A \in T_0(f_A, A^{-1}(\Omega))$, $u^A \in u_0^A + W_0^{1,\infty}(A^{-1}(\Omega))$, and $\int_{A^{-1}(\Omega)} f_A(\nabla u^A)dy = \int_\Omega f(\nabla u)dx < +\infty$. Therefore, by the particular case above considered, we conclude that $u^A = u_0^A$, that is $u = u_0$. ∎

By Lemma 10.7.9 we deduce the following representation result.

Theorem 10.7.10. *Let f be a Borel function as in (10.1.1) satisfying (10.7.29), $\overline{F}_0$ be given by (10.6.2), and $T_0(f, \cdot)$ by (10.6.5). Then*

$$\overline{F}_0(u_0, \Omega, u) = \begin{cases} \int_\Omega f(\nabla u_0)dx & \text{if } u = u_0 \text{ a.e. in } \Omega \\ +\infty & \text{otherwise} \end{cases}$$

for every $\Omega \in \mathcal{A}_0$, $u_0 \in T_0(f, \Omega)$, $u \in L^1(\Omega)$.

Proof. Trivial by Lemma 10.7.9. ∎

§10.8 Applications to Minimum Problems

In this section we apply the relaxation results of the present chapter to the study of some classes of minimum problems.

If f is as in (10.1.1), and $p \in [1, +\infty]$, we assume that f satisfies the following coerciveness conditions

$$(10.8.1) \qquad \begin{cases} |z|^p \leq f(z) \text{ for every } z \in \mathbf{R}^n & \text{if } p \in [1, +\infty[\\ \text{dom} f \text{ is bounded} & \text{if } p = +\infty. \end{cases}$$

We observe that, by using (1.3.3), conditions in (10.8.1) imply that

$$(10.8.2) \qquad \begin{cases} |z|^p \leq f^{**}(z) \text{ for every } z \in \mathbf{R}^n & \text{if } p \in [1, +\infty[\\ \text{dom} f^{**} \text{ is bounded} & \text{if } p = +\infty. \end{cases}$$

We start with the case of Neumann minimum problems.

Theorem 10.8.1. *Let $p \in [1, +\infty]$, and f be a Borel function as in (10.1.1) satisfying (10.1.11)÷(10.1.13), and (10.8.1). Let $\Omega \in \mathcal{A}_0$ be convex, $\lambda \in]0, +\infty[$, and $r \in]1, p^*[$. Then (10.8.2) holds. Moreover,*
i) if $p \in]1, +\infty]$, and $\beta \in L^{p'}(\Omega)$, then

$$(10.8.3) \quad \inf \left\{ \int_\Omega f(\nabla u)dx + \lambda \int_\Omega |u|^r dx + \int_\Omega \beta u dx : u \in W^{1,\infty}(\Omega) \right\} =$$

$$= \min \left\{ \int_\Omega f^{**}(\nabla u)dx + \lambda \int_\Omega |u|^r dx + \int_\Omega \beta u dx : u \in W^{1,p}(\Omega) \right\},$$

the minimizing sequences of the functional in the left-hand side of (10.8.3) are compact in $L^p(\Omega)$, and their converging subsequences converge to solutions of the right-hand side of (10.8.3),
ii) if $p = 1$, and $\beta \in L^{r'}(\Omega)$, then

$$(10.8.4) \quad \inf \left\{ \int_\Omega f(\nabla u)dx + \lambda \int_\Omega |u|^r dx + \int_\Omega \beta u dx : u \in W^{1,\infty}(\Omega) \right\} =$$

$$= \min \left\{ \int_\Omega f^{**}(\nabla u)dx + \int_\Omega (f^{**})^\infty (\nabla^s u)d|D^s u| + \right.$$

$$\left. + \lambda \int_\Omega |u|^r dx + \int_\Omega \beta u dx : u \in BV(\Omega) \right\},$$

the minimizing sequences of the functional in the left-hand side of (10.8.4) are compact in $L^r(\Omega)$, and their converging subsequences converge to solutions of the right-hand side of (10.8.4).

Proof. We first prove i).

Let $F(\Omega, \cdot)$ be given by (10.1.2), let us set $s = \max\{p, r\}$, and prove that

$$(10.8.5) \quad \mathrm{sc}^-(L^s(\Omega))F(\Omega, u) = \begin{cases} \int_\Omega f^{**}(\nabla u)dx & \text{if } u \in W^{1,p}(\Omega) \\ +\infty & \text{if } u \in L^s(\Omega) \setminus W^{1,p}(\Omega) \end{cases}$$

for every $u \in L^s(\Omega)$.

By Theorem 10.5.2 and Proposition 10.5.3 it follows that

$$(10.8.6) \quad \mathrm{sc}^-(L^1(\Omega))F(\Omega, u) \leq \begin{cases} \int_\Omega f^{**}(\nabla u)dx & \text{if } u \in W^{1,p}(\Omega) \\ +\infty & \text{if } u \in L^1(\Omega) \setminus W^{1,p}(\Omega) \end{cases}$$

for every $u \in L^1(\Omega)$.

On the other side, if $u \in L^s(\Omega)$ is such that $\mathrm{sc}^-(L^1(\Omega))F(\Omega, u) < +\infty$, let $\{u_h\} \subseteq W^{1,\infty}_{\mathrm{loc}}(\mathbf{R}^n)$ be such that $u_h \to u$ in $L^1(\Omega)$, and

$$\mathrm{sc}^-(L^1(\Omega))F(\Omega, u) = \liminf_{h \to +\infty} \int_\Omega f(\nabla u_h)dx.$$

Then, by (10.8.1), Lemma 4.4.2, and the Rellich-Kondrachov Compactness Theorem we conclude that $u_h \to u$ in $L^s(\Omega)$, from which it follows that

$$(10.8.7) \quad \mathrm{sc}^-(L^1(\Omega))F(\Omega, u) \geq \mathrm{sc}^-(L^s(\Omega))F(\Omega, u) \text{ for every } u \in L^s(\Omega).$$

Now, if $u \in L^s(\Omega)$ is such that $\mathrm{sc}^-(L^s(\Omega))F(\Omega, u) < +\infty$, let $\{u_h\} \subseteq W^{1,\infty}_{\mathrm{loc}}(\mathbf{R}^n)$ be such that $u_h \to u$ in $L^s(\Omega)$, and

$$(10.8.8) \quad \mathrm{sc}^-(L^s(\Omega))F(\Omega, u) = \liminf_{h \to +\infty} \int_\Omega f(\nabla u_h)dx.$$

Then, again by (10.8.1), and Lemma 4.4.2, we conclude that $u_h \to u$ in weak-$W^{1,p}(\Omega)$ (weak*-$W^{1,\infty}(\Omega)$ if $p = +\infty$), from which it follows that $u \in W^{1,p}(\Omega)$. Consequently, by (10.8.8), (1.3.2), and Theorem 7.4.6 applied to f^{**}, we conclude that

$$\mathrm{sc}^-(L^s(\Omega))F(\Omega, u) = \liminf_{h \to +\infty} \int_\Omega f(\nabla u_h)dx \geq$$

$$\geq \liminf_{h \to +\infty} \int_\Omega f^{**}(\nabla u_h)dx \geq \int_\Omega f^{**}(\nabla u)dx,$$

from which, together with (10.8.6) and (10.8.7), (10.8.5) follows.

By (10.8.5) and Proposition 3.5.2, once we observe that the functional $u \in L^s(\Omega) \mapsto \lambda \int_\Omega |u|^r dx + \int_\Omega \beta u dx$ is $L^s(\Omega)$-continuous, we immediately obtain that

$$(10.8.9) \quad \mathrm{sc}^-(L^s(\Omega))\left\{ F(\Omega, u) + \lambda \int_\Omega |u|^r dx + \int_\Omega \beta u dx \right\} =$$

$$= \begin{cases} \int_\Omega f^{**}(\nabla u)dx + \lambda \int_\Omega |u|^r dx + \int_\Omega \beta u dx & \text{if } u \in W^{1,p}(\Omega) \\ +\infty & \text{if } u \in L^s(\Omega) \setminus W^{1,p}(\Omega) \end{cases}$$

for every $u \in L^s(\Omega)$.

In conclusion, let us prove that the functional $u \in L^s(\Omega) \mapsto F(\Omega, u) + \lambda \int_\Omega |u|^r dx + \int_\Omega \beta u dx$ is coercive.

To do this, let us consider only the case in which $p \in]1, +\infty[$, the one in which $p = +\infty$ being similar.

In this case, since (10.8.1) implies that

$$F(\Omega, u) + \lambda \int_\Omega |u|^r dx + \int_\Omega \beta u dx \geq$$

$$\geq \||\nabla u|\|^p_{L^p(\Omega)} + \lambda \|u\|^r_{L^r(\Omega)} - \|\beta\|_{L^{p'}(\Omega)} \|u\|_{L^p(\Omega)} \text{ for every } u \in W^{1,p}(\Omega),$$

and since every $u \in L^s(\Omega)$ satisfying $F(\Omega, u) < +\infty$ actually is in $W^{1,p}(\Omega)$, then $\{u \in L^s(\Omega) : F(\Omega, u) + \lambda \int_\Omega |u|^r dx + \int_\Omega \beta u dx \leq c\} \subseteq \{u \in W^{1,p}(\Omega) : \||\nabla u|\|^p_{L^p(\Omega)} + \lambda \|u\|^r_{L^r(\Omega)} - \|\beta\|_{L^{p'}(\Omega)} \|u\|_{W^{1,p}(\Omega)} \leq c\}$ for every $c \in \mathbf{R}$. Consequently, the desired coerciveness follows from Proposition 4.4.3.

By the coerciveness of $u \in L^s(\Omega) \mapsto F(\Omega, u) + \lambda \int_\Omega |u|^r dx + \int_\Omega \beta u dx$, and (10.8.9) the assumptions of Theorem 3.5.6 are fulfilled with $U = L^s(\Omega)$, and the proof follows from Theorem 3.5.6, once we observe that obviously the left-hand side of (10.8.3) is finite, and that, being Ω convex, every $u \in W^{1,\infty}(\Omega)$ can be thought as the restriction to Ω of a function in $W^{1,\infty}_{\text{loc}}(\mathbf{R}^n)$.

Let us now prove ii).

In this case the proof follows the same outlines of the one for i), with the obvious changes.

Clearly in this case $s = r$, and an argument similar to the one exploited above, but with Rellich-Kondrachov Compactness Theorem replaced by Theorem 4.2.11, yields

$$\text{sc}^-(L^r(\Omega)) \left\{ F(\Omega, u) + \lambda \int_\Omega |u|^r dx + \int_\Omega \beta u dx \right\} =$$

$$= \begin{cases} \int_\Omega f^{**}(\nabla u)dx + \int_\Omega (f^{**})^\infty(\nabla^s u)d|D^s u| + \\ \qquad\qquad + \lambda \int_\Omega |u|^r dx + \int_\Omega \beta u dx & \text{if } u \in BV(\Omega) \\ +\infty & \text{if } u \in L^r(\Omega) \setminus BV(\Omega) \end{cases}$$

for every $u \in L^r(\Omega)$.

Analogously, Proposition 4.4.1 provides the coerciveness of $u \in L^r(\Omega) \mapsto F(\Omega, u) + \lambda \int_\Omega |u|^r dx + \int_\Omega \beta u dx$, and the proof completes as in case i). ∎

By Theorem 10.8.1 we deduce the following corollary.

Corollary 10.8.2. Let $p \in [1, +\infty]$, $g: \mathbf{R}^n \to [0, +\infty[$ be continuous, $C \subseteq \mathbf{R}^n$ be convex, and assume that

$$\begin{cases} |z|^p \leq g(z) \text{ for every } z \in \mathbf{R}^n & \text{if } p \in [1, +\infty[\\ C \text{ is bounded} & \text{if } p = +\infty. \end{cases}$$

Let $\Omega \in \mathcal{A}_0$ be convex, $\lambda \in]0, +\infty[$, and $r \in]1, p^*[$. Then

$$\begin{cases} |z|^p \leq (g + I_C)^{**}(z) \text{ for every } z \in \mathbf{R}^n & \text{if } p \in [1, +\infty[\\ \mathrm{dom}(g + I_C)^{**} \text{ is bounded} & \text{if } p = +\infty. \end{cases}$$

Moreover,
i) if $p \in]1, +\infty]$, and $\beta \in L^{p'}(\Omega)$, then

$$(10.8.10) \qquad \inf \left\{ \int_\Omega g(\nabla u) dx + \lambda \int_\Omega |u|^r dx + \int_\Omega \beta u dx : \right.$$

$$\left. u \in W^{1,\infty}(\Omega), \ \nabla u(x) \in C \text{ for a.e. } x \in \Omega \right\} =$$

$$= \min \left\{ \int_\Omega (g + I_C)^{**}(\nabla u) dx + \lambda \int_\Omega |u|^r dx + \int_\Omega \beta u dx : u \in W^{1,p}(\Omega) \right\},$$

the minimizing sequences of the functional in the left-hand side of (10.8.10) are compact in $L^p(\Omega)$, and their converging subsequences converge to solutions of the right-hand side of (10.8.10),
ii) if $p = 1$, and $\beta \in L^{r'}(\Omega)$, then

$$(10.8.11) \qquad \inf \left\{ \int_\Omega g(\nabla u) dx + \lambda \int_\Omega |u|^r dx + \int_\Omega \beta u dx : \right.$$

$$\left. u \in W^{1,\infty}(\Omega), \ \nabla u(x) \in C \text{ for a.e. } x \in \Omega \right\} =$$

$$= \min \left\{ \int_\Omega (g + I_C)^{**}(\nabla u) dx + \int_\Omega ((g + I_C)^{**})^\infty (\nabla^s u) d|D^s u| + \right.$$

$$\left. + \lambda \int_\Omega |u|^r dx + \int_\Omega \beta u dx : u \in BV(\Omega) \right\},$$

the minimizing sequences of the functional in the left-hand side of (10.8.11) are compact in $L^r(\Omega)$, and their converging subsequences converge to solutions of the right-hand side of (10.8.11).

Proof. Follows from Theorem 10.8.1 applied with $f = g + I_C$, once we observe that $g + I_C$ satisfies conditions (10.1.11)÷(10.1.13), the last two being fulfilled by exploiting the uniform continuity of g on the bounded subsets of $\mathbf{R}^n$. ∎

We now come to Dirichlet minimum problems.

Theorem 10.8.3. Let $p \in [1, +\infty]$, and f be a Borel function as in (10.1.1) satisfying (10.1.11)÷(10.1.13), (10.7.1), and (10.8.1). Let $\Omega \in \mathcal{A}_0$, $T_1(f, \Omega)$ be given by (10.6.6), and $u_0 \in T_1(f, \Omega)$. Then (10.8.2) holds. Moreover,
i) if $p \in]n, +\infty]$, and $\beta \in L^1(\Omega)$, then

$$(10.8.12) \qquad \inf\left\{ \int_\Omega f(\nabla u)dx + \int_\Omega \beta u dx : u \in u_0 + W_0^{1,\infty}(\Omega) \right\} =$$

$$= \min\left\{ \int_\Omega f^{**}(\nabla u)dx + \int_\Omega \beta u dx : u \in u_0 + W_0^{1,p}(\Omega) \right\},$$

the minimizing sequences of the functional in the left-hand side of (10.8.12) are compact in $L^\infty(\Omega)$, and their converging subsequences converge to solutions of the right-hand side of (10.8.12),
ii) if $p \in]1, n]$, Ω is also convex, $\beta \in L^p(\Omega)$, and there exists $x_0 \in \Omega$ such that $T[x_0]u_0 - u_0$ is positively 1-homogeneous, then (10.8.12) holds, the minimizing sequences of the functional in the left-hand side of (10.8.12) are compact in $L^p(\Omega)$, and their converging subsequences converge to solutions of the right-hand side of (10.8.12),
iii) if $p = 1$, Ω is also convex, $\lambda \in]0, +\infty[$, $r \in]1, 1^*[$, $\beta \in L^{r'}(\Omega)$, and there exists $x_0 \in \Omega$ such that $T[x_0]u_0 - u_0$ is positively 1-homogeneous, then

$$(10.8.13) \qquad \inf\left\{ \int_\Omega f(\nabla u)dx + \lambda \int_\Omega |u|^r dx + \int_\Omega \beta u dx : \right.$$

$$u \in u_0 + W_0^{1,\infty}(\Omega) \right\} = \min\left\{ \int_\Omega f^{**}(\nabla u)dx + \int_\Omega (f^{**})^\infty(\nabla^s u)d|D^s u| + \right.$$

$$+ \int_{\partial\Omega} (f^{**})^\infty((u_0 - u)\mathbf{n}_\Omega)d\mathcal{H}^{n-1} + \lambda \int_\Omega |u|^r dx + \int_\Omega \beta u dx : u \in BV(\Omega) \right\},$$

the minimizing sequences of the functional in the left-hand side of (10.8.13) are compact in $L^r(\Omega)$, and their converging subsequences converge to solutions of the right-hand side of (10.8.13).

Proof. We first prove i).

Let $F_0(\Omega, u_0, \cdot)$ be given by (10.6.1). Then, an argument similar to the one exploited to get (10.8.5) in the proof of Theorem 10.8.1, but with Theorem 10.5.2 replaced by Theorem 10.7.4, yields

$$(10.8.14) \qquad \mathrm{sc}^-(L^\infty(\Omega))\left\{ F_0(\Omega, u_0, u) + \int_\Omega \beta u dx \right\} =$$

$$= \begin{cases} \int_\Omega f^{**}(\nabla u)dx + \int_\Omega \beta u dx & \text{if } u \in u_0 + W_0^{1,p}(\Omega) \\ +\infty & \text{if } u \in L^\infty(\Omega) \setminus (u_0 + W^{1,p}(\Omega)) \end{cases}$$

$$\text{for every } u \in L^\infty(\Omega).$$

Moreover, (10.8.1), Sobolev Imbedding Theorem, and Proposition 4.4.4 prove that the functional $u \in L^\infty(\Omega) \mapsto F_0(\Omega, u_0, u) + \int_\Omega \beta u dx$ is coercive.

Because of this and (10.8.14), the assumptions of Theorem 3.5.6 are fulfilled with $U = L^\infty(\Omega)$, and the proof follows from Theorem 3.5.6, once we observe that obviously the left-hand side of (10.8.12) is finite.

The proof of case ii) follows the same outlines of the one of i), with the obvious changes. In particular, by considering relaxation processes in L^p spaces in place of L^∞ ones, and by replacing Theorem 10.7.4 with Theorem 10.7.6.

Finally, the proof of case iii) follows the same outlines of the one of i) with the obvious changes, by considering relaxation processes in L^r spaces in place of L^∞ ones, and by replacing Theorem 10.7.4 with Theorem 10.7.6. By using Theorem 10.7.6, one first proves that

$$
\mathrm{sc}^-(L^r(\Omega))\left\{ F_0(\Omega, u_0, u) + \lambda \int_\Omega |u|^r dx + \int_\Omega \beta u dx \right\} =
$$

$$
= \begin{cases} \int_\Omega f^{**}(\nabla u)dx + \int_\Omega (f^{**})^\infty (\nabla^s u)d|D^s u| + \\ \quad + \int_{\partial\Omega}(f^{**})^\infty((u_0 - u)\mathbf{n}_\Omega)d\mathcal{H}^{n-1} + \lambda \int_\Omega |u|^r dx + \int_\Omega \beta u dx \\ \hspace{5cm} \text{if } u \in BV(\Omega) \\ +\infty \hspace{3.6cm} \text{if } u \in L^r(\Omega) \setminus BV(\Omega), \end{cases}
$$

and then, by exploiting Proposition 4.4.1, that the functional $u \in L^r(\Omega) \mapsto F_0(\Omega, u_0, u) + \lambda \int_\Omega |u|^r dx + \int_\Omega \beta u dx$ is coercive.

Because of this, the proof follows from an application of Theorem 3.5.6 with $U = L^r(\Omega)$, once we observe that obviously the left-hand side of (10.8.13) is finite. ∎

By Theorem 10.8.3 we deduce the following corollary.

Corollary 10.8.4. *Let* $g : \mathbf{R}^n \to [0, +\infty[$ *be continuous, and* $C \subseteq \mathbf{R}^n$ *be convex with* $\mathrm{int}(C) \neq \emptyset$*, and assume that*

$$
\begin{cases} |z|^p \leq g(z) \text{ for every } z \in \mathbf{R}^n & \text{if } p \in [1, +\infty[\\ C \text{ is bounded} & \text{if } p = +\infty. \end{cases}
$$

Let $\Omega \in \mathcal{A}_0$*,* $T_1(g + I_C, \Omega)$ *be given by (10.6.6), and* $u_0 \in T_1(g + I_C, \Omega)$*. Then*

$$
\begin{cases} |z|^p \leq (g + I_C)^{**}(z) \text{ for every } z \in \mathbf{R}^n & \text{if } p \in [1, +\infty[\\ \mathrm{dom}(g + I_C)^{**} \text{ is bounded} & \text{if } p = +\infty. \end{cases}
$$

Moreover,

i) if $p \in]n, +\infty]$*, and* $\beta \in L^1(\Omega)$*, then*

$$
(10.8.15) \qquad \inf\left\{ \int_\Omega g(\nabla u)dx + \int_\Omega \beta u dx : \right.
$$

$$u \in u_0 + W_0^{1,\infty}(\Omega), \ \nabla u(x) \in C \text{ for a.e. } x \in \Omega \Big\} =$$

$$= \min \Big\{ \int_\Omega (g + I_C)^{**}(\nabla u)dx + \int_\Omega \beta u dx : u \in u_0 + W_0^{1,p}(\Omega) \Big\},$$

the minimizing sequences of the functional in the left-hand side of (10.8.15) are compact in $L^\infty(\Omega)$, and their converging subsequences converge to solutions of the right-hand side of (10.8.15),

ii) if $p \in \ ]1, n]$, Ω is also convex, $\beta \in L^{p'}(\Omega)$, and there exists $x_0 \in \Omega$ such that $T[x_0]u_0 - u_0$ is positively 1-homogeneous, then (10.8.15) holds, the minimizing sequences of the functional in the left-hand side of (10.8.15) are compact in $L^p(\Omega)$, and their converging subsequences converge to solutions of the right-hand side of (10.8.15),

iii) if $p = 1$, $\lambda \in \ ]0, +\infty[$, $r \in \ ]1, 1^*[$, Ω is also convex, $\beta \in L^{r'}(\Omega)$, and there exists $x_0 \in \Omega$ such that $T[x_0]u_0 - u_0$ is positively 1-homogeneous, then

$$(10.8.16) \qquad \inf \Big\{ \int_\Omega g(\nabla u)dx + \lambda \int_\Omega |u|^r dx + \int_\Omega \beta u dx :$$

$$u \in u_0 + W_0^{1,\infty}(\Omega), \ \nabla u(x) \in C \text{ for a.e. } x \in \Omega \Big\} =$$

$$= \min \Big\{ \int_\Omega (g + I_C)^{**}(\nabla u)dx + \int_\Omega ((g + I_C)^{**})^\infty(\nabla^s u)d|D^s u| +$$

$$+ \int_{\partial\Omega} ((g + I_C)^{**})^\infty((u_0 - u)\mathbf{n}_\Omega)d\mathcal{H}^{n-1} + \lambda \int_\Omega |u|^r dx + \int_\Omega \beta u dx :$$

$$u \in BV(\Omega) \Big\},$$

the minimizing sequences of the functional in the left-hand side of (10.8.16) are compact in $L^r(\Omega)$, and their converging subsequences converge to solutions of the right-hand side of (10.8.16).

Proof. Follows from Theorem 10.8.3 applied with $f = g + I_C$. ∎

Theorem 10.8.5. Let $f : \mathbf{R}^n \to [0, \infty]$ be a Borel function with $\text{int}(\text{dom} f) = \emptyset$, $\Omega \in \mathcal{A}_0$, $T_0(f, \Omega)$ be given by (10.6.5), $u_0 \in T_0(f, \Omega)$, and $\beta \in L^1(\Omega)$. Then u_0 is the only function in $u_0 + W_0^{1,\infty}(\Omega)$ that makes the functional $u \in u_0 + W_0^{1,\infty}(\Omega) \mapsto \int_\Omega f(\nabla u)dx + \int_\Omega \beta u dx$ finite, and

$$\inf \Big\{ \int_\Omega f(\nabla u)dx + \int_\Omega \beta u dx : u \in u_0 + W_0^{1,\infty}(\Omega) \Big\} =$$

$$= \int_\Omega f(\nabla u_0)dx + \int_\Omega \beta u_0 dx.$$

Proof. Follows from Lemma 10.7.9. ∎

Corollary 10.8.6. *Let $p \in [1, +\infty]$, $g: \mathbf{R}^n \to [0, +\infty[$ be continuous, $C \subseteq \mathbf{R}^n$ be convex with $\mathrm{int}(C) = \emptyset$, $\Omega \in \mathcal{A}_0$, $T_0(I_C, \Omega)$ be given by (10.6.5), $u_0 \in T_0(I_C, \Omega)$, and $\beta \in L^1(\Omega)$. Then u_0 is the only function in $u_0 + W_0^{1,\infty}(\Omega)$ that fulfils the constraint $\nabla u(x) \in C$ for a.e. $x \in \Omega$, and*

$$\inf \left\{ \int_\Omega g(\nabla u) dx + \int_\Omega \beta u \, dx : u \in u_0 + W_0^{1,\infty}(\Omega), \right.$$

$$\left. \nabla u(x) \in C \text{ for a.e. } x \in \Omega \right\} = \int_\Omega g(\nabla u_0) dx + \int_\Omega \beta u_0 dx.$$

Proof. Follows from Lemma 10.7.9. ∎

§10.9 Additional Remarks on Integral Representation on the Whole Space of Lipschitz Functions

Let f be a Borel function as in (10.1.1), F be defined by (10.1.2), and $F^{(\infty)}$ by (10.4.1).

In the present section we deepen the study of $F^{(\infty)}$, and, in particular, of its integral representation properties on the whole $W^{1,\infty}$ spaces.

First of all, we start to discuss on the lower semicontinuity properties of $F^{(\infty)}$.

Example 10.9.1. Let f be given by Example 1.4.2. Then f fulfils (10.1.11)÷(10.1.13).

Let us prove that, given a $\Omega \in \mathcal{A}_0$, $F^{(\infty)}(\Omega, \cdot)$ is not even strongly $W^{1,\infty}(\Omega)$-lower semicontinuous.

To see this take $\overline{z} = (0, \overline{b})$, with $\overline{b} > 0$, and $\{z_h\} \subseteq]0, +\infty[^2$ such that $z_h \to \overline{z}$. Then $u_{z_h} \to u_{\overline{z}}$ in $W^{1,\infty}(\Omega)$, and by (10.4.7) of Theorem 10.4.1 and Proposition 10.4.3, we get that

$$(10.9.1) \quad F^{(\infty)}(\Omega, u_{\overline{z}}) \geq \mathrm{co}(\mathrm{sc}^- f)(\overline{z}) \mathcal{L}^n(\Omega) > \liminf_{h \to +\infty} \mathrm{co}(\mathrm{sc}^- f)(z_h) \mathcal{L}^n(\Omega).$$

On the other side, since we have that

$$F^{(\infty)}(\Omega, u_{z_h}) \leq f(z_h) \mathcal{L}^n(\Omega) < +\infty \text{ for every } h \in \mathbf{N},$$

by (10.9.1), (10.4.6) of Theorem 10.4.1, and Proposition 10.4.3, we conclude that

$$F^{(\infty)}(\Omega, u_{\overline{z}}) > \liminf_{h \to +\infty} F^{(\infty)}(\Omega, u_{z_h}).$$

We now prove that in some cases the inequalities in Theorem 10.4.1 can be strict. Actually, even being the assumptions of Theorem 10.4.1

fulfilled, one can have $+\infty = F^{(\infty)}(\Omega, u) > \int_\Omega \mathrm{co}(\mathrm{sc}^- f)(\nabla u)dx$ for some regular $\Omega \in \mathcal{A}_0$, $u \in C^\infty(\mathbf{R}^n)$.

Example 10.9.2. Let $n = 2$, and let f be defined by

$$f: (z_1, z_2) \in \mathbf{R}^2 \mapsto \begin{cases} +\infty & \text{if } z_1 \leq 0 \\ \frac{1}{z_1} - e^{z_2^2} & \text{if } 0 < z_1 \leq e^{-z_2^2} \\ 0 & \text{if } z_1 > e^{-z_2^2}. \end{cases}$$

Then f is continuous, and satisfies $(10.1.11) \div (10.1.13)$. Moreover it is clear that

$$(10.9.2) \quad \mathrm{co}(\mathrm{sc}^- f)(z_1, z_2) = \begin{cases} +\infty & \text{if } z_1 \leq 0 \\ 0 & \text{if } z_1 > 0 \end{cases} \text{ for every } (z_1, z_2) \in \mathbf{R}^2.$$

In addition, let us also observe that

$$(10.9.3) \qquad\qquad (f + I_{Q_m(0)})^{**}(z_1, z_2) =$$

$$= \begin{cases} +\infty & \text{if } z_1 \leq 0 \text{ or } z_1 > m \text{ or } |z_2| > m \\ \frac{1}{z_1} - e^{m^2} & \text{if } 0 < z_1 \leq e^{-m^2} \text{ and } -m \leq z_2 \leq m \\ 0 & \text{if } e^{-m^2} < z_1 \leq m \text{ and } -m \leq z_2 \leq m \end{cases}$$

for every $m \in \mathbf{N}$, $(z_1, z_2) \in \mathbf{R}^2$.

Let $\Omega =]0, 1[\times] - 1, 1[$, and $u: (x_1, x_2) \in \mathbf{R}^2 \mapsto x_1^2/2$. Then, by Theorem 10.4.1, Remark 10.4.2, and $(10.9.3)$, it follows that $F^{(\infty)}(\Omega, u) = +\infty$ whilst, by $(10.9.2)$, it results $\int_\Omega \mathrm{co}(\mathrm{sc}^- f)(\nabla u)dx = 0$.

We now propose some sufficient conditions ensuring the validity of $(10.4.3)$ and $(10.4.5)$ of Theorem 10.4.1, without any finiteness restriction. More precisely that

$$(10.9.4) \qquad\qquad F^{(\infty)}(\Omega, u) = \int_\Omega \mathrm{co}(\mathrm{sc}^- f)(\nabla u)dx$$

for every $\Omega \in \mathcal{A}_0$ convex, $u \in W^{1,\infty}(\Omega)$,

or, if $\mathrm{int}(\mathrm{dom} f) \neq \emptyset$, that

$$(10.9.5) \qquad\qquad F^{(\infty)}(\Omega, u) = \int_\Omega \mathrm{co}(\mathrm{sc}^- f)(\nabla u)dx$$

for every $\Omega \in \mathcal{A}_0$, $u \in W^{1,\infty}_{\mathrm{loc}}(\mathbf{R}^n)$.

Proposition 10.9.3. *Let f be a Borel function as in $(10.1.1)$ satisfying $(10.1.11) \div (10.1.13)$, and let $F^{(\infty)}$ be given by $(10.4.1)$. Let $\Omega \in \mathcal{A}_0$, Ω also* convex if $\mathrm{int}(\mathrm{dom} f) \neq \emptyset$, let $u \in W_{\mathrm{loc}}^{1,\infty}(\mathbf{R}^n)$, and assume that one of the following conditions is fulfilled

(10.9.6) $$\int_\Omega f(\nabla u)dx < +\infty,$$

(10.9.7)   there exists $K \subseteq \mathrm{ri}(\mathrm{dom} f)$ compact such that

$$\nabla u(x) \in K \text{ for a.e. } x \in K,$$

(10.9.8) $$\int_\Omega \mathrm{co}(\mathrm{sc}^- f)(\nabla u)dx = +\infty.$$

Then

$$F^{(\infty)}(\Omega, u) = \int_\Omega \mathrm{co}(\mathrm{sc}^- f)(\nabla u)dx.$$

Proof. If (10.9.6) holds, by (10.4.1) it results $F^{(\infty)}(\Omega, u) < +\infty$, and the proposition follows from Theorem 10.4.1, and Proposition 10.4.3.

If (10.9.7) holds, then (10.1.12) yields (10.9.6), and the proof follows.

If (10.9.8) holds, the proof follows from Theorem 10.4.1, and Proposition 10.4.3. ∎

Proposition 10.9.4. *Let f be a Borel function as in (10.1.1) satisfying (10.1.11)÷(10.1.13), and let $F^{(\infty)}$ be given by (10.4.1). Assume that $\mathrm{dom} f$ is bounded. Then (10.9.4) holds.*

If in addition $\mathrm{int}(\mathrm{dom} f) \neq \emptyset$, then (10.9.5) too holds.

Proof. Let us first observe that, if $\mathrm{dom} f$ is bounded, then $f + I_{Q_m}(0) = f$ for every $m \in \mathbf{N}$ sufficiently large, and therefore, by Theorem 10.4.1, and Remark 10.4.2, that $F^{(\infty)}(\Omega, u) = \int_\Omega f^{**}(\nabla u)dx$ for every $\Omega \in \mathcal{A}_0$ convex, $u \in W^{1,\infty}(\Omega)$ or, if $\mathrm{int}(\mathrm{dom} f) \neq \emptyset$, for every $\Omega \in \mathcal{A}_0$, $u \in W_{\mathrm{loc}}^{1,\infty}(\mathbf{R}^n)$.

Because of this, and by Corollary 1.4.14 the proof follows. ∎

Lemma 10.9.5. *Let $f: \mathbf{R}^n \to [0, +\infty]$ be bounded on the bounded subsets of $\mathrm{dom} f$, and satisfying (10.1.11). Then for every open set A it results that*

$$\mathrm{dom}(\mathrm{co}(\mathrm{sc}^- f)) \cap A \subseteq \mathrm{dom}(f + I_A)^{**}.$$

Proof. Let us preliminarily prove that the boundedness of f on the bounded subsets of $\mathrm{dom} f$ implies that

(10.9.9) $$\overline{\mathrm{dom} f} \cap A \subseteq \mathrm{dom}(f + I_A)^{**}.$$

To do this, we observe that if $z \in \overline{\mathrm{dom}f} \cap A$, and $\{z_h\} \subseteq \mathrm{dom}f \cap A$ is such that $z_h \to z$, then by the lower semicontinuity of $(f + I_A)^{**}$, (1.3.2), and the boundedness of f on the bounded subsets of $\mathrm{dom}f$, we infer that

$$(f + I_A)^{**}(z) \le \liminf_{h \to +\infty}(f + I_A)^{**}(z_h) \le \liminf_{h \to +\infty}(f + I_A)(z_h) =$$

$$= \liminf_{h \to +\infty} f(z_h) < +\infty \text{ for every } z \in \overline{\mathrm{dom}f} \cap A,$$

from which inclusion in (10.9.9) follows.

At this point, by (1.4.1) of Proposition 1.4.1, (1.3.9), the convexity of $\mathrm{dom}f$, and (10.9.9), we conclude that

$$\mathrm{dom}(\mathrm{co}(\mathrm{sc}^- f)) \cap A \subseteq \mathrm{dom}f^{**} \cap A \subseteq \overline{\mathrm{dom}f} \cap A \subseteq \mathrm{dom}(f + I_A)^{**},$$

which proves the lemma. ∎

Theorem 10.9.6. *Let f be a Borel function as in (10.1.1) satisfying (10.1.11), (10.1.13), and let $F^{(\infty)}$ be given by (10.4.1). Assume that f is bounded on the bounded subsets of $\mathrm{dom}f$. Then (10.9.4) holds.*

If in addition $\mathrm{int}(\mathrm{dom}f) \ne \emptyset$, then (10.9.5) too holds.

Proof. Let us prove (10.9.4), the proof of (10.9.5) being similar.

It is clear that, by our assumptions on f, condition (10.1.12) too follows.

Let $\Omega \in \mathcal{A}_0$ be convex, $u \in W^{1,\infty}(\Omega)$. Then it is clear that, by Theorem 10.4.1 and Proposition 10.4.3, we have to treat only the case in which $F^{(\infty)}(\Omega, u) = +\infty$. If this is the case, let $m_0 > \||\nabla u|\|_{L^\infty(\Omega)}$. Then, by Theorem 10.4.1 and Remark 10.4.2, we get that $\int_\Omega (f + I_{Q_{m_0}(0)})^{**}(\nabla u)dx = +\infty$ from which, taking into account the boundedness of f on the bounded subsets of $\mathrm{dom}f$, we conclude that

$$(10.9.10) \qquad \text{there exists } E \in \mathcal{L}_n(\Omega) \text{ with } \mathcal{L}^n(E) > 0 \text{ such that}$$

$$\nabla u(x) \notin \mathrm{dom}(f + I_{Q_{m_0}(0)})^{**} \text{ for a.e. } x \in E.$$

By (10.9.10), and Lemma 10.9.5 applied with $A = Q_{m_0}(0)$, we deduce that $\nabla u(x) \notin \mathrm{dom}(\mathrm{co}(\mathrm{sc}^- f))$ for a.e. $x \in E$. This implies that $\int_\Omega \mathrm{co}(\mathrm{sc}^- f)(\nabla u)dx = +\infty$, from which (10.9.4) follows. ∎

By Theorem 10.9.6 we deduce the following corollaries.

Corollary 10.9.7. *Let f be a Borel function as in (10.1.1) satisfying (10.1.11), (10.1.13), and let $F^{(\infty)}$ be given by (10.4.1). Assume that $\mathrm{dom}f$ is closed, and that f is upper semicontinuous. Then (10.9.4) holds.*

If in addition $\mathrm{int}(\mathrm{dom}f) \ne \emptyset$, then (10.9.5) too holds.

Proof. Follows from Theorem 10.9.6. ∎

Corollary 10.9.8. *Let* $g: \mathbf{R}^n \to [0, +\infty[$ *be continuous,* C *be a convex subset of* $\mathbf{R}^n$, *and let* $F^{(\infty)}$ *be given by (10.4.1) with* $f = g + I_C$. *Then (10.9.4) holds.*

If in addition $\mathrm{int}(C) \neq \emptyset$, *then (10.9.5) too holds.*

Proof. Follows from Theorem 10.9.6, once we observe that $g + I_C$ satisfies (10.1.11)÷(10.1.13). ∎

Corollary 10.9.9. *Let* f *be a Borel function as in (10.1.1) satisfying (10.1.11)÷(10.1.13), and let* $F^{(\infty)}$ *be given by (10.4.1). Assume that* $\mathrm{dom} f$ *is an affine set. Then (10.9.4) holds.*

If $\mathrm{dom} f = \mathbf{R}^n$, *then (10.9.5) holds.*

Proof. Follows from (10.1.12) and Theorem 10.9.6. ∎

The following result shows that Example 10.9.2 needs to be settled at least in dimension two.

Proposition 10.9.10. *Let* $n = 1$, f *be a Borel function as in (10.1.1) satisfying (10.1.11)÷(10.1.13), and let* $F^{(\infty)}$ *be given by (10.4.1). Then*

$$(10.9.11) \qquad F^{(\infty)}(\Omega, u) = \int_\Omega \mathrm{co}(\mathrm{sc}^- f)(u')dx$$

for every $\Omega \in \mathcal{A}_0$, $u \in W^{1,\infty}_{\mathrm{loc}}(\mathbf{R})$.

Proof. It is clear that we can assume that $\mathrm{int}(\mathrm{dom} f) \neq \emptyset$, so that $\mathrm{dom} f$ turns out to be an interval.

If $\mathrm{dom} f$ is a bounded interval, the proof follows from Proposition 10.9.4.

If $\mathrm{dom} f = \mathbf{R}$, the proof follows from Corollary 10.9.9, therefore we have to treat only the case in which $\mathrm{dom} f$ is an unbounded interval with one real endpoint, say for example $\mathrm{dom} f =]a, +\infty[$, or $\mathrm{dom} f = [a, +\infty[$ for some $a \in \mathbf{R}$.

Let us prove that

$$(10.9.12) \qquad (f + I_{Q_m(0)})^{**}(z) \leq \mathrm{co}(\mathrm{sc}^- f)(z) + f(z_0) + 1$$

for every $z_0 > a$, $m > |a| + |z_0| + 1$, $z \in]a, z_0[$.

To do this let $z_0 > a$, $m > |a| + |z_0| + 1$, $z \in]a, z_0[$. Then by Theorem 1.2.6 there exist $z_1, z_2 \in \mathrm{dom} f$ with $z_1 \leq z$, $t \in [0,1]$ such that $z = tz_1 + (1-t)z_2$, and

$$(10.9.13) \qquad tf(z_1) + (1-t)f(z_2) < \mathrm{co} f(z) + 1.$$

Since $a, z_0 \in Q_m(0)$, and $z_1 \in [a, z]$, it is clear that $z_1 \in Q_m(0)$, and we treat separately the cases in which $z_2 \in Q_m(0)$ and $z_2 \notin Q_m(0)$.

If $z_2 \in Q_m(0)$, by (1.3.2), Theorem 1.2.6, (10.9.13), and Proposition 1.4.1, we have that

$$(f + I_{Q_m(0)})^{**}(z) \le \operatorname{co}(f + I_{Q_m(0)})(z) \le$$

$$\le t(f + I_{Q_m(0)})(z_1) + (1 - t)(f + I_{Q_m(0)})(z_2) =$$

$$= tf(z_1) + (1 - t)f(z_2) < \operatorname{co}f(z) + 1 = \operatorname{co}(\operatorname{sc}^- f)(z) + 1,$$

from which (10.9.12) follows.

If $z_2 \notin Q_m(0)$, let $s \in [0,1]$ be such that $z = sz_1 + (1 - s)z_0$. Let us consider separately the two cases in which $sf(z_1) + (1 - s)f(z_0) \le tf(z_1) + (1 - t)f(z_2)$, and $sf(z_1) + (1 - s)f(z_0) > tf(z_1) + (1 - t)f(z_2)$.

If $sf(z_1) + (1 - s)f(z_0) \le tf(z_1) + (1 - t)f(z_2)$, by (1.3.2), Theorem 1.2.6, (10.9.13), and Proposition 1.4.1 we have that

$$(f + I_{Q_m(0)})^{**}(z) \le \operatorname{co}(f + I_{Q_m(0)})(z) \le$$

$$\le s(f + I_{Q_m(0)})(z_1) + (1 - s)(f + I_{Q_m(0)})(z_0) =$$

$$= sf(z_1) + (1-s)f(z_0) \le tf(z_1) + (1-t)f(z_2) < \operatorname{co}f(z) + 1 = \operatorname{co}(\operatorname{sc}^- f)(z) + 1,$$

from which (10.9.12) follows.

If $sf(z_1) + (1 - s)f(z_0) > tf(z_1) + (1 - t)f(z_2)$, by (1.3.2), Theorem 1.2.6, and (10.9.13) we have that

$$(10.9.14) \qquad (f + I_{Q_m(0)})^{**}(z) \le \operatorname{co}(f + I_{Q_m(0)})(z) \le$$

$$\le s(f + I_{Q_m(0)})(z_1) + (1 - s)(f + I_{Q_m(0)})(z_0) = sf(z_1) + (1 - s)f(z_0) =$$

$$= tf(z_1) + (1 - t)f(z_2) + sf(z_1) + (1 - s)f(z_0) - (tf(z_1) + (1 - t)f(z_2)) <$$

$$< \operatorname{co}f(z) + 1 + sf(z_1) + (1 - s)f(z_0) - (tf(z_1) + (1 - t)f(z_2)).$$

We now observe that $tf(z_1) + (1-t)f(z_2)$ is the value at z of the affine function α satisfying $\alpha(z_1) = f(z_1)$, and $\alpha(z_2) = f(z_2)$, whilst $sf(z_1) + (1-s)f(z_0)$ is the one at z of the affine function β satisfying $\beta(z_1) = f(z_1)$, and $\beta(z_0) = f(z_0)$. Therefore, once we observe that $\beta(z_1) = \alpha(z_1)$, and that $a(z_0) \ge 0$, we obtain that

$$(10.9.15) \qquad sf(z_1) + (1 - s)f(z_0) - (tf(z_1) + (1 - t)f(z_2)) =$$

$$= \beta(z) - \alpha(z) \le \beta(z_0) - \alpha(z_0) \le f(z_0).$$

By (10.9.14), (10.9.15), and Proposition 1.4.1 we conclude that

$$(f + I_{Q_m(0)})^{**}(z) \le \operatorname{co}f(z) + 1 + f(z_0) = \operatorname{co}(\operatorname{sc}^- f)(z) + 1 + f(z_0),$$

from which (10.9.12) follows also in this case.

Let us observe now that, if $z_0 > a$, $m > |a| + |z_0| + 1$, by the lower semicontinuity of $(f + I_{Q_m(0)})^{**}$, (10.9.12), the convexity of $\mathrm{co}(\mathrm{sc}^- f)$, and (1.4.1) of Proposition 1.4.1, it results that

$$(10.9.16) \quad (f + I_{Q_m(0)})^{**}(a) \leq \liminf_{t \to 1-}(f + I_{Q_m(0)})^{**}(ta + (1-t)z_0) \leq$$

$$\leq \limsup_{t \to 1-}\{t\,\mathrm{co}(\mathrm{sc}^- f)(a) + (1-t)f(z_0)\} + f(z_0) + 1 \leq \mathrm{co}(\mathrm{sc}^- f)(a) + f(z_0) + 1,$$

whilst by Proposition 1.4.1, and (1.2.6) it clearly follows that

$$(10.9.17) \quad (f + I_{Q_m(0)})^{**}(z) \leq \mathrm{co}(\mathrm{sc}^- f)(z) + f(z_0) + 1 \text{ for every } z < a.$$

Hence by (10.9.12), (10.9.16), and (10.9.17) we conclude that

$$(10.9.18) \qquad (f + I_{Q_m(0)})^{**}(z) \leq \mathrm{co}(\mathrm{sc}^- f)(z) + f(z_0) + 1$$

$$\text{for every } z_0 > a, \ m > |a| + |z_0| + 1, \ z \in \,]-\infty, z_0[.$$

In conclusion, if $\Omega \in \mathcal{A}_0$, $u \in W^{1,\infty}_{\mathrm{loc}}(\mathbf{R})$ with $F^{(\infty)}(\Omega, u) = +\infty$, and $z_0 > \|u'\|_{L^\infty(\Omega)}$, we deduce from Theorem 10.4.1, Remark 10.4.2, and from the monotonicity properties of $\{\int_\Omega (f + I_{Q_m(0)})^{**}(u')dx\}$, that $\int_\Omega (f + I_{Q_m(0)})^{**}(u')dx = +\infty$ for every $m \in \mathbf{N}$.

Because of this, and (10.9.18), we thus obtain that $\int_\Omega \mathrm{co}(\mathrm{sc}^- f)(u')dx = +\infty$, from which, together with Theorem 10.4.1, and Proposition 10.4.3, (10.9.11) follows. ■

For every $\Omega \in \mathcal{A}_0$ let $F(\Omega, \cdot)$ be defined by (10.1.2). Then the above results can be applied to study the relationship between $F^{(\infty)}(\Omega, \cdot)$ and the greatest sequentially weak*-$W^{1,\infty}(\Omega)$-lower semicontinuous functional less than or equal to $F(\Omega, \cdot)$. For every $u \in W^{1,\infty}(\Omega)$ we denote by $\overline{F}^{(\infty)}(\Omega, u)$ the value of such functional in u.

More precisely, under different sets of assumptions, we prove that

$$(10.9.19) \qquad F^{(\infty)}(\Omega, u) = \overline{F}^{(\infty)}(\Omega, u) = \int_\Omega f^{**}(\nabla u)dx$$

$$\text{for every } \Omega \in \mathcal{A}_0 \text{ convex}, \ u \in W^{1,\infty}(\Omega),$$

or

$$(10.9.20) \qquad F^{(\infty)}(\Omega, u) = \overline{F}^{(\infty)}(\Omega, u) = \int_\Omega f^{**}(\nabla u)dx$$

$$\text{for every } \Omega \in \mathcal{A}_0, \ u \in W^{1,\infty}_{\mathrm{loc}}(\mathbf{R}^n).$$

Let us preliminarily observe that, by using also Theorem 7.4.6, it follows that

$$(10.9.21) \qquad \int_\Omega f^{**}(\nabla u)dx \leq \overline{F}^{(\infty)}(\Omega, u) \leq F^{(\infty)}(\Omega, u)$$

for every $\Omega \in \mathcal{A}_0$, $u \in W^{1,\infty}(\Omega)$.

Theorem 10.9.11. *Let f be a Borel function as in (10.1.1) satisfying (10.1.11)÷(10.1.13). Assume that $\mathrm{dom}f$ is an affine set. Then (10.9.19) holds.*

If $\mathrm{dom}f = \mathbf{R}^n$, then (10.9.20) too holds.

Proof. Follows by (10.9.21), Corollary 10.9.9, and Proposition 1.4.8. ∎

Theorem 10.9.12. *Let f be a Borel function as in (10.1.1) satisfying (10.1.11)÷(10.1.13). Assume that $\mathrm{dom}f$ is bounded. Then (10.9.19) holds.*

If in addition $\mathrm{int}(\mathrm{dom}f) \neq \emptyset$, then (10.9.20) too holds.

Proof. Follows from (10.9.21), Proposition 10.9.4, and Corollary 1.4.14. ∎

Theorem 10.9.13. *Let f be a Borel function as in (10.1.1) satisfying (10.1.11), (10.1.13). Assume that f is bounded on the bounded subsets of $\mathrm{dom}f$, and that one of the following conditions is fulfilled*

$$\lim_{z \to \infty} \frac{f(z)}{|z|} = +\infty,$$

for every $z_0 \in \mathrm{rb}(\mathrm{co}(\mathrm{dom}f))$ there exists a non-trivial supporting

hyperplane to $\mathrm{co}(\mathrm{dom}f)$ containing z_0 having a bounded intersection

with $\mathrm{rb}(\mathrm{co}(\mathrm{dom}f))$.

Then (10.9.19) holds.

If in addition $\mathrm{int}(\mathrm{dom}f) \neq \emptyset$, then (10.9.20) too holds.

Proof. Follows from (10.9.21), Theorem 10.9.6, and Proposition 1.4.8 or Theorem 1.4.13. ∎

Theorem 10.9.14. *Let $n = 1$, f be a Borel function as in (10.1.1) satisfying (10.1.11)÷(10.1.13). Then (10.9.20) holds.*

Proof. Follows from (10.9.21), Proposition 10.9.10, and Corollary 1.4.16. ∎

Chapter 11

Cut-off Functions
and Partitions of Unity

In the present chapter, and in the next three, we study the homogenization process for some classes of unbounded integral functionals of the calculus of variations.

This chapter has a rather technical nature, and is preparatory to the next ones, where the full process will be analyzed. Here, we just discuss the construction of some special cut-off functions and partitions of unity on which the analysis carried out in the next chapters will depend deeply.

We also want point out here that in homogenization theory both sequences of discrete parameters (denoted by h) and continuous ones (denoted by ε) are traditionally used. Generally, the sequences of discrete parameters are assumed to be diverging in order to give the idea of the thickening of the materials that mix together. On the other side, continuous parameters are assumed to be vanishing in order to recall that the size of the zones occupied by the single materials becomes smaller and smaller.

We will use both the types of parameters. For sake of simplicity, we use sequences of discrete parameters for technical or intermediate results and the continuous parameters for the main theorems, to make them independent of the choice of sequences.

§11.1 Cut-off Functions

Let f be an integrand of the following type

$$(11.1.1) \quad \begin{cases} f \colon (x, z) \in \mathbf{R}^n \times \mathbf{R}^n \mapsto f(x, z) \in [0, +\infty] \\ f \ (\mathcal{L}_n(\mathbf{R}^n) \times \mathcal{B}(\mathbf{R}^n))\text{-measurable} \\ f \ Y\text{-periodic in the } x \text{ variable, convex in the } z \text{ one,} \end{cases}$$

and let us introduce for every $q \in [1, +\infty]$, the function $\tilde{f}^q_{\text{hom}}$ defined by

(11.1.2) $\tilde{f}^q_{\text{hom}} : z \in \mathbf{R}^n \mapsto$

$$\inf \left\{ \int_Y f(y, z + \nabla v) dy : v \in W^{1,q}_{\text{per}}(Y) \cap L^\infty(Y) \right\}.$$

It is clear that, for every $q \in [1, +\infty]$, $\tilde{f}^q_{\text{hom}}$ turns out to be convex. In order to prove our results, we assume that

(11.1.3) $0 \in \text{int}(\text{dom} \tilde{f}^q_{\text{hom}})$.

Then there exists $\delta \in]0, 1[$ such that

(11.1.4) $B_{2\delta}(0) \subseteq \text{int}(\text{dom} \tilde{f}^q_{\text{hom}})$,

and, for every $j \in \{1, \ldots, n\}$, there exist $w_j^+, w_j^- \in W^{1,q}_{\text{per}}(Y) \cap L^\infty(Y)$ such that

(11.1.5) $f(\cdot, \delta \mathbf{e}_j + \nabla w_j^+(\cdot)) \in L^1(Y), \ f(\cdot, \delta \mathbf{e}_j + \nabla w_j^-(\cdot)) \in L^1(Y)$.

Lemma 11.1.1. *Let f be as in (11.1.1), $q \in [1, +\infty]$, and let $\tilde{f}^q_{\text{hom}}$ be given by (11.1.2). Assume that (11.1.3) holds. Let $\delta \in]0, 1[$ satisfy (11.1.4). Then, for every $\{r_h\} \subseteq]0, +\infty[$ strictly increasing and diverging, $\Omega \in \mathcal{A}_0$, and any compact subset K of Ω there exist $\{\psi_h\} \subseteq W^{1,q}(\mathbf{R}^n) \cap L^\infty(\mathbf{R}^n)$, $\psi \in W^{1,q}(\mathbf{R}^n) \cap L^\infty(\mathbf{R}^n)$, and $c_f \in]0, +\infty[$ (c_f depending only on n, f, q, and δ) such that*

$$\psi_h = \psi = 0 \text{ a.e. in } \mathbf{R}^n \setminus \Omega \text{ for every } h \in \mathbf{N},$$

(11.1.6) $0 \leq \psi_h \leq 1 \text{ a.e. in } \Omega \text{ for every } h \in \mathbf{N},$

(11.1.7) $\psi_h = 1 \text{ a.e. in } K \text{ for every } h \in \mathbf{N},$

(11.1.8) $\psi_h \to \psi \text{ in } L^\infty(\Omega) \text{ as } h \text{ diverges},$

(11.1.9) $\limsup_{h \to +\infty} \int_A f\left(r_h x, \frac{\delta \text{dist}(K, \partial\Omega)}{64 n 3^n \sqrt{n}} \nabla \psi_h\right) dx \leq c_f \mathcal{L}^n(A)$

for every $A \in \mathcal{A}_0$.

Proof. We first consider the case in which $\Omega = Q_r(0)$ and $K = \overline{Q_\rho(0)}$.

Let us fix $j \in \{1,\ldots,n\}$, and let u_j^+ and u_j^- be two affine functions such that $\nabla u_j^+ = \delta e_j$, $\nabla u_j^- = -\delta e_j$ and

(11.1.10)
$$\begin{cases} u_j^+ = 0 \text{ on the first face of } Q_r(0) \text{ in the direction of } e_j, \\ u_j^- = 0 \text{ on the second face of } Q_r(0) \text{ in the direction of } e_j. \end{cases}$$

Since $Q_r(0)$ and $Q_\rho(0)$ have the same centre, it turns out that

(11.1.11)
$$\begin{cases} u_j^+ = \tfrac{\delta}{2}(r - \rho) \text{ on the first face of } Q_\rho(0) \\ \qquad\qquad\qquad\qquad\qquad \text{in the direction of } e_j, \\ u_j^- = \tfrac{\delta}{2}(r - \rho) \text{ on the second face of } Q_\rho(0) \\ \qquad\qquad\qquad\qquad\qquad \text{in the direction of } e_j. \end{cases}$$

Let w_j^+, $w_j^- \in W_{\mathrm{per}}^{1,q}(Y) \cap L^\infty(Y)$ satisfy (11.1.5) and, for every $h \in \mathbf{N}$, $u_{j,h}^+$, $u_{j,h}^-$ be defined by $u_{j,h}^+ = u_j^+ + \frac{1}{r_h} w_j^+ (r_h \cdot)$, $u_{j,h}^- = u_j^- + \frac{1}{r_h} w_j^- (r_h \cdot)$. Then, it results that

(11.1.12) $u_{j,h}^+ \to u_j^+$, $u_{j,h}^- \to u_j^-$ in $L^\infty(Q_r(0))$,

and, by (11.1.5),

(11.1.13) $\displaystyle \lim_{h \to +\infty} \int_A f(r_h x, \nabla u_{j,h}^+) dx = \mathcal{L}^n(A) \int_Y f(y, \delta e_j + \nabla w_j^+) dy,$

$$\lim_{h \to +\infty} \int_A f(r_h x, \nabla u_{j,h}^-) dx = \mathcal{L}^n(A) \int_Y f(y, -\delta e_j + \nabla w_j^-) dy$$

for every $A \in \mathcal{A}_0$.

Set $M = \max\{\max_{j \in \{1,\ldots,n\}} \|w_j^+\|_{L^\infty(Y)}, \max_{j \in \{1,\ldots,n\}} \|w_j^-\|_{L^\infty(Y)}\} + 1$, $h_0 = [\frac{8M}{\delta(r-\rho)}] + 1$, and let $\{\chi_h\} \subseteq C^1(\mathbf{R})$ be such that

(11.1.14)
$$\begin{cases} \chi_h(t) = 0 \text{ for every } t \in]-\infty, \frac{M}{h}[, \\ \chi_h(t) = \tfrac{\delta}{2}(r - \rho) \text{ for every } t \in]\frac{\delta}{2}(r-\rho) - \frac{M}{h}, +\infty[\\ \chi_h \text{ affine in }]2\frac{M}{h}, \frac{\delta}{2}(r-\rho) - 2\frac{M}{h}[\\ 0 \le \chi_h'(t) \le 2 \text{ for every } t \in \mathbf{R} \end{cases}$$

for every $h \ge h_0$.

It results that

(11.1.15) $\chi_h \to \chi_\infty$ in $L^\infty(\mathbf{R})$,

where χ_∞ is the function on $\mathbf{R}$ defined by

(11.1.16) $\chi_\infty(t) = \begin{cases} 0 & \text{if } t \in]-\infty, 0[\\ t & \text{if } t \in [0, \frac{\delta}{2}(r - \rho)] \\ \frac{\delta}{2}(r - \rho) & \text{if } t \in]\frac{\delta}{2}(r - \rho), +\infty[. \end{cases}$

For every $h \in \mathbf{N}$, $\chi_h(u_{j,h}^+)$, $\chi_h(u_{j,h}^-)$, $\chi_\infty(u^j)$, and $\chi_\infty(v^j)$ are in $W_{\mathrm{loc}}^{1,q}(\mathbf{R}^n) \cap L^\infty(\mathbf{R}^n)$. From $(11.1.10) \div (11.1.12)$, $(11.1.14)$, and $(11.1.15)$, it follows that

$$(11.1.17) \quad 0 \leq \chi_h(u_{j,h}^+) \leq \frac{\delta}{2}(r - \rho), \ 0 \leq \chi_h(u_{j,h}^-) \leq \frac{\delta}{2}(r - \rho) \ \text{a.e. in } \mathbf{R}^n$$

$$\text{for every } h \geq h_0,$$

$$(11.1.18) \quad \begin{cases} \text{the trace of } \chi_h(u_{j,h}^+) \text{ on the first face of } Q_r(0) \\ \qquad\qquad\qquad\qquad\qquad \text{in the direction of } \mathbf{e}_j \text{ is } 0, \\ \text{the trace of } \chi_h(u_{j,h}^-) \text{on the second face of } Q_r(0) \\ \qquad\qquad\qquad\qquad\qquad \text{in the direction of } \mathbf{e}_j \text{ is } 0 \end{cases}$$

$$\text{for every } h \geq h_0$$

$$(11.1.19) \quad \chi_h(u_{j,h}^+) = \frac{\delta}{2}(r - \rho), \ \chi_h(u_{j,h}^-) = \frac{\delta}{2}(r - \rho) \ \text{a.e. in } Q_\rho(0)$$

$$\text{for every } h \geq h_0,$$

and

$$(11.1.20) \quad \chi_h(u_{j,h}^+) \to \chi_\infty(u_j^+), \ \chi_h(u_{j,h}^-) \to \chi_\infty(u_j^-) \ \text{in } L^\infty(Q_r(0)).$$

Moreover, it results that

$$(11.1.21) \qquad \limsup_{h \to +\infty} \int_A f\left(r_h x, \frac{1}{2}\nabla(\chi_h(u_{j,h}^+))\right) dx \leq$$

$$\leq \left(\int_Y f(y, \delta \mathbf{e}_j + \nabla w_j^+) dy + \int_Y f(y, 0) dy\right) \mathcal{L}^n(A),$$

$$\limsup_{h \to +\infty} \int_A f\left(r_h x, \frac{1}{2}\nabla(\chi_h(u_{j,h}^-))\right) dx \leq$$

$$\leq \left(\int_Y f(y, -\delta \mathbf{e}_j + \nabla w_j^-) dy + \int_Y f(y, 0) dy\right) \mathcal{L}^n(A)$$

$$\text{for every } A \in \mathcal{A}_0.$$

In fact, because of the convexity of f, and by $(11.1.14)$, it results that (we prove only the first statement in $(11.1.21)$, the other being similar)

$$(11.1.22) \quad f\left(r_h x, \frac{1}{2}\nabla(\chi_h(u_{j,h}^+))(x)\right) = f\left(r_h x, \frac{1}{2}\chi_h'(u_{j,h}^+(x))\nabla u_{j,h}^+(x)\right) \leq$$

$$\leq \frac{1}{2}\chi_h'(u_{j,h}^+(x))f(r_hx,\nabla u_{j,h}^+(x)) + \left(1 - \frac{1}{2}\chi_h'(u_{j,h}^+(x))\right)f(r_hx,0) \leq$$

$$\leq f(r_hx,\nabla u_{j,h}^+(x)) + f(r_hx,0) \text{ for a.e. } x \in \mathbf{R}^n, \text{ and every } h \geq h_0,$$

therefore (11.1.21) follows by combining (11.1.22) with (11.1.13), and by recalling that $f(\cdot,0) \in L^1_{\text{loc}}(\mathbf{R}^n)$.

Set now

$$\psi_h(x) = \frac{1}{(\frac{\delta}{2})^{2n}(r-\rho)^{2n}} \prod_{j=1}^n \chi_h(u_{j,h}^+(x))\chi_h(u_{j,h}^-(x))$$

for a.e. $x \in \mathbf{R}^n$, and every $h \in \mathbf{N}$,

$$\psi(x) = \frac{1}{(\frac{\delta}{2})^{2n}(r-\rho)^{2n}} \prod_{j=1}^n \chi_\infty(u_j^+(x))\chi_\infty(u_j^-(x)) \text{ for a.e. } x \in \mathbf{R}^n.$$

It is obvious that, ψ_h and ψ are in $W^{1,q}_{\text{loc}}(\mathbf{R}^n) \cap L^\infty(\mathbf{R}^n)$ for every $h \in \mathbf{N}$. Moreover, from (11.1.10) and (11.1.16)÷(11.1.20) it follows that

(11.1.23) $\psi_h \in W_0^{1,q}(Q_r(0)), \ \psi \in W_0^{1,q}(Q_r(0))$ for every $h \geq h_0$,

(11.1.24) $0 \leq \psi_h \leq 1$ a.e. in $Q_r(0)$ for every $h \geq h_0$,

(11.1.25) $\psi_h = 1$ a.e. in $Q_\rho(0)$ for every $h \geq h_0$,

(11.1.26) $\psi_h \to \psi$ in $L^\infty(Q_r(0))$.

Furthermore, it results that

(11.1.27) $\limsup_{h\to+\infty} \int_A f\left(r_hx, \frac{\delta(r-\rho)}{8n}\nabla\psi_h\right) dx \leq c\mathcal{L}^n(A)$

for every $A \in \mathcal{A}_0$,

where

(11.1.28) $c = \sum_{j=1}^n \left\{ \int_Y f(y,\delta e_j + \nabla w_j^+)dy + \int_Y f(y,-\delta e_j + \nabla w_j^-)dy \right\} +$

$$+(3n+1)\int_Y f(y,0)dy.$$

In fact, once we define for every $h \geq h_0$, $j \in \{1, \ldots, n\}$, and a.e.

$x \in \mathbf{R}^n$ $\lambda_h^j(x) = \dfrac{\prod_{i=1, i \neq j}^n \chi_h(u_{j,h}^+(x)) \chi_h(u_{j,h}^-(x))}{(\frac{\delta}{2})^{2n-2}(r-\rho)^{2n-2}n}$, we have that $0 \leq \lambda_h^j(x) \leq \frac{1}{n}$,

and, consequently, $\sum_{j=1}^n \lambda_h^j(x) \leq 1$. Therefore by the convexity properties of f, and (11.1.17), it results that

$$f\left(r_h x, \frac{\delta(r-\rho)}{8n} \nabla \psi_h(x)\right) =$$

$$= f\left(r_h x, \sum_{j=1}^n \lambda_h^j(x) \left(\frac{\chi_h(u_{j,h}^+(x))}{4\frac{\delta}{2}(r-\rho)} \nabla(\chi_h(u_{j,h}^-))(x) + \right.\right.$$

$$\left.\left. + \frac{\chi_h(u_{j,h}^-(x))}{4\frac{\delta}{2}(r-\rho)} \nabla(\chi_h(u_{j,h}^+))(x)\right)\right) \leq$$

$$\leq \sum_{j=1}^n f\left(r_h x, \frac{1}{2}\frac{\chi_h(u_{j,h}^+(x))}{\frac{\delta}{2}(r-\rho)} \frac{1}{2}\nabla(\chi_h(u_{j,h}^-))(x) + \right.$$

$$\left. + \frac{1}{2}\frac{\chi_h(u_{j,h}^-(x))}{\frac{\delta}{2}(r-\rho)} \frac{1}{2}\nabla(\chi_h(u_{j,h}^+))(x)\right) + f(r_h x, 0) \leq$$

$$\leq \sum_{j=1}^n f\left(r_h x, \frac{1}{2}\nabla(\chi_h(u_{j,h}^-))(x)\right) + \sum_{j=1}^n f\left(r_h x, \frac{1}{2}\nabla(\chi_h(u_{j,h}^+))(x)\right) +$$

$$+ \sum_{j=1}^n f(r_h x, 0) + f(r_h x, 0) \text{ for a.e. } x \in \mathbf{R}^n, \text{ and every } h \geq h_0.$$

In conclusion, (11.1.27) follows by combining the above inequalities with (11.1.21), and recalling that $f(\cdot, 0) \in L^1_{\text{loc}}(\mathbf{R}^n)$.

Consider now the general case.

Let $\mathcal{R} = \{Q_\rho^j\}_{j \in \mathbf{N}}$ be a partition of $\mathbf{R}^n$ into half open cubes with faces parallel to the coordinate planes, and sidelength $\rho = \frac{\text{dist}(K, \partial\Omega)}{2\sqrt{n}}$. Let us observe that it is not restrictive to assume the existence of $m \in \mathbf{N}$ such that $Q_\rho^j \cap K \neq \emptyset$ if and only if $j \in \{1, \ldots, m\}$.

Let $r = \frac{3}{2}\rho = \frac{3}{4}\frac{\text{dist}(K, \partial\Omega)}{\sqrt{n}}$ and, for every $j \in \mathbf{N}$, let Q_r^j be an open cube with faces parallel to the coordinate planes, centred as Q_ρ^j, and with sidelength equal to r.

Because of (11.1.23)÷(11.1.27), for every $j \in \{1, \ldots, m\}$ there exist $\{\psi_h^j\} \subseteq W_0^{1,q}(Q_r^j)$, and $\psi^j \in W_0^{1,q}(Q_r^j)$ such that

(11.1.29) $0 \leq \psi_h^j \leq 1$ a.e. in Q_r^j for every $h \in \mathbf{N}$,

(11.1.30) $\psi_h^j = 1$ a.e. in Q_ρ^j for every $h \in \mathbf{N}$,

(11.1.31) $$\psi_h^j \to \psi^j \text{ in } L^\infty(Q_r^j),$$

(11.1.32) $$\limsup_{h \to +\infty} \int_A f\left(r_h x, \frac{\delta(r-\rho)}{8n} \nabla \psi_h^j\right) dx \leq c\mathcal{L}^n(A)$$

$$\text{for every } A \in \mathcal{A}_0,$$

where c is given by (11.1.28).

Let $\chi \in C^1(\mathbf{R})$ satisfy

$$\begin{cases} \chi(t) = 0 \text{ for every } t \in \,] -\infty, 0] \\ \chi(t) = 1 \text{ for every } t \in [1, +\infty[\\ 0 \leq \chi'(t) \leq 2 \text{ for every } t \in \mathbf{R}, \end{cases}$$

and, for $h \in \mathbf{N}$, let ψ_h and ψ be defined by

$$\psi_h = \chi\left(\sum_{j=1}^m \psi_h^j\right) \quad \psi = \chi\left(\sum_{j=1}^m \psi^j\right).$$

It is obvious that, for every $h \in \mathbf{N}$, ψ_h and ψ belong to $W_0^{1,q}(\Omega) \cap L^\infty(\Omega)$ and that (11.1.6) holds. Moreover (11.1.8) follows from (11.1.29) and (11.1.31).

Since $K \subseteq \cup_{j=1}^m \overline{Q_\rho^j}$, from (11.1.30) it follows that $\sum_{j=1}^m \psi_h^j(x) \geq 1$ a.e. in K for every $h \in \mathbf{N}$. Consequently, equality (11.1.7) holds.

Finally, we prove (11.1.9).

Let $A \in \mathcal{A}_0$. It is not restrictive to assume the existence of a positive integer $s \geq m$ such that $A \subseteq \cup_{i=1}^s Q_\rho^i$.

On the other hand, for every fixed $i \in \{1, \ldots, s\}$, let c_i be the number of the cubes in $\{Q_r^j\}_{j \in \{1, \ldots, m\}}$ that have nonempty intersection with Q_ρ^i, and let $j_1(i), \ldots, j_{c_i}(i) \in \{1, \ldots, m\}$ be such that $Q_r^j \cap Q_\rho^i \neq \emptyset$ if and only if $j \in \{j_1(i), \ldots, j_{c_i}(i)\}$. Clearly for every $i \in \{1, \ldots, s\}$, it turns out that $c_i \leq 3^n$.

Consequently (11.1.1), (11.1.32), and the properties of χ provide that

$$\limsup_{h \to +\infty} \int_A f\left(r_h x, \frac{\delta \text{dist}(K, \partial\Omega)}{64n3^n\sqrt{n}} \nabla \psi_h\right) dx =$$

$$= \limsup_{h \to +\infty} \int_A f\left(r_h x, \frac{1}{2}\chi'\left(\sum_{j=1}^m \psi_h^j\right) \frac{\delta \text{dist}(K, \partial\Omega)}{32n3^n\sqrt{n}} \sum_{j=1}^m \nabla \psi_h^j\right) dx \leq$$

$$\leq \limsup_{h \to +\infty} \int_A \frac{1}{2}\chi'\left(\sum_{j=1}^m \psi_h^j\right) f\left(r_h x, \frac{\delta \text{dist}(K, \partial\Omega)}{32n3^n\sqrt{n}} \sum_{j=1}^m \nabla \psi_h^j\right) dx +$$

$$+ \limsup_{h\to+\infty} \int_A \left(1 - \frac{1}{2}\chi'\left(\sum_{j=1}^m \psi_h^j\right)\right) f(r_h x, 0)\, dx \le$$

$$\le \limsup_{h\to+\infty} \sum_{i=1}^s \int_{Q_\rho^i \cap A} f\left(r_h x, \frac{\delta \operatorname{dist}(K, \partial\Omega)}{32 n 3^n \sqrt{n}} \sum_{j=1}^m \nabla\psi_h^j\right) dx +$$

$$+ \mathcal{L}^n(A) \int_Y f(y, 0)\, dy =$$

$$= \limsup_{h\to+\infty} \sum_{i=1}^s \int_{Q_\rho^i \cap A} f\left(r_h x, \frac{\delta \operatorname{dist}(K, \partial\Omega)}{32 n 3^n \sqrt{n}} \sum_{k=1}^{c_i} \nabla\psi_h^{j_k(i)}\right) dx +$$

$$+ \mathcal{L}^n(A) \int_Y f(y, 0)\, dy \le$$

$$\le \limsup_{h\to+\infty} \sum_{i=1}^s \frac{c_i}{3^n} \int_{Q_\rho^i \cap A} f\left(r_h x, \sum_{k=1}^{c_i} \frac{\delta \operatorname{dist}(K, \partial\Omega)}{c_i 32 n \sqrt{n}} \nabla\psi_h^{j_k(i)}\right) dx +$$

$$+ \sum_{i=1}^s \left(1 - \frac{c_i}{3^n}\right) \lim_{h\to+\infty} \int_{Q_\rho^i \cap A} f(r_h x, 0)\, dx + \mathcal{L}^n(A) \int_Y f(y, 0)\, dy \le$$

$$\le \frac{1}{3^n} \limsup_{h\to+\infty} \sum_{i=1}^s \sum_{k=1}^{c_i} \int_{Q_\rho^i \cap A} f\left(r_h x, \frac{\delta \operatorname{dist}(K, \partial\Omega)}{32 n \sqrt{n}} \nabla\psi_h^{j_k(i)}\right) dx +$$

$$+ 2\mathcal{L}^n(A) \int_Y f(y, 0)\, dy =$$

$$= \frac{1}{3^n} \limsup_{h\to+\infty} \sum_{i=1}^s \sum_{k=1}^{c_i} \int_{Q_\rho^i \cap A} f\left(r_h x, \frac{\delta(r - \rho)}{8n} \nabla\psi_h^{j_k(i)}\right) dx +$$

$$+ 2\mathcal{L}^n(A) \int_Y f(y, 0)\, dy \le$$

$$\le c \sum_{i=1}^s \mathcal{L}^n(Q_\rho^i \cap A) + 2\mathcal{L}^n(A) \int_Y f(y, 0)\, dy = \left(c + 2\int_Y f(y, 0)\, dy\right) \mathcal{L}^n(A),$$

where c is given by (11.1.28).

Because of this, inequality (11.1.9) follows. This completes the proof. ∎

Remark 11.1.2. We point out that under the assumptions of Lemma 11.1.1, if δ satisfies (11.1.4) and, for every $j \in \{1,\ldots,n\}$, w_j^+, $w_j^- \in W_{\mathrm{per}}^{1,q}(Y) \cap L^\infty(Y)$ satisfy (11.1.5), the constant c_f in (11.1.9) is given by

$$c_f = \sum_{j=1}^n \left(\int_Y f(y, \delta \mathbf{e}_j + Dw_j^+)\, dy + \int_Y f(y, -\delta \mathbf{e}_j + Dw_j^-)\, dy\right) +$$

$$+3(n+1) \int_Y f(y,0)dy.$$

§11.2 Partitions of Unity

For every f as in (11.1.1) and $z_0 \in \mathbf{R}^n$ we introduce the function $f^{(z_0)}$ given by

$$(11.2.1) \qquad f^{(z_0)} \colon (x,z) \in \mathbf{R}^n \times \mathbf{R}^n \mapsto f(x,z) + f(x, 2z_0 - z).$$

It is clear that, for fixed $z_0 \in \mathbf{R}^n$, $f^{(z_0)}(x, \cdot)$ is symmetric with respect to z_0 for a.e. $x \in \mathbf{R}^n$, and that

$$f(x,z) \leq f^{(z_0)}(x,z) \text{ for a.e. } x \in \mathbf{R}^n, \text{ and every } z \in \mathbf{R}^n.$$

For every $q \in [1, +\infty]$ and $z_0 \in \mathbf{R}^n$, we set

$$(11.2.2) \qquad \widetilde{C}^q(z_0) = \mathrm{dom}\,\widetilde{f^{(z_0)}}^q_{\mathrm{hom}} = \left\{ z \in \mathbf{R}^n : \text{ there exists} \right.$$

$$\left. v \in W^{1,q}_{\mathrm{per}}(Y) \cap L^\infty(Y) \text{ with } \int_Y f^{(z_0)}(y, z + \nabla v)dy < +\infty \right\}.$$

It is clear that, for every $q \in [1, +\infty]$ and $z_0 \in \mathbf{R}^n$, $\widetilde{C}^q(z_0)$ is convex. In the following, given $z_0 \in \mathbf{R}^n$, we assume that

$$(11.2.3) \qquad \mathrm{int}(\widetilde{C}^q(z_0)) \neq \emptyset.$$

We also observe that (11.2.3) with $z_0 = 0$ implies (11.1.3).

Proposition 11.2.1. *Let f be as in (11.1.1), $q \in [1, +\infty]$, $z_0 \in \mathbf{R}^n$, and $\widetilde{C}^q(z_0)$ be defined in (11.2.2). Assume that (11.2.3) holds. Then there exists $\delta \in {]}0,1{[}$ such that*

$$(11.2.4) \qquad B_{2\delta}(z_0) \subset\subset \mathrm{int}(\widetilde{C}^q(z_0)),$$

and

$$(11.2.5) \qquad \int_Y f(y, z_0)dy < +\infty.$$

Proof. Since $\widetilde{C}^q(z_0)$ is symmetric with respect to z_0 and convex, assumption (11.2.3) provides the existence of $\delta \in {]}0,1{[}$ for which (11.2.4) holds.

By (11.2.4) it follows trivially that $z_0 \in \widetilde{C}^q(z_0)$. Consequently there exists $v_0 \in W^{1,q}_{per}(Y) \cap L^\infty(Y)$ such that $\int_Y f^{(z_0)}(y, z_0 + \nabla v_0) dy < +\infty$. Because of this, (11.2.5) follows, once we observe that (11.1.1) implies that

$$f(y, z_0) = f\left(y, \frac{1}{2}(z_0 + \nabla v_0) + \frac{1}{2}(z_0 - \nabla v_0)\right) \leq$$

$$\leq \frac{1}{2}\{f(y, z_0 + \nabla v_0) + f(y, z_0 - \nabla v_0)\} = \frac{1}{2}f^{(z_0)}(y, z_0 + \nabla v_0)$$

$$\text{for a.e. } y \in Y. \qquad \blacksquare$$

Let $\varepsilon \in]0, +\infty[$, and $P_1, \ldots, P_m \subseteq \mathbf{R}^n$. For every $i \in \{1, \ldots, m\}$, denote by $\nu_\varepsilon(P_1, \ldots, P_m)(P_i)$ the number of the elements in $\{P_1, \ldots, P_m\}$ whose distance from P_i is less than ε. Moreover, set

$$(11.2.6) \qquad \sigma_\varepsilon(P_1, \ldots, P_m) = \sup_{i \in \{1, \ldots, m\}} \nu_\varepsilon(P_1, \ldots, P_m)(P_i).$$

It is obvious that $\sigma_\varepsilon(P_1, \ldots, P_m) \in \{1, \ldots, m\}$.

Finally, for every $i \in \{1, \ldots, m\}$, set $P^+_{i,\varepsilon} = (P_i)^+_\varepsilon$, $P^-_{i,\varepsilon} = (P_i)^-_\varepsilon$.

Lemma 11.2.2. *Let f be as in (11.1.1), $z_0 = 0$, $f^{(0)}$ be given by (11.2.1), $q \in [1, +\infty]$, and $\widetilde{C}^q(0)$ be defined in (11.2.2). Assume that (11.2.3) holds. Let $\delta \in]0, 1[$ satisfy (11.2.4). Let $\{\Omega_1, \ldots, \Omega_m\}$ be a finite family of bounded disjoint open subsets of $\mathbf{R}^n$, and $\varepsilon \in]0 + \infty[$ be such that $\Omega^-_{j,\varepsilon} \neq \emptyset$ for every $j \in \{1, \ldots, m\}$. Then, for every $j \in \{1, \ldots, m\}$ there exist $\{\gamma^{\varepsilon,j}_h\} \subseteq W^{1,q}(\mathbf{R}^n) \cap L^\infty(\mathbf{R}^n)$, $\gamma^{\varepsilon,j} \in W^{1,q}(\mathbf{R}^n) \cap L^\infty(\mathbf{R}^n)$ such that*

$$(11.2.7) \qquad \gamma^{\varepsilon,j}_h = \gamma^{\varepsilon,j} = 0 \text{ a.e. in } \mathbf{R}^n \setminus \Omega^+_{j,\varepsilon} \text{ for every } h \in \mathbf{N},$$

$$(11.2.8) \qquad 0 \leq \gamma^{\varepsilon,j}_h \leq 1 \text{ a.e. in } \cup^m_{i=1} \Omega_i \text{ for every } h \in \mathbf{N},$$

$$(11.2.9) \qquad \gamma^{\varepsilon,j}_h = 1 \text{ a.e. in } \Omega^-_{j,\varepsilon} \text{ for every } h \in \mathbf{N},$$

$$(11.2.10) \qquad \sum^m_{i=1} \gamma^{\varepsilon,i}_h = 1 \text{ a.e. in } \cup^m_{i=1} \Omega_i \text{ for every } h \in \mathbf{N},$$

$$(11.2.11) \qquad \gamma^{\varepsilon,j}_h \to \gamma^{\varepsilon,j} \text{ in } L^\infty(\cup^m_{i=1}\Omega_i),$$

$$(11.2.12) \quad \limsup_{h \to +\infty} \int_A f^{(0)}\left(hx, \frac{\delta\varepsilon}{256n3^n \sqrt{n}\sigma_\varepsilon(\Omega_1, \ldots, \Omega_m)} \nabla\gamma^{\varepsilon,j}_h\right) dx \leq$$

$$\leq 2 \left(c_f + 2 \int_Y f(y,0) dy \right) \mathcal{L}^n(A) \text{ for every } A \in \mathcal{A} \left(\cup_{i=1}^m \Omega_i \right),$$

where c_f is defined in Remark 11.1.2.

Proof. For every $j \in \{1, \dots, m\}$ let $\{\psi_h^{\varepsilon,j}\}$, and $\psi^{\varepsilon,j}$ be given by Lemma 11.1.1 applied to $f^{(0)}$ with $\Omega = \Omega_{j,\varepsilon}^+$, and $K = \Omega_{j,\frac{\varepsilon}{2}}^+$. Then it results that

$$\sum_{j=1}^m \psi_h^{\varepsilon,j}(x) \geq 1, \ \sum_{j=1}^m \psi^{\varepsilon,j}(x) \geq 1 \text{ for a.e. } x \in \cup_{j=1}^m \Omega_{j,\frac{\varepsilon}{2}}^+, \text{ and every } h \in \mathbf{N}.$$

Let A_ε be an open set with Lipschitz boundary such that $\cup_{j=1}^m \Omega_j \subset\subset A_\varepsilon \subset\subset \cup_{j=1}^m \Omega_{j,\frac{\varepsilon}{2}}^+$, and, for every $h \in \mathbf{N}$, let ϑ_h^ε, $\vartheta^\varepsilon \in W_{\text{loc}}^{1,q}(\mathbf{R}^n) \cap L_{\text{loc}}^\infty(\mathbf{R}^n)$ such that

$$(11.2.13) \quad \begin{cases} \vartheta_h^\varepsilon(x) = \sum_{i=1}^m \psi_h^{\varepsilon,i}(x), \ \vartheta^\varepsilon(x) = \sum_{i=1}^m \psi^{\varepsilon,i}(x) \text{ for a.e. } x \in A_\varepsilon, \\ \vartheta_h^\varepsilon(x) \geq 1, \ \vartheta^\varepsilon(x) \geq 1 \text{ for a.e. } x \in \mathbf{R}^n. \end{cases}$$

For every $j \in \{1, \dots, m\}$, $h \in \mathbf{N}$ let $\gamma_h^{\varepsilon,j}$, and $\gamma^{\varepsilon,j}$ be the functions defined by

$$(11.2.14) \quad \gamma_h^{\varepsilon,j} = \frac{\psi_h^{\varepsilon,j}}{\vartheta_h^\varepsilon}, \quad \gamma^{\varepsilon,j} = \frac{\psi^{\varepsilon,j}}{\vartheta^\varepsilon}.$$

Then, by Lemma 11.1.1, and (11.2.13), the functions in (11.2.14) satisfy conditions (11.2.7)÷(11.2.11).

To prove (11.2.12) let us fix $j \in \{1, \dots, m\}$, set $\sigma_\varepsilon = \sigma_\varepsilon(\Omega_1, \dots, \Omega_m)$, and let, for every $h \in \mathbf{N}$ and a.e. $x \in \cup_{i=1}^m \Omega_i$, $\lambda_h^\varepsilon(x) = \frac{1}{2 \sum_{i=1}^m \psi_h^{\varepsilon,i}(x)}$,

$$\lambda_h^{\varepsilon,j}(x) = \frac{\psi_h^{\varepsilon,j}(x)}{2(\sum_{i=1}^m \psi_h^{\varepsilon,i}(x))^2}.$$

Because of (11.2.14) and (11.2.13), of the convexity of f, of the symmetry properties of $f^{(0)}$, and by observing that, for every $h \in \mathbf{N}$ and a.e. $x \in \cup_{i=1}^m \Omega_i$, $0 \leq \lambda_h^\varepsilon(x) \leq \frac{1}{2}$ and $0 \leq \lambda_h^{\varepsilon,j}(x) \leq \frac{1}{2}$, it results that

$$(11.2.15) \quad f^{(0)} \left(hx, \frac{\delta\varepsilon}{256n3^n \sqrt{n}\sigma_\varepsilon} \nabla\gamma_h^{\varepsilon,j}(x) \right) =$$

$$= f^{(0)} \left(hx, \lambda_h^\varepsilon(x) \frac{\delta\varepsilon}{128n3^n \sqrt{n}\sigma_\varepsilon} \nabla\psi_h^{\varepsilon,j}(x) - \right.$$

$$\left. - \lambda_h^{\varepsilon,j}(x) \frac{\delta\varepsilon}{128n3^n \sqrt{n}\sigma_\varepsilon} \sum_{i=1}^m \nabla\psi_h^{\varepsilon,i}(x) \right) \leq$$

$$\leq \lambda_h^\varepsilon(x) f^{(0)} \left(hx, \frac{\delta\varepsilon}{128n3^n \sqrt{n}\sigma_\varepsilon} \nabla\psi_h^{\varepsilon,j}(x) \right) +$$

$$+\lambda_h^{\varepsilon,j}(x)f^{(0)}\left(hx,\frac{\delta\varepsilon}{128n3^n\sqrt{n}}\sum_{i=1}^m\frac{1}{\sigma_\varepsilon}\nabla\psi_h^{\varepsilon,i}(x)\right)+$$

$$+\left(1-\lambda_h^\varepsilon(x)-\lambda_h^{\varepsilon,j}(x)\right)f^{(0)}(hx,0)\le$$

$$\le f^{(0)}\left(hx,\frac{\delta\varepsilon}{128n3^n\sqrt{n}\sigma_\varepsilon}\nabla\psi_h^{\varepsilon,j}(x)\right)+$$

$$+f^{(0)}\left(hx,\frac{\delta\varepsilon}{128n3^n\sqrt{n}}\sum_{i=1}^m\frac{1}{\sigma_\varepsilon}\nabla\psi_h^{\varepsilon,i}(x)\right)+2f(hx,0)$$

for a.e. $x\in\cup_{i=1}^m\Omega_i$, and every $h\in\mathbf{N}$.

Let us consider separately the first two terms after the last inequality in (11.2.15). Since $\frac{1}{\sigma_\varepsilon}\in\,]0,1]$, the convexity of f provides that

(11.2.16) $$f^{(0)}\left(hx,\frac{\delta\varepsilon}{128n3^n\sqrt{n}\sigma_\varepsilon}\nabla\psi_h^{\varepsilon,j}(x)\right)\le$$

$$\le f^{(0)}\left(hx,\frac{\delta\varepsilon/2}{64n3^n\sqrt{n}}\nabla\psi_h^{\varepsilon,j}(x)\right)+2f(hx,0)$$

for a.e. $x\in\cup_{i=1}^m\Omega_i$, and every $h\in\mathbf{N}$.

On the other hand, for a fixed $l\in\{1,\dots,m\}$, the number of the sets Ω_i, with $i\in\{1,\dots,m\}$, such that $\mathrm{dist}(\Omega_i,\Omega_l)<\varepsilon$ is less than or equal to σ_ε. Let $\{\Omega_{i_1},\dots,\Omega_{i_{\sigma_\varepsilon}}\}$ be a subset of $\{\Omega_1,\dots,\Omega_m\}$ containing all the sets Ω_i that satisfy $\mathrm{dist}(\Omega_i,\Omega_l)<\varepsilon$. Consequently, from the convexity of f, it follows that

(11.2.17) $$f^{(0)}\left(hx,\frac{\delta\varepsilon}{128n3^n\sqrt{n}}\sum_{i=1}^m\frac{1}{\sigma_\varepsilon}\nabla\psi_h^{\varepsilon,i}(x)\right)=$$

$$=f^{(0)}\left(hx,\frac{\delta\varepsilon}{128n3^n\sqrt{n}}\sum_{k=1}^{\sigma_\varepsilon}\frac{1}{\sigma_\varepsilon}\nabla\psi_h^{\varepsilon,i_k}(x)\right)\le$$

$$\le\sum_{k=1}^{\sigma_\varepsilon}\frac{1}{\sigma_\varepsilon}f^{(0)}\left(hx,\frac{\delta\varepsilon/2}{64n3^n\sqrt{n}}\nabla\psi_h^{\varepsilon,i_k}(x)\right)\quad\text{for a.e. }x\in\Omega_l,\text{ and every }h\in\mathbf{N}.$$

Let now A be as in (11.2.12). Then by combining Lemma 11.1.1 applied to $f^{(0)}$ with (11.2.15)÷(11.2.17), by using (11.2.5), and by recalling that $\mathrm{dist}(\Omega_{j,\varepsilon}^+,\Omega_{j,\frac{\varepsilon}{2}}^+)=\frac{\varepsilon}{2}$, we obtain that

$$\limsup_{h\to+\infty}\int_{A\cap\Omega_l}f^{(0)}\left(hx,\frac{\delta\varepsilon}{256n3^n\sqrt{n}}\frac{1}{\sigma_\varepsilon}\nabla\gamma_h^{\varepsilon,j}\right)dx\le$$

$$\le 4\mathcal{L}^n(A\cap\Omega_l)\int_Y f(y,0)dy+c_f\mathcal{L}^n(A\cap\Omega_l)+\sum_{i=1}^{\sigma_\varepsilon}\frac{1}{\sigma_\varepsilon}c_f\mathcal{L}^n(A\cap\Omega_l)=$$

$$=2\left(2\int_Y f(y,0)dy+c_f\right)\mathcal{L}^n(A\cap\Omega_l)\quad\text{for every }l\in\{1,\dots,m\},$$

from which (11.2.12) easily follows. ∎

Chapter 12

Homogenization
of Unbounded Functionals

In this chapter we analyze the homogenization process as $\varepsilon \to 0^+$ for energy functionals of the kind $u \mapsto \int_\Omega f(\frac{x}{\varepsilon}, \nabla u)dx$, where the densities f are actually unbounded, and satisfy

(12.0.1) $\quad \begin{cases} f \colon (x,z) \in \mathbf{R}^n \times \mathbf{R}^n \mapsto f(x,z) \in [0,+\infty] \\ f \ (\mathcal{L}_n(\mathbf{R}^n) \times \mathcal{B}(\mathbf{R}^n))\text{-measurable} \\ f \ Y\text{-periodic in the } x \text{ variable, convex in the } z \text{ one.} \end{cases}$

In the same order of ideas of [CCDAG1] and [CCDAG2], we develop here a general study of the homogenization of integral energies with densities as in (12.0.1), but under high coerciveness assumptions due to the fact that (12.0.1) do not involve, as x varies, any kind of control on the sets where the partial functions $f(x,\cdot)$ take the value $+\infty$, not even on the behaviour of $f(x,\cdot)$ itself near the boundary of such sets.

On the contrary, in the next chapter, we will treat cases in which some controls on the above quantities are assumed, and less restrictive coerciveness assumptions are needed.

Energies of the above type appear in the treatment of various problems of applied mathematics, as recalled in Chapter 6. Because of this, we develop homogenization processes for the treatment of various classes of minimum problems, for example of Dirichlet, Neumann, and mixed type, with boundary conditions that look to be natural in the problems suggested by the models recalled in §6.5. These homogenization processes also provide an answer to a conjecture stated in [BLP, §17 of Chapter 1].

§12.1 Notations and Basic Results

Let f be as in (12.0.1).

We recall that, for every $q \in [1, +\infty]$, the function $\tilde{f}^q_{\text{hom}}$ is defined in Chapter 11 by

$$(12.1.1) \quad \tilde{f}^q_{\text{hom}} : z \in \mathbf{R}^n \mapsto \inf\left\{ \int_Y f(y, z + \nabla v)dy : v \in W^{1,q}_{\text{per}}(Y) \cap L^\infty(Y) \right\},$$

and that it turns out to be convex.

For every $r \in]0, +\infty[$, $q \in [1, +\infty]$, $\{r_h\} \subseteq [0, +\infty[$, $\Omega \in \mathcal{A}_0$, $\Gamma \subseteq \partial\Omega$, and $u_0 \in W^{1,1}_{\text{loc}}(\mathbf{R}^n)$ we define the following functionals on $L^\infty_{\text{loc}}(\mathbf{R}^n)$

$$(12.1.2) \quad F_r(\Omega, \cdot) : u \in L^\infty_{\text{loc}}(\mathbf{R}^n) \mapsto$$

$$\begin{cases} \int_\Omega f(rx, \nabla u)dx & \text{if } u \in W^{1,q}_{\text{loc}}(\mathbf{R}^n) \cap L^\infty_{\text{loc}}(\mathbf{R}^n) \\ +\infty & \text{otherwise,} \end{cases}$$

$$(12.1.3) \quad F_r(\Omega, \Gamma, u_0, \cdot) : u \in L^\infty_{\text{loc}}(\mathbf{R}^n) \mapsto$$

$$\begin{cases} \int_\Omega f(rx, \nabla u)dx & \text{if } u \in u_0 + W^{1,q}_{0,\Gamma}(\Omega) \cap L^\infty_{\text{loc}}(\mathbf{R}^n) \\ +\infty & \text{otherwise,} \end{cases}$$

and set

$$(12.1.4) \quad \begin{cases} \widetilde{F}'(\Omega, \cdot) : u \in L^\infty_{\text{loc}}(\mathbf{R}^n) \mapsto \Gamma^-(L^\infty(\Omega)) \liminf_{h \to +\infty} F_{r_h}(\Omega, u) \\ \\ \widetilde{F}''(\Omega, \cdot) : u \in L^\infty_{\text{loc}}(\mathbf{R}^n) \mapsto \Gamma^-(L^\infty(\Omega)) \limsup_{h \to +\infty} F_{r_h}(\Omega, u), \end{cases}$$

$$(12.1.5) \quad \begin{cases} \widetilde{F}'(\Omega, \Gamma, u_0, \cdot) : u \in L^\infty_{\text{loc}}(\mathbf{R}^n) \mapsto \\ \qquad\qquad \Gamma^-(L^\infty(\Omega)) \liminf_{h \to +\infty} F_{r_h}(\Omega, \Gamma, u_0, u) \\ \\ \widetilde{F}''(\Omega, \Gamma, u_0, \cdot) : u \in L^\infty_{\text{loc}}(\mathbf{R}^n) \mapsto \\ \qquad\qquad \Gamma^-(L^\infty(\Omega)) \limsup_{h \to +\infty} F_{r_h}(\Omega, \Gamma, u_0, u). \end{cases}$$

Moreover, we also set

$$(12.1.6) \quad \begin{cases} F'(\Omega, \cdot) : u \in L^\infty_{\text{loc}}(\mathbf{R}^n) \mapsto \Gamma^-(L^\infty(\Omega)) \liminf_{h \to +\infty} F_h(\Omega, u) \\ \\ F''(\Omega, \cdot) : u \in L^\infty_{\text{loc}}(\mathbf{R}^n) \mapsto \Gamma^-(L^\infty(\Omega)) \limsup_{h \to +\infty} F_h(\Omega, u). \end{cases}$$

It is clear that all the functionals in (12.1.2)÷(12.1.6) depend also on q even if, for the sake of simplicity, we omit an explicit indication of it.

Nevertheless, we point out that the index q measures the regularity of the admissible configurations, and that the dependence on it in the corresponding homogenization results may be true. In fact, it is well known that a Lavrentiev phenomenon may appear, and even survive the homogenization processes (cf. [CEDA1], [CESC], [DDG]).

Because of (12.0.1) and of Proposition 3.4.1 it follows that

$$(12.1.7) \qquad \widetilde{F}'(\cdot, u), \ \widetilde{F}''(\cdot, u) \text{ are increasing}$$

for every $u \in L^\infty_{loc}(\mathbf{R}^n)$, and every $\{r_h\} \subseteq [0, +\infty[$,

and

$$(12.1.8) \qquad \widetilde{F}'(\Omega, \cdot), \ \widetilde{F}''(\Omega, \cdot) \text{ are convex}$$

for every $\Omega \in \mathcal{A}_0$, and every $\{r_h\} \subseteq [0, +\infty[$.

Moreover, the following properties are straightaway verified.

$$(12.1.9) \qquad \widetilde{F}'(\Omega, u_1) = \widetilde{F}'(\Omega, u_2), \ \widetilde{F}''(\Omega, u_1) = \widetilde{F}''(\Omega, u_2)$$

whenever $\{r_h\} \subseteq [0, +\infty[, \ \Omega \in \mathcal{A}_0$

$u_1, u_2 \in L^\infty_{loc}(\mathbf{R}^n)$ satisfy $u_1 = u_2$ a.e. in Ω,

$$(12.1.10) \qquad \widetilde{F}'(\Omega, u+c) = \widetilde{F}'(\Omega, u), \ \widetilde{F}''(\Omega, u+c) = \widetilde{F}''(\Omega, u)$$

for every $\{r_h\} \subseteq [0, +\infty[, \ \Omega \in \mathcal{A}_0, \ u \in L^\infty_{loc}(\mathbf{R}^n), \ c \in \mathbf{R}$.

Proposition 12.1.1. Let f be as in (12.0.1), $q \in [1, +\infty]$, and let $\widetilde{F}'$ and $\widetilde{F}''$ be defined in (12.1.4). Then

$$\widetilde{F}'_-(\Omega \quad x_0, T[x_0]u) - \widetilde{F}'_-(\Omega, u), \ \widetilde{F}''_-(\Omega - x_0, T[x_0]u) = \widetilde{F}''_-(\Omega, u)$$

for every $\{r_h\} \subseteq]0, +\infty[$ increasing and diverging, $\Omega \in \mathcal{A}_0$,

$$x_0 \in \mathbf{R}^n, \ u \in C^0(\mathbf{R}^n).$$

Proof. We prove the first equality, the second being analogous.

Let $\{r_h\}, \Omega, x_0, u$ be as above. Let us prove that

$$(12.1.11) \qquad \widetilde{F}'_-(\Omega - x_0, T[x_0]u) \geq \widetilde{F}'_-(\Omega, u)$$

To do this, let us assume that the left-hand side of (12.1.11) is finite. Let us take $O, B \in \mathcal{A}(\Omega)$ with $O \subset\subset B \subset\subset \Omega$. Then, there exists $\{u_h\} \subseteq W^{1,q}_{loc}(\mathbf{R}^n) \cap L^\infty_{loc}(\mathbf{R}^n)$ such that $u_h \to T[x_0]u$ in $L^\infty(B - x_0)$ and

$$\widetilde{F}'(B - x_0, T[x_0]u) = \liminf_{k \to +\infty} \int_{B-x_0} f(r_h x, \nabla u_h)dx.$$

For every $h \in \mathbf{N}$ let $m_h \in \mathbf{Z}^n$ be such that $\frac{m_h}{r_h} \to x_0$. Then by performing in the above integrals the change of variables $y = x + \frac{m_h}{r_h}$, and by exploiting the periodicity properties of f, we obtain that

$$(12.1.12) \qquad \widetilde{F}'(B - x_0, T[x_0]u) =$$

$$= \liminf_{h \to +\infty} \int_{B + \frac{m_h}{r_h} - x_0} f\left(r_h\left(y - \frac{m_h}{r_h} \right), \nabla_x u_h \left(y - \frac{m_h}{r_h} \right) \right) dy =$$

$$= \liminf_{h \to +\infty} \int_{B + \frac{m_h}{r_h} - x_0} f\left(r_h y, \nabla_y \left(T\left[-\frac{m_h}{r_h} \right] u_h \right)(y) \right) dy.$$

We now observe that $O \subseteq B + \frac{m_h}{r_h} - x_0$ provided h is large enough, and that, because of the continuity of u, it turns out that $T[-\frac{m_h}{r_h}]u_h \to u$ in $L^\infty(O)$. Consequently, by (12.1.12) we infer that

$$\widetilde{F}'_-(\Omega - x_0, T[x_0]u) \geq \liminf_{h \to +\infty} \int_O f\left(r_h y, \nabla_y \left(T\left[-\frac{m_h}{r_h} \right] u_h \right)(y) \right) dy \geq$$

$$\geq \widetilde{F}'(O, u) \text{ for every } O \subset\subset \Omega,$$

from which (12.1.11) follows.

By symmetry, the reverse inequality to (12.1.11) follows. This completes the proof. ∎

Lemma 12.1.2. *Let f be as in (12.0.1), $q \in [1, +\infty]$, and let F', F'', $\widetilde{F}'$, and $\widetilde{F}''$ be defined in (12.1.6) and (12.1.4). Then*

$$(12.1.13) \qquad F'_-(\Omega, u) \leq \widetilde{F}'_-(\Omega, u), \ \widetilde{F}''_-(\Omega, u) \leq F''_-(\Omega, u)$$

for every $\{r_h\} \subseteq \,]0, +\infty[$ diverging, $\Omega \in \mathcal{A}_0$, $u \in C^0(\mathbf{R}^n)$.

Proof. Let $\{r_h\}$, Ω, u be as in (12.1.13) and set, for every $h \in \mathbf{N}$, $k_h = [r_h]$. Then $\lim_{h \to +\infty} \frac{k_h}{r_h} = 1$.

In order to prove the first inequality in (12.1.13) we observe that we can obviously assume that $\widetilde{F}'_-(\Omega, u) < +\infty$ so that, if $\Omega'' \in \mathcal{A}_0$ satisfies $\Omega'' \subset\subset \Omega$, there exist $\{h_j\} \subseteq \mathbf{N}$ strictly increasing, and $\{u_h\} \subseteq W^{1,q}_{loc}(\mathbf{R}^n) \cap L^\infty_{loc}(\mathbf{R}^n)$ such that $u_h \to u$ in $L^\infty(\Omega'')$, and

$$(12.1.14) \qquad \widetilde{F}'(\Omega'', u) \geq \lim_{j \to +\infty} \int_{\Omega''} f(r_{h_j} x, \nabla u_{h_j}) dx.$$

Let $\Omega' \in \mathcal{A}_0$ with $\Omega' \subset\subset \Omega''$. For every $j \in \mathbf{N}$ we perform in the integrals in (12.1.14) the change of variable $x = \frac{k_{h_j}}{r_{h_j}} y$, set $v_h = \frac{r_h}{k_h} u_h(\frac{k_h}{r_h} \cdot)$, and observe that, provided h is large enough, $\Omega' \subseteq \frac{r_h}{k_h} \Omega''$. Because of this, and by the continuity of u, we have that $v_h \to u$ in $L^\infty(\Omega')$ and, by (12.1.14) and (3.2.5), that

$$(12.1.15) \qquad \widetilde{F}'_-(\Omega, u) \geq$$

$$\geq \lim_{j \to +\infty} \left(\frac{k_{h_j}}{r_{h_j}}\right)^n \int_{\frac{r_{h_j}}{k_{h_j}}\Omega''} f\left(k_{h_j}y, \nabla_x u_{h_j}\left(\frac{k_{h_j}}{r_{h_j}}y\right)\right) dy \geq$$

$$\geq \liminf_{j \to +\infty} \int_{\Omega'} f(k_{h_j}y, \nabla_y v_{h_j}) dy \geq \Gamma^-(L^\infty(\Omega')) \liminf_{h \to +\infty} F_{k_h}(\Omega', u) \geq$$

$$\geq F'(\Omega', u).$$

By (12.1.15) we deduce the first inequality in (12.1.13) as Ω' increases to Ω.

In order to prove the second inequality in (12.1.13), we can assume that $F''(\Omega, u) < +\infty$ so that, because of (3.2.5), $\sup\{\Gamma^-(L^\infty(A)) \limsup_{h \to +\infty} F_{k_h}(A, u) : A \subset\subset \Omega\} < +\infty$.

Let Ω'', Ω' be as before. Then there exists $\{u_h\} \subseteq W^{1,q}_{\text{loc}}(\mathbf{R}^n) \cap L^\infty_{\text{loc}}(\mathbf{R}^n)$ with $u_h \to u$ in $L^\infty(\Omega'')$, and

$$(12.1.16) \quad \Gamma^-(L^\infty(\Omega'')) \limsup_{h \to +\infty} F_{k_h}(\Omega'', u) \geq \limsup_{h \to +\infty} \int_{\Omega''} f(k_h x, \nabla u_h) dx.$$

For every $h \in \mathbf{N}$ we perform in the integrals in (12.1.16) the change of variable $x = \frac{r_h}{k_h}y$, set $v_h = \frac{k_h}{r_h}u_h(\frac{r_h}{k_h}\cdot)$, and observe that, provided h is large enough, $\frac{k_h}{r_h}\Omega' \subseteq \Omega''$. Because of this, and by the continuity of u, we have that $v_h \to u$ in $L^\infty(\Omega')$, and by (12.1.16) and (3.2.5), that

$$(12.1.17) \qquad \widetilde{F}'(\Omega', u) \leq \limsup_{h \to +\infty} \int_{\Omega'} f(r_h y, \nabla_y v_h) dy \leq$$

$$\leq \limsup_{h \to +\infty} \left(\frac{r_h}{k_h}\right)^n \int_{\frac{k_h}{r_h}\Omega'} f(k_h x, \nabla_x u_h) dx \leq$$

$$\leq \sup\left\{\Gamma^-(L^\infty(A)) \limsup_{h \to +\infty} F_{k_h}(A, u) : A \subset\subset \Omega\right\} \leq F''_-(\Omega, u).$$

By (12.1.17) we deduce the second inequality in (12.1.13) as Ω' increases to Ω. ∎

As usual in homogenization problems, we introduce for every $q \in [1, +\infty]$, the function f^q_{hom} defined by

$$(12.1.18) \qquad f^q_{\text{hom}}: z \in \mathbf{R}^n \mapsto \inf\left\{\int_Y f(y, z + \nabla v) dy : v \in W^{1,q}_{\text{per}}(Y)\right\}.$$

Then it is clear that for every $q \in [1, +\infty]$, f^q_{hom} is convex, that $f^q_{\text{hom}} \leq \tilde{f}^q_{\text{hom}}$, and that

(12.1.19) $\qquad\qquad f^q_{\text{hom}} = \tilde{f}^q_{\text{hom}}$ provided $q \in]n, +\infty]$,

where $\tilde{f}^q_{\text{hom}}$ is given by (12.1.1).

We point out that sometimes we improperly refer to the definition of f^q_{hom} in (12.1.18) as to the *homogenization formula*.

For every $q \in [1, +\infty]$, and $z_0 \in \mathbf{R}^n$, we also set

(12.1.20) $\qquad C^q(z_0) = \text{dom}(f^{(z_0)})^q_{\text{hom}} = \Big\{ z \in \mathbf{R}^n : \text{ there exists}$

$$v \in W^{1,q}_{\text{per}}(Y) \text{ with } \int_Y f^{(z_0)}(y, z + \nabla v) dy < +\infty \Big\}.$$

Then it is clear that for every $q \in [1, +\infty]$ and $z_0 \in \mathbf{R}^n$, $C^q(z_0)$ is convex, that $\tilde{C}^q(z_0) \subseteq C^q(z_0)$, and that

(12.1.21) $\qquad\qquad C^q(z_0) = \tilde{C}^q(z_0)$ provided $q \in]n, +\infty]$,

where $\tilde{C}^q(z_0)$ is defined in (11.2.2).

The next result collects some properties of the functions defined by (12.1.18). In it we assume that

(12.1.22) $\begin{cases} |z|^p \leq f(x, z) \text{ for a.e. } x \in \mathbf{R}^n \text{ and every } z \in \mathbf{R}^n \\ \qquad\qquad\qquad\qquad\qquad\qquad\qquad\qquad\quad \text{if } p \in [1, +\infty[\\ \text{dom} f(x, \cdot) \subseteq B_R(0) \text{ for a.e. } x \in \mathbf{R}^n \qquad \text{if } p = +\infty \end{cases}$

and that

(12.1.23) $\qquad\qquad f(x, \cdot)$ is lower semicontinuous for a.e. $x \in \mathbf{R}^n$,

for some $R > 0$.

Proposition 12.1.3. *Let f be as in* (12.0.1), $q \in [1, +\infty]$, *and let f^q_{hom} be defined in* (12.1.18). *Then f^q_{hom} is convex. Let now $p \in [1, +\infty]$, $q \in [p, +\infty]$, and assume that* (12.1.22) *holds. Then f^q_{hom} satisfies*

(12.1.24) $\begin{cases} |z|^p \leq f^q_{\text{hom}}(z) \text{ for every } z \in \mathbf{R}^n & \text{if } p \in [1, +\infty[\\ \text{dom} f^q_{\text{hom}} \subseteq B_R(0) & \text{if } p = +\infty. \end{cases}$

Finally, if $p \in]1, +\infty]$, $q = p$ and (12.1.23) *holds, then f^q_{hom} is also lower semicontinuous, and*

$$f^p_{\text{hom}}(z) = \min\Big\{ \int_Y f(y, z + \nabla v) dy : v \in W^{1,p}_{\text{per}}(Y) \Big\} \text{ for every } z \in \mathbf{R}^n.$$

Proof. The convexity of f_{hom}^q has already been observed.

Let us prove (12.1.24). To this aim, we first assume that $p < +\infty$, and observe that the Trace Theorem for Sobolev Functions and Jensen's inequality imply that

$$(12.1.25) \qquad |z|^p = \left| \int_Y z \, dy + \int_{\partial Y} \gamma_Y v \mathbf{n}_Y d\mathcal{H}^{n-1} \right|^p =$$

$$= \left| \int_Y (z + \nabla v) dy \right|^p \leq \int_Y |z + \nabla v|^p dy \quad \text{for every } v \in W_{\text{per}}^{1,1}(Y).$$

Then (12.1.25) and (12.1.22) yield that

$$|z|^p = \min \left\{ \int_Y |z + \nabla v|^p dy : v \in W_{\text{per}}^{1,q}(Y) \right\} \leq$$

$$\leq \inf \left\{ \int_Y f(y, z + \nabla v) dy : v \in W_{\text{per}}^{1,q}(Y) \right\} = f_{\text{hom}}^q(z) \quad \text{for every } z \in \mathbf{R}^n,$$

from which the first estimate in (12.1.24) follows.

When $p = +\infty$, then also $q = +\infty$. Let $z \in \text{dom} f_{\text{hom}}^{+\infty}$, and $w \in W_{\text{per}}^{1,\infty}(Y)$ be such that $z + \nabla w(x) \in \text{dom} f(x, \cdot)$ for a.e. $x \in Y$. Then, again by (12.1.25) with $p = 1$, and (12.1.22), we obtain that

$$|z| = \min \left\{ \int_Y |z + \nabla v| dy : v \in W_{\text{per}}^{1,\infty}(Y) \right\} \leq \int_Y |z + \nabla w| dy \leq R$$

$$\text{for every } z \in \text{dom} f_{\text{hom}}^{+\infty},$$

that is the second estimate in (12.1.24).

Let us assume now that $q = p \in]1, +\infty]$. Let $z \in \mathbf{R}^n$, and $\{z_h\} \subseteq \mathbf{R}^n$ be such that $z_h \to z$ and $\liminf_{h \to +\infty} f_{\text{hom}}^p(z_h) < +\infty$. Then there exists $\{h_k\} \subseteq \mathbf{N}$ strictly increasing such that, for every $k \in \mathbf{N}$, there is $v_k \in W_{\text{per}}^{1,p}(Y)$ with $\int_Y v_k dy = 0$, and $\lim_{k \to +\infty} \int_Y f(y, z_{h_k} + \nabla v_k) dy = \liminf_{h \to +\infty} f_{\text{hom}}^p(z_h) < +\infty$. Because of this, (12.1.22) and Theorem 4.3.19, $\{v_k\}$ turns out to be bounded in $W^{1,p}(Y)$. Consequently, by Proposition 4.5.1, it follows that there exists $v \in W_{\text{per}}^{1,p}(Y)$ such that, up to subsequences, $v_k \to v$ in weak-$W^{1,p}(Y)$ (in weak*-$W^{1,\infty}(Y)$ if $p = +\infty$). Then, by (12.0.1), (12.1.23), and Theorem 5.2.2 we obtain that

$$f_{\text{hom}}^p(z) \leq \int_Y f(y, z + \nabla v) dy \leq$$

$$\leq \lim_{k \to +\infty} \int_Y f(y, z_{h_k} + \nabla v_k) dy = \liminf_{h \to +\infty} f_{\text{hom}}^p(z_h),$$

from which the lower semicontinuity of f_{hom}^p follows.

In conclusion, an argument similar to the one just exploited yields that, for every $z \in \mathbf{R}^n$, the infimum in the definition of f_{hom}^p is attained. ∎

§12.2 Some Properties of Γ-Limits

Let f be as in (12.0.1), $q \in [1, +\infty]$.

For every $h \in \mathbf{N}$ let F_h be defined by (12.1.2). In this section we study some measure theoretic properties of the Γ-limits in (12.1.4) for u fixed in $L_{\text{loc}}^\infty(\mathbf{R}^n)$, and investigate the relationships between the limits in (12.1.4) and (12.1.5).

To this purpose we consider $\tilde{f}_{\text{hom}}^q$ given by (12.1.1), assume that

$$(12.2.1) \qquad \qquad \text{int}(\text{dom} \tilde{f}_{\text{hom}}^q) \neq \emptyset.$$

Proposition 12.2.1. *Let f be as in (12.0.1), $q \in [1, +\infty]$, $\{r_h\} \subseteq]0, +\infty[$ be strictly increasing and diverging, $\tilde{F}'$, $\tilde{F}''$ be defined in (12.1.4), and Ω, Ω_1, $\Omega_2 \in \mathcal{A}_0$.*

If $\Omega_1 \cap \Omega_2 = \emptyset$, and $\Omega_1 \cup \Omega_2 \subseteq \Omega$, then

$$(12.2.2) \qquad \tilde{F}'_-(\Omega, u) \geq \tilde{F}'_-(\Omega_1, u) + \tilde{F}'_-(\Omega_2, u) \ \text{for every } u \in L_{\text{loc}}^\infty(\mathbf{R}^n).$$

If $\Omega \subseteq \Omega_1 \cup \Omega_2$, and (12.2.1) holds, then

$$(12.2.3) \qquad \tilde{F}''_-(\Omega, u) \leq \tilde{F}''_-(\Omega_1, u) + \tilde{F}''_-(\Omega_2, u) \ \text{for every } u \in L_{\text{loc}}^\infty(\mathbf{R}^n).$$

Proof. Inequality (12.2.2) follows directly from the definition of $\tilde{F}'_-$.

To prove (12.2.3) we can assume that (11.1.3) holds, otherwise, taken $z_0 \in \text{int}(\text{dom} \tilde{f}_{\text{hom}}^q)$, it suffices to replace f with $f(\cdot, z_0 + \cdot)$. Moreover, it also suffices to consider the case in which $\Omega \subset\subset \Omega_1 \cup \Omega_2$, and to prove that

$$(12.2.4) \qquad \tilde{F}''(\Omega, u) \leq \tilde{F}''(\Omega_1, u) + \tilde{F}''(\Omega_2, u) \ \text{for every } u \in L_{\text{loc}}^\infty(\mathbf{R}^n).$$

Fix $u \in L_{\text{loc}}^\infty(\mathbf{R}^n)$, and assume that the right-hand side of (12.2.4) is finite. Consequently, for $i = 1, 2$, there exists $\{u_h^{(i)}\} \subseteq W_{\text{loc}}^{1,q}(\mathbf{R}^n) \cap L_{\text{loc}}^\infty(\mathbf{R}^n)$ such that $u_h^{(i)} \to u$ in $L^\infty(\Omega_i)$ and

$$(12.2.5) \qquad \limsup_{h \to +\infty} \int_{\Omega_i} f\left(r_h x, \nabla u_h^{(i)}\right) dx \leq \tilde{F}''(\Omega_i, u).$$

Since $\Omega \subset\subset \Omega_1 \cup \Omega_2$, there exists $A_1 \subset\subset \Omega_1$ such that $\Omega \subset\subset A_1 \cup \Omega_2$.

Let $\{\psi_h\}$ be given by Lemma 11.1.1 applied to Ω_1 and $K = \overline{A_1}$, and let $\{w_h\} \subseteq W_{\text{loc}}^{1,q}(\mathbf{R}^n) \cap L_{\text{loc}}^\infty(\mathbf{R}^n)$ be defined by

$$w_h = \psi_h \left(u_h^{(1)} + \varepsilon_h\right) + (1 - \psi_h) u_h^{(2)},$$

where $\varepsilon_h = \|u_h^{(2)} - u_h^{(1)}\|_{L^\infty(\Omega_1 \cap \Omega_2)}$ for every $h \in \mathbf{N}$.

Then it is clear that $w_h \to u$ in $L^\infty(\Omega)$.

Fix $t \in [0, 1[$. Then, by making use of the convexity properties of f, and by recalling that $\Omega \setminus A_1 \subset\subset \Omega_2$, it results that

$$(12.2.6) \qquad \widetilde{F}''(\Omega, tu) \le \limsup_{h \to +\infty} \int_\Omega f(r_h x, t\nabla w_h) dx =$$

$$= \limsup_{h \to +\infty} \int_\Omega f\left(r_h x, t\left(\psi_h \nabla u_h^{(1)} + (1 - \psi_h)\nabla u_h^{(2)} + \right.\right.$$

$$\left.\left. \left(u_h^{(1)} + \varepsilon_h - u_h^{(2)}\right)\nabla \psi_h\right)\right) dx \le$$

$$\le \limsup_{h \to +\infty} \left\{ t \int_\Omega f\left(r_h x, \psi_h \nabla u_h^{(1)} + (1 - \psi_h)\nabla u_h^{(2)}\right) dx + \right.$$

$$\left. + (1 - t) \int_\Omega f\left(r_h x, \frac{t}{1 - t}\left(u_h^{(1)} + \varepsilon_h - u_h^{(2)}\right)\nabla \psi_h\right) dx \right\} \le$$

$$\le \limsup_{h \to +\infty} \int_\Omega \psi_h(x) f\left(r_h x, \nabla u_h^{(1)}\right) dx +$$

$$+ \limsup_{h \to +\infty} \int_\Omega (1 - \psi_h(x)) f\left(r_h x, \nabla u_h^{(2)}\right) dx +$$

$$+ (1 - t)\limsup_{h \to +\infty} \int_\Omega f\left(r_h x, \frac{t}{1 - t}\left(u_h^{(1)} + \varepsilon_h - u_h^{(2)}\right)\nabla \psi_h\right) dx \le$$

$$\le \limsup_{h \to +\infty} \int_{\Omega_1} f\left(r_h x, \nabla u_h^{(1)}\right) dx + \limsup_{h \to +\infty} \int_{\Omega_2} f\left(r_h x, \nabla u_h^{(2)}\right) dx +$$

$$+ (1 - t)\limsup_{h \to +\infty} \int_{\Omega \cap (\Omega_1 \setminus \overline{A_1})} f\left(r_h x, \frac{t}{1 - t}\left(u_h^{(1)} + \varepsilon_h - u_h^{(2)}\right)\nabla \psi_h\right) dx +$$

$$+ (1 - t)\limsup_{h \to +\infty} \int_{\Omega \setminus (\Omega_1 \setminus \overline{A_1})} f(r_h x, 0) dx.$$

On the other hand, since $\Omega \cap (\Omega_1 \setminus \overline{A_1}) \subset\subset \Omega_1 \cap \Omega_2$, it results that $u_h^{(1)} + \varepsilon_h - u_h^{(2)} \to 0$ in $L^\infty(\Omega \cap (\Omega_1 \setminus \overline{A_1}))$, and $u_h^{(1)} + \varepsilon_h - u_h^{(2)} \ge 0$ a.e. in $\Omega \cap (\Omega_1 \setminus \overline{A_1})$. Consequently, there exists $h_t \in \mathbf{N}$ such that

$$\frac{t}{1 - t}\left(u_h^{(1)}(x) + \varepsilon_h - u_h^{(2)}(x)\right) \in \left[0, \frac{\delta \text{dist}(A_1, \partial \Omega_1)}{64 n 3^n \sqrt{n}}\right[$$

for a.e. $x \in \Omega \cap (\Omega_1 \setminus \overline{A_1})$, and every $h \ge h_t$,

where $\delta \in]0, 1[$ satisfies (11.2.4). Because of this, we get that for a.e. $x \in \Omega \cap (\Omega_1 \setminus \overline{A_1})$, and every $h \ge h_t$ the vector $\frac{t}{1 - t}(u_h^{(1)}(x) + \varepsilon_h - u_h^{(2)}(x))\nabla \psi_h(x)$

is a convex combination of 0 and $\frac{\delta \text{dist}(A_1, \partial\Omega_1)}{64n3^n\sqrt{n}}\nabla\psi_h(x)$. Therefore, by the convexity of f in the second group of variables, it follows that

$$(12.2.7) \qquad \int_{\Omega\cap(\Omega_1\setminus\overline{A_1})} f\left(r_h x, \frac{t}{1-t}\left(u_h^{(1)} + \varepsilon_h - u_h^{(2)}\right)\nabla\psi_h\right) dx \leq$$

$$\leq \int_{\Omega\cap(\Omega_1\setminus\overline{A_1})} f(r_h x, 0)dx +$$

$$+\int_{\Omega\cap(\Omega_1\setminus\overline{A_1})} f^{(0)}\left(r_h x, \frac{\delta \text{dist}(A_1, \partial\Omega_1)}{64n3^n\sqrt{n}}\nabla\psi_h\right) dx \text{ for every } h \geq h_t.$$

from which, by making use of (11.1.10) of Lemma 11.1.1, we infer that

$$(12.2.8) \quad \limsup_{h\to+\infty} \int_{\Omega\cap(\Omega_1\setminus\overline{A_1})} f\left(r_h x, \frac{t}{1-t}\left(u_h^{(1)} + \varepsilon_h - u_h^{(2)}\right)\nabla\psi_h\right) dx \leq$$

$$\leq \left(c_f + \int_Y f(y,0)dy\right)\mathcal{L}^n\left(\Omega\cap(\Omega_1\setminus\overline{A_1})\right) \leq \left(c_f + \int_Y f(y,0)dy\right)\mathcal{L}^n(\Omega),$$

where c_f is defined in Remark 11.1.2.

By combining (12.2.6) with (11.2.5), (12.2.5), and (12.2.8), it results that

$$(12.2.9) \qquad \widetilde{F}''(\Omega, tu) \leq \widetilde{F}''(\Omega_1, u) + \widetilde{F}''(\Omega_2, u) +$$

$$+(1-t)\left(c_f + \int_Y f(y,0)dy\right)\mathcal{L}^n(\Omega) + (1-t)\mathcal{L}^n(\Omega)\int_Y f(y,0)dy$$

$$\text{for every } t \in [0, 1[.$$

Finally, passing to the limit in (12.2.9) as t tends to 1^-, and making use of Proposition 3.3.2, inequality (12.2.3) follows. ∎

Proposition 12.2.2. *Let f be as in (12.0.1), $q \in [1, +\infty]$, $\{r_h\} \subseteq]0, +\infty[$ be strictly increasing and diverging, $\widetilde{F}'$, $\widetilde{F}''$ be defined in (12.1.4), $\Omega \in \mathcal{A}_0$, and $\widetilde{F}'(\Omega, \partial\Omega, 0, \cdot)$, $\widetilde{F}''(\Omega, \partial\Omega, 0, \cdot)$ be defined in (12.1.5). Assume that (11.1.3) holds. Then*

$$(12.2.10) \qquad \widetilde{F}'(\Omega, u) = \widetilde{F}'_-(\Omega, u) = \widetilde{F}'(\Omega, \partial\Omega, 0, u),$$

$$\widetilde{F}''(\Omega, u) = \widetilde{F}''_-(\Omega, u) = \widetilde{F}''(\Omega, \partial\Omega, 0, u)$$

for every $u \in L^\infty_{\text{loc}}(\mathbf{R}^n) \cap C^0(\overline{\Omega})$ such that $u = 0$ on $\partial\Omega$.

Proof. Let u be as in (12.2.10). We prove (12.2.10) for $\widetilde{F}''(\Omega, \cdot)$, $\widetilde{F}''_-(\Omega, \cdot)$, and $\widetilde{F}''(\Omega, \partial\Omega, 0, \cdot)$, the proof for $\widetilde{F}'(\Omega, \cdot)$, $\widetilde{F}'_-(\Omega, \cdot)$, and $\widetilde{F}'(\Omega, \partial\Omega, 0, \cdot)$ being analogous.

Let $\{\varepsilon_k\}$ be a decreasing sequence of positive numbers converging to zero, and $\{\chi_k\}$ be the sequence of functions defined by

$$(12.2.11) \qquad \chi_k \colon t \in \mathbf{R} \mapsto \min\{t + \varepsilon_k, \max\{t - \varepsilon_k, 0\}\}.$$

For every $k \in \mathbf{N}$ let $A_k, \Omega_k \in \mathcal{A}_0$ be such that $A_k \subset\subset A_{k+1}$, $\cup_{k=1}^{+\infty} A_k = \Omega$, Ω_k has Lipschitz boundary, $\Omega_k \subset\subset \Omega_{k+1}$, $A_k \subset\subset \Omega_k \subset\subset \Omega$ and

$$(12.2.12) \qquad \sup_{x \in \Omega \setminus A_k} |u(x)| < \frac{\varepsilon_k}{2}.$$

Let us prove that

$$(12.2.13) \qquad \widetilde{F}''(\Omega, \partial\Omega, 0, u) \le \widetilde{F}''_-(\Omega, u).$$

To do this assume that $\widetilde{F}''(\Omega, u) < +\infty$. Then, for every $k \in \mathbf{N}$, there exists $\{u_h^{(k)}\} \subseteq W_{\mathrm{loc}}^{1,q}(\mathbf{R}^n) \cap L_{\mathrm{loc}}^{\infty}(\mathbf{R}^n)$ such that $u_h^{(k)} \to u$ in $L^{\infty}(\Omega_k)$, and

$$\limsup_{h \to +\infty} \int_{\Omega_k} f\left(r_h x, \nabla u_h^{(k)}\right) dx \le \widetilde{F}''(\Omega_k, u).$$

For every $k \in \mathbf{N}$ let $s_k \in \mathbf{N}$ be such that $s_k > k$, $s_{k+1} > s_k$,

$$(12.2.14) \qquad \int_{\Omega_k} f\left(r_h x, \nabla u_h^{(k)}\right) dx \le \widetilde{F}''(\Omega_k, u) + \frac{1}{k} \text{ for every } h \ge s_k,$$

and

$$(12.2.15) \qquad \left\| u - u_h^{(k)} \right\|_{L^{\infty}(\Omega_k)} \le \frac{\varepsilon_k}{2} \text{ for every } h \ge s_k.$$

For $h \ge s_1$ set $k_h = \max\{k \in \mathbf{N} : s_k \le h\}$, and define u_h and $\tilde{u}_h$ by $u_h = u_h^{k_h}$, $\tilde{u}_h = \chi_{k_h}(u_h)$. Then, for every $h \ge s_1$, $u_h \in W_{\mathrm{loc}}^{1,q}(\mathbf{R}^n) \cap L_{\mathrm{loc}}^{\infty}(\mathbf{R}^n)$ and by (12.2.11), (12.2.12), and (12.2.15), we infer that $\tilde{u}_h \in W_0^{1,q}(\Omega_{k_h}) \cap L_{\mathrm{loc}}^{\infty}(\mathbf{R}^n)$. Let us denote again by $\tilde{u}_h$ the zero extension of $\tilde{u}_h$ from Ω_{k_h} to $\mathbf{R}^n$.

By (12.2.11), (12.2.15) and (12.2.12) it turns out that

$$(12.2.16) \qquad |\tilde{u}_h(x) - u(x)| \le |\tilde{u}_h(x) - u_h(x)| + |u_h(x) - u(x)| \le$$

$$\le \varepsilon_{k_h} + \frac{\varepsilon_{k_h}}{2} = \frac{3}{2}\varepsilon_{k_h} \text{ for a.e. } x \in \Omega_{k_h} \text{ and every } h \ge s_1,$$

and, by (12.2.12), that

$$(12.2.17) \qquad |\tilde{u}_h(x) - u(x)| = |u(x)| \le \frac{1}{2}\varepsilon_{k_h}$$

for a.e. $x \in \Omega \setminus \Omega_{k_h}$, and every $h \geq s_1$.

Consequently, from (12.2.16) and (12.2.17), it follows that

(12.2.18) $$\tilde{u}_h \to u \text{ in } L^\infty(\Omega).$$

Now let B_1, $B_2 \in \mathcal{A}_0$ be such that $B_1 \subset\subset B_2 \subset\subset \Omega_{k_h}$ for h sufficiently large. Let $\{\psi_h\}$ be given by Lemma 11.1.1 applied to B_2 and $K = \overline{B_1}$, and let $\{w_h\} \subseteq W_0^{1,q}(\Omega) \cap L^\infty_{loc}(\mathbf{R}^n)$ be defined by

$$w_h = \psi_h(u_h + \varepsilon_{k_h}) + (1 - \psi_h)\tilde{u}_h.$$

Then obviously $w_h \to u$ in $L^\infty(\Omega)$.

By making use of the convexity of f, it results that

$$\int_\Omega f(r_h x, t \nabla w_h) dx \leq$$

$$\leq t \int_\Omega \psi_h(x) f(r_h x, \nabla u_h) dx + t \int_\Omega (1 - \psi_h(x)) f(r_h x, \nabla \tilde{u}_h) dx +$$

$$+ (1 - t) \int_\Omega f\left(r_h x, \frac{t}{1-t}(u_h + \varepsilon_{k_h} - \tilde{u}_h)\nabla \psi_h\right) dx \leq$$

$$\leq \int_{\Omega_{k_h}} \psi_h(x) f(r_h x, \nabla u_h) dx + \int_{\Omega_{k_h}} (1 - \psi_h(x)) f(r_h x, \nabla \tilde{u}_h) dx +$$

$$+ \int_{\Omega \setminus \Omega_{k_h}} f(r_h x, 0) dx + (1 - t) \int_\Omega f\left(r_h x, \frac{t}{1-t}(u_h + \varepsilon_{k_h} - \tilde{u}_h)\nabla \psi_h\right) dx \leq$$

$$\leq \int_{\Omega_{k_h}} \psi_h(x) f(r_h x, \nabla u_h) dx + \int_{\Omega_{k_h}} (1 - \psi_h(x)) \chi'_{k_h}(u_h(x)) f(r_h x, \nabla u_h) dx +$$

$$+ \int_{\Omega_{k_h}} (1 - \psi_h(x)) \left(1 - \chi'_{k_h}(u_h(x))\right) f(r_h x, 0) dx + \int_{\Omega \setminus \Omega_{k_h}} f(r_h x, 0) dx +$$

$$+ (1 - t) \int_\Omega f\left(r_h x, \frac{t}{1-t}(u_h + \varepsilon_{k_h} - \tilde{u}_h)\nabla \psi_h\right) dx \leq$$

$$\leq \int_{\Omega_{k_h}} f(r_h x, \nabla u_h) dx + 2 \int_{\Omega \setminus B_1} f(r_h x, 0) dx +$$

$$+ (1 - t) \int_\Omega f\left(r_h x, \frac{t}{1-t}(u_h + \varepsilon_{k_h} - \tilde{u}_h)\nabla \psi_h\right) dx$$

for every $h \in \mathbf{N}$ sufficiently large, $t \in [0, 1[$.

Hence because of (12.2.14), we conclude that

$$(12.2.19) \qquad \int_\Omega f(r_h x, t\nabla w_h)dx \leq$$

$$\leq \widetilde{F}''(\Omega_{k_h}, u) + \frac{1}{k_h} + 2\int_{\Omega\setminus B_1} f(r_h x, 0)dx +$$

$$+(1-t)\int_\Omega f\left(r_h x, \frac{t}{1-t}(u_h + \varepsilon_{k_h} - \tilde{u}_h)\nabla\psi_h\right)dx \leq$$

$$\leq \widetilde{F}''_-(\Omega, u) + \frac{1}{k_h} + 2\int_{\Omega\setminus B_1} f(r_h x, 0)dx +$$

$$+(1-t)\left\{\int_{B_2} f\left(r_h x, \frac{t}{1-t}(u_h + \varepsilon_{k_h} - \tilde{u}_h)\nabla\psi_h\right)dx + \int_{\Omega\setminus B_2} f(r_h x, 0)dx\right\}$$

for every $h \in \mathbf{N}$ sufficiently large, $t \in [0, 1[$.

On the other hand, because of (12.2.15), (12.2.18), and again the inclusion $B_2 \subset\subset \Omega_{k_h}$ for every h large enough, it results that $u_h + \varepsilon_{k_h} - \tilde{u}_h \to 0$ in $L^\infty(B_2)$, and that $u_h + \varepsilon_{k_h} - \tilde{u}_h \geq 0$ a.e. in B_2. Consequently, by making use of (11.1.10) of Lemma 11.1.1, and by arguing as in the proof of (12.2.8), it is easy to verify that

$$(12.2.20) \qquad \limsup_{h\to+\infty} \int_{B_2} f\left(r_h x, \frac{t}{1-t}(u_h + \varepsilon_{k_h} - \tilde{u}_h)\nabla\psi_h\right)dx \leq$$

$$\leq \left(c_f + \int_Y f(y,0)dy\right)\mathcal{L}^n(B_2) \leq \left(c_f + \int_Y f(y,0)dy\right)\mathcal{L}^n(\Omega)$$

for every $t \in [0, 1[$,

where c_f is defined in Remark 11.1.2.

Passing to the limit in (12.2.19) as h tends to infinity, because of (11.2.5) and (12.2.20), it results that

$$(12.2.21) \qquad \widetilde{F}''(\Omega, \partial\Omega, 0, tu) \leq$$

$$\leq \widetilde{F}''_-(\Omega, u) + 2\mathcal{L}^n(\Omega\setminus B_1)\int_Y f(y,0)dy + (1-t)\left(c_f + 2\int_Y f(y,0)dy\right)\mathcal{L}^n(\Omega)$$

for every $t \in [0, 1[$.

Finally, letting t increase to 1 in (12.2.21), by Proposition 3.3.2, and again (11.2.5), we conclude that

$$\widetilde{F}''(\Omega, \partial\Omega, 0, u) \leq \widetilde{F}''_-(\Omega, u) + 2\mathcal{L}^n(\Omega\setminus B_1)\int_Y f(y,0)dy,$$

from which (12.2.13) follows as B_1 increases to Ω.

On the other hand, since it is always true that

$$(12.2.22) \qquad \widetilde{F}''_-(\Omega, u) \leq \widetilde{F}''(\Omega, u) \leq \widetilde{F}''(\Omega, \partial\Omega, 0, u),$$

the proof follows from (12.2.13) and (12.2.22). ∎

Proposition 12.2.3. *Let f be as in (12.0.1), $q \in [1, +\infty]$, $\{r_h\} \subseteq]0, +\infty[$ be strictly increasing and diverging, $\widetilde{F}'$, $\widetilde{F}''$ be defined in (12.1.4), $\Omega \in \mathcal{A}_0$, $\Gamma \subseteq \partial\Omega$, and $\widetilde{F}'(\Omega, \Gamma, 0, \cdot)$, $\widetilde{F}''(\Omega, \Gamma, 0, \cdot)$ be defined in (12.1.5). Assume that (11.1.3) holds. Then*

$$(12.2.23) \qquad \widetilde{F}'(\Omega, u) = \widetilde{F}'(\Omega, \Gamma, 0, u), \ \widetilde{F}''(\Omega, u) = \widetilde{F}''(\Omega, \Gamma, 0, u)$$

for every $u \in L^\infty_{\mathrm{loc}}(\mathbf{R}^n) \cap C^0(\overline{\Omega})$ such that $u = 0$ on Γ.

Proof. The proof follows the outlines of the one of Proposition 12.2.2.

Let u be as in (12.2.23). Let us prove that

$$(12.2.24) \qquad \widetilde{F}''(\Omega, \Gamma, 0, u) \leq \widetilde{F}''(\Omega, u).$$

To do this we can assume that $\widetilde{F}''(\Omega, u) < +\infty$, so that there exists $\{u_h\} \subseteq W^{1,q}_{\mathrm{loc}}(\mathbf{R}^n) \cap L^\infty_{\mathrm{loc}}(\mathbf{R}^n)$ such that $u_h \to u$ in $L^\infty(\Omega)$, and

$$\limsup_{h \to +\infty} \int_\Omega f(r_h x, \nabla u_h) dx \leq \widetilde{F}''(\Omega, u).$$

For every $h \in \mathbf{N}$ let $\varepsilon_h = 2\|u_h - u\|_{L^\infty(\Omega)}$, and χ_h be as in (12.2.11). Then $\chi_h(u_h) \in W^{1,q}_{0,\Gamma}(\Omega) \cap L^\infty_{\mathrm{loc}}(\mathbf{R}^n)$ for every $h \in \mathbf{N}$. In fact, if $h \in \mathbf{N}$, it is clear that $\chi_h(u_h) \in W^{1,q}_{\mathrm{loc}}(\mathbf{R}^n) \cap L^\infty_{\mathrm{loc}}(\mathbf{R}^n)$. Moreover by the continuity of u in $\partial\Omega$ it follows that the set $I_h = \{x \in \Omega : |u(x)| < \frac{1}{2}\varepsilon_h\}$ is a neighborhood of Γ in Ω, consequently we have that

$$|u_h(x)| \leq |u_h(x) - u(x)| + |u(x)| \leq \|u_h - u\|_{L^\infty(\Omega)} + \frac{1}{2}\varepsilon_h = \varepsilon_h \ \text{a.e. in } I_h,$$

that is $\chi_h(u_h) = 0$ a.e. in I_h, and therefore $\chi_h(u_h) \in W^{1,q}_{0,\Gamma}(\Omega) \cap L^\infty_{\mathrm{loc}}(\mathbf{R}^n)$. Finally it is clear that $\chi_h(u_h) \to u$ in $L^\infty(\Omega)$.

Let B_1, B_2 be two open sets such that $B_1 \subset\subset B_2 \subset\subset \Omega$, and let $\{\psi_h\}$ and $\{w_h\}$ be as in Proposition 12.2.2. Then (12.2.24) follows from the same arguments used in the proof of (12.2.13) in Proposition 12.2.2.

By (12.2.24) and the obvious inequality

$$\widetilde{F}''(\Omega, u) \leq \widetilde{F}''(\Omega, \Gamma, 0, u),$$

the right-hand side of (12.2.23) follows. This completes the proof. ∎

§12.3 Finiteness Conditions

Let f be a function satisfying (12.0.1), $p \in [1, +\infty]$, and F'' be the functional defined in (12.1.6). In this section we give sufficient conditions on Ω and u in order to get finiteness of $F''(\Omega, u)$.

For every $u = \sum_{j=1}^{m}(u_{z_j} + s_j)\chi_{P_j} \in PA(\mathbf{R}^n)$ we set

$$\sigma(u) = \max_{j \in \{1,\dots,m\}} \text{card}\left\{ i \in \{1,\dots,m\} : \overline{P_i} \cap \overline{P_j} \neq \emptyset \right\}.$$

Lemma 12.3.1. Let f be as in (12.0.1), $z_0 = 0$, $q \in [1, +\infty]$, $\widetilde{C}^q(0)$ and $\tilde{f}_{\text{hom}}^q$ be defined in (11.2.2) and (12.1.1) respectively, and F'' in (12.1.6). Assume that (11.2.3) holds. Let $\delta \in \,]0, 1[$ satisfy (11.2.4). Then

$$(12.3.1) \qquad F''(\Omega, tu) \le t \int_\Omega \tilde{f}_{\text{hom}}^q(\nabla u)dx + (1-t)\mathcal{L}^n(\Omega)\int_Y f(y, 0)dy$$

for every $\Omega \in \mathcal{A}_0$, $u \in PA(\mathbf{R}^n)$,

$$t \in \left[0, \frac{\delta}{256n3^n\sqrt{n}\sigma(u)^2\left(2\|\nabla u\|_{L^\infty(\Omega)} + 1\right) + \delta} \right].$$

Proof. Let Ω, $u = \sum_{i=1}^{m}(u_{z_j} + s_j)\chi_{P_j}$, t be as in (12.3.1), and set, for every $j \in \{1,\dots,m\}$, $\Omega_j = \Omega \cap \text{int}(P_j)$.

In order to prove (12.3.1), let us assume that

$$(12.3.2) \qquad \sum_{j=1}^{m} \tilde{f}_{\text{hom}}^q(z_j)\mathcal{L}^n(\Omega_j) = \int_\Omega \tilde{f}_{\text{hom}}^q(\nabla u)dx < +\infty.$$

Inequality (12.3.2) provides that $z_j \in \text{dom}\tilde{f}_{\text{hom}}^q$ for every $j \in \{1,\dots, m\}$. Hence, for every fixed $\theta \in \,]0, +\infty[$ and $j \in \{1,\dots,m\}$, there exists $v^j \in W_{\text{per}}^{1,q}(Y) \cap L^\infty(Y)$ such that

$$\int_Y f(y, z_j + \nabla v^j)dy \le \tilde{f}_{\text{hom}}^q(z_j) + \theta.$$

Whence, for every $j \in \{1,\dots,m\}$, by setting $v_h^j = \frac{1}{h}v^j(h\cdot)$ for every $h \in \mathbf{N}$, it follows that

$$(12.3.3) \quad \lim_{h \to +\infty} \int_{\Omega \cap \Omega_{j,\varepsilon}^+} f\left(hx, z_j + \nabla v_h^j\right)dx \le \mathcal{L}^n(\Omega \cap \Omega_{j,\varepsilon}^+)(\tilde{f}_{\text{hom}}^q(z_j) + \theta).$$

For every $\varepsilon > 0$ sufficiently small, and $j \in \{1,\dots,m\}$, let $\{\gamma_h^{\varepsilon,j}\}$ and $\gamma^{\varepsilon,j}$ be given by Lemma 11.2.2, and let, for every $h \in \mathbf{N}$,

$$w_h^\varepsilon = \sum_{j=1}^{m}(u_{z_j} + s_j + v_h^j)\gamma_h^{\varepsilon,j}.$$

Then, because of Lemma 11.2.2, it results that

$$(12.3.4) \qquad w_h^\varepsilon \to w_\varepsilon = \sum_{j=1}^{m}(u_{z_j}+s_j)\gamma^{\varepsilon,j} \text{ in } L^\infty(\Omega) \text{ and a.e. in } \Omega$$

for every $\varepsilon > 0$ sufficiently small.

By (12.3.4), the convexity properties of f, Lemma 11.2.2, and by re-calling that $\sum_{j=1}^{m}\nabla\gamma_h^{\varepsilon,j} = 0$ a.e. in Ω, it results that

$$(12.3.5) \qquad F''(\Omega, tw_\varepsilon) \leq \limsup_{h\to+\infty} \int_\Omega f(hx, t\nabla w_h^\varepsilon)dx =$$

$$= \limsup_{h\to+\infty} \int_\Omega f\left(hx, t\sum_{j=1}^{m}(z_j+\nabla v_h^j)\gamma_h^{\varepsilon,j}+ \right.$$

$$\left. +(1-t)\frac{t}{1-t}\sum_{j=1}^{m}(u_{z_j}+s_j+v_h^j)\nabla\gamma_h^{\varepsilon,j} \right)dx \leq$$

$$\leq t\limsup_{h\to+\infty} \int_\Omega f\left(hx, \sum_{j=1}^{m}(z_j + \nabla v_h^j)\gamma_h^{\varepsilon,j} \right)dx+$$

$$+(1-t)\limsup_{h\to+\infty} \int_\Omega f\left(hx, \frac{t}{1-t}\sum_{j=1}^{m}(u_{z_j} + s_j + v_h^j)\nabla\gamma_h^{\varepsilon,j} \right)dx \leq$$

$$\leq t\sum_{j=1}^{m}\limsup_{h\to+\infty} \int_\Omega \gamma_h^{\varepsilon,j}(x)f\left(hx, z_j + \nabla v_h^j \right)dx+$$

$$+(1-t)\limsup_{h\to+\infty} \int_\Omega f\left(hx, \frac{t}{1-t}\sum_{j=1}^{m}(u_{z_j} + s_j + v_h^j - u)\nabla\gamma_h^{\varepsilon,j} \right)dx \leq$$

$$\leq t\sum_{j=1}^{m}\limsup_{h\to+\infty} \int_{\Omega\cap\Omega_{j,\varepsilon}^+} f\left(hx, z_j + \nabla v_h^j \right)dx+$$

$$+(1-t)\limsup_{h\to+\infty} \sum_{i=1}^{m}\int_{\Omega_{i,\varepsilon}^-} f(hx, 0)dx+$$

$$+(1-t)\limsup_{h\to+\infty} \int_{\Omega\setminus\cup_{i=1}^{m}\Omega_{i,\varepsilon}^-} f\left(hx, \frac{t}{1-t}\sum_{j=1}^{m}(u_{z_j} + s_j + v_h^j - u)\nabla\gamma_h^{\varepsilon,j} \right)dx \leq$$

$$\leq t\sum_{j=1}^{m}\limsup_{h\to+\infty} \int_{\Omega\cap\Omega_{j,\varepsilon}^+} f\left(hx, z_j + \nabla v_h^j \right)dx+$$

$$+(1-t)\limsup_{h\to+\infty}\sum_{i=1}^{m}\int_{\Omega_{i,\varepsilon}^-}f(hx,0)\,dx+$$

$$+(1-t)\sum_{i=1}^{m}\limsup_{h\to+\infty}\int_{\Omega_i\setminus\Omega_{i,\varepsilon}^-}f\left(hx,\frac{t}{1-t}\sum_{j=1}^{m}(u_{z_j}+s_j+v_h^j-u)\nabla\gamma_h^{\varepsilon,j}\right)dx$$

for every $\varepsilon>0$ sufficiently small.

On the other hand, let $\sigma_\varepsilon=\sigma_\varepsilon(\Omega_1,\ldots,\Omega_m)$ be given by (11.2.6). Let us observe that, for a fixed $\varepsilon>0$ sufficiently small and $i\in\{1,\ldots,m\}$, the number of the sets Ω_j such that $\mathrm{dist}(\Omega_i,\Omega_j)<\varepsilon$ is less than or equal to σ_ε. Let $\{\Omega_{j_1},\ldots,\Omega_{j_{\sigma_\varepsilon}}\}$ be a subset of $\{\Omega_1,\ldots,\Omega_m\}$ containing all the sets Ω_j satisfying $\mathrm{dist}(\Omega_i,\Omega_j)<\varepsilon$. Consequently, as regards the last term in (12.3.5), it results that

$$(12.3.6)\qquad \int_{\Omega_i\setminus\Omega_{i,\varepsilon}^-}f\left(hx,\frac{t}{1-t}\sum_{j=1}^{m}(u_{z_j}+s_j+v_h^j-u)\nabla\gamma_h^{\varepsilon,j}\right)dx=$$

$$=\int_{\Omega_i\setminus\Omega_{i,\varepsilon}^-}f\left(hx,\sum_{k=1}^{\sigma_\varepsilon}\frac{1}{\sigma_\varepsilon}\sigma_\varepsilon\frac{t}{1-t}(u_{z_{j_k}}+s_{j_k}+v_h^{j_k}-u)\nabla\gamma_h^{\varepsilon,j_k}\right)dx\le$$

$$\le\sum_{k=1}^{\sigma_\varepsilon}\frac{1}{\sigma_\varepsilon}\int_{\Omega_i\setminus\Omega_{i,\varepsilon}^-}f\left(hx,\sigma_\varepsilon\frac{t}{1-t}(u_{z_{j_k}}+s_{j_k}+v_h^{j_k}-u)\nabla\gamma_h^{\varepsilon,j_k}\right)dx\le$$

$$\le\sum_{k=1}^{\sigma_\varepsilon}\frac{1}{\sigma_\varepsilon}\int_{(\Omega_i\setminus\Omega_{i,\varepsilon}^-)\cap\Omega_{j,\varepsilon}^+}f\left(hx,\sigma_\varepsilon\frac{t}{1-t}(u_{z_{j_k}}+s_{j_k}+v_h^{j_k}-u)\nabla\gamma_h^{\varepsilon,j_k}\right)dx+$$

$$+\int_{\Omega_i\setminus\Omega_{i,\varepsilon}^-}f(hx,0)dx$$

for every $i\in\{1,\ldots,m\}$, $\varepsilon>0$ sufficiently small, $h\in\mathbf{N}$.

We now observe that there exists $\varepsilon(u)\in\,]0,+\infty[$ such that

$$\sigma_\varepsilon(\Omega_1,\ldots,\Omega_m)\le\sigma(u)\quad\text{for every }\varepsilon\in\,]0,\varepsilon(u)[\,.$$

Fix $\varepsilon\in\,]0,\varepsilon(u)[$. Then, since

$$\left\|\sigma_\varepsilon\frac{t}{1-t}\left(u_{z_j}+s_j+v_h^j-u\right)\right\|_{L^\infty(\Omega\cap\Omega_{j,\varepsilon}^+)}\le\sigma_\varepsilon\frac{t}{1-t}(2\|\nabla u\|_{L^\infty(\Omega)}+1)\varepsilon$$

for every h sufficiently large, $j\in\{1,\ldots,m\}$,

our choice of t provides that

$$\left\|\sigma_\varepsilon\frac{t}{t-t}\left(u_{z_j}+s_j+v_h^j-u\right)\right\|_{L^\infty(\Omega\cap\Omega_{j,\varepsilon}^+)}\le\frac{\delta\varepsilon}{256n3^n\sqrt{n}\sigma_\varepsilon}$$

for every h sufficiently large, $j \in \{1, \ldots, m\}$,

i.e.

$$(12.3.7) \quad \sigma_\varepsilon \frac{t}{1-t} \left(u_{z_j}(x) + s_j + v_h^j(x) - u(x) \right) \in$$

$$\left] -\frac{\delta \varepsilon}{256n3^n \sqrt{n\sigma_\varepsilon}}, \frac{\delta \varepsilon}{256n3^n \sqrt{n\sigma_\varepsilon}} \right[$$

for a.e. $x \in (\Omega_i \setminus \Omega_{i,\varepsilon}^-) \cap \Omega_{j,\varepsilon}^+$, every h sufficiently large, $j \in \{1, \ldots, m\}$.

Then (12.3.7), an argument similar to the one used to get (12.2.7), the convexity properties of f, and Lemma 11.2.2 provide that

$$(12.3.8) \quad \limsup_{h \to +\infty} \sum_{k=1}^{\sigma_\varepsilon} \frac{1}{\sigma_\varepsilon} \int_{(\Omega_i \setminus \Omega_{i,\varepsilon}^-) \cap \Omega_{j,\varepsilon}^+} f \left(hx, \right.$$

$$\left. \sigma_\varepsilon \frac{t}{1-t} (u_{z_{j_k}} + s_{j_k} + v_h^{j_k} - u) \nabla \gamma_h^{\varepsilon, j_k} \right) dx \leq$$

$$\leq \limsup_{h \to +\infty} \sum_{k=1}^{\sigma_\varepsilon} \frac{1}{\sigma_\varepsilon} \int_{(\Omega_i \setminus \Omega_{i,\varepsilon}^-) \cap \Omega_{j,\varepsilon}^+} f^{(0)} \left(hx, \frac{\delta \varepsilon}{256n3^n \sqrt{n\sigma_\varepsilon}} \nabla \gamma_h^{\varepsilon, j_k} \right) dx \leq$$

$$\leq 2 \left(c_f + 2 \int_Y f(y,0) dy \right) \mathcal{L}^n (\Omega_i \setminus \Omega_{i,\varepsilon}^-) \quad \text{for every } \varepsilon \in {]0, \varepsilon(u)[},$$

where $f^{(0)}$ is defined by (11.2.1), and c_f by Remark 11.1.2.

By combining (12.3.5) with (12.3.3), (12.3.6), (12.3.8), and by making use of (11.2.5) and of the periodicity of $f(\cdot, 0)$, it then results that

$$(12.3.9) \quad F''(\Omega, tw_\varepsilon) \leq$$

$$\leq t \sum_{j=1}^m \mathcal{L}^n \left(\Omega \cap \Omega_{j,\varepsilon}^+ \right) \left(\tilde{f}_{\text{hom}}^q (z_j) + \theta \right) + (1-t) \sum_{i=1}^m \mathcal{L}^n \left(\Omega_{i,\varepsilon}^- \right) \int_Y f(y,0) dy +$$

$$+ (1-t) \sum_{i=1}^m \mathcal{L}^n \left(\Omega_i \setminus \Omega_{i,\varepsilon}^- \right) \int_Y f(y,0) dy +$$

$$+ (1-t) 2 \sum_{i=1}^m \mathcal{L}^n \left(\Omega_i \setminus \Omega_{i,\varepsilon}^- \right) \left(c_f + 2 \int_Y f(y,0) dy \right)$$

for every $\varepsilon \in {]0, \varepsilon(u)[}$.

Observe now that, because of Lemma 11.2.2,

$$\|w_\varepsilon - u\|_{L^\infty(\Omega)} = \left\| \sum_{j=1}^m (u_{z_j} + s_j - u) \gamma^{\varepsilon, j} \right\|_{L^\infty(\Omega)} \leq$$

$$\leq \sum_{j=1}^{m} \left\| (u_{z_j} + s_j - u)\gamma^{\varepsilon,j} \right\|_{L^\infty(\Omega \cap \Omega_{j,\varepsilon}^+)} \leq m \left(2 \left\| \nabla u \right\|_{L^\infty(\Omega)} + 1 \right) \varepsilon$$

$$\text{for every } \varepsilon \in]0, \varepsilon(u)[,$$

and consequently that

$$(12.3.10) \qquad\qquad w_\varepsilon \to u \text{ in } L^\infty(\Omega).$$

Then, because of Proposition 3.3.2, (12.3.10), and (12.3.9), it results that

$$(12.3.11) \qquad\qquad F''(\Omega, tu) \leq \liminf_{\varepsilon \to 0} F''(\Omega, tw_\varepsilon) \leq$$

$$\leq \sum_{j=1}^{m} \mathcal{L}^n \left(\Omega \cap \overline{\Omega_j} \right) \left(\tilde{f}^q_{\text{hom}}(z_j) + \theta \right) + (1-t)\mathcal{L}^n(\Omega) \int_Y f(y, 0)dy.$$

By passing to the limit in (12.3.11) as θ tends to 0^+, and recalling that $\sum_{j=1}^{m} \mathcal{L}^n(\Omega \cap \overline{\Omega_j})\tilde{f}^q_{\text{hom}}(z_j) = \int_\Omega \tilde{f}^q_{\text{hom}}(\nabla u)dx$, inequality (12.3.1) follows. ∎

We can now prove the finiteness result.

Proposition 12.3.2. Let f be as in (12.0.1), $z_0 = 0$, $q \in [1, +\infty]$, $\tilde{C}^q(0)$ be defined in (11.2.2), and F'' in (12.1.6). Assume that (11.2.3) holds. Let $\delta \in]0, +\infty[$ satisfy (11.2.4). Then there exist $r \in]0, \delta[$, and $c \in]0, +\infty[$ such that

$$(12.3.12) \qquad\qquad F''_-(\Omega, u) \leq c\mathcal{L}^n(\Omega)$$

for every $\Omega \in \mathcal{A}_0$, $u \in W^{1,\infty}_{\text{loc}}(\mathbf{R}^n)$ such that $\|\nabla u\|_{L^\infty(\Omega)} \leq r$.

Proof. Let $\Omega \in \mathcal{A}_0$, and Q be an open cube with $\Omega \subset\subset Q$.

Let $r \in]0, +\infty[$ to be specified later, and $u \in W^{1,\infty}_{\text{loc}}(\mathbf{R}^n)$ such that $\|\nabla u\|_{L^\infty(\Omega)} \leq r$.

Because of (12.1.9), it is not restrictive to assume u equal to 0 in $\mathbf{R}^n \setminus Q$.

Let $S_1, \ldots, S_l \subseteq \mathbf{R}^n \setminus Q$ be polyhedral sets with pairwise disjoint interiors such that $\mathcal{L}^n((\mathbf{R}^n \setminus Q) \setminus \cup_{j=1}^{l} S_j) = 0$, and let $P_1, \ldots, P_m \subseteq Q$ be n-simplexes with pairwise disjoint interiors such that $Q = \cup_{j=1}^{m} P_j$. For every $h \in \mathbf{N}$ let $P_1^h, \ldots, P_{m^h}^h$ be the n-simplexes obtained by taking the $\frac{1}{h}$-replies of $P_1, \ldots, P_m$ repeated $\frac{1}{h}Q$-periodically so that $Q = \cup_{j=1}^{m^h} P_j^h$.

For every $h \in \mathbf{N}$ let $u_h \in PA(\mathbf{R}^n)$ be such that u_h is affine on each n-simplex of $\{P_1^h, \ldots, P_{m^h}^h\}$, equal to u on the vertices of the elements of $\{P_1^h, \ldots, P_{m^h}^h\}$ and equal to 0 in each element of $\{S_1, \ldots, S_l\}$. Then, since for every $h \in \mathbf{N}$ and $j \in \{1, \ldots, m^h\}$, P_j^h intersects at most m^n elements of $\{P_1^h, \ldots, P_{m^h}^h\}$, we immediately obtain that

$$(12.3.13) \qquad\qquad \sigma(u_h) \leq m^n + l \text{ for every } h \in \mathbf{N},$$

(12.3.14) $u_h \to u$ in $L^\infty(\mathbf{R}^n)$,

(12.3.15) $\|\nabla u_h\|_{L^\infty(\Omega)} \le \bar{c}\|\nabla u\|_{L^\infty(\Omega)} \le \bar{c}r$ for every $h \in \mathbf{N}$,

where $\bar{c}$ is in $[1, +\infty[$ and depends only on n.

Since, because of (12.3.15) and (12.3.13), it results that

$$\frac{\delta}{256n3^n\sqrt{n}(m^n + l)^2(2\delta + 1) + \delta} \le$$

$$\le \frac{\delta}{256n3^n\sqrt{n}\sigma^2(u_h)\left(2\left\|\frac{\delta}{\bar{c}r}\nabla u_h\right\|_{L^\infty(\Omega)} + 1\right) + \delta} \text{ for every } h \in \mathbf{N},$$

Lemma 12.3.1 provides that

(12.3.16) $F''\left(\Omega, t\frac{\delta}{\bar{c}r}u_h\right) \le$

$$\le t\int_\Omega \tilde{f}^q_{\text{hom}}\left(\frac{\delta}{\bar{c}r}\nabla u_h\right)dx + (1-t)\mathcal{L}^n(\Omega)\int_Y f(y,0)dy$$

for every $h \in \mathbf{N}$, $t \in \left[0, \dfrac{\delta}{256n3^n\sqrt{n}(m^n + l)^2(2\delta + 1) + \delta}\right]$,

where $\tilde{f}^q_{\text{hom}}$ is given by (12.1.1).

In (12.3.16) it is possible to take $t = \frac{\bar{c}r}{\delta}$ if and only if

(12.3.17) $r \le \dfrac{\delta^2}{\bar{c}(256n3^n\sqrt{n}(m^n + l)^2(2\delta + 1) + \delta)}$,

furthermore, since clearly $\bar{c} \ge 1$, it results that $r \in]0, \delta[$.

By choosing r as in (12.3.17), from (12.3.16) written with $t = \frac{\bar{c}r}{\delta}$ it then follows that

(12.3.18) $F''(\Omega, u_h) \le$

$$\le \frac{\bar{c}r}{\delta}\int_\Omega \tilde{f}^q_{\text{hom}}\left(\frac{\delta}{\bar{c}r}\nabla u_h\right)dx + (1 - \frac{\bar{c}r}{\delta})\mathcal{L}^n(\Omega)\int_Y f(y,0)dy \text{ for every } h \in \mathbf{N}.$$

We now observe that, by (12.3.15) it results

(12.3.19) $\left\|\frac{\delta}{\bar{c}r}\nabla u_h\right\|_{L^\infty(\Omega)} \le \delta$ for every $h \in \mathbf{N}$, and that, by using (12.0.1) and (11.2.4), $\tilde{f}^q_{\text{hom}}$ turns out to be bounded in $B_\delta(0)$. Consequently, by (12.3.18) and (12.3.19) it follows that

$$(12.3.20) \qquad F''(\Omega, u_h) \le \left(\max_{\overline{B_\delta(0)}} \tilde{f}^q_{\text{hom}} + \int_Y f(y,0)dy \right) \mathcal{L}^n(\Omega).$$

Finally (12.3.14), Proposition 3.3.2, (12.3.20), and (11.2.5) provide (12.3.12) with r satisfying (12.3.17) and c deduced by (12.3.20). ∎

§12.4 Representation on Linear Functions

Let f be as in (12.0.1), $q \in [1, +\infty]$, and F', F'' be defined in (12.1.6). In this section we prove that, for every bounded open set Ω, $F'(\Omega, \cdot) = F''(\Omega, \cdot)$ on the class of the linear functions, and give a representation result for their common value.

Lemma 12.4.1. *Let f be as in (12.0.1), $q \in [1, +\infty]$, $\tilde{f}^q_{\text{hom}}$ be defined in (12.1.1), and F'' in (12.1.6). Then*

$$(12.4.1) \qquad F''(\Omega, u_z) \le \mathcal{L}^n(\Omega) \text{sc}^- \tilde{f}^q_{\text{hom}}(z) \text{ for every } \Omega \in \mathcal{A}_0, \ z \in \mathbf{R}^n.$$

Proof. Fix $\Omega \in \mathcal{A}_0$, and $z \in \mathbf{R}^n$.

In order to prove (12.4.1), we can assume that $\text{sc}^- \tilde{f}^q_{\text{hom}}(z) < +\infty$. Then, for every $\varepsilon \in]0, +\infty[$, there exist $z_\varepsilon \in \mathbf{R}^n$ and $v_\varepsilon \in W^{1,q}_{\text{per}}(Y) \cap L^\infty_{\text{loc}}(\mathbf{R}^n)$ satisfying $z_\varepsilon \to z$ as $\varepsilon \to 0$, and

$$(12.4.2) \qquad \text{sc}^- \tilde{f}^q_{\text{hom}}(z) + 2\varepsilon \ge \tilde{f}^q_{\text{hom}}(z_\varepsilon) + \varepsilon \ge \int_Y f(y, z_\varepsilon + \nabla v_\varepsilon) \, dy.$$

For every $\varepsilon \in]0, +\infty[$ let $\{v_h\} \subseteq W^{1,q}_{\text{loc}}(\mathbf{R}^n) \cap L^\infty_{\text{loc}}(\mathbf{R}^n)$ be defined by $v_h = \frac{1}{h} v_\varepsilon(h\cdot)$. It is obvious that $v_h \to 0$ in $L^\infty(\Omega)$, consequently, because of (12.0.1) and (12.4.2), it results that

$$(12.4.3) \qquad F''(\Omega, u_{z_\varepsilon}) \le \limsup_{h \to +\infty} \int_\Omega f(hx, z_\varepsilon + \nabla v_h) dx =$$

$$= \mathcal{L}^n(\Omega) \int_Y f(y, z_\varepsilon + \nabla v_\varepsilon) dy \le \mathcal{L}^n(\Omega) \left(\text{sc}^- \tilde{f}^q_{\text{hom}}(z) + 2\varepsilon \right)$$

for every $\varepsilon > 0$.

Inequality (12.4.1) now follows from Proposition 3.3.2 as ε tends to 0^+ in (12.4.3). ∎

To prove the reverse inequality of (12.4.1) with F'' replaced by F', we need some technical lemmas.

Lemma 12.4.2. *Let f be as in (12.0.1), $q \in [1, +\infty]$, and F' be defined in (12.1.6). Assume that $f(\cdot, 0) \in L^1_{loc}(\mathbf{R}^n)$. Then*

$$F'(\Omega, tu) \le tF'(\Omega, u) + (1-t)\mathcal{L}^n(\Omega) \int_Y f(y, 0)dy$$

for every $\Omega \in \mathcal{A}_0$, $u \in L^\infty_{loc}(\mathbf{R}^n)$, $t \in [0,1]$.

Moreover similar inequalities hold for F'', F'_-, F''_- in place of F'.

Proof. The proof follows trivially from (12.1.8), and the obvious inequality

$$F'(\Omega, 0) \le \liminf_{h \to +\infty} \int_\Omega f(hx, 0)dx = \mathcal{L}^n(\Omega) \int_Y f(y, 0)dy. \qquad \blacksquare$$

Lemma 12.4.3. *Let f be as in (12.0.1), $q \in [1, +\infty]$, and F' be defined in (12.1.6). Then*

$$\frac{1}{r_1^n}F'(x_1 + r_1 Y, u_z) = \frac{1}{r_2^n}F'(x_2 + r_2 Y, u_z)$$

for every x_1, $x_2 \in \mathbf{R}^n$, r_1, $r_2 \in \,]0, +\infty[$, $z \in \mathbf{R}^n$.

Proof. Let $x_1 \in \mathbf{R}^n$, $r_1, r_2 \in \,]0, +\infty[$, $z \in \mathbf{R}^n$ be as above, and let $s_1 < r_1$, $s_2 > r_2$.

In addition, let $\{m_h\} \subseteq \mathbf{Z}^n$ be such that

$$(12.4.4) \qquad \frac{(m_h)_i}{h} \ge (x_1)_i \text{ for every } i \in \{1, \dots, n\}, \qquad \lim_{h \to +\infty} \frac{m_h}{h} = x_1,$$

let $\{k_h\} \subseteq \mathbf{N}$ satisfy

$$(12.4.5) \qquad \frac{k_h}{h} \le \frac{s_1}{s_2} \text{ for every } h \in \mathbf{N}, \qquad \lim_{h \to +\infty} \frac{k_h}{h} = \frac{s_1}{s_2},$$

and let $\{n_h\} \subseteq \mathbf{Z}^n$ be such that

$$(12.4.6) \qquad \frac{(n_h)_i}{k_h} \le (x_2)_i \text{ for every } i \in \{1, \dots, n\}, \qquad \lim_{h \to +\infty} \frac{n_h}{k_h} = x_2.$$

Let us prove that

$$(12.4.7) \qquad \frac{1}{r_1^n}F'(x_1 + r_1 Y, u_z) \ge \frac{1}{r_2^n}F'(x_2 + r_2 Y, u_z).$$

To do this we can assume that $\frac{1}{r_1^n}F'(x_1 + r_1 Y, u_z) < +\infty$, so that there exists $\{u_h\} \subseteq W^{1,q}_{loc}(\mathbf{R}^n) \cap L^\infty_{loc}(\mathbf{R}^n)$ with $u_h \to u_z$ in $L^\infty(x_1 + r_1 Y)$, and

$$(12.4.8) \qquad F'(x_1 + r_1 Y, u_z) \ge \liminf_{h \to +\infty} \int_{x_1 + r_1 Y} f(hx, \nabla u_h)dx.$$

By (12.4.8), and (12.4.4) we thus have that

$$(12.4.9) \qquad F'(x_1 + r_1Y, u_z) \geq \liminf_{h \to +\infty} \int_{\frac{m_h}{h} + s_1Y} f(hx, \nabla u_h) dx = .$$

$$= \liminf_{h \to +\infty} \int_{s_1Y} f\left(hx + m_h, \nabla u_h \left(x + \frac{m_h}{h}\right)\right) dx =$$

$$= \liminf_{h \to +\infty} \int_{s_1Y} f(hx, \nabla v_h) dx,$$

where $v_h = u_h(\cdot + \frac{m_h}{h})$ obviously satisfies $v_h \to u_z + z \cdot x_1$ in $L^\infty(s_1Y)$.

We now observe that (12.4.5) yields

$$(12.4.10) \quad \int_{s_1Y} f(hx, \nabla v_h) dx \geq \left(\frac{k_h}{h}\right)^n \int_{\frac{h}{k_h} s_1 Y} f\left(k_h y, \nabla v_h \left(\frac{k_h}{h} y\right)\right) dy \geq$$

$$\geq \left(\frac{k_h}{h}\right)^n \int_{s_2Y} f(k_h y, \nabla w_h) dy \text{ for every } h \in \mathbf{N},$$

where $w_h = \frac{h}{k_h} v_h(\frac{k_h}{h} \cdot)$ obviously satisfies $w_h \to u_z + \frac{s_2}{s_1} z \cdot x_1$ in $L^\infty(s_2Y)$.

Finally, by (12.4.6) we infer that

$$(12.4.11) \qquad\qquad \liminf_{h \to +\infty} \int_{s_2Y} f(k_h y, \nabla w_h) dy =$$

$$= \liminf_{h \to +\infty} \int_{s_2Y} f\left(k_h \left(y + \frac{n_h}{k_h}\right), \nabla w_h\right) dy =$$

$$= \liminf_{h \to +\infty} \int_{\frac{n_h}{k_h} + s_2Y} f\left(k_h y, \nabla w_h \left(y - \frac{n_h}{k_h}\right)\right) dy \geq$$

$$\geq \liminf_{h \to +\infty} \int_{x_2 + r_2Y} f(k_h y, \nabla z_h) dy \geq F'\left(x_2 + r_2Y, u_z + \frac{s_2}{s_1} z \cdot x_1 - z \cdot x_2\right),$$

where $z_h = w_h(\cdot - \frac{n_h}{k_h})$ satisfies $z_h \to u_z + \frac{s_2}{s_1} z \cdot x_1 - z \cdot x_2$ in $L^\infty(x_2 + r_2Y)$.

In conclusion, by (12.4.9)÷(12.4.11), and (12.4.5) we deduce that

$$F'(x_1 + r_1Y, u_z) \geq \left(\frac{s_1}{s_2}\right)^n F'\left(x_2 + r_2Y, u_z + \frac{s_2}{s_1} z \cdot x_1 - z \cdot x_2\right) =$$

$$= \left(\frac{s_1}{s_2}\right)^n F'(x_2 + r_2Y, u_z),$$

from which inequality (12.4.7) follows as $s_1 \to r_1^-$, and $s_2 \to r_2^-$.

By exchanging the roles of x_1 and x_2, and of r_1 and r_2 in (12.4.7), the proof follows. ∎

Lemma 12.4.4. *Let f be as in (12.0.1), $q \in [1, +\infty]$, and $\tilde{f}_{hom}^q$ be defined in (12.1.1). Then*

$$\tilde{f}_{hom}^q(z) = \inf \left\{ \int_Y f(hx, z + \nabla v)dx : v \in W_{per}^{1,q}(Y) \cap L^\infty(Y) \right\}$$

for every $z \in \mathbf{R}^n$, $h \in \mathbf{N}$.

Proof. Let $z \in \mathbf{R}^n$, and $h \in \mathbf{N}$.

Let us first prove that

$$(12.4.12) \quad \tilde{f}_{hom}^q(z) \leq \inf \left\{ \int_Y f(hx, z + \nabla v)dx : v \in W_{per}^{1,q}(Y) \cap L^\infty(Y) \right\}.$$

Let $v \in W_{per}^{1,q}(Y) \cap L^\infty(Y)$, and define v_h as

$$v_h(x_1, \ldots, x_n) = \frac{1}{h^n} \sum_{i_1, \ldots, i_n = 0}^{h-1} v\left(x_1 + \frac{i_1}{h}, \ldots, x_n + \frac{i_n}{h}\right).$$

Then $v_h \in W_{loc}^{1,q}(\mathbf{R}^n) \cap L^\infty(Y)$, and is $\frac{1}{h}Y$-periodic.

Then, by using the convexity and periodicity properties of f, by performing the change of variables $y = x + \frac{(i_1, \ldots, i_n)}{h}$, and by exploiting the Y-periodicity of v, we have that

$$(12.4.13) \qquad\qquad \int_Y f(hx, z + \nabla v_h)dx \leq$$

$$\leq \frac{1}{h^n} \sum_{i_1, \ldots, i_n = 0}^{h-1} \int_Y f\left(hx, z + \nabla v\left(x + \frac{(i_1, \ldots, i_n)}{h}\right)\right)dx =$$

$$= \frac{1}{h^n} \sum_{i_1, \ldots, i_n = 0}^{h-1} \int_{\frac{(i_1, \ldots, i_n)}{h} + Y} f\left(hx - (i_1, \ldots, i_n), z + \nabla v(x)\right)dx =$$

$$= \frac{1}{h^n} \sum_{i_1, \ldots, i_n = 0}^{h-1} \int_{\frac{(i_1, \ldots, i_n)}{h} + Y} f(hx, z + \nabla v(x))dx =$$

$$= \frac{1}{h^n} \sum_{i_1, \ldots, i_n = 0}^{h-1} \int_Y f(hx, z + \nabla v(x))dx = \int_Y f(hx, z + \nabla v)dx.$$

We now observe that $O_{1/h}v_h$ is actually Y-periodic, therefore, by the $\frac{1}{h}Y$-periodicity of v_h, and the Y-periodicity properties of f, we conclude that

$$(12.4.14) \qquad\qquad \tilde{f}_{hom}^q(z) \leq \int_Y f(y, z + \nabla(O_{1/h}v_h))dy =$$

$$= \int_Y f\left(y, z + \nabla_x v_h\left(\frac{y}{h}\right)\right) dy = h^n \int_{\frac{1}{h}Y} f(hx, z + \nabla v_h(x)) dx =$$

$$= \sum_{i_1,\ldots,i_n=0}^{h-1} \int_{\frac{1}{h}Y} f(hx - (i_1,\ldots,i_n), z + \nabla v_h(x)) dx =$$

$$= \sum_{i_1,\ldots,i_n=0}^{h-1} \int_{\frac{(i_1,\ldots,i_n)}{h}+\frac{1}{h}Y} f\left(hx, z + \nabla v_h\left(x + \frac{(i_1,\ldots,i_n)}{h}\right)\right) dx =$$

$$= \sum_{i_1,\ldots,i_n=0}^{h-1} \int_{\frac{(i_1,\ldots,i_n)}{h}+\frac{1}{h}Y} f(hx, z + \nabla v_h(x)) dx = \int_Y f(hx, z + \nabla v_h) dx.$$

In conclusion, by (12.4.14), and (12.4.13), we deduce (12.4.12). We now prove that

$$(12.4.15) \quad \inf\left\{ \int_Y f(hx, z + \nabla v) dx : v \in W^{1,q}_{\mathrm{per}}(Y) \cap L^\infty(Y) \right\} \le \tilde{f}^q_{\mathrm{hom}}(z).$$

Let $v \in W^{1,q}_{\mathrm{per}}(Y) \cap L^\infty(Y)$. Then $O_{1/h}v$ is Y-periodic, and the periodicity properties of f provide that

$$\inf\left\{ \int_Y f(hx, z + \nabla v) dx : v \in W^{1,q}_{\mathrm{per}}(Y) \cap L^\infty(Y) \right\} \le$$

$$\le \int_Y f(hx, z + \nabla_x(O_{1/h}v)(x)) dx = \int_Y f(hx, z + \nabla_y v(hx)) dx =$$

$$= \frac{1}{h^n} \int_{hY} f(y, z + \nabla_y v(y)) dy =$$

$$= \frac{1}{h^n} \sum_{i_1,\ldots,i_n=0}^{h-1} \int_{(i_1,\ldots,i_n)+Y} f(y, z + \nabla_y v(y)) dy =$$

$$= \frac{1}{h^n} \sum_{i_1,\ldots,i_n=0}^{h-1} \int_Y f(y - (i_1,\ldots,i_n), z + \nabla_y v(y - (i_1,\ldots,i_n))) dy =$$

$$= \int_Y f(y, z + \nabla v) dy,$$

from which (12.4.15) follows.

By (12.4.12) and (12.4.15) the proof follows. ∎

Lemma 12.4.5. Let f be as in (12.0.1), $z_0 = 0$, $q \in [1, +\infty]$, $\tilde{C}^q(0)$ and $\tilde{f}^q_{\mathrm{hom}}$ be defined in (11.2.2) and (12.1.1) respectively, and F' in (12.1.6).

Assume that (11.2.3) holds. Let $z \in \mathbf{R}^n$ be such that $F'(]-1,2[^n, u_z) < +\infty$. Then

$$\tilde{f}^q_{\mathrm{hom}}(tz) < +\infty \text{ for every } t \in [0,1[.$$

Proof. By the above assumptions there exist $\{v_h\} \subseteq W^{1,q}_{\mathrm{loc}}(\mathbf{R}^n) \cap L^\infty_{\mathrm{loc}}(\mathbf{R}^n)$ and $\{h_k\} \subseteq \mathbf{N}$ strictly increasing such that $v_h \to u_z$ in $L^\infty(]-1,2[^n)$, and

$$(12.4.16) \qquad \int_{]-1,2[^n} f(h_k x, \nabla v_{h_k}) \, dx < +\infty \text{ for every } k \in \mathbf{N}.$$

Since (11.2.3) with $z_0 = 0$ implies (11.1.3), for a fixed $\eta \in]0,1[$, let $\{\psi_h\}$ be given by Lemma 11.1.1 applied to $r_h = h$ for every $h \in \mathbf{N}$, $]-\eta, 1+\eta[^n$ and $K = [0,1]^n$. Because of (11.1.7) and (11.1.8) of Lemma 11.1.1, it then results that for a.e. $x \in \mathbf{R}^n$ and every $h \in \mathbf{N}$ the sum $\sum_{i \in Z^n} \psi_h(x+i)$ is actually extended only to a finite set of indices i, and that

$$(12.4.17) \qquad \sum_{i \in Z^n} \psi_h(x+i) \geq 1 \text{ for a.e. } x \in \mathbf{R}^n, \text{ and every } h \in \mathbf{N}.$$

For every $h \in \mathbf{N}$ let $\tilde{\psi}_h$ be defined by

$$\tilde{\psi}_h(x) = \frac{\psi_h(x)}{\sum_{j \in Z^n} \psi_h(x+j)} \text{ for a.e. } x \in \mathbf{R}^n.$$

Then, for every $h \in \mathbf{N}$, $\tilde{\psi}_h \in W^{1,q}(\mathbf{R}^n) \cap L^\infty(\mathbf{R}^n)$, $\tilde{\psi}_h = 0$ a.e. in $\mathbf{R}^n \setminus]-\eta, 1+\eta[^n$, and $0 \leq \tilde{\psi}_h \leq 1$ a.e. in $\mathbf{R}^n$. Moreover,

$$(12.4.18) \qquad \sum_{i \in Z^n} \tilde{\psi}_h(x+i) = \sum_{i \in Z^n} \frac{\psi_h(x+i)}{\sum_{j \in Z^n} \psi_h(x+j+i)} =$$

$$= \sum_{i \in Z^n} \frac{\psi_h(x+i)}{\sum_{j \in Z^n} \psi_h(x+j)} = 1 \text{ for a.e. } x \in \mathbf{R}^n, \text{ and every } h \in \mathbf{N}.$$

Let now $\{u_h\}$ be the sequence of functions defined by

$$u_h(x) = u_z(x) + \sum_{i \in Z^n} (v_h(x+i) - u_z(x+i)) \tilde{\psi}_h(x+i)$$

for a.e. $x \in \mathbf{R}^n$, and every $h \in \mathbf{N}$.

Then, by using the properties of $\{\tilde{\psi}_h\}$, it is easy to verify that, for every $h \in \mathbf{N}$, the above sums are extended only to a finite set of indices i, and, consequently, that $u_h \in W^{1,q}_{\mathrm{loc}}(\mathbf{R}^n) \cap L^\infty_{\mathrm{loc}}(\mathbf{R}^n)$. Furthermore it also results that

$$(12.4.19) \qquad (u_h - u_z) \in W^{1,q}_{\mathrm{per}}(Y) \cap L^\infty(Y) \text{ for every } h \in \mathbf{N}.$$

In fact

$$(u_h - u_z)(x + \mathbf{e}_j) = \sum_{i \in Z^n} (v_h(x + \mathbf{e}_j + i) - u_z(x + \mathbf{e}_j + i)) \, \tilde{\psi}_h(x + \mathbf{e}_j + i) =$$

$$= \sum_{i \in Z^n} (v_h(x + i) - u_z(x + i)) \, \tilde{\psi}_h(x + i) = (u_h - u_z)(x)$$

for a.e. $x \in \mathbf{R}^n$, and $j \in \{1, \ldots, n\}$.

Let now $t \in [0, 1[$. Let us prove that there exists $k_t \in \mathbf{N}$ such that

(12.4.20) $$\int_Y f\left(h_{k_t} x, tz + t\nabla(u_{h_{k_t}} - u_z)\right) dx < +\infty.$$

In fact, because of (12.4.18) and of the convexity properties of f, it results that

(12.4.21) $$\int_Y f(h_k x, tz + t\nabla(u_{h_k} - u_z)) dx =$$

$$= \int_Y f\left(h_k x, tz + t \sum_{i \in Z^n} \tilde{\psi}_{h_k}(x+i) \nabla(v_{h_k} - u_z)(x+i) + \right.$$

$$\left. + t \sum_{i \in Z^n} (v_{h_k} - u_z)(x+i) \nabla \tilde{\psi}_{h_k}(x+i) \right) dx =$$

$$= \int_Y f\left(h_k x, t \sum_{i \in Z^n} \tilde{\psi}_{h_k}(x+i) \nabla v_{h_k}(x+i) + \right.$$

$$\left. + t \sum_{i \in Z^n} (v_{h_k} - u_z)(x+i) \nabla \tilde{\psi}_{h_k}(x+i) \right) dx \le$$

$$\le t \int_Y f\left(h_k x, \sum_{i \in Z^n} \tilde{\psi}_{h_k}(x+i) \nabla v_{h_k}(x+i) \right) dx +$$

$$+ (1-t) \int_Y f\left(h_k x, \frac{t}{1-t} \sum_{i \in Z^n} (v_{h_k} - u_z)(x+i) \nabla \tilde{\psi}_{h_k}(x+i) \right) dx$$

for every $k \in \mathbf{N}$.

To estimate the last two integrals in (12.4.21) set $I = \{i \in Z^n :$ $(Y+i) \cap]-\eta, 1+\eta[^n \ne \emptyset\}$. Then I has 3^n elements,

(12.4.22) $$\mathcal{L}^n\left(]-1, 2[^n \setminus \cup_{i \in I} (Y+i)\right) = 0,$$

and, because of (12.4.18),

(12.4.23) $$\sum_{i\in I} \tilde{\psi}_h(x+i) = 1 \text{ for a.e. } x \in Y, \text{ and every } h \in \mathbf{N}.$$

The convexity and periodicity properties of f, (12.4.23), (12.4.22), and (12.4.16) provide the finiteness of the first integral in the last term of (12.4.21). In fact

(12.4.24) $$\int_Y f\left(h_k x, \sum_{i\in Z^n} \tilde{\psi}_{h_k}(x+i)\nabla v_{h_k}(x+i)\right) dx =$$

$$= \int_Y f\left(h_k x, \sum_{i\in I} \tilde{\psi}_{h_k}(x+i)\nabla v_{h_k}(x+i)\right) dx \le$$

$$\le \sum_{i\in I} \int_Y \tilde{\psi}_{h_k}(x+i) f\left(h_k x, \nabla v_{h_k}(x+i)\right) dx \le$$

$$\le \sum_{i\in I} \int_{Y+i} f\left(h_k(y-i), \nabla v_{h_k}(y)\right) dy = \sum_{i\in I} \int_{Y+i} f\left(h_k y, \nabla v_{h_k}\right) dy =$$

$$= \int_{]-1,2[^n} f(h_k y, \nabla v_{h_k}) dy < +\infty$$

for every $k \in \mathbf{N}$.

In order to treat the last integral in (12.4.21), for a.e. $x \in Y$, every $k \in \mathbf{N}$, and $i \in I$ let us set $\lambda_k^{(i)}(x) = \frac{1}{2\sum_{j\in Z^n} \psi_{h_k}(x+i+j)}$, $\mu_k^{(i)}(x) = \frac{\psi_{h_k}(x+i)}{2(\sum_{j\in Z^n} \psi_{h_k}(x+i+j))^2}$, and observe that, by (12.4.17), it results $0 \le \lambda_k^{(i)}(x) \le \frac{1}{2}$, $0 \le \mu_k^{(i)}(x) \le \frac{1}{2}$ for a.e. $x \in Y$, every $k \in \mathbf{N}$ and $i \in I$. Then the convexity properties of f provide that

(12.4.25) $$\int_Y f\left(h_k x, \frac{t}{1-t}\sum_{i\in Z^n}(v_{h_k}-u_z)(x+i)\nabla\tilde{\psi}_{h_k}(x+i)\right) dx =$$

$$= \int_Y f\left(h_k x, 3^n\frac{t}{1-t}\sum_{i\in I}\frac{1}{3^n}(v_{h_k}-u_z)(x+i)\nabla\tilde{\psi}_{h_k}(x+i)\right) dx \le$$

$$\le \sum_{i\in I}\frac{1}{3^n}\int_Y f\left(h_k x, 3^n\frac{t}{1-t}(v_{h_k}-u_z)(x+i)\nabla\tilde{\psi}_{h_k}(x+i)\right) dx \le$$

$$\le \sum_{i\in I}\int_Y f\left(h_k x, 3^n\frac{t}{1-t}\left(2\lambda_k^{(i)}(x)(v_{h_k}-u_z)(x+i)\nabla\psi_{h_k}(x+i)-\right.\right.$$

$$2\mu_k^{(i)}(x)(v_{h_k}-u_z)(x+i)\sum_{j\in Z^n}\nabla\psi_{h_k}(x+i+j)+\left(1-\lambda_k^{(i)}(x)-\mu_k^{(i)}(x)\right)0\Big)\Big)dx\le$$

$$\le\sum_{i\in I}\int_Y f\left(h_kx,3^n2\frac{t}{1-t}(v_{h_k}-u_z)(x+i)\nabla\psi_{h_k}(x+i)\right)dx+$$

$$+\sum_{i\in I}\int_Y f\left(h_kx,-3^n2\frac{t}{1-t}(v_{h_k}-u_z)(x+i)\sum_{j\in Z^n}\nabla\psi_{h_k}(x+i+j)\right)dx+$$

$$+3^n\int_Y f(h_kx,0)dx \text{ for every } k\in\mathbf{N}.$$

Consider now the first term in the last sum of (12.4.25).

Fix $i\in I$. Then the periodicity properties of f provide that

$$(12.4.26)\qquad \int_Y f\left(h_kx,3^n2\frac{t}{1-t}(v_{h_k}-u_z)(x+i)\nabla\psi_{h_k}(x+i)\right)dx=$$

$$=\int_{Y+i} f\left(h_k(y-i),3^n2\frac{t}{1-t}(v_{h_k}-u_z)(y)\nabla\psi_{h_k}(y)\right)dy=$$

$$=\int_{Y+i} f\left(h_ky,3^n2\frac{t}{1-t}(v_{h_k}-u_z)\nabla\psi_{h_k}\right)dy \text{ for every } k\in\mathbf{N}.$$

Since $Y+i\subseteq]-1,2[^n$, and $v_h\to u_z$ in $L^\infty(]-1,2[^n)$, it results that there exists $k_{\eta,t}\in\mathbf{N}$ such that

$$(12.4.27)\qquad 3^n2\frac{t}{1-t}(v_{h_k}-u_z)(y)\in\left]-\frac{\delta\eta}{64n3^n\sqrt{n}},\frac{\delta\eta}{64n3^n\sqrt{n}}\right[$$

$$\text{for a.e. } y\in Y+i \text{ and every } k\ge k_{\eta,t},$$

where δ is given by (11.2.4). Consequently, an argument similar to the one utilized to get (12.2.7), together with (11.1.10) of Lemma 11.1.1, yields

$$(12.4.28)\qquad \int_{Y+i} f\left(h_ky,3^n2\frac{t}{1-t}(v_{h_k}-u_z)\nabla\psi_{h_k}\right)dy\le$$

$$\le\int_{Y+i} f^{(0)}\left(h_ky,\frac{\delta\eta}{64n3^n\sqrt{n}}\nabla\psi_{h_k}\right)dy<+\infty \text{ for every } k\ge k_{\eta,t},$$

where $f^{(0)}$ is defined by (11.2.1).

By combining (12.4.26) with (12.4.28), the finiteness of the first term in the last sum of (12.4.25) is obtained, i.e.

$$(12.4.29)\quad \sum_{i\in I}\int_Y f\left(h_kx,3^n2\frac{t}{1-t}(v_{h_k}(x+i)-u_z(x+i))\right)$$

$$\nabla\psi_{h_k}(x+i)\Big)dx < +\infty \text{ for every } k \geq k_{\eta,t}.$$

Consider, now, the second term in the last sum of (12.4.25).
Fix $i \in I$. Then the periodicity properties of f provide that

$$(12.4.30) \quad \int_Y f\left(h_k x, -3^n 2\frac{t}{1-t}(v_{h_k}-u_z)(x+i)\sum_{j\in Z^n}\nabla\psi_{h_k}(x+i+j)\right)dx =$$

$$= \int_{Y+i} f\left(h_k(y-i), -3^n 2\frac{t}{1-t}(v_{h_k}-u_z)(y)\sum_{j\in Z^n}\nabla\psi_{h_k}(y+j)\right)dy =$$

$$= \int_{Y+i} f\left(h_k y, -3^n 2\frac{t}{1-t}(v_{h_k}-u_z)(y)\sum_{j\in Z^n}\nabla\psi_{h_k}(y+j)\right)dy$$

for every $k \in \mathbf{N}$.

Let $J_i = \{j \in \mathbf{Z}^n : (Y+i)\cap]-\eta-j, 1+\eta-j[^n\neq\emptyset\}$. It is obvious that J_i has 3^n elements. Consequently, (12.4.30) and the convexity properties of f imply that

$$(12.4.31) \quad \int_Y f\left(h_k x, -3^n 2\frac{t}{1-t}(v_{h_k}-u_z)(x+i)\sum_{j\in Z^n}\nabla\psi_{h_k}(x+i+j)\right)dx =$$

$$= \int_{Y+i} f\left(h_k y, -3^n 2\frac{t}{1-t}(v_{h_k}-u_z)(y)3^n\sum_{j\in J_i}\frac{1}{3^n}\nabla\psi_{h_k}(y+j)\right)dy \leq$$

$$\leq \sum_{j\in J_i}\int_{Y+i} f\left(h_k y, -9^n 2\frac{t}{1-t}(v_{h_k}-u_z)(y)\nabla\psi_{h_k}(y+j)\right)dy$$

for every $k \in \mathbf{N}$.

By arguing as in (12.4.27) and (12.4.28), by using also the periodicity properties of f, and (11.1.10) of Lemma 11.1.1, it follows that

$$(12.4.32) \quad \sum_{j\in J_i}\int_{Y+i} f\left(h_k y, -9^n 2\frac{t}{1-t}(v_{h_k}-u_z)\nabla\psi_{h_k}(y+j)\right)dy \leq$$

$$\leq \sum_{j\in J_i}\int_{Y+i} f^{(0)}\left(h_k y, \frac{\delta\eta}{64n3^n\sqrt{n}}\nabla\psi_{h_k}(y+j)\right)dy =$$

$$= \sum_{j\in J_i}\int_{Y+i+j} f^{(0)}\left(h_k(y-j), \frac{\delta\eta}{64n3^n\sqrt{n}}\nabla\psi_{h_k}(y)\right)dy =$$

$$= \sum_{j \in J_i} \int_{Y+i+j} f^{(0)} \left(h_k y, \frac{\delta \eta}{64 n 3^n \sqrt{n}} \nabla \psi_{h_k} \right) dy < +\infty$$

for every k sufficiently large.

In conclusion, by combining (12.4.31) with (12.4.32), we obtain that

$$(12.4.33) \quad \sum_{i \in I} \int_Y f \left(h_k x, -3^n 2 \frac{t}{1-t} (v_{h_k} - u_z)(x+i) \right.$$

$$\left. \sum_{j \in Z^n} \nabla \psi_{h_k}(x+i+j) \right) dx < +\infty \text{ for every } k \text{ sufficiently large.}$$

Inequality (12.4.20) now follows by combining (12.4.21) with (12.4.24), (12.4.25), (12.4.29), and (12.4.33).

Finally Lemma 12.4.4, (12.4.19), and (12.4.20) provide the lemma. ∎

We now prove the reverse inequality of (12.4.1) with F'' replaced by F'.

Proposition 12.4.6. *Let* f *be as in* (12.0.1), $z_0 = 0$, $q \in [1, +\infty]$, $\tilde{C}^q(0)$ *and* $\tilde{f}^q_{\mathrm{hom}}$ *be defined in* (11.2.2) *and* (12.1.1) *respectively, and* F' *in* (12.1.6). *Assume that* (11.2.3) *holds. Then*

$$(12.4.34) \quad \mathcal{L}^n(\Omega)\mathrm{sc}^- \tilde{f}^q_{\mathrm{hom}}(z) \leq F'(\Omega, u_z) \text{ for every } \Omega \in \mathcal{A}_0, \ z \in \mathbf{R}^n.$$

Proof. Fix $z \in \mathbf{R}^n$. Let us first consider the case $\Omega = Y$.

Clearly we can assume that

$$(12.4.35) \quad F'(Y, u_z) < +\infty.$$

Consequently Lemma 12.4.3 yields

$$(12.4.36) \quad F'(]-1, 2[^n, u_z) < +\infty.$$

Fix $t \in \ ]0, 1[$. Then (12.4.36) and Lemma 12.4.5 provide that $\tilde{f}^q_{\mathrm{hom}}(tz)$ $< +\infty$, whence there exists $v \in W^{1,q}_{\mathrm{per}}(Y) \cap L^\infty(Y)$ such that $\int_Y f(y, tz + \nabla v) dy < +\infty$.

For every $h \in \mathbf{N}$ let us set $v_h = \frac{1}{h} v(h \cdot)$. Then $v_h \in W^{1,q}_{\mathrm{per}}(Y) \cap L^\infty_{\mathrm{loc}}(\mathbf{R}^n)$ for every $h \in \mathbf{N}$, and $v_h \to 0$ in $L^\infty(Y)$.

On the other hand, (11.2.5), (12.4.35), and Lemma 12.4.2 provide that $F'(Y, tu_z) < +\infty$. Hence, there exist $\{u_h\} \subseteq W^{1,q}_{\mathrm{loc}}(\mathbf{R}^n) \cap L^\infty_{\mathrm{loc}}(\mathbf{R}^n)$ and $\{h_k\} \subseteq \mathbf{N}$ strictly increasing such that $u_h \to tu_z$ in $L^\infty(Y)$ and

$$(12.4.37) \quad F'(Y, tu_z) = \lim_{k \to +\infty} \int_Y f(h_k x, \nabla u_{h_k}) dx.$$

Let now Ω', Ω'' be open sets such that $\Omega' \subset\subset \Omega'' \subset\subset Y$, $\{\psi_h\}$ be given by Lemma 11.1.1 applied to $r_h = h$ for every $h \in \mathbf{N}$, Ω'' and $K = \overline{\Omega'}$, and, for every $k \in \mathbf{N}$, let $w_k \in W^{1,q}_{\mathrm{loc}}(\mathbf{R}^n)Y) \cap L^\infty_{\mathrm{loc}}(\mathbf{R}^n)$ defined by $w_k = \psi_{h_k} u_{h_k} + (1 - \psi_{h_k})(v_{h_k} + tu_z)$.

It is obvious that $w_k - tu_z \in W^{1,q}_{\mathrm{per}}(Y) \cap L^\infty(Y)$ for every $k \in \mathbf{N}$, and $w_k \to tu_z$ in $L^\infty(Y)$. Consequently, because of Lemma 12.4.4, and of the convexity properties of f it results that

$$(12.4.38) \qquad \tilde{f}^q_{\mathrm{hom}}(t^2 z) =$$

$$= \inf \left\{ \int_Y f\left(h_k x, t^2 z + \nabla v\right) dx : v \in W^{1,q}_{\mathrm{per}}(Y) \cap L^\infty(Y) \right\} \le$$

$$\le \int_Y f\left(h_k x, t^2 z + \nabla\left(t(w_k - tu_z)\right)\right) dx = \int_Y f(h_k x, t\nabla w_k) dx \le$$

$$\le t \int_Y f(h_k x, \psi_{h_k} \nabla u_{h_k} + (1 - \psi_{h_k})(\nabla v_{h_k} + tz)) dx +$$

$$+ (1-t) \int_Y f\left(h_k x, \frac{t}{1-t}(u_{h_k} - v_{h_k} - tu_z)\nabla\psi_{h_k}\right) dx \le$$

$$\le t \int_Y f(h_k x, \nabla u_{h_k}) dx + t \int_{Y \setminus \overline{\Omega'}} f(h_k x, \nabla v(h_k x) + tz) dx +$$

$$+ (1-t) \int_Y f\left(h_k x, \frac{t}{1-t}(u_{h_k} - v_{h_k} - tu_z)\nabla\psi_{h_k}\right) dx$$

for every $k \in \mathbf{N}$.

On the other hand, it turns out that $u_{h_k} - v_{h_k} - tu_z \to 0$ in $L^\infty(Y)$. Consequently, by an argument similar to the one used to get (12.2.7), there exists $k_t \in \mathbf{N}$ such that, for every $k > k_t$,

$$(12.4.39) \qquad \int_Y f\left(h_k x, \frac{t}{1-t}(u_{h_k} - v_{h_k} - tu_z)\nabla\psi_{h_k}\right) dx \le$$

$$\le \int_Y f^{(0)}\left(h_k x, \frac{\delta \mathrm{dist}(\Omega', \partial\Omega'')}{64 n 3^n \sqrt{n}}\nabla\psi_{h_k}\right) dx,$$

where $f^{(0)}$ is defined by (11.2.1).

Then (11.1.10) of Lemma 11.1.1 and (12.4.39) provide that

$$(12.4.40) \qquad \limsup_{k \to +\infty} \int_Y f\left(h_k x, \frac{t}{1-t}(u_{h_k} - v_{h_k} - tu_z)\nabla\psi_{h_k}\right) dx \le$$

$$\le c_f \mathcal{L}^n(Y) = c_f,$$

where c_f is defined in Remark 11.1.2.

By passing to the limit in (12.4.38) as k tends to infinity, (12.4.37) and (12.4.40) provide that

$$\tilde{f}^q_{\text{hom}}(t^2 z) \leq F'(Y, tu_z) + \mathcal{L}^n (Y \setminus \Omega') \int_Y f(y, \nabla v + tz) \, dy + (1 - t)c_f,$$

from which, by using also Lemma 12.4.2, it follows that

(12.4.41) $\tilde{f}^q_{\text{hom}}(t^2 z) \leq$

$$\leq F'(Y, u_z) + (1-t) \int_Y f(y, 0) dy + \mathcal{L}^n(Y \setminus \Omega') \int_Y f(y, \nabla v + tz) dy + (1-t)c_f.$$

Inequality (12.4.41) holds for all $t \in [0, 1[$ and $\Omega' \subset\subset Y$. Therefore, as Ω' increases to Y, and t converges to 1^- in (12.4.41), by (11.2.4) it results that

(12.4.42) $\text{sc}^- \tilde{f}^q_{\text{hom}}(z) \leq \liminf_{t \to 1^-} \tilde{f}^q_{\text{hom}}(t^2 z) \leq F'(Y, u_z).$

Consider now the general case in which Ω is a bounded open set.

For every $k \in \mathbf{N}$ let $Q_1^k, \ldots, Q_{m_k}^k, B_1^k, \ldots, B_{m_k}^k$ be cubes with faces parallel to the coordinate planes such that $Q_i^k \cap Q_j^k = \emptyset$ if $i \neq j$, $\cup_{j=1}^{m_k} Q_j^k \subseteq \Omega$, $B_j^k \subset\subset Q_j^k$ for every $j \in \{1, \ldots, m_k\}$, and

(12.4.43) $\mathcal{L}^n \left(\Omega \setminus \cup_{j=1}^{m_k} Q_j^k \right) < \dfrac{1}{k}, \; \mathcal{L}^n \left(Q_j^k \setminus B_j^k \right) < \dfrac{1}{m_k k}$

for every $j \in \{1, \ldots, m_k\}$.

From (12.2.2) of Proposition 12.2.1 it follows that

(12.4.44) $F'(\Omega, u_z) \geq F'_-(\Omega, u_z) \geq F'_-(\cup_{j=1}^{m_k} Q_j^k, u_z) \geq$

$$\geq \sum_{j=1}^{m_k} F'_-(Q_j^k, u_z) \geq \sum_{j=1}^{m_k} F'(B_j^k, u_z) \text{ for every } k \in \mathbf{N}.$$

On the other hand, Lemma 12.4.3 and (12.4.42) provide that

(12.4.45) $F'(B_j^k, u_z) = \mathcal{L}^n \left(B_j^k \right) F'(Y, u_z) \geq \mathcal{L}^n \left(B_j^k \right) \text{sc}^- \tilde{f}^q_{\text{hom}}(z)$

for every $j \in \{1, \ldots, m_k\}$, $k \in \mathbf{N}$.

Therefore, by combining (12.4.44) with (12.4.45) and (12.4.43), it results that

(12.4.46) $F'(\Omega, u_z) \geq \sum_{j=1}^{m_k} \mathcal{L}^n \left(B_j^k \right) \text{sc}^- \tilde{f}^q_{\text{hom}}(z) =$

$$= \mathcal{L}^n \left(\cup_{j=1}^{m_k} B_j^k \right) \text{sc}^- \tilde{f}^q_{\text{hom}}(z) \geq \left(\mathcal{L}^n(\Omega) - \dfrac{2}{k} \right) \text{sc}^- \tilde{f}^q_{\text{hom}}(z)$$

for every $k \in \mathbf{N}$.

As k tends to infinity in (12.4.46), inequality (12.4.34) follows. ∎

Combining Lemma 12.4.1 with Proposition 12.4.6, the result below follows.

Proposition 12.4.7. *Let f be as in (12.0.1), $z_0 = 0$, $q \in [1, +\infty]$, $\tilde{C}^q(0)$ and $\tilde{f}^q_{\text{hom}}$ be defined in (11.2.2) and (12.1.1) respectively, and F', F'' in (12.1.6). Assume that (11.2.3) holds. Then*

$$F'(\Omega, u_z) = F''(\Omega, u_z) = F'_-(\Omega, u_z) = F''_-(\Omega, u_z) = \mathcal{L}^n(\Omega)\text{sc}^- \tilde{f}^q_{\text{hom}}(z)$$

for every $\Omega \in \mathcal{A}_0$, $z \in \mathbf{R}^n$.

§12.5 A Blow-up Condition

In this section we prove that the functional F'_- defined by means of (12.1.6) satisfies a blow-up condition.

Lemma 12.5.1. *Let f be as in (12.0.1), $q \in [1, +\infty]$, and F' be defined in (12.1.6). Then*

$$F'_-\big(t\Omega, O_{1/t}u\big) = t^n F'_-(\Omega, u) \text{ for every } \Omega \in \mathcal{A}_0, \ u \in C^0(\mathbf{R}^n), \ t \in]0, +\infty[.$$

Proof. Proof. Let Ω, u, t be as above. Let us prove that

$$(12.5.1) \qquad\qquad F'_-\big(t\Omega, O_{1/t}u\big) \geq t^n F'_-(\Omega, u).$$

We can assume the left-hand side of (12.5.1) to be finite, so that for every $A \subset\subset \Omega$ there exist $\{h_k\} \subseteq \mathbf{N}$ strictly increasing, and $\{u_h\} \subseteq W^{1,q}_{\text{loc}}(\mathbf{R}^n) \cap L^\infty_{\text{loc}}(\mathbf{R}^n)$ with $u_h \to O_{1/t}u$ in $L^\infty(tA)$, and

$$F'_-(t\Omega, O_{1/t}u) \geq \lim_{k\to+\infty} \int_{tA} f(h_k x, \nabla u_{h_k})dx.$$

By performing in the last inequality the change of variable $x = ty$, we deduce that $O_t u_h \to u$ in $L^\infty(A)$, and, by (3.2.5), that

$$(12.5.2) \qquad F'_-(t\Omega, O_{1/t}u) \geq t^n \lim_{k\to+\infty} \int_A f(th_k x, \nabla O_t u_{h_k})dy \geq$$

$$\geq t^n \Gamma^-(L^\infty(A)) \liminf_{h\to+\infty} F_{th}(A, u) \text{ for every } A \in \mathcal{A}_0 \text{ with } A \subset\subset \Omega.$$

At this point, by (12.5.2), Lemma 12.1.2, and (3.2.5), we infer (12.5.1).

By symmetry, the reverse inequality of (12.5.1) follows. This completes the proof of the lemma. ∎

Proposition 12.5.2. *Let f be as in (12.0.1), $q \in [1, +\infty]$, and F' be defined in (12.1.6). Then*

$$(12.5.3) \quad \limsup_{r\to 0^+} \frac{1}{r^n} F'_-(Q_r(x_0), u) \geq F'_-(Q_1(x_0), u(x_0) + \nabla u(x_0) \cdot (\cdot - x_0))$$

for a.e. $x_0 \in \mathbf{R}^n$, and every $u \in W^{1,\infty}_{\mathrm{loc}}(\mathbf{R}^n)$.

Proof. Let x_0, u be as in (12.5.3). Because of (12.1.10), Proposition 12.1.1, and Lemma 12.5.1 it results that

(12.5.4)

$$\limsup_{r \to 0^+} \frac{1}{r^n} F'_-(Q_r(x_0), u) = \limsup_{r \to 0^+} \frac{1}{r^n} F'_-(Q_r(0), T[x_0](u - u(x_0))) \geq$$

$$\geq \liminf_{r \to +0^+} F'_-(Q_1(0), O_r T[x_0](u - u(x_0))) =$$

$$= \liminf_{r \to +0^+} F'_-(Q_1(x_0), T[-x_0]O_r T[x_0](u - u(x_0))).$$

We now recall that

$$T[-x_0]O_r T[x_0](u - u(x_0)) \to \nabla u(x_0)(\cdot - x_0) \text{ in } L^\infty(Q_1(x_0)) \text{ as } r \to 0^+.$$

Then (12.5.3) follows from (12.5.4), the $L^\infty(Q_1(x_0))$-lower semicontinuity of $F'_-(Q_1(x_0), \cdot)$, and (12.1.10). ∎

§12.6 Representation Results

In this section we prove some integral representation result for the Γ-limits of the functionals in (12.1.3).

Proposition 12.6.1. Let f be as in (12.0.1), $z_0 = 0$, $q \in [1, +\infty]$, $\widetilde{C}^q(0)$ and $\widetilde{f}^q_{\mathrm{hom}}$ be defined in (11.2.2) and (12.1.1) respectively, and F', F'' in (12.1.6). Assume that (11.2.3) holds. Then

$$F'_-(\Omega, u) = F''_-(\Omega, u) = \int_\Omega \mathrm{sc}^- \widetilde{f}^q_{\mathrm{hom}}(\nabla u) dx$$

for every $\Omega \in \mathcal{A}_0$, $u \in \cup_{s>n} W^{1,s}_{\mathrm{loc}}(\mathbf{R}^n)$.

Proof. Let $\{h_k\} \subseteq \mathbf{N}$ be strictly increasing. Then Proposition 3.4.3 provides the existence of $\{h_{k_j}\} \subseteq \{h_k\}$ such that

(12.6.1)

$$\sup \left\{ \Gamma^-(L^\infty(A)) \liminf_{j \to +\infty} F_{h_{k_j}}(A, u) : A \subset\subset \Omega \right\} =$$

$$= \sup \left\{ \Gamma^-(L^\infty(A)) \limsup_{j \to +\infty} F_{h_{k_j}}(A, u) : A \subset\subset \Omega \right\}$$

for every $\Omega \in \mathcal{A}_0, u \in L^\infty_{\mathrm{loc}}(\mathbf{R}^n)$.

Let now $p \in]n, +\infty]$. Then $W^{1,p}_{\mathrm{loc}}(\mathbf{R}^n) \subseteq C^0(\mathbf{R}^n) \subseteq L^\infty_{\mathrm{loc}}(\mathbf{R}^n)$. Consequently, for every $\Omega \in \mathcal{A}_0$, we can consider the functional $G(\Omega, \cdot)$ defined in $W^{1,p}_{\text{loc}}(\mathbf{R}^n)$ that to every u assigns the value in (12.6.1), and prove that G fulfils the assumptions of Theorem 9.3.8.

In fact (9.3.8) is trivial, (9.2.5) follows from (12.1.10), (9.3.31) from Proposition 12.1.1, (9.3.32) from (12.1.7) and Proposition 12.2.1. Moreover, (9.3.33) comes from (3.2.5), Proposition 12.5.2, Proposition 12.4.7 and (12.1.10), and (9.3.34) from (12.1.8).

In order to verify (12.3.3) we preliminarily observe that Proposition 12.4.7 yields that f_G in Theorem 9.3.8 agrees with $\text{sc}^- \tilde{f}^q_{\text{hom}}$, and that, by Proposition 11.2.1, we have that $0 \in \tilde{C}^q(0) \subseteq \text{dom} \tilde{f}^q_{\text{hom}} \subseteq \text{dom} \, \text{sc}^- \tilde{f}^q_{\text{hom}}$. Therefore (12.3.3) with $z_0 = 0$ follows from Proposition 12.3.2, and (3.2.5). Moreover (9.3.7) too holds, in fact, given $\Omega \in \mathcal{A}_0$, and an open set with Lipschitz boundary A such that $A \subset\subset \Omega$, Proposition 3.3.2 yields that the functionals $\Gamma^-(L^\infty(A)) \liminf_{j \to +\infty} F_{h_{k_j}}(A, \cdot)$, and $\Gamma^-(L^\infty(A)) \limsup_{j \to +\infty} F_{h_{k_j}}(A, \cdot)$ are $W^{1,p}(\Omega)$ $(\cap_{q \in [1, +\infty[} W^{1,q}(\Omega)$ if $p = +\infty)$ -lower semicontinuous in $W^{1,p}_{\text{loc}}(\mathbf{R}^n)$, and hence that so is also $G(\Omega, \cdot)$, since it agrees with the last upper bound of the family of such functionals obtained letting A vary with the above properties.

Consequently, by Theorem 9.3.8, Proposition 12.4.7, and (3.2.5) it follows that

$$G(\Omega, u) = \int_\Omega \text{sc}^- \tilde{f}^q_{\text{hom}}(\nabla u) dx \text{ for every } \Omega \in \mathcal{A}_0, u \in W^{1,p}_{\text{loc}}(\mathbf{R}^n).$$

Then we have proved that

for every $\{h_k\} \subseteq \mathbf{N}$ strictly increasing there exists $\{h_{k_j}\} \subseteq \{h_k\}$ such that

$$\sup \left\{ \Gamma^-(L^\infty(A)) \liminf_{j \to +\infty} F_{h_{k_j}}(A, u) : A \subset\subset \Omega \right\} =$$

$$= \sup \left\{ \Gamma^-(L^\infty(A)) \limsup_{j \to +\infty} F_{h_{k_j}}(A, u) : A \subset\subset \Omega \right\} =$$

$$= \int_\Omega \text{sc}^- \tilde{f}^q_{\text{hom}}(\nabla u) dx \text{ for every } \Omega \in \mathcal{A}_0, \ u \in W^{1,p}_{\text{loc}}(\mathbf{R}^n).$$

Because of this, and by Proposition 3.4.2, we thus have that

$$F'_-(\Omega, u) = F''_-(\Omega, u) = \int_\Omega \text{sc}^- \tilde{f}^q_{\text{hom}}(\nabla u) dx$$

$$\text{for every } \Omega \in \mathcal{A}_0, \ u \in W^{1,p}_{\text{loc}}(\mathbf{R}^n),$$

from which, as p varies in $]n, +\infty]$, the proof follows. ∎

The following representation result in the Dirichlet case holds.

Theorem 12.6.2. Let f be as in (12.0.1), $z_0 \in \mathbf{R}^n$, $c \in \mathbf{R}$, $q \in [1, +\infty]$, $\Omega \in \mathcal{A}_0$, $\widetilde{C}^q(z_0)$ and $\tilde{f}^q_{\text{hom}}$ be defined in (11.2.2) and (12.1.1) respectively, and $F_r(\Omega, \partial\Omega, u_{z_0} + c, \cdot)$ in (12.1.3) for every $r \in]0, +\infty[$. Assume that (11.2.3) holds. Then

$$\Gamma^-(L^\infty(\Omega)) \liminf_{\varepsilon \to 0^+} F_{1/\varepsilon}(\Omega, \partial\Omega, u_{z_0} + c, u) =$$

$$= \Gamma^-(L^\infty(\Omega)) \limsup_{\varepsilon \to 0^+} F_{1/\varepsilon}(\Omega, \partial\Omega, u_{z_0} + c, u) =$$

$$= \int_\Omega \text{sc}^- \tilde{f}^q_{\text{hom}}(\nabla u) dx \text{ for every } u \in u_{z_0} + c + \cup_{s>n} W^{1,s}_0(\Omega).$$

Proof. Let u be as above, and let $\{\varepsilon_h\} \subseteq]0, +\infty[$ be strictly decreasing and converging to 0.

Let F' and F'' be defined in (12.1.6), and $\widetilde{F}'$, $\widetilde{F}''$, $\widetilde{F}'(\Omega, \partial\Omega, u_{z_0} + c, \cdot)$ and $\widetilde{F}''(\Omega, \partial\Omega, u_{z_0} + c, \cdot)$ by (12.1.4) and (12.1.5) with $r_h = 1/\varepsilon_h$ for every $h \in \mathbf{N}$. Then, from Proposition 12.6.1, Lemma 12.1.2, and Proposition 12.2.2 applied to the function $(x, z) \in \mathbf{R}^n \times \mathbf{R}^n \mapsto f(x, z_0 + z)$ we obtain that

$$\int_\Omega \text{sc}^- \tilde{f}^q_{\text{hom}}(\nabla u) dx = F'_-(\Omega, u) \leq \widetilde{F}'_-(\Omega, u) = \widetilde{F}'(\Omega, \partial\Omega, u_{z_0} + c, u) \leq$$

$$\leq \widetilde{F}''(\Omega, \partial\Omega, u_{z_0} + c, u) = \widetilde{F}''_-(\Omega, u) \leq F''_-(\Omega, u) = \int_\Omega \text{sc}^- \tilde{f}^q_{\text{hom}}(\nabla u) dx$$

$$\text{for every } u \in u_{z_0} + c + \cup_{s>n} W^{1,s}_0(\Omega),$$

from which, together with Proposition 3.2.3 and Proposition 3.2.6, the proof follows. ∎

Regarding the Neumann case, the following result holds.

Theorem 12.6.3. Let f be as in (12.0.1), $z_0 \in \mathbf{R}^n$, $q \in [1, +\infty]$, $\widetilde{C}^q(z_0)$ and $\tilde{f}^q_{\text{hom}}$ be defined in (11.2.2) and (12.1.1) respectively, and F_r in (12.1.2) for every $r \in]0, +\infty[$. Assume that (11.2.3) holds. Then

$$\Gamma^-(L^\infty(\Omega)) \liminf_{\varepsilon \to 0^+} F_{1/\varepsilon}(\Omega, u) = \Gamma^-(L^\infty(\Omega)) \limsup_{\varepsilon \to 0^+} F_{1/\varepsilon}(\Omega, u) =$$

$$= \int_\Omega \text{sc}^- \tilde{f}^q_{\text{hom}}(\nabla u) dx \text{ for every } \Omega \in \mathcal{A}_0 \text{ convex}, u \in \cup_{s>n} W^{1,s}_{\text{loc}}(\mathbf{R}^n).$$

Proof. Let us first consider the case in which $z_0 = 0$.

Let $p \in]n, +\infty]$.

Let F' and F'' be given by (12.1.6). Then, by Lemma 12.1.2, Proposition 3.2.3, and Proposition 3.2.6, it follows that

$$F'_-(\Omega, u) \leq \sup \left\{ \Gamma^-(L^\infty(A)) \liminf_{\varepsilon \to 0^+} F_{1/\varepsilon}(A, u) : A \subset\subset \Omega \right\} \leq$$

$$\leq \sup\left\{\Gamma^-(L^\infty(A))\limsup_{\varepsilon\to0^+} F_{1/\varepsilon}(A,u) : A \subset\subset \Omega\right\} \leq F''_-(\Omega,u)$$

for every $\Omega \in \mathcal{A}_0$, $u \in C^0(\mathbf{R}^n)$.

Consequently, by making use of Proposition 12.6.1, we infer that

$$(12.6.2) \qquad \sup\left\{\Gamma^-(L^\infty(A))\liminf_{\varepsilon\to0^+} F_{1/\varepsilon}(A,u) : A \subset\subset \Omega\right\} =$$

$$= \sup\left\{\Gamma^-(L^\infty(A))\limsup_{\varepsilon\to0^+} F_{1/\varepsilon}(A,u) : A \subset\subset \Omega\right\} = \int_\Omega \mathrm{sc}^-\tilde{f}^q_{\mathrm{hom}}(\nabla u)dx$$

for every $\Omega \in \mathcal{A}_0$, $u \in W^{1,p}_{\mathrm{loc}}(\mathbf{R}^n)$.

To complete the proof, let us verify that $\Gamma^-(L^\infty(\cdot))\liminf_{\varepsilon\to0^+} F_{1/\varepsilon}$ and $\Gamma^-(L^\infty(\cdot))\limsup_{\varepsilon\to0^+} F_{1/\varepsilon}$ fulfil the assumptions of Proposition 2.7.4 with $\mathcal{O} = \mathcal{A}_0$, and $U = W^{1,p}_{\mathrm{loc}}(\mathbf{R}^n)$.

By (12.1.7) they are increasing. Moreover, the continuity of the elements of $W^{1,p}_{\mathrm{loc}}(\mathbf{R}^n)$ implies that $T[-x_0]O_tT[x_0]u \to u$ uniformly in Ω as $t \to 1^-$ for every $\Omega \in \mathcal{A}_0$, $x_0 \in \mathbf{R}^n$, $u \in W^{1,p}_{\mathrm{loc}}(\mathbf{R}^n)$, consequently, by Proposition 3.3.2, (2.7.2) follows. Finally, because of (12.6.2), (2.7.3) too holds. Consequently, Proposition 2.7.4 applies, and (12.6.2) yields

$$\Gamma^-(L^\infty(\Omega))\liminf_{\varepsilon\to0^+} F_{1/\varepsilon}(\Omega,u) =$$

$$= \Gamma^-(L^\infty(\Omega))\limsup_{\varepsilon\to0^+} F_{1/\varepsilon}(\Omega,u) = \int_\Omega \mathrm{sc}^-\tilde{f}^q_{\mathrm{hom}}(\nabla u)dx$$

for every $\Omega \in \mathcal{A}_0$ convex, $u \in W^{1,p}_{\mathrm{loc}}(\mathbf{R}^n)$,

from which, letting p vary in $]n, +\infty]$, the proof follows if $z_0 = 0$.

Finally, if $z_0 \neq 0$, the theorem follows from the above considered particular case applied to the function $(x,z) \in \mathbf{R}^n \times \mathbf{R}^n \mapsto f(x,z_0+z)$. ∎

By Theorem 12.6.3 we deduce the following result concerning the mixed problem case.

Theorem 12.6.4. Let f be as in (12.0.1), $z_0 \in \mathbf{R}^n$, $c \in \mathbf{R}$, $q \in [1,+\infty]$, $\Omega \in \mathcal{A}_0$ be convex, $\Gamma \subseteq \partial\Omega$, $\tilde{C}^q(z_0)$ and $\tilde{f}^q_{\mathrm{hom}}$ be defined in (11.2.2) and (12.1.1), and $F_r(\Omega,\Gamma,u_{z_0}+c,\cdot)$ in (12.1.3) for every $r \in]0,+\infty[$. Assume that (11.2.3) holds. Then

$$\Gamma^-(L^\infty(\Omega))\liminf_{\varepsilon\to0^+} F_{1/\varepsilon}(\Omega,\Gamma,u_{z_0}+c,u) =$$

$$= \Gamma^-(L^\infty(\Omega))\limsup_{\varepsilon\to0^+} F_{1/\varepsilon}(\Omega,\Gamma,u_{z_0}+c,u) = \int_\Omega \mathrm{sc}^-\tilde{f}^q_{\mathrm{hom}}(\nabla u)dx$$

for every $u \in \cup_{s>n} W_{\text{loc}}^{1,s}(\mathbf{R}^n)$ such that $u = u_{z_0} + c$ on Γ.

Proof. Let Ω, u be as above, and let $\{\varepsilon_h\} \subseteq]0, +\infty[$ be strictly decreasing and converging to 0.

Let $\widetilde{F}'$, $\widetilde{F}''$, $\widetilde{F}'(\Omega, \Gamma, u_{z_0} + c, \cdot)$ and $\widetilde{F}''(\Omega, \Gamma, u_{z_0} + c, \cdot)$ be defined by (12.1.4) and (12.1.5) with $r_h = 1/\varepsilon_h$ for every $h \in \mathbf{N}$. Then, from Theorem 12.6.3, Proposition 3.2.6, and Proposition 12.2.3 applied to the function $(x, z) \in \mathbf{R}^n \times \mathbf{R}^n \mapsto f(x, z_0 + z)$ we obtain that

$$\int_\Omega \text{sc}^- \tilde{f}_{\text{hom}}^q (\nabla u) dx = \Gamma^-(L^\infty(\Omega)) \liminf_{\varepsilon \to 0^+} F_{1/\varepsilon}(\Omega, u) \leq \widetilde{F}'(\Omega, u) \leq$$

$$\leq \widetilde{F}'(\Omega, \Gamma, u_{z_0} + c, u) \leq \widetilde{F}''(\Omega, \Gamma, u_{z_0} + c, u) \leq \widetilde{F}''(\Omega, u),$$

from which, together with Proposition 3.2.6, Proposition 3.2.3, and Theorem 12.6.3, we conclude that

$$\int_\Omega \text{sc}^- \tilde{f}_{\text{hom}}^q (\nabla u) dx = \Gamma^-(L^\infty(\Omega)) \liminf_{\varepsilon \to 0^+} F_{1/\varepsilon}(\Omega, \Gamma, u_{z_0} + c, u) \leq$$

$$\leq \Gamma^-(L^\infty(\Omega)) \limsup_{\varepsilon \to 0^+} F_{1/\varepsilon}(\Omega, \Gamma, u_{z_0} + c, u) \leq$$

$$\leq \Gamma^-(L^\infty(\Omega)) \limsup_{\varepsilon \to 0^+} F_{1/\varepsilon}(\Omega, u) = \int_\Omega \text{sc}^- \tilde{f}_{\text{hom}}^q (\nabla u) dx.$$

This completes the proof. ∎

§12.7 Applications to the Convergence of Minima and of Minimizers

In this section we apply the theorems of the previous one to deduce convergence results for minima and minimizers of some classes of variational problems.

To do this, we take f as in (12.0.1), $p \in]1, +\infty]$, $q \in [p, +\infty]$, and assume that (12.1.22) holds. Moreover, if $C^q(z_0)$ is given by (12.1.20) for every $z_0 \in \mathbf{R}^n$, we also assume that

(12.7.1) $$\text{int}(C^q(z_0) \neq \emptyset$$

for some $z_0 \in \mathbf{R}^n$.

If f_{hom}^q is defined in (12.1.18), then Proposition 12.1.3 yields

(12.7.2) $$\begin{cases} |z|^p \leq f_{\text{hom}}^q(z) \text{ for every } z \in \mathbf{R}^n & \text{if } p \in [1, +\infty[\\ \text{dom} f_{\text{hom}}^q \subseteq B_R(0) & \text{if } p = +\infty. \end{cases}$$

We start with the case of Dirichlet minimum problems.

Theorem 12.7.1. *Let f be as in (12.0.1), $p \in \,]n, +\infty]$, $q \in [p, +\infty]$, $z_0 \in \mathbf{R}^n$, $C^q(z_0)$ and f_{hom}^q be defined in (12.1.20) and (12.1.18) respectively. Assume that (12.1.22) and (12.7.1) hold. For every $\varepsilon > 0$, $\Omega \in \mathcal{A}_0$, $\beta \in L^1(\Omega)$, and $c \in \mathbf{R}$ let*

$$(12.7.3) \qquad\qquad i_\varepsilon^0(\Omega, \beta) =$$

$$= \inf \left\{ \int_\Omega f\left(\frac{x}{\varepsilon}, \nabla u\right) dx + \int_\Omega \beta u \, dx : u \in u_{z_0} + c + W_0^{1,q}(\Omega) \right\},$$

$$m_{\mathrm{hom}}^0(\Omega, \beta) =$$

$$= \min \left\{ \int_\Omega \mathrm{sc}^- f_{\mathrm{hom}}^q(\nabla u) dx + \int_\Omega \beta u \, dx : u \in u_{z_0} + c + W_0^{1,p}(\Omega) \right\},$$

and let $\{u_\varepsilon\}_{\varepsilon>0} \subseteq u_{z_0} + c + W_0^{1,q}(\Omega)$ be such that

$$\lim_{\varepsilon \to 0^+} \left(\int_\Omega f\left(\frac{x}{\varepsilon}, \nabla u_\varepsilon\right) dx + \int_\Omega \beta u_\varepsilon dx - i_\varepsilon^0(\Omega, \beta) \right) = 0.$$

Then f_{hom}^q is convex and satisfies (12.7.2), $\{i_\varepsilon^0(\Omega, \beta)\}_{\varepsilon>0}$ converges as $\varepsilon \to 0^+$ to $m_{\mathrm{hom}}^0(\Omega, \beta)$, $\{u_\varepsilon\}_{\varepsilon>0}$ has cluster points in $L^\infty(\Omega)$ as $\varepsilon \to 0^+$, and every such point is a solution of $m_{\mathrm{hom}}^0(\Omega, \beta)$.

Moreover, if $q = p$ and (12.1.23) too holds, then $\mathrm{sc}^- f_{\mathrm{hom}}^p = f_{\mathrm{hom}}^p$, for every $z \in \mathbf{R}^n$ the infimum in the definition of $f_{\mathrm{hom}}^p(z)$ is attained, problems in (12.7.3) have solutions, and for every $\varepsilon > 0$ one can take as u_ε a solution of $i_\varepsilon^0(\Omega, \beta)$.

Proof. The properties of f_{hom}^q follow from (12.0.1) and Proposition 12.1.3.

Let Ω, β, c be as above, and, for every $\varepsilon > 0$, let $F_{1/\varepsilon}(\Omega, \partial\Omega, u_{z_0} + c, \cdot)$ be defined by (12.1.3).

First of all, we prove that the limit below exists, and that

$$(12.7.4) \qquad \Gamma^-(L^\infty(\Omega)) \lim_{\varepsilon \to 0^+} F_{1/\varepsilon}(\Omega, \partial\Omega, u_{z_0} + c, u) =$$

$$= \begin{cases} \int_\Omega \mathrm{sc}^- f_{\mathrm{hom}}^q(\nabla u) dx & \text{if } u \in u_{z_0} + c + W_0^{1,p}(\Omega) \\ +\infty & \text{if } u \in L^\infty(\Omega) \setminus (u_{z_0} + c + W_0^{1,p}(\Omega)) \end{cases}$$

$$\text{for every } u \in L^\infty(\Omega).$$

By (12.1.21), Theorem 12.6.2, and (12.1.19) it follows that

$$(12.7.5) \qquad \Gamma^-(L^\infty(\Omega)) \limsup_{\varepsilon \to 0^+} F_{1/\varepsilon}(\Omega, \partial\Omega, u_{z_0} + c, u) \leq$$

$$\leq \begin{cases} \int_\Omega \mathrm{sc}^- f_{\mathrm{hom}}^q(\nabla u) dx & \text{if } u \in u_{z_0} + c + W_0^{1,p}(\Omega) \\ +\infty & \text{if } u \in L^\infty(\Omega) \setminus (u_{z_0} + c + W_0^{1,p}(\Omega)) \end{cases}$$

$$\text{for every } u \in L^\infty(\Omega).$$

On the other side, if $u \in L^\infty(\Omega)$ is such that $\Gamma^-(L^\infty(\Omega)) \liminf_{\varepsilon \to 0^+}$ $F_{1/\varepsilon}(\Omega, \partial\Omega, u_{z_0} + c, u) < +\infty$, then, by Proposition 3.2.6, there exists $\varepsilon_h \to$ 0 such that $\Gamma^-(L^\infty(\Omega)) \liminf_{h \to +\infty} F_{\varepsilon_h}(\Omega, \partial\Omega, u_{z_0} + c, u) < +\infty$.

Let $\{u_h\} \subseteq u_{z_0} + c + W_0^{1,q}(\Omega)$ be such that $u_h \to u$ in $L^\infty(\Omega)$, and

$$\Gamma^-(L^\infty(\Omega)) \liminf_{h \to +\infty} F_{\varepsilon_h}(\Omega, \partial\Omega, u_{z_0} + c, u) = \liminf_{h \to +\infty} \int_\Omega f\left(\frac{x}{\varepsilon_h} \nabla u_h\right) dx.$$

Then, by (12.1.22), and the Rellich-Kondrachov Compactness Theorem, we conclude that $u \in u_{z_0} + c + W_0^{1,p}(\Omega)$, and therefore, by Theorem 12.6.2, and (12.1.19), that

$$\Gamma^-(L^\infty(\Omega)) \liminf_{\varepsilon \to 0^+} F_{1/\varepsilon}(\Omega, \partial\Omega, u_{z_0} + c, u) \geq \int_\Omega \text{sc}^- f_{\text{hom}}^q(\nabla u) dx,$$

from which, together with (12.7.5), (12.7.4) follows.

By (12.7.4), and Proposition 3.2.2, once we observe that the functional $u \in L^\infty(\Omega) \mapsto \int_\Omega \beta u dx$ is $L^\infty(\Omega)$-continuous, we immediately obtain that

$$(12.7.6) \qquad \Gamma^-(L^\infty(\Omega)) \lim_{\varepsilon \to 0^+} \left\{ F_{1/\varepsilon}(\Omega, \partial\Omega, u_{z_0} + c, u) + \int_\Omega \beta u dx \right\} =$$

$$= \begin{cases} \int_\Omega \text{sc}^- f_{\text{hom}}^q(\nabla u) dx + \int_\Omega \beta u dx & \text{if } u \in u_{z_0} + c + W_0^{1,p}(\Omega) \\ +\infty & \text{if } u \in L^\infty(\Omega) \setminus (u_{z_0} + c + W_0^{1,p}(\Omega)) \end{cases}$$

$$\text{for every } u \in L^\infty(\Omega).$$

Let us now prove that the functionals $u \in L^\infty(\Omega) \mapsto F_{1/\varepsilon}(\Omega, \partial\Omega, u_{z_0} + c, u) + \int_\Omega \beta u dx$ are equi-coercive.

To do this, let us consider only the case in which $p \in]1, +\infty[$, the one in which $p = +\infty$ being similar.

In this case, since (12.1.22), and Sobolev Imbedding Theorem imply that

$$F_{1/\varepsilon}(\Omega, \partial\Omega, u_{z_0} + c, u) + \int_\Omega \beta u dx \geq$$

$$\geq \|\nabla u\|_{L^p(\Omega)}^p - \|\beta\|_{L^1(\Omega)} \|u\|_{L^\infty(\Omega)} \geq \|\nabla u\|_{L^p(\Omega)}^p - C\|\beta\|_{L^1(\Omega)} \|u\|_{W^{1,p}(\Omega)}$$

$$\text{for every } \varepsilon > 0, \ u \in u_{z_0} + c + W_0^{1,p}(\Omega),$$

for some $C \geq 0$ not depending on h, and since every $u \in L^\infty(\Omega)$ satisfying $F_{1/\varepsilon}(\Omega, \partial\Omega, u_{z_0} + c, u) < +\infty$ actually is in $u_{z_0} + c + W_0^{1,p}(\Omega)$, then $\{u \in L^\infty(\Omega) : F_{1/\varepsilon}(\Omega, \partial\Omega, u_{z_0} + c, u) + \int_\Omega \beta u dx \leq \lambda\} \subseteq u_{z_0} + c + \{u \in W_0^{1,p}(\Omega) : \|\nabla u\|_{L^p(\Omega)}^p - C\|\beta\|_{L^1(\Omega)} \|u\|_{W^{1,p}(\Omega)} \leq \lambda\}$ for every $\lambda \in \mathbf{R}, \varepsilon > 0$. Consequently, the desired coerciveness follows from Proposition 4.4.4.

By (12.7.6), and the equi-coerciveness of $u \in L^\infty(\Omega) \mapsto F_{1/\varepsilon}(\Omega, \partial\Omega,$ $u_{z_0} + c, u) + \int_\Omega \beta u dx$, the assumptions of Theorem 3.3.8 are fulfilled with $U = L^\infty(\Omega)$, and the proof follows from Theorem 3.3.8, once we observe that obviously

$$\limsup_{\varepsilon \to 0^+} i_0^\varepsilon(\Omega, \beta) \le \mathcal{L}^n(\Omega) \int_Y f(y, z_0) dy + \|\beta\|_{L^{p'}(\Omega)} \|u_{z_0} + c\|_{L^p(\Omega)} < +\infty. \quad \blacksquare$$

We now treat the case of Neumann minimum problems.

Theorem 12.7.2. *Let f be as in (12.0.1), $p \in]n, +\infty]$, $q \in [p, +\infty]$, $z_0 \in \mathbf{R}^n$, $C^q(z_0)$ and f_{hom}^q be defined in (12.1.20) and (12.1.18) respectively. Assume that (12.1.22) and (12.7.1) hold. For every $\varepsilon > 0$, $\Omega \in \mathcal{A}_0$ convex, $\lambda \in]0, +\infty[$, $r \in]1, +\infty[$, and $\mu \in \mathcal{M}(\overline{\Omega})$ let*

$$(12.7.7) \qquad\qquad i_\varepsilon(\Omega, \lambda, \mu) =$$

$$= \inf \left\{ \int_\Omega f\left(\frac{x}{\varepsilon}, \nabla u\right) dx + \lambda \int_\Omega |u|^r dx + \int_{\overline{\Omega}} u d\mu : u \in W^{1,q}(\Omega) \right\},$$

$$m_{\mathrm{hom}}(\Omega, \lambda, \mu) =$$

$$= \min \left\{ \int_\Omega \mathrm{sc}^- f_{\mathrm{hom}}^q(\nabla u) dx + \lambda \int_\Omega |u|^r dx + \int_{\overline{\Omega}} u d\mu : u \in W^{1,p}(\Omega) \right\},$$

and let $\{u_\varepsilon\}_{\varepsilon > 0} \subseteq W^{1,q}(\Omega)$ be such that

$$\lim_{\varepsilon \to 0^+} \left(\int_\Omega f\left(\frac{x}{\varepsilon}, \nabla u_\varepsilon\right) dx + \lambda \int_\Omega |u_\varepsilon|^r dx + \int_{\overline{\Omega}} u_\varepsilon d\mu - i_\varepsilon(\Omega, \lambda, \mu) \right) = 0.$$

Then f_{hom}^q is convex and satisfies (12.7.2), $\{i_\varepsilon(\Omega, \lambda, \mu)\}_{\varepsilon > 0}$ converges as $\varepsilon \to 0^+$ to $m_{\mathrm{hom}}(\Omega, \lambda, \mu)$, $\{u_\varepsilon\}_{\varepsilon > 0}$ has cluster points in $L^\infty(\Omega)$ as $\varepsilon \to 0^+$, and every such point is a solution of $m_{\mathrm{hom}}(\Omega, \lambda, \mu)$.

Moreover, if $q = p$ and (12.1.23) too holds, then $\mathrm{sc}^- f_{\mathrm{hom}}^p = f_{\mathrm{hom}}^p$, for every $z \in \mathbf{R}^n$ the infimum in the definition of $f_{\mathrm{hom}}^p(z)$ is attained, problems in (12.7.7) have solutions, and for every $\varepsilon > 0$ one can take as u_ε a solution of $i_\varepsilon(\Omega, \lambda, \mu)$.

Proof. The properties of f_{hom}^q follow from (12.0.1) and Proposition 12.1.3.

Let Ω, λ, r, μ be as above, and, for every $\varepsilon > 0$, let $F_{1/\varepsilon}(\Omega, \cdot)$ be defined by (12.1.2).

First of all, let us set

$$B(\Omega, \cdot) : u \in L^\infty(\Omega) \mapsto \begin{cases} \int_{\overline{\Omega}} u d\mu & \text{if } u \in C^0(\overline{\Omega}) \\ +\infty & \text{if } u \in L^\infty(\Omega) \setminus C^0(\overline{\Omega}). \end{cases}$$

Then, by exploiting Theorem 12.6.3, and the $L^\infty(\Omega)$-continuity of the restriction of $B(\Omega, \cdot)$ to $W^{1,p}(\Omega)$, it is easy to prove that

$$(12.7.8) \qquad \Gamma^-(L^\infty(\Omega)) \lim_{\varepsilon \to 0^+} \left\{ F_{1/\varepsilon}(\Omega, u) + B(\Omega, u) \right\} =$$

$$= \int_\Omega \mathrm{sc}^- f_{\mathrm{hom}}^q(\nabla u) dx + \int_{\overline{\Omega}} u d\mu \quad \text{for every } u \in W^{1,p}(\Omega).$$

Then, by (12.7.8), an argument similar to the one exploited in the proof of Theorem 12.7.1, Proposition 2.5.1, and the Rellich-Kondrachov Compactness Theorem it follows that the limit below exists, and that

$$(12.7.9) \qquad \Gamma^-(L^\infty(\Omega)) \lim_{\varepsilon \to 0^+} \left\{ F_{1/\varepsilon}(\Omega, u) + \lambda \int_\Omega |u|^r dx + B(\Omega, u) \right\} =$$

$$= \begin{cases} \int_\Omega \mathrm{sc}^- f_{\mathrm{hom}}^q(\nabla u) dx + \lambda \int_\Omega |u|^r dx + \int_{\overline{\Omega}} u d\mu & \text{if } u \in W^{1,p}(\Omega) \\ +\infty & \text{if } u \in L^\infty(\Omega) \setminus W^{1,p}(\Omega) \end{cases}$$

$$\text{for every } u \in L^\infty(\Omega).$$

Let us now prove that the functionals $u \in L^\infty(\Omega) \mapsto F_{1/\varepsilon}(\Omega, u) + \lambda \int_\Omega |u|^r dx + B(\Omega, u)$ are equi-coercive.

To do this, we first recall that, since Ω has Lipschitz boundary, and $p \in]n, +\infty]$, by (12.1.22), and Sobolev Imbedding Theorem there exists $C \in [0, +\infty[$ such that

$$F_{1/\varepsilon}(\Omega, u) + \lambda \int_\Omega |u|^r dx + B(\Omega, u) \geq$$

$$\geq \|\|\nabla u\|\|_{L^p(\Omega)}^p + \lambda \|u\|_{L^r(\Omega)}^r - |\mu|(\overline{\Omega})\|u\|_{L^\infty(\Omega)} \geq$$

$$\geq \|\|\nabla u\|\|_{L^p(\Omega)}^p + \lambda \|u\|_{L^r(\Omega)}^r - C|\mu|(\overline{\Omega})\|u\|_{W^{1,p}(\Omega)}$$

$$\text{for every } \varepsilon > 0, \text{ and } u \in W^{1,p}(\Omega).$$

Therefore, once we recall that every $u \in L^\infty(\Omega)$ satisfying $F_{1/\varepsilon}(\Omega, u) < +\infty$ actually is in $W^{1,p}(\Omega)$, we obtain that $\{ u \in L^\infty(\Omega) : F_{1/\varepsilon}(\Omega, u) + \lambda \int_\Omega |u|^r dx + B(\Omega, u) \leq c \} \subseteq \{ u \in W^{1,p}(\Omega) : \|\|\nabla u\|\|_{L^p(\Omega)}^p + \lambda \|u\|_{L^r(\Omega)}^r - C|\mu|(\overline{\Omega})\|u\|_{W^{1,p}(\Omega)} \leq c \}$ for every $c \in \mathbf{R}$, $\varepsilon > 0$. Consequently, the desired coerciveness follows from Proposition 4.4.3.

When $p = +\infty$, the same result follows from (12.1.22), and Proposition 4.4.3, once we observe that

$$F_{1/\varepsilon}(\Omega, u) + \lambda \int_\Omega |u|^r dx + B(\Omega, u) \geq \lambda \int_\Omega |u|^r dx - |\mu|(\overline{\Omega})\|u\|_{L^\infty(\Omega)} \geq$$

$$\geq \lambda \|u\|_{L^r(\Omega)}^r - |\mu|(\overline{\Omega})\|u\|_{W^{1,\infty}(\Omega)}$$

$$\text{for every } \varepsilon > 0, \text{ and } u \in W^{1,\infty}(\Omega).$$

By (12.7.9), and the equi-coerciveness of $u \in L^{\infty}(\Omega) \mapsto F_{1/\varepsilon}(\Omega, u) + \lambda \int_{\Omega} |u|^r dx + B(\Omega, u)$, the assumptions of Theorem 3.3.8 are fulfilled with $U = L^{\infty}(\Omega)$, and the proof follows from Theorem 3.3.8, once we observe that obviously

$$\limsup_{\varepsilon \to 0^+} i_{\varepsilon}(\Omega, \lambda, \mu) \le$$

$$\le \mathcal{L}^n(\Omega) \int_Y f(y, z_0) dy + \lambda \int_{\Omega} |u_{z_0}|^r dx + \int_{\overline{\Omega}} u_{z_0} d\mu < +\infty. \qquad \blacksquare$$

The following result deals with another case of Neumann minimum problems.

Theorem 12.7.3. *Let f be as in (12.0.1), $p \in]n, +\infty]$, $q \in [p, +\infty]$, $z_0 \in \mathbf{R}^n$, $C^q(z_0)$ and f^q_{hom} be defined in (12.1.20) and (12.1.18) respectively. Assume that (12.1.22) and (12.7.1) hold. For every $\varepsilon > 0$, $\Omega \in \mathcal{A}_0$ convex, and $\mu \in \mathcal{M}(\overline{\Omega})$ such that $\mu(\overline{\Omega}) = 0$ let*

$$(12.7.10) \quad i_{\varepsilon}(\Omega, \mu) = \inf\left\{ \int_{\Omega} f\left(\frac{x}{\varepsilon}, \nabla u\right) dx + \int_{\overline{\Omega}} u d\mu : u \in W^{1,q}(\Omega) \right\},$$

$$m_{\mathrm{hom}}(\Omega, \mu) = \min\left\{ \int_{\Omega} \mathrm{sc}^- f^q_{\mathrm{hom}}(\nabla u) dx + \int_{\overline{\Omega}} u d\mu : u \in W^{1,p}(\Omega) \right\},$$

and let $\{u_{\varepsilon}\}_{\varepsilon > 0} \subseteq W^{1,q}(\Omega)$ be such that

$$\lim_{\varepsilon \to 0^+} \left(\int_{\Omega} f\left(\frac{x}{\varepsilon}, \nabla u_{\varepsilon}\right) dx + \int_{\overline{\Omega}} u_{\varepsilon} d\mu - i_{\varepsilon}(\Omega, \mu) \right) = 0.$$

Then f^q_{hom} is convex and satisfies (12.7.2), $\{i_{\varepsilon}(\Omega, \mu)\}_{\varepsilon > 0}$ converges as $\varepsilon \to 0^+$ to $m_{\mathrm{hom}}(\Omega, \mu)$, $\{u_{\varepsilon} - \int_{\Omega} u_{\varepsilon} dx\}_{\varepsilon > 0}$ has cluster points in $L^{\infty}(\Omega)$ as $\varepsilon \to 0^+$, and every such point is a solution of $m_{\mathrm{hom}}(\Omega, \mu)$.

Moreover, if $q = p$ and (12.1.23) too holds, then $\mathrm{sc}^- f^p_{\mathrm{hom}} = f^p_{\mathrm{hom}}$, for every $z \in \mathbf{R}^n$ the infimum in the definition of $f^p_{\mathrm{hom}}(z)$ is attained, problems in (12.7.10) have solutions, and for every $\varepsilon > 0$ one can take u_{ε} as a minimizer of $i_{\varepsilon}(\Omega, \mu)$.

Proof. Let Ω, μ be as above. Then the theorem follows by arguing as in the proof of Theorem 12.7.2, once we observe that the condition $\mu(\overline{\Omega}) = 0$ yields

$$i_{\varepsilon}(\Omega, \mu) = \inf\left\{ \int_{\Omega} f\left(\frac{x}{\varepsilon}, \nabla u\right) dx + \int_{\overline{\Omega}} u d\mu : u \in W^{1,q}(\Omega), \int_{\Omega} u dx = 0 \right\}$$

$$\text{for every } \varepsilon > 0,$$

$$m_{\text{hom}}(\Omega, \mu) =$$

$$= \min \left\{ \int_\Omega \text{sc}^- f_{\text{hom}}^q (\nabla u) dx + \int_\Omega u d\mu : u \in W^{1,p}(\Omega), \int_\Omega u dx = 0 \right\},$$

and that by (12.1.22), Theorem 4.3.19, and Proposition 4.4.3, the above functionals are equi-coercive in $L^\infty(\Omega)$. ∎

Finally, the result below is concerned with mixed minimum problems.

Theorem 12.7.4. Let f be as in (12.0.1), $p \in]n, +\infty]$, $q \in [p, +\infty]$, $z_0 \in \mathbf{R}^n$, $C^q(z_0)$ and f_{hom}^q be defined in (12.1.20) and (12.1.18) respectively. Assume that (12.1.22) and (12.7.1) hold. For every $\varepsilon > 0$, $\Omega \in \mathcal{A}_0$ convex, $\emptyset \neq \Gamma \subseteq \partial\Omega$, $c \in \mathbf{R}$, and $\mu \in \mathcal{M}(\overline{\Omega})$ let

(12.7.11)
$$i_\varepsilon(\Omega, \Gamma, \mu) =$$

$$= \inf \left\{ \int_\Omega f \left(\frac{x}{\varepsilon}, \nabla u \right) dx + \int_\Omega u d\mu : u \in u_{z_0} + c + W_{0,\Gamma}^{1,q}(\Omega) \right\},$$

$$m_{\text{hom}}(\Omega, \Gamma, \mu) =$$

$$= \min \left\{ \int_\Omega \text{sc}^- f_{\text{hom}}^q (\nabla u) dx + \int_\Omega u d\mu : u \in u_{z_0} + c + W_{0,\Gamma}^{1,p}(\Omega) \right\},$$

and let $\{u_\varepsilon\}_{\varepsilon>0} \subseteq u_{z_0} + c + W_{0,\Gamma}^{1,q}(\Omega)$ be such that

$$\lim_{\varepsilon \to 0^+} \left(\int_\Omega f \left(\frac{x}{\varepsilon}, \nabla u_h \right) dx + \int_\Omega u_h d\mu - i_\varepsilon(\Omega, \Gamma, \mu) \right) = 0.$$

Then f_{hom}^q is convex and satisfies (12.7.2), $\{i_\varepsilon(\Omega, \Gamma, \mu)\}_{\varepsilon>0}$ converges as $\varepsilon \to 0^+$ to $m_{\text{hom}}(\Omega, \Gamma, \mu)$, $\{u_\varepsilon\}_{\varepsilon>0}$ has cluster points in $L^\infty(\Omega)$ as $\varepsilon \to 0^+$, and every such point is a solution of $m_{\text{hom}}(\Omega, \Gamma, \mu)$.

Moreover, if $q = p$ and (12.1.23) too holds, then $\text{sc}^- f_{\text{hom}}^p = f_{\text{hom}}^p$, for every $z \in \mathbf{R}^n$ the infimum in the definition of $f_{\text{hom}}^p(z)$ is attained, problems in (12.7.11) have solutions, and for every $\varepsilon > 0$ one can take as u_ε a solution of $i_\varepsilon(\Omega, \Gamma, \mu)$.

Proof. The proof follows the same outlines of the one of Theorem 12.7.1 with the necessary changes.

In particular, if $F_{1/\varepsilon}$ is defined in (12.1.3) for every $\varepsilon > 0$ and B is the functional introduced in the proof of Theorem 12.7.1, by considering Theorem 12.6.4 in place of Theorem 12.6.3, one first proves that

$$\Gamma^-(L^\infty(\Omega)) \lim_{\varepsilon \to 0^+} \left\{ F_{1/\varepsilon}(\Omega, \Gamma, u_{z_0} + c, u) + B(\Omega, u) \right\} =$$

$$= \begin{cases} \int_\Omega \text{sc}^- f_{\text{hom}}^q (\nabla u) dx + \int_{\overline{\Omega}} u d\mu & \text{if } u \in u_{z_0} + c + W_{0,\Gamma}^{1,p}(\Omega) \\ +\infty & \text{if } u \in L^\infty(\Omega) \setminus u_{z_0} + c + W_{0,\Gamma}^{1,p}(\Omega) \end{cases}$$

for every $u \in L^\infty(\Omega)$,

and then, by using Proposition 4.4.3, that the functionals $u \in L^\infty(\Omega) \mapsto F_{1/\varepsilon}(\Omega, \Gamma, u_{z_0} + c, u) + B(\Omega, u)$ are equi-coercive. ∎

Chapter 13

Homogenization of Unbounded Functionals with Special Constraints

In this chapter we examine the homogenization process for unbounded integral functionals when the constraints on the admissible deformations are not allowed to oscillate freely. We consider essentially the extreme case in which they are fixed, and the intermediate one in which they can oscillate, but with some restrictions.

In both the cases it is possible for us to prove results sharper than those of the previous chapter, and settle the homogenization process in the two settings of Sobolev and BV spaces.

§13.1 Homogenization with Fixed Constraints: the Case of Neumann Boundary Conditions

In this section we start the study of homogenization problems when the constraints are fixed.

Thus, if f is as in (12.0.1), we assume that

$$(13.1.1) \qquad \mathrm{dom} f(x, \cdot) = C \text{ for a.e. } x \in \mathbf{R}^n$$

for some convex set C, not necessarily bounded, satisfying

$$(13.1.2) \qquad \mathrm{int}(C) \neq \emptyset,$$

and that the following mild summability condition in the space variable

$$(13.1.3) \qquad f(\cdot, z) \in L^1(Y) \text{ for every } z \in C$$

is fulfilled.

In this setting we are able to carry out the homogenization processes for Neumann, Dirichlet, and mixed problems under weak coerciveness assumptions. We refer to [CCDAG3] and [CCDAG4] for additional references on the subject.

Let f be as in (12.0.1). For every $r \in]0, +\infty[$, $q \in [1, +\infty]$, $\{r_h\} \subseteq [0, +\infty[$, $\Omega \in \mathcal{A}_0$, $\Gamma \subseteq \partial\Omega$, and $u_0 \in W^{1,1}_{\mathrm{loc}}(\mathbf{R}^n)$ we define the following functionals on $L^1_{\mathrm{loc}}(\mathbf{R}^n)$

$$(13.1.4) \quad G_r(\Omega, \cdot) : u \in L^1_{\mathrm{loc}}(\mathbf{R}^n) \mapsto \begin{cases} \int_\Omega f(rx, \nabla u)dx & \text{if } u \in W^{1,q}_{\mathrm{loc}}(\mathbf{R}^n) \\ +\infty & \text{otherwise,} \end{cases}$$

$$(13.1.5) \quad G_r(\Omega, \Gamma, u_0, \cdot) : u \in L^1_{\mathrm{loc}}(\mathbf{R}^n) \mapsto$$

$$\begin{cases} \int_\Omega f(rx, \nabla u)dx & \text{if } u \in u_0 + W^{1,q}_{0,\Gamma}(\Omega) \\ +\infty & \text{otherwise,} \end{cases}$$

and set

$$(13.1.6) \quad \begin{cases} \widetilde{G}'(\Omega, \cdot) : u \in L^1_{\mathrm{loc}}(\mathbf{R}^n) \mapsto \Gamma^-(L^1(\Omega)) \liminf_{h \to +\infty} G_{r_h}(\Omega, u) \\[2mm] \widetilde{G}''(\Omega, \cdot) : u \in L^1_{\mathrm{loc}}(\mathbf{R}^n) \mapsto \Gamma^-(L^1(\Omega)) \limsup_{h \to +\infty} G_{r_h}(\Omega, u), \end{cases}$$

$$(13.1.7) \quad \begin{cases} \widetilde{G}'(\Omega, \Gamma, u_0, \cdot) : u \in L^1_{\mathrm{loc}}(\mathbf{R}^n) \mapsto \\ \qquad\qquad \Gamma^-(L^1(\Omega)) \liminf_{h \to +\infty} G_{r_h}(\Omega, \Gamma, u_0, u) \\[2mm] \widetilde{G}''(\Omega, \Gamma, u_0, \cdot) : u \in L^1_{\mathrm{loc}}(\mathbf{R}^n) \mapsto \\ \qquad\qquad \Gamma^-(L^1(\Omega)) \limsup_{h \to +\infty} G_{r_h}(\Omega, \Gamma, u_0, u). \end{cases}$$

Moreover, we also set

$$(13.1.8) \quad \begin{cases} G'(\Omega, \cdot) : u \in L^1_{\mathrm{loc}}(\mathbf{R}^n) \mapsto \Gamma^-(L^1(\Omega)) \liminf_{h \to +\infty} G_h(\Omega, u) \\[2mm] G''(\Omega, \cdot) : u \in L^1_{\mathrm{loc}}(\mathbf{R}^n) \mapsto \Gamma^-(L^1(\Omega)) \limsup_{h \to +\infty} G_h(\Omega, u), \end{cases}$$

Because of (12.0.1) and of Proposition 3.4.1 it follows that

$$(13.1.9) \qquad \widetilde{G}'(\cdot, u), \ \widetilde{G}''(\cdot, u) \text{ are increasing}$$

for every $u \in L^1_{\mathrm{loc}}(\mathbf{R}^n)$, and every $\{r_h\} \subseteq [0, +\infty[$,

and

$$(13.1.10) \qquad \widetilde{G}'(\Omega, \cdot), \ \widetilde{G}''(\Omega, \cdot) \text{ are convex}$$

for every $\Omega \in \mathcal{A}_0$, and every $\{r_h\} \subseteq [0, +\infty[$.

Moreover, by using arguments analogous to those exploited in the proof of Proposition 12.1.1, it turns out that

$$(13.1.11) \quad \widetilde{G}'_-(\Omega - x_0, T[x_0]u) = \widetilde{G}'_-(\Omega, u), \quad \widetilde{G}''_-(\Omega - x_0, T[x_0]u) = \widetilde{G}''_-(\Omega, u)$$

for every $\{r_h\} \subseteq]0, +\infty[$ increasing and diverging, $\Omega \in \mathcal{A}_0$,

$$x_0 \in \mathbf{R}^n, \ u \in L^1_{\text{loc}}(\mathbf{R}^n).$$

Lemma 13.1.1. *Let f be as in (12.0.1), $q \in [1, +\infty]$, and let G', G'', $\widetilde{G}'$, and $\widetilde{G}''$ be defined in (13.1.8), and (13.1.6). Then*

$$G'_-(\Omega, u) \leq \widetilde{G}'_-(\Omega, u), \ \widetilde{G}''_-(\Omega, u) \leq G''_-(\Omega, u)$$

for every $\{r_h\} \subseteq]0, +\infty[$ diverging, $\Omega \in \mathcal{A}_0$, $u \in L^1_{\text{loc}}(\mathbf{R}^n)$.

Proof. Follows as the one of Lemma 12.1.2. Actually, it is even simpler, because the consideration of L^1-convergence allows to drop the continuity assumptions on the limit functions required in Lemma 12.1.2. ∎

In the present section we represent the limits defined in (13.1.8).
In the following result we assume that

$$(13.1.12) \qquad \phi(z) \leq f(x, z) \leq a(x) + M\phi(z)$$

for a.e. $x \in \mathbf{R}^n$, and every $z \in \mathbf{R}^n$

for some $\phi: \mathbf{R}^n \to [0, +\infty]$, $a \in L^1_{\text{loc}}(\mathbf{R}^n)$ Y-periodic, $M \geq 0$.

Lemma 13.1.2. *Let f be as in (12.0.1), $q \in [1, +\infty]$, $\{r_h\} \subseteq [0, +\infty[$ be increasing and diverging, and $\widetilde{G}'$ be defined in (13.1.8). Assume that*
i) $C \subseteq \mathbf{R}^n$ is convex satisfies (13.1.1) and (13.1.3), $0 \in \text{ri}(C)$, and $\Omega \in \mathcal{A}_0$, $u \in W^{1,\infty}_{\text{loc}}(\mathbf{R}^n)$ are such that $\widetilde{G}'_-(\Omega, u) < +\infty$,
or that
ii) f satisfies (13.1.12) for some $\phi: \mathbf{R}^n \to [0, +\infty[$ convex with $0 \in \text{ri}(\text{dom}\phi)$, $a \in L^1_{\text{loc}}(\mathbf{R}^n)$ Y-periodic, $M \geq 0$, and $\Omega \in \mathcal{A}_0$, $u \in W^{1,1}_{\text{loc}}(\mathbf{R}^n)$ are such that $\widetilde{G}'_-(\Omega, u) < +\infty$.
Then, for every $t \in [0, 1[$, the integrals $\{\int f(r_h x, t\nabla u)dx\}$ are equi-absolutely continuous in Ω.

Proof. Let us first prove the lemma under the assumptions in i).
Since $\widetilde{G}'_-(\Omega, u) < +\infty$, fixed $A \in \mathcal{A}_0$ with $A \subset\subset \Omega$, by (13.1.1) there exists $\{u_k\} \subseteq W^{1,q}_{\text{loc}}(\mathbf{R}^n)$ such that $u_k \to u$ in $L^1(A)$, and

$$(13.1.13) \qquad \text{for every } k \in \mathbf{N}, \ \nabla u_k(x) \in C \text{ for a.e. } x \in A.$$

By (13.1.13), and an argument similar to the one exploited in the proof of Lemma 10.1.2, we obtain that $\nabla u(x) \in \overline{C}$ for a.e. $x \in A$, from which, letting A increase to Ω, we conclude that

$$(13.1.14) \qquad \nabla u(x) \in \overline{C} \text{ for a.e. } x \in \Omega.$$

We now fix $t \in [0,1[$, and observe that, since $0 \in \mathrm{ri}(C)$ and $\nabla u \in (L^\infty(\Omega))^n$, (13.1.14), the convexity of C, and Proposition 1.1.5 provide the existence of $z_1, \ldots, z_m \in \mathrm{ri}(C)$ such that $t\nabla u(x) \in \mathrm{co}(\{z_1, \ldots, z_m\})$ for a.e. $x \in \Omega$. Consequently, by the convexity of f, we deduce that

$$f(r_h x, t\nabla u(x)) \le \sum_{j=1}^m f(r_h x, \overline{z}_j) \text{ for a.e. } x \in \Omega, \text{ every } h \in \mathbf{N},$$

from which, together with (13.1.3) and the weak convergence in $L^1(\Omega)$ of $\{f(r_h\cdot, \overline{z}_j)\}$, the lemma under assumptions in i) follows.

Let us now assume that ii) holds. Then, fixed $A \in \mathcal{A}_0$ with $A \subset\subset \Omega$, there exist $\{h_k\} \subseteq \mathbf{N}$ strictly increasing, and $\{u_k\} \subseteq W^{1,q}_{\mathrm{loc}}(\mathbf{R}^n)$ such that $u_k \to u$ in $L^1(A)$, and

$$\liminf_{k \to +\infty} \int_A f(r_{h_k} x, \nabla u_k) dx \le \widetilde{G}'(A, u) \le \widetilde{G}'_-(\Omega, u) < +\infty,$$

from which, making use of the left-hand side of (13.1.12), and of the $L^1(A)$-lower semicontinuity of $v \in W^{1,1}_{\mathrm{loc}}(\mathbf{R}^n) \mapsto \int_A \mathrm{sc}^- \phi(\nabla v) dx$ ensured by Theorem 7.4.6, it turns out that

$$\int_A \mathrm{sc}^- \phi(\nabla u) dx \le \liminf_{k \to +\infty} \int_A \mathrm{sc}^- \phi(\nabla u_k) dx \le \liminf_{k \to +\infty} \int_A \phi(\nabla u_k) dx \le$$

$$\le \liminf_{k \to +\infty} \int_A f(r_{h_k} x, \nabla u_k) dx \le \widetilde{G}'_-(\Omega, u) < +\infty$$

for every $A \in \mathcal{A}_0$ with $A \subset\subset \Omega$,

and therefore that

$$(13.1.15) \qquad \int_\Omega \mathrm{sc}^- \phi(\nabla u) dx < +\infty.$$

Let us now fix $t \in [0,1[$. By (13.1.15), once we observe that $\mathrm{dom}\phi$ convex, $0 \in \mathrm{ri}(\mathrm{dom}\phi)$, and $\mathrm{ri}(\mathrm{dom}\phi) = \mathrm{ri}(\mathrm{dom\,sc}^- \phi)$, we get that $t\nabla u(x) \in \mathrm{ri}(\mathrm{dom\,sc}^- \phi)$ for a.e. $x \in \Omega$, and consequently, by the convexity of ϕ, that $\mathrm{sc}^- \phi(t\nabla u(x)) = \phi(t\nabla u(x))$ for a.e. $x \in \Omega$. Because of this, the right-hand side of (13.1.12), and again the convexity of ϕ provide that

$$f(r_h x, t\nabla u(x)) \le a(r_h x) + M\phi(t\nabla u(x)) = a(r_h x) + M\mathrm{sc}^- \phi(t\nabla u(x)) \le$$

$$\leq a(r_h x) + Mtsc^- \phi(\nabla u(x)) + M(1-t)\phi(0)$$

for a.e. $x \in \Omega$, and every $h \in \mathbf{N}$,

from which, together with (13.1.15), the weak convergence in $L^1(\Omega)$ of $\{a(r_h \cdot)\}$, and the finiteness of $\phi(0)$, the lemma follows also under assumptions in ii). ∎

Lemma 13.1.3. *Let f be as in (12.0.1), $q \in [1, +\infty]$, and f^q_{hom} be defined in (12.1.18). Let $C \subseteq \mathbf{R}^n$ be such that (13.1.1), and (13.1.3) hold. Then*

$$C \subseteq \text{dom}(f^q_{\text{hom}}) \subseteq \overline{C}.$$

Proof. By (13.1.3) it follows trivially that

$$f^q_{\text{hom}}(z) \leq \int_Y f(y, z)dy < +\infty \text{ for every } z \in C,$$

from which the left-hand side inequality follows.

Let now $z \in \text{dom}(f^q_{\text{hom}})$. Then there exists $v \in W^{1,q}_{\text{per}}(Y)$ such that $\int_Y f(y, z + \nabla v)dy < +\infty$. Consequently, by (13.1.1), it follows that

$$(13.1.16) \qquad z + \nabla v(y) \in C \text{ for a.e. } y \in Y$$

Since $\overline{C}$ is closed and convex, there exist two families $\{\alpha_\theta\}_{\theta \in \mathcal{T}} \subseteq \mathbf{R}^n$, and $\{\beta_\theta\}_{\theta \in \mathcal{T}} \subseteq \mathbf{R}$ such that $\zeta \in \overline{C}$ if and only if $\alpha_\theta \cdot \zeta + \beta_\theta \geq 0$ for every $\theta \in \mathcal{T}$. Therefore, by (13.1.16), we obtain that

$$(13.1.17) \qquad \alpha_\theta \cdot \int_Y (z + \nabla v)dy + \beta_\theta \geq 0 \text{ for every } \theta \subset \mathcal{T}.$$

By (13.1.17), the Gauss-Green Theorem, and the Y-periodicity of v we deduce that

$$\alpha_\theta \cdot z + \alpha_\theta \cdot \int_{\partial Y} \gamma_Y v \mathbf{n}_Y d\mathcal{H}^{n-1} + \beta_\theta = \alpha_\theta \cdot z + \beta_\theta \geq 0 \text{ for every } \theta \in \mathcal{T},$$

from which we conclude that

$$\text{dom}(f^q_{\text{hom}}) \subseteq \overline{C}. \qquad ∎$$

Proposition 13.1.4. *Let f be as in (12.0.1), $q \in [1, +\infty]$, and let f^q_{hom} be defined in (12.1.18), and $\tilde{f}^q_{\text{hom}}$ in (12.1.1). Let $C \subseteq \mathbf{R}^n$ be convex such that (13.1.1), and (13.1.3) hold. Then*

$$sc^- f^q_{\text{hom}}(z) = sc^- \tilde{f}^q_{\text{hom}}(z) \text{ for every } z \in \mathbf{R}^n.$$

Proof. Since $f_{\text{hom}}^q \leq \tilde{f}_{\text{hom}}^q$, it is clear that

$$(13.1.18) \qquad \text{sc}^- f_{\text{hom}}^q(z) \leq \text{sc}^- \tilde{f}_{\text{hom}}^q(z) \text{ for every } z \in \mathbf{R}^n.$$

To prove the reverse inequality we first take $z \in \text{ri}(C)$. Then by Lemma 13.1.3, the convexity of f_{hom}^q, and Theorem 1.1.17 it follows that $\text{sc}^- f_{\text{hom}}^q(z) = f_{\text{hom}}^q(z)$. Because of this, we can assume that $f_{\text{hom}}^q(z) < +\infty$, so that for every $\varepsilon > 0$ there exists $u \in W_{\text{per}}^{1,q}(Y)$ such that

$$f_{\text{hom}}^q(z) + \varepsilon \geq \int_Y f(y, z + \nabla u) dy.$$

For every $k \in \mathbf{N}$ set $u_k = \max\{\min\{u, k\}, -k\}$. Then $u_k \in W_{\text{per}}^{1,q}(Y) \cap L^\infty(Y)$, and

$$f_{\text{hom}}^q(z) + \varepsilon \geq \int_Y f(y, z + \nabla u_k) dy - \int_{\{y \in Y : |u(y)| \geq k\}} f(y, z) dy \geq$$

$$\geq \tilde{f}_{\text{hom}}^q(z) - \int_{\{y \in Y : |u(y)| \geq k\}} f(y, z) dy,$$

from which, letting first k diverge and then ε go to 0, and by taking into account (13.1.3), we conclude that

$$(13.1.19) \qquad f_{\text{hom}}^q(z) \geq \tilde{f}_{\text{hom}}^q(z) \geq \text{sc}^- \tilde{f}_{\text{hom}}^q(z) \text{ for every } z \in \text{ri}(C).$$

By the convexity of f_{hom}^q, Proposition 1.3.1, (1.3.8) of Proposition 1.3.2, and (13.1.19) we thus have that

$$(13.1.20) \qquad \text{sc}^- f_{\text{hom}}^q(z) = \lim_{t \to 1^-} f_{\text{hom}}^q(tz + (1-t)z_0) \geq$$

$$\geq \liminf_{t \to 1^-} \text{sc}^- \tilde{f}_{\text{hom}}^q(tz + (1-t)z_0) \geq \text{sc}^- \tilde{f}_{\text{hom}}^q(z) \text{ for every } z_0 \in \text{ri}(C), \ z \in \overline{C}.$$

In addition, by Lemma 13.1.3, it follows that $f_{\text{hom}}^q(z) = +\infty$ for every $z \in \mathbf{R}^n \setminus \overline{C}$, from which we conclude that

$$(13.1.21) \qquad \text{sc}^- f_{\text{hom}}^q(z) = +\infty \text{ for every } z \in \mathbf{R}^n \setminus \overline{C}.$$

By (13.1.18), (13.1.20), and (13.1.21), the proof follows. ∎

In the sequel, given f as in (12.0.1), we consider, for every $q \in [1, +\infty]$, $h \in \mathbf{N}$, and $\Omega \in \mathcal{A}_0$, the following functionals on $L_{\text{loc}}^\infty(\mathbf{R}^n)$

$$F_h(\Omega, \cdot) : u \in L_{\text{loc}}^\infty(\mathbf{R}^n) \mapsto G_h(\Omega, u),$$

where, for every $h \in \mathbf{N}$, G_h is defined in (13.1.4), and their limits

$$(13.1.22) \begin{cases} F'(\Omega, \cdot): u \in L^\infty_{\text{loc}}(\mathbf{R}^n) \mapsto \Gamma^-(L^\infty(\Omega)) \liminf_{h \to +\infty} F_h(\Omega, u) \\ F''(\Omega, \cdot): u \in L^\infty_{\text{loc}}(\mathbf{R}^n) \mapsto \Gamma^-(L^\infty(\Omega)) \limsup_{h \to +\infty} F_h(\Omega, u). \end{cases}$$

Lemma 13.1.5. Let f be as in (12.0.1), $q \in [1, +\infty]$, G', G'' be defined in (13.1.8), and F', F'' in (13.1.22). Then,

i) if $C \subseteq \mathbf{R}^n$ is convex, and satisfies (13.1.1) and (13.1.3), it results that

$$F'(\Omega, u) = G'(\Omega, u) \text{ for every } \Omega \in \mathcal{A}_0, \ u \in W^{1,\infty}_{\text{loc}}(\mathbf{R}^n),$$

ii) if f satisfies (13.1.12) for some $\phi: \mathbf{R}^n \to [0, +\infty]$ convex, $a \in L^1_{\text{loc}}(\mathbf{R}^n)$ Y-periodic, and $M \geq 0$, it results that

$$F''(\Omega, u) = G''(\Omega, u) \text{ for every } \Omega \in \mathcal{A}_0, \ u \in W^{1,1}_{\text{loc}}(\mathbf{R}^n) \cap L^\infty(\Omega).$$

Proof. We prove only part i) of the lemma, the proof of the other being similar.

First of all, let us observe that it is not restrictive to assume that

$$(13.1.23) \qquad\qquad 0 \in \text{ri}(C),$$

otherwise, taken $z_0 \in \text{ri}(C)$, we only have to consider the function $(x, z) \in \mathbf{R}^n \times \mathbf{R}^n \mapsto f(x, z_0 + z)$ in place of f.

Let $\Omega \in \mathcal{A}_0$, $u \in W^{1,\infty}_{\text{loc}}(\mathbf{R}^n)$. Let us first prove that

$$(13.1.24) \qquad\qquad F'(\Omega, u) \leq G'(\Omega, u).$$

To do this, we can assume that $G'(\Omega, u) < +\infty$, and observe that, by (13.1.23), (13.1.1), and (13.1.3) it follows that $G'(\Omega, 0) < +\infty$ too.

Let $t \in [0, 1[$. Then the finiteness of $G'(\Omega, u)$ and of $G'(\Omega, 0)$, and (13.1.10) yield that $G'(\Omega, tu) < +\infty$ too. Consequently, there exists $\{v_{t,h}\} \subseteq L^1_{\text{loc}}(\mathbf{R}^n)$ such that $v_{t,h} \to tu$ in $L^1(\Omega)$, and

$$G'(\Omega, tu) = \liminf_{h \to +\infty} G_h(\Omega, v_{t,h}).$$

Then it is easy to produce $\{h_k(t)\} \subseteq \mathbf{N}$ strictly increasing such that, setting for every $k \in \mathbf{N}$ $u_{t,k} = v_{t,h_k(t)}$, it results that $u_{t,k} \in W^{1,q}_{\text{loc}}(\mathbf{R}^n)$ for every $k \in \mathbf{N}$, $u_{t,k} \to tu$ in $L^1(\Omega)$, in measure in Ω and a.e. in Ω, and

$$(13.1.25) \qquad G'(\Omega, tu) = \liminf_{k \to +\infty} \int_\Omega f(h_k(t)x, \nabla u_{t,k}) dx.$$

For every $\varepsilon > 0$ let $\vartheta_\varepsilon \in C^1(\mathbf{R})$ be such that $0 \leq \vartheta'_\varepsilon \leq 1$, and

$$\vartheta_\varepsilon(s) = \begin{cases} -2\varepsilon & \text{if } s < -3\varepsilon \\ s & \text{if } -\varepsilon \leq s \leq \varepsilon \\ 2\varepsilon & \text{if } s > 3\varepsilon, \end{cases}$$

and set, for every $k \in \mathbf{N}$, $w_{t,k} = tu + \vartheta_\varepsilon(u_{t,k} - tu)$.

It is clear that $w_{t,k} \in W^{1,q}_{loc}(\mathbf{R}^n) \cap L^\infty_{loc}(\mathbf{R}^n)$ for every $k \in \mathbf{N}$, and that $w_{t,k} \to tu$ in $L^\infty(\Omega)$. Moreover, by the convexity of f, (13.1.25), Lemma 13.1.2, and (13.1.10), we have

$$(13.1.26) \qquad F'(\Omega, tu) \le \liminf_{k \to +\infty} \int_\Omega f(h_k(t)x, \nabla w_{t,k})dx \le$$

$$\le \limsup_{k \to +\infty} \int_\Omega \vartheta'_\varepsilon(u_{t,k} - tu)f(h_k(t)x, \nabla u_{t,k})dx +$$

$$+ \limsup_{k \to +\infty} \int_{\{x \in \Omega: |u_{t,k}(x) - tu(x)| > \varepsilon\}} f(h_k(t)x, t\nabla u)dx \le$$

$$\le G'(\Omega, tu) + \rho_\varepsilon \le tG'(\Omega, u) + (1 - t)G'(\Omega, 0) + \rho_\varepsilon$$

$$\text{for every } t \in [0, 1[, \ \varepsilon > 0,$$

where $\rho_\varepsilon \ge 0$ for every $\varepsilon > 0$, and $\lim_{\varepsilon \to 0} \rho_\varepsilon = 0$.

In conclusion, once we observe that $tu \to u$ in $L^1(\Omega)$ as $t \to 1^-$, by (13.1.26) and Proposition 3.3.2, letting first ε go to 0, and then t increase to 1, (13.1.24) follows.

Because of (13.1.24), the proof follows, being obvious that $G'(\Omega, u) \le F'(\Omega, u)$. ∎

Proposition 13.1.6. *Let f be as in (12.0.1), $q \in [1, +\infty]$, f^q_{hom} be defined in (12.1.18), and G' and G'' in (13.1.8). Let $C \subseteq \mathbf{R}^n$ be convex such that (13.1.1)÷(13.1.3) hold. Then f^q_{hom} is convex, and*

$$G'_-(\Omega, u) = G''_-(\Omega, u) = \int_\Omega sc^- f^q_{hom}(\nabla u)dx + \int_\Omega (sc^- f^q_{hom})^\infty(\nabla^s u)d|D^s u|$$

for every $\Omega \in \mathcal{A}_0, \ u \in BV(\mathbf{R}^n)$.

Proof. As usual, it is not restrictive to assume that $0 \in int(C)$.

Let F' and F'' be defined in (13.1.22), $\tilde{f}^q_{hom}$ in (12.1.1), and, for every $z_0 \in \mathbf{R}^n$, $\tilde{C}^q(z_0)$ in (11.2.2). Then by i) of Lemma 13.1.5 we get that

$$(13.1.27) \qquad F'(\Omega, u) = G'(\Omega, u) \le G''(\Omega, u) \le F''(\Omega, u)$$

$$\text{for every } \Omega \in \mathcal{A}_0, \ u \in W^{1,\infty}_{loc}(\mathbf{R}^n).$$

We now observe that (12.0.1) implies the convexity of f^q_{hom}, and that, by using (13.1.3), it is easy to prove that $C \cap (-C) \subseteq \tilde{C}^q(0)$, from which, together with (13.1.2), we infer that $int(\tilde{C}^q(0)) \ne \emptyset$. Hence, by Proposition 12.6.1, we obtain that

$$(13.1.28) \qquad F'_-(\Omega, u) = F''_-(\Omega, u) = \int_\Omega sc^- \tilde{f}^q_{hom}(\nabla u)dx$$

$$\text{for every } \Omega \in \mathcal{A}_0, \ u \in W^{1,\infty}_{\text{loc}}(\mathbf{R}^n).$$

On the other side, Proposition 13.1.4 holds. Then because of (13.1.27), (13.1.28), and Proposition 13.1.4 we conclude that

$$(13.1.29) \qquad G'_-(\Omega, u) = G''_-(\Omega, u) = \int_\Omega \text{sc}^- f^q_{\text{hom}}(\nabla u) dx$$

$$\text{for every } \Omega \in \mathcal{A}_0, \ u \in W^{1,\infty}_{\text{loc}}(\mathbf{R}^n).$$

In conclusion, by (13.1.11), (13.1.10), Proposition 3.3.2, and (13.1.29), a double application of the first part of Proposition 8.4.2 to $\text{sc}^- f^q_{\text{hom}}$ $\mathcal{E}_0 = \mathcal{E} = \mathcal{A}_0$, and F equal to the restriction of G'_- to $\mathcal{A}_0 \times C^\infty(\mathbf{R}^n)$, and to $\text{sc}^- f^q_{\text{hom}}$, $\mathcal{E}_0 = \mathcal{E} = \mathcal{A}_0$, and F equal to the restriction of G''_- to $\mathcal{A}_0 \times C^\infty(\mathbf{R}^n)$ completes the proof. ■

Theorem 13.1.7. *Let f be as in (12.0.1), $q \in [1, +\infty]$, f^q_{hom} be defined in (12.1.18), and G_r in (13.1.4) for every $r \in]0, +\infty[$. Let $C \subseteq \mathbf{R}^n$ be convex such that (13.1.1)$\div$(13.1.3) hold. Then f^q_{hom} is convex, and*

$$\Gamma^-(L^1(\Omega)) \liminf_{\varepsilon \to 0^+} G_{1/\varepsilon}(\Omega, u) = \Gamma^-(L^1(\Omega)) \limsup_{\varepsilon \to 0^+} G_{1/\varepsilon}(\Omega, u) =$$

$$= \int_\Omega \text{sc}^- f^q_{\text{hom}}(\nabla u) dx + \int_\Omega (\text{sc}^- f^q_{\text{hom}})^\infty (\nabla^s u) d|D^s u|$$

for every $\Omega \in \mathcal{A}_0$ convex, $u \in BV(\Omega)$.

Proof. Let G' and G'' be given by (13.1.8). Then, by Lemma 13.1.1, Proposition 3.2.3, and Proposition 3.2.6, it follows that

$$G'_-(\Omega, u) \leq \sup \left\{ \Gamma^-(L^1(A)) \liminf_{\varepsilon \to 0^+} G_{1/\varepsilon}(A, u) : A \subset\subset \Omega \right\} \leq$$

$$\leq \sup \left\{ \Gamma^-(L^1(A)) \limsup_{\varepsilon \to 0^+} G_{1/\varepsilon}(A, u) : A \subset\subset \Omega \right\} \leq G''_-(\Omega, u)$$

$$\text{for every } \Omega \in \mathcal{A}_0, \ u \in L^1_{\text{loc}}(\mathbf{R}^n).$$

Consequently, by making use of Proposition 13.1.6, we infer that

$$(13.1.30) \qquad \sup \left\{ \Gamma^-(L^1(A)) \liminf_{\varepsilon \to 0^+} G_{1/\varepsilon}(A, u) : A \subset\subset \Omega \right\} =$$

$$= \sup \left\{ \Gamma^-(L^1(A)) \limsup_{\varepsilon \to 0^+} G_{1/\varepsilon}(A, u) : A \subset\subset \Omega \right\} =$$

$$= \int_\Omega \text{sc}^- f^q_{\text{hom}}(\nabla u) dx + \int_\Omega (\text{sc}^- f^q_{\text{hom}})^\infty (\nabla^s u) d|D^s u|$$

for every $\Omega \in \mathcal{A}_0$, $u \in BV(\mathbf{R}^n)$.

To complete the proof, let us verify that $\Gamma^-(L^1(\cdot)) \liminf_{\varepsilon \to 0^+} G_{1/\varepsilon}$ and $\Gamma^-(L^1(\cdot)) \limsup_{\varepsilon \to 0^+} G_{1/\varepsilon}$ fulfil the assumptions of Proposition 2.7.4 with $\mathcal{O} = \mathcal{A}_0$, and $U = BV(\mathbf{R}^n)$.

By (13.1.9) they are increasing. Moreover, since $T[-x_0]O_t T[x_0]u \to u$ in $L^1(\Omega)$ as $t \to 1^-$ for every $\Omega \in \mathcal{A}_0$, $x_0 \in \mathbf{R}^n$, $u \in BV(\mathbf{R}^n)$, by Proposition 3.3.2, (2.7.2) follows. Finally, because of (13.1.30), (2.7.3) too holds. Consequently, Proposition 2.7.4 applies, and (13.1.30) yields

$$\Gamma^-(L^1(\Omega)) \liminf_{\varepsilon \to 0^+} G_{1/\varepsilon}(\Omega, u) = \Gamma^-(L^1(\Omega)) \limsup_{\varepsilon \to 0^+} G_{1/\varepsilon}(\Omega, u) =$$

$$= \int_\Omega \mathrm{sc}^- f^q_{\mathrm{hom}}(\nabla u)dx + \int_\Omega (\mathrm{sc}^- f^q_{\mathrm{hom}})^\infty (\nabla^s u)d|D^s u|$$

for every $\Omega \in \mathcal{A}_0$ convex, $u \in BV(\mathbf{R}^n)$,

from which the proof follows, once we recall that, due to the smoothness of $\partial\Omega$, the null extension of an element of $BV(\Omega)$ is in $BV(\mathbf{R}^n)$. ∎

§13.2 Homogenization with Fixed Constraints: the Case of Dirichlet Boundary Conditions

In this section we want to prove identity between the limits in (13.1.7) when $\Gamma = \partial\Omega$, and an integral representation result for their common value.

Lemma 13.2.1. Let f be as in (12.0.1), $q \in [1, +\infty]$, $\{r_h\} \subseteq [0, +\infty[$ be increasing and diverging, $\Omega \in \mathcal{A}_0$, $z_0 \in \mathbf{R}^n$ such that $\int_Y f(y, z_0)dy < +\infty$ and let $\widetilde{G}'$ be defined in (13.1.6), and $\widetilde{G}'(\Omega, \partial\Omega, \cdot, \cdot)$ in (13.1.7). Then

$$\widetilde{G}'(\Omega', u) - \mathcal{L}^n(\Omega' \setminus \Omega) \int_Y f(y, z_0)dy \leq \widetilde{G}'(\Omega, \partial\Omega, u_{z_0} + c, u)$$

for every $\Omega' \in \mathcal{A}_0$ with $\Omega \subset\subset \Omega'$, $c \in \mathbf{R}$,

$$u \in L^1_{\mathrm{loc}}(\mathbf{R}^n) \text{ with } u = u_{z_0} + c \text{ a.e. in } \Omega' \setminus \Omega.$$

Proof. Let Ω', c, u be as above. We can obviously assume that $\widetilde{G}'(\Omega, \partial\Omega, u_{z_0} + c, u) < +\infty$, so that there exists $\{u_h\} \subseteq u_{z_0} + c + W_0^{1,q}(\Omega)$ such that $u_h \to u$ in $L^1(\Omega)$, and

(13.2.1) $$\widetilde{G}'(\Omega, \partial\Omega, u_{z_0} + c, u) = \liminf_{h \to +\infty} \int_\Omega f(r_h x, \nabla u_h)dx.$$

It is obvious that, for every $h \in \mathbf{N}$, u_h can be thought as an element of $u_{z_0} + c + W_0^{1,q}(\Omega')$ once we extend it by $u_{z_0} + c$ out of Ω. Therefore $u_h \to u$ in $L^1(\Omega')$, and by Theorem 2.2.9 and (13.2.1), it follows that

$$\widetilde{G}'(\Omega', u) \leq \liminf_{h \to +\infty} \int_{\Omega'} f(r_h x, \nabla u_h)dx \leq$$

$$\leq \widetilde{G}'(\Omega, \partial\Omega, u_{z_0} + c, u) + \limsup_{h \to +\infty} \int_{\Omega' \setminus \Omega} f(r_h x, z_0) dx =$$

$$= \widetilde{G}'(\Omega, \partial\Omega, u_{z_0} + c, u) + \mathcal{L}^n(\Omega' \setminus \Omega) \int_Y f(y, z_0) dy,$$

which proves the lemma. ∎

To prove the next result, it is convenient to introduce for every $u_0 \in W^{1,1}_{\mathrm{loc}}(\mathbf{R}^n)$ and $q \in [1, +\infty]$, the functional

(13.2.2) $$G_{\mathrm{hom}}(\Omega, u_0, \cdot) : u \in L^1_{\mathrm{loc}}(\mathbf{R}^n) \mapsto$$

$$\begin{cases} \int_\Omega \mathrm{sc}^- f^q_{\mathrm{hom}}(\nabla u) dx & \text{if } u \in u_0 + W^{1,\infty}_0(\Omega) \\ +\infty & \text{otherwise}, \end{cases}$$

where f^q_{hom} is defined in (12.1.18).

Lemma 13.2.2. *Let f be as in (12.0.1), $q \in [1, +\infty]$, $\{r_h\} \subseteq [0, +\infty[$ be increasing and diverging, f^q_{hom} be defined in (12.1.18), $\widetilde{G}'$ and $\widetilde{G}''$ in (13.1.7), and G_{hom} in (13.2.2). Let $C \subseteq \mathbf{R}^n$ be convex such that $(13.1.1) \div (13.1.3)$ hold. Then f^q_{hom} is convex and*

$$\int_\Omega \mathrm{sc}^- f^q_{\mathrm{hom}}(\nabla u) dx + \int_\Omega (\mathrm{sc}^- f^q_{\mathrm{hom}})^\infty (\nabla^s u) d|D^s u| +$$

$$+ \int_{\partial\Omega} (\mathrm{sc}^- f^q_{\mathrm{hom}})^\infty ((u_{z_0} + c - \gamma_\Omega u)\mathbf{n}_\Omega) d\mathcal{H}^{n-1} \leq \widetilde{G}'(\Omega, \partial\Omega, u_{z_0} + c, u) \leq$$

$$\leq \widetilde{G}''(\Omega, \partial\Omega, u_{z_0} + c, u) \leq \mathrm{sc}^-(L^1(\Omega)) G_{\mathrm{hom}}(\Omega, u_{z_0} + c, u)$$

for every $\Omega \in \mathcal{A}_0$ with Lipschitz boundary, $z_0 \in C$, $c \in \mathbf{R}$, $u \in BV(\Omega)$.

Proof. Let $\tilde{f}^q_{\mathrm{hom}}$, and, for every $z_0 \in \mathbf{R}^n$, $\widetilde{C}^q(z_0)$ be defined in (12.1.1), and (11.2.2) respectively. For every $h \in \mathbf{N}$, $\Omega \in \mathcal{A}_0$, and $u_0 \in W^{1,1}_{\mathrm{loc}}(\mathbf{R}^n)$ we define the following functionals on $L^\infty_{\mathrm{loc}}(\mathbf{R}^n)$

$$F_h(\Omega, \partial\Omega, u_0, \cdot) : u \in L^\infty_{\mathrm{loc}}(\mathbf{R}^n) \mapsto G_{r_h}(\Omega, \partial\Omega, u_0, u),$$

and their limits

$$\begin{cases} F'(\Omega, \partial\Omega, u_0, \cdot) : u \in L^\infty_{\mathrm{loc}}(\mathbf{R}^n) \mapsto \\ \qquad\qquad\qquad \Gamma^-(L^\infty(\Omega)) \liminf_{h \to +\infty} F_h(\Omega, \partial\Omega, u_0, u) \\ F''(\Omega, \partial\Omega, u_0, \cdot) : u \in L^\infty_{\mathrm{loc}}(\mathbf{R}^n) \mapsto \\ \qquad\qquad\qquad \Gamma^-(L^\infty(\Omega)) \limsup_{h \to +\infty} F_h(\Omega, \partial\Omega, u_0, u). \end{cases}$$

We now observe that, since $C \cap (2z_0 - C) \subseteq \widetilde{C}^q(z_0)$, then $\mathrm{int}(\widetilde{C}^q(z_0)) \neq \emptyset$ for every $z_0 \in \mathrm{int}(C)$, therefore by Theorem 12.6.2, and Proposition 13.1.4 we conclude that

$$(13.2.3) \qquad \widetilde{G}''(\Omega, \partial\Omega, u_{z_0} + c, u) \leq$$

$$\leq F''(\Omega, \partial\Omega, u_{z_0} + c, u) = \int_\Omega \mathrm{sc}^- \widetilde{f}^q_{\mathrm{hom}}(\nabla u)dx = \int_\Omega \mathrm{sc}^- f^q_{\mathrm{hom}}(\nabla u)dx$$

for every $\Omega \in \mathcal{A}_0$, $z_0 \in \mathrm{int}(C)$, $c \in \mathbf{R}$, $u \in u_{z_0} + c + W_0^{1,\infty}(\Omega)$.

Let now $\Omega \in \mathcal{A}_0$ have Lipschitz boundary, $z_0 \in C$, $c \in \mathbf{R}$, and $u \in BV(\Omega)$. Let us extend u in $\mathbf{R}^n \setminus \Omega$, and call again with u such extension, by defining $u = u_{z_0} + c$ in $\mathbf{R}^n \setminus \Omega$, so that $u \in BV_{\mathrm{loc}}(\mathbf{R}^n)$. Let $\{u_h\} \subseteq u_{z_0} + c + W_0^{1,\infty}(\Omega)$ be such that $u_h \to u$ in $L^1(\Omega)$, and

$$(13.2.4) \quad \mathrm{sc}^-(L^1(\Omega))G_{\mathrm{hom}}(\Omega, u_{z_0} + c, u) \geq \lim_{h \to +\infty} \int_\Omega \mathrm{sc}^- f^q_{\mathrm{hom}}(\nabla u_h)dx.$$

Then, by Proposition 13.1.6, Lemma 13.1.1, Lemma 13.2.1, Proposition 3.3.2, (13.2.3), and (13.2.4), we have that

$$(13.2.5) \qquad \int_{\Omega'} \mathrm{sc}^- f^q_{\mathrm{hom}}(\nabla u)dx + \int_{\Omega'} (\mathrm{sc}^- f^q_{\mathrm{hom}})^\infty(\nabla^s u)d|D^s u| -$$

$$-\mathcal{L}^n(\Omega' \setminus \Omega) \int_Y f(y, z_0)dy \leq$$

$$\leq \widetilde{G}'(\Omega, \partial\Omega, u_{z_0} + c, u) \leq \widetilde{G}''(\Omega, \partial\Omega, u_{z_0} + c, u) \leq$$

$$\leq \liminf_{h \to +\infty} \widetilde{G}''(\Omega, \partial\Omega, u_{z_0} + c, u_h) \leq \liminf_{h \to +\infty} \int_\Omega \mathrm{sc}^- f^q_{\mathrm{hom}}(\nabla u_h)dx \leq$$

$$\leq \mathrm{sc}^-(L^1(\Omega))G_{\mathrm{hom}}(\Omega, u_{z_0} + c, u) \text{ for every } \Omega' \in \mathcal{A}_0 \text{ with } \Omega \subset\subset \Omega'.$$

In conclusion, once we observe that, as in (12.3.5), it follows that

$$(13.2.6) \qquad \int_{\Omega'} \mathrm{sc}^- f^q_{\mathrm{hom}}(\nabla u)dx + \int_{\Omega'} (\mathrm{sc}^- f^q_{\mathrm{hom}})^\infty(\nabla^s u)d|D^s u| =$$

$$= \int_\Omega \mathrm{sc}^- f^q_{\mathrm{hom}}(\nabla u)dx + \mathcal{L}^n(\Omega' \setminus \Omega)\mathrm{sc}^- f^q_{\mathrm{hom}}(z_0) +$$

$$+ \int_\Omega (\mathrm{sc}^- f^q_{\mathrm{hom}})^\infty(\nabla^s u)d|D^s u| + \int_{\partial\Omega} (\mathrm{sc}^- f^q_{\mathrm{hom}})^\infty((u_{z_0} + c - \gamma_\Omega u)\mathbf{n}_\Omega)d\mathcal{H}^{n-1}$$

for every $\Omega' \in \mathcal{A}_0$ with $\Omega \subset\subset \Omega'$,

and that $\mathrm{sc}^- f^q_{\mathrm{hom}}(z_0) \leq f^q_{\mathrm{hom}}(z_0) < +\infty$ by Lemma 13.1.3, the proof follows from (13.2.5), (13.2.6), and (13.1.3) letting Ω' decrease to Ω. ∎

Theorem 13.2.3. *Let f be as in (12.0.1), $q \in [1, +\infty]$, f_{hom}^q be defined in (12.1.18), and G_r in (13.1.5) for every $r \in \,]0, +\infty[$. Let $C \subseteq \mathbf{R}^n$ be convex such that (13.1.1)÷(13.1.3) hold. Then f_{hom}^q is convex and*

$$\Gamma^-(L^1(\Omega)) \liminf_{\varepsilon \to 0^+} G_{1/\varepsilon}(\Omega, \partial\Omega, u_{z_0} + c, u) =$$

$$= \Gamma^-(L^1(\Omega)) \limsup_{\varepsilon \to 0^+} G_{1/\varepsilon}(\Omega, \partial\Omega, u_{z_0} + c, u) =$$

$$= \int_\Omega \mathrm{sc}^- f_{\text{hom}}^q(\nabla u)dx + \int_\Omega (\mathrm{sc}^- f_{\text{hom}}^q)^\infty(\nabla^s u)d|D^s u| +$$

$$+ \int_{\partial\Omega} (\mathrm{sc}^- f_{\text{hom}}^q)^\infty((u_{z_0} + c - \gamma_\Omega u)\mathbf{n}_\Omega)d\mathcal{H}^{n-1}$$

for every $\Omega \in \mathcal{A}_0$ convex, $z_0 \in \mathrm{int}(C)$, $c \in \mathbf{R}$, $u \in BV(\Omega)$.

Proof. Follows from Lemma 13.2.2, Theorem 10.7.6, Proposition 3.2.6, and Proposition 3.2.3. ∎

§13.3 Homogenization with Fixed Constraints: the Case of Mixed Boundary Conditions

Let f be as in (12.0.1), $\Omega \in \mathcal{A}_0$ with Lipschitz boundary, $\Gamma \subseteq \partial\Omega$, $z_0 \in \mathbf{R}^n$, and $c \in \mathbf{R}$.

This section is devoted to prove identity between the limits $\Gamma^-(L^1(\Omega))$ $\liminf_{\varepsilon \to 0^+} G_{1/\varepsilon}(\Omega, \Gamma, u_{z_0} + c, \cdot)$ and $\Gamma^-(L^1(\Omega)) \limsup_{\varepsilon \to 0^+} G_{1/\varepsilon}(\Omega, \Gamma, u_{z_0} + c, \cdot)$, and an integral representation result for their common value.

We do this when (13.1.12) is fulfilled for some $\phi \colon \mathbf{R}^n \to [0, +\infty]$ convex such that $\mathrm{int}(\mathrm{dom}\phi) \neq \emptyset$, $a \in L^1_{\mathrm{loc}}(\mathbf{R}^n)$ Y-periodic, and $M \geq 0$, and when $q = 1$. We point out explicitly that (13.1.12) implies that $\mathrm{dom}f(x, \cdot) = \mathrm{dom}\phi$ for a.e. $x \in \mathbf{R}^n$, and that the choice $q = 1$ is natural since again (13.1.12) selects the space of the functions that make the functionals in (13.1.5) finite as the one of the elements $u \in u_{z_0} + c + W^{1,1}_{0,\Gamma}(\Omega)$ for which $\int_\Omega \phi(\nabla u)dx < +\infty$.

Theorem 13.3.1. *Let f be as in (12.0.1), f_{hom}^1 be defined in (12.1.18) with $q = 1$, and G_r in (13.1.5) again with $q = 1$ for every $r \in \,]1, +\infty[$. Assume that (13.1.12) holds with $\phi \colon \mathbf{R}^n \to [0, +\infty]$ convex and satisfying $\mathrm{int}(\mathrm{dom}\phi) \neq \emptyset$, $a \in L^1_{\mathrm{loc}}(\mathbf{R}^n)$ Y-periodic, and $M \geq 0$. Then f_{hom}^1 is convex and*

$$\Gamma^-(L^1(\Omega)) \liminf_{\varepsilon \to 0^+} G_{1/\varepsilon}(\Omega, \Gamma, u_{z_0} + c, u) =$$

$$= \Gamma^-(L^1(\Omega)) \limsup_{\varepsilon \to 0^+} G_{1/\varepsilon}(\Omega, \Gamma, u_{z_0} + c, u) = \int_\Omega \mathrm{sc}^- f_{\text{hom}}^1(\nabla u)dx$$

for every $\Omega \in \mathcal{A}_0$ convex, $\Gamma \subseteq \partial\Omega$, $z_0 \in \mathrm{int}(\mathrm{dom}\phi)$, $c \in \mathbf{R}$,

$$u \in u_{z_0} + c + W^{1,1}_{0,\Gamma}(\Omega).$$

Proof. Let Ω, Γ, z_0, and c be as above.

It is clear that

(13.3.1) $\Gamma^-(L^1(\Omega)) \liminf_{\varepsilon \to 0^+} G_{1/\varepsilon}(\Omega, u) \leq$

$$\leq \Gamma^-(L^1(\Omega)) \liminf_{\varepsilon \to 0^+} G_{1/\varepsilon}(\Omega, \Gamma, u_{z_0} + c, u) \text{ for every } u \in L^1_{loc}(\mathbf{R}^n).$$

Let now $\{r_h\} \subseteq [0, +\infty[$ be increasing and diverging, and let $G''(\Omega, \cdot)$, $\widetilde{G}''(\Omega, \Gamma, \cdot, \cdot)$ be given by (13.1.8) and (13.1.7) respectively with $q = 1$.

Let us first assume that $z_0 = 0$. Then

(13.3.2) $0 \in \text{int}(\text{dom}\phi),$

and let us prove that

(13.3.3) $\widetilde{G}''(\Omega, \Gamma, c, u) \leq G''_-(\Omega, u)$ for every $u \in c + W^{1,1}_{0,\Gamma}(\Omega) \cap L^\infty(\Omega).$

To do this, let u as in (13.3.3). Let us assume that $G''_-(\Omega, u) < +\infty$. Then by ii) of Lemma 13.1.5, and Lemma 12.1.2, given $A \subset\subset \Omega$, there exists $\{u_h\} \subseteq W^{1,1}_{loc}(\mathbf{R}^n)$ such that $u_h \to u$ in $L^\infty(A)$, and

(13.3.4) $\limsup_{h \to +\infty} \int_A f(r_h x, \nabla u_h) dx \leq G''(A, u).$

By (13.3.4), the left-hand side of (13.1.12), and since $G''_-(\Omega, u) < +\infty$, it follows that $\nabla u_h(x) \in \text{dom}\phi$ for every $h \in \mathbf{N}$ sufficiently large, and a.e. $x \in A$.

Let $B \in \mathcal{A}_0$ with $B \subset\subset A$, and let $\psi \in C^\infty_0(A)$ be such that $0 \leq \psi(x) \leq 1$ for every $x \in A$, and $\psi(x) = 1$ for every $x \in B$.

For every $h \in \mathbf{N}$ let w_h be defined by $w_h = \psi_h + (1 - \psi)u$. Then obviously $w_h \in c + W^{1,1}_{0,\Gamma}(\Omega)$ for every $h \in \mathbf{N}$, and $w_h \to u$ in $L^\infty(\Omega)$.

Let now $t \in [0, 1[$. Then, by making use of the convexity properties of f, it results that

$$\int_\Omega f(r_h x, t\nabla w_h) dx \leq$$

$$\leq t \int_\Omega f(r_h x, \psi \nabla u_h + (1 - \psi)\nabla u) dx +$$

$$+ (1 - t) \int_\Omega f\left(r_h x, \frac{t}{1-t}(u_h - u)\nabla\psi\right) dx \leq$$

$$\leq t \int_\Omega \psi(x) f(r_h x, \nabla u_h) dx + t \int_\Omega (1 - \psi(x)) f(r_h x, \nabla u) dx +$$

$$+(1-t)\int_\Omega f\left(r_h x, \frac{t}{1-t}(u_h-u)\nabla\psi\right)dx \le$$

$$\le \int_A f(r_h x, \nabla u_h)dx + \int_{\Omega\setminus B} f(r_h x, \nabla u)dx+$$

$$+(1-t)\int_\Omega f\left(r_h x, \frac{t}{1-t}(u_h-u)\nabla\psi\right)dx \text{ for every } h \in \mathbf{N}.$$

Hence, because of (13.3.4), we get that

$$(13.3.5) \qquad \widetilde{G}''(\Omega,\Gamma,c,tu) \le \limsup_{h\to+\infty}\int_\Omega f(r_h x, t\nabla w_h)dx \le$$

$$\le G''_-(\Omega,u) + \limsup_{h\to+\infty}\int_{\Omega\setminus B} f(r_h x, \nabla u)dx+$$

$$+(1-t)\limsup_{h\to+\infty}\int_\Omega f\left(r_h x, \frac{t}{1-t}(u_h-u)\nabla\psi\right) dx \text{ for every } t \in [0,1[.$$

We now observe that the finiteness of $G''_-(\Omega,u)$, and ii) of Lemma 13.1.2 yield that

$$(13.3.6) \qquad \limsup_{h\to+\infty}\int_{\Omega\setminus B} f(r_h x, \nabla u)dx = \rho_B$$

for some $\rho_B \in [0,+\infty[$ decreasing to 0 as B increases to Ω.

Moreover, let us fix $r \in {]}0,\mathrm{dist}(0,\partial\mathrm{dom}\phi)[$. Then, since obviously $\|(u_h-u)|\nabla\psi|\|_{L^\infty(\Omega)} \to 0$, by using (13.3.2) and the properties of ψ, it results that

$$(13.3.7) \qquad \text{for every } t \in [0,1[\text{ there exists } h_t \in \mathbf{N} \text{ such that}$$

$$\frac{t}{1-t}(u_h(x)-u(x))\nabla\psi(x) \in B_{\frac{r}{\sqrt{n}}}(0) \subseteq \mathrm{dom}\phi$$

$$\text{for a.e. } x \in \Omega \text{ and every } h \in \mathbf{N}\cap[h_t,+\infty[.$$

Consequently, denoted by $\overline{z}_1,\ldots,\overline{z}_{2^n}$ the vertices of the cube centred in 0 and with sidelength $\frac{2r}{\sqrt{n}}$, by (13.3.7), the convexity properties of f, and (13.1.12) it is easy to verify that

$$(13.3.8) \qquad \limsup_{h\to+\infty}\int_\Omega f\left(r_h x, \frac{t}{1-t}(u_h-u)\nabla\psi\right) dx \le$$

$$\le \mathcal{L}^n(\Omega)\sum_{j=1}^{2^n}\int_Y f(y,\overline{z}_j)dy < +\infty \text{ for every } t \in [0,1[.$$

Passing to the limit in (13.3.5) as t increases to 1, by (13.3.5), (13.3.6), (13.3.8), and Proposition 3.3.2, it results that

$$\widetilde{G}''(\Omega, \Gamma, c, u) \leq \liminf_{t \to 1^-} \widetilde{G}''(\Omega, \Gamma, c, tu) \leq G''_-(\Omega, u) + \rho_B$$

for every $B \in \mathcal{A}_0$ with $B \subset\subset \Omega$,

from which, letting B increase to Ω, (13.3.3) follows.

Again under assumption (13.3.2), let us now prove that

(13.3.9) $G''(\Omega, \Gamma, c, u) \leq G''_-(\Omega, u)$ for every $u \in c + W^{1,1}_{0,\Gamma}(\Omega)$.

To do this, let u be as in (13.3.9), and, for every $k \in \mathbf{N}$, let $T_k u$ be the truncation of u at level k.

It is clear that, since $u \in c + W^{1,1}_{0,\Gamma}(\Omega)$, then $T_k u \in c + W^{1,1}_{0,\Gamma}(\Omega) \cap L^\infty(\Omega)$ for every $k \in \mathbf{N}$ sufficiently large. Moreover

(13.3.10) $$\limsup_{k \to +\infty} G''_-(\Omega, T_k u) \leq G''_-(\Omega, u).$$

In fact, if $k \in \mathbf{N}$, and if $G''_-(\Omega, u) < +\infty$, let $A \subset\subset \Omega$, and $\{u_h\} \subseteq W^{1,1}_{loc}(\mathbf{R}^n)$ be such that $u_h \to u$ in $L^1(A)$, and

$$G''(A, u) = \limsup_{h \to +\infty} \int_A f(hx, \nabla u_h) dx.$$

Then

$$G''(A, T_k u) \leq \limsup_{h \to +\infty} \int_A f(hx, \nabla T_k u_h) dx \leq$$

$$\leq G''_-(\Omega, u) + \limsup_{h \to +\infty} \int_{\{x \in A : |u_h(x)| \geq k\}} f(hx, 0) dx.$$

Now it is clear that $\mathcal{L}^n(\{x \in A : |u_h(x)| \geq k\}) \leq \frac{1}{k} \int_A |u_h| dx$ for every $h \in \mathbf{N}$. Consequently, by (13.3.2), (13.1.12), and the equi-absolute continuity of the integrals $\int_\cdot f(r_h x, 0) dx$, it turns out that

$$\limsup_{k \to +\infty} \limsup_{h \to +\infty} \int_{\{x \in A : |u_h(x)| \geq k\}} f(hx, 0) dx = 0,$$

from which (13.3.10) follows letting also A increase to Ω.

By Proposition 3.3.2, (13.3.3), and (13.3.10), inequality (13.3.9) follows once we observe that $T_k u \to u$ in $L^1(\Omega)$.

In conclusion, if (13.3.2) is dropped, taken $z_0 \in \text{int}(\text{dom}\phi)$, we only have to apply (13.3.3) with f replaced by $(x, z) \in \mathbf{R}^n \times \mathbf{R}^n \mapsto f(x, z_0 + z)$, thus getting

(13.3.11) $$\widetilde{G}''(\Omega, \Gamma, u_{z_0} + c, u) \leq G''_-(\Omega, u)$$

for every $\Omega \in \mathcal{A}_0$, $\Gamma \subseteq \partial\Omega$, $z_0 \in \text{int}(\text{dom}\phi)$, $c \in \mathbf{R}$, $u \in u_{z_0} + c + W_{0,\Gamma}^{1,1}(\Omega)$.

By (13.3.1), (13.3.11), and Theorem 13.1.7 the proof follows. ∎

§13.4 Homogenization with Fixed Constraints: Applications to the Convergence of Minima and of Minimizers

In this section, by using the theorems of the previous ones, we obtain convergence results for minima and minimizers of some classes of variational problems both in BV and Sobolev spaces.

We start with the case of Neumann minimum problems in BV spaces.

Theorem 13.4.1. *Let f be as in (12.0.1), $C \subseteq \mathbf{R}^n$ be convex, $q \in [1, +\infty]$, and let f_{hom}^q be defined in (12.1.18). Assume that (12.1.22) with $p = 1$, (13.1.1), (13.1.2) and (13.1.3) hold. For every $\varepsilon > 0$, $\Omega \in \mathcal{A}_0$ convex, $\lambda \in]0, +\infty[$, $r \in]1, 1^*[$, and $\beta \in L^{r'}(\Omega)$ let*

$$i_\varepsilon(\Omega, \lambda, \beta) =$$

$$= \inf\left\{ \int_\Omega f\left(\frac{x}{\varepsilon}, \nabla u\right) dx + \lambda \int_\Omega |u|^r dx + \int_\Omega \beta u \, dx : u \in W^{1,q}(\Omega) \right\},$$

$$m_{\text{hom}}(\Omega, \lambda, \beta) =$$

$$= \min\left\{ \int_\Omega \text{sc}^- f_{\text{hom}}^q(\nabla u) dx + \int_\Omega (\text{sc}^- f_{\text{hom}}^q)^\infty(\nabla^s u) d|D^s u| + \right.$$

$$\left. + \lambda \int_\Omega |u|^r dx + \int_\Omega \beta u \, dx : u \in BV(\Omega) \right\},$$

and let $\{u_\varepsilon\}_{\varepsilon > 0} \subseteq W^{1,q}(\Omega)$ be such that

$$\lim_{\varepsilon \to 0^+} \left(\int_\Omega f\left(\frac{x}{\varepsilon}, \nabla u_\varepsilon\right) dx + \lambda \int_\Omega |u_\varepsilon|^r dx + \int_\Omega \beta u_\varepsilon dx - i_\varepsilon(\Omega, \lambda, \beta) \right) = 0.$$

Then f_{hom}^q is convex and satisfies (12.7.2) with $p = 1$, $\{i_\varepsilon(\Omega, \lambda, \beta)\}_{\varepsilon > 0}$ converges as $\varepsilon \to 0^+$ to $m_{\text{hom}}(\Omega, \lambda, \beta)$, $\{u_\varepsilon\}_{\varepsilon > 0}$ has cluster points in $L^r(\Omega)$ as $\varepsilon \to 0^+$, and every such point is a solution of $m_{\text{hom}}(\Omega, \lambda, \beta)$.

Proof. The properties of f_{hom}^q follow from (12.0.1) and Proposition 12.1.3.

Let Ω, λ, r, β be as above, and let, for every $\varepsilon > 0$, $G_{1/\varepsilon}$ be defined in (13.1.4). Let us prove that

(13.4.1) $$\Gamma^-(L^r(\Omega)) \lim_{\varepsilon \to 0^+} G_{1/\varepsilon}(\Omega, u) =$$

$$= \begin{cases} \int_\Omega \text{sc}^- f_{\text{hom}}^q(\nabla u) dx + \int_\Omega (\text{sc}^- f_{\text{hom}}^q)^\infty(\nabla^s u) d|D^s u| \\ \qquad\qquad\qquad\qquad\qquad\qquad \text{if } u \in BV(\Omega) \\ +\infty \qquad\qquad\qquad\qquad\qquad\quad \text{if } u \in L^r(\Omega) \setminus BV(\Omega) \end{cases}$$

for every $u \in L^r(\Omega)$.

To do this, we first take $u \in BV(\Omega)$ such that $\int_\Omega \mathrm{sc}^- f^q_{\mathrm{hom}}(\nabla u) dx + \int_\Omega (\mathrm{sc}^- f^q_{\mathrm{hom}})^\infty (\nabla^s u) d|D^s u| < +\infty$, and $\{\varepsilon_h\} \subseteq]0, +\infty[$ with $\varepsilon_h \to 0$. Then, by Theorem 13.1.7, and Proposition 3.2.3, it follows that there exists $\{u_h\} \subseteq W^{1,q}_{\mathrm{loc}}(\mathbf{R}^n)$ such that $u_h \to u$ in $L^1(\Omega)$, and

$$\limsup_{h \to +\infty} \int_\Omega f\left(\frac{x}{\varepsilon_h}, \nabla u_h\right) dx \le$$

$$\le \int_\Omega \mathrm{sc}^- f^q_{\mathrm{hom}}(\nabla u) dx + \int_\Omega (\mathrm{sc}^- f^q_{\mathrm{hom}})^\infty (\nabla^s u) d|D^s u|.$$

Consequently, by (12.1.22) with $p = 1$, and Theorem 4.2.11, we obtain that $u_h \to u$ in $L^r(\Omega)$, and that

$$\Gamma^-(L^r(\Omega)) \limsup_{h \to +\infty} G_{1/\varepsilon_h}(\Omega, u) \le$$

$$\le \int_\Omega \mathrm{sc}^- f^q_{\mathrm{hom}}(\nabla u) dx + \int_\Omega (\mathrm{sc}^- f^q_{\mathrm{hom}})^\infty (\nabla^s u) d|D^s u|,$$

from which, together with Proposition 3.2.3, we conclude that

$$(13.4.2) \qquad \Gamma^-(L^r(\Omega)) \limsup_{\varepsilon \to 0^+} G_{1/\varepsilon}(\Omega, u) \le$$

$$\le \begin{cases} \int_\Omega \mathrm{sc}^- f^q_{\mathrm{hom}}(\nabla u) dx + \int_\Omega (\mathrm{sc}^- f^q_{\mathrm{hom}})^\infty (\nabla^s u) d|D^s u| \\ \hspace{6cm} \text{if } u \in BV(\Omega) \\ +\infty \hspace{4.5cm} \text{if } u \in L^r(\Omega) \setminus BV(\Omega) \end{cases}$$

for every $u \in L^r(\Omega)$.

On the other side, if $u \in L^r(\Omega)$ is such that $\Gamma^-(L^r(\Omega)) \liminf_{\varepsilon \to 0^+} G_{1/\varepsilon}(\Omega, u) < +\infty$, by virtue of Proposition 3.2.6, let $\{\varepsilon_h\} \subseteq]0, +\infty[$ with $\varepsilon_h \to 0$, and $\{u_h\} \subseteq W^{1,q}_{\mathrm{loc}}(\mathbf{R}^n)$ be such that $u_h \to u$ in $L^r(\Omega)$, and

$$\Gamma^-(L^r(\Omega)) \liminf_{\varepsilon \to 0^+} G_{1/\varepsilon}(\Omega, u) = \liminf_{h \to +\infty} \int_\Omega f\left(\frac{x}{\varepsilon_h}, \nabla u_h\right) dx.$$

Then, by (12.1.22) with $p = 1$, we infer that $\{u_h\}$ is bounded in $BV(\Omega)$, and hence that $u \in BV(\Omega)$.

Because of this, and of Theorem 13.1.7, we thus conclude that

$$\Gamma^-(L^r(\Omega)) \liminf_{\varepsilon \to 0^+} G_{1/\varepsilon}(\Omega, u) \ge$$

$$\ge \begin{cases} \int_\Omega \mathrm{sc}^- f^q_{\mathrm{hom}}(\nabla u) dx + \int_\Omega (\mathrm{sc}^- f^q_{\mathrm{hom}})^\infty (\nabla^s u) d|D^s u| \\ \hspace{6cm} \text{if } u \in BV(\Omega) \\ +\infty \hspace{4.5cm} \text{if } u \in L^r(\Omega) \setminus BV(\Omega) \end{cases}$$

for every $u \in L^r(\Omega)$,

from which, together with (13.4.2), equality (13.4.1) follows.

Moreover, by (13.4.1), and Proposition 3.5.2, once we observe that the functional $u \in L^r(\Omega) \mapsto \lambda \int_\Omega |u|^r dx + \int_\Omega \beta u dx$ is $L^r(\Omega)$-continuous, we obtain that

$$(13.4.3) \quad \Gamma^-(L^r(\Omega)) \lim_{\varepsilon \to 0^+} \left\{ G_{1/\varepsilon}(\Omega, u) + \lambda \int_\Omega |u|^r dx + \int_\Omega \beta u dx \right\} =$$

$$= \begin{cases} \int_\Omega \mathrm{sc}^- f_{\mathrm{hom}}^q(\nabla u) dx + \int_\Omega (\mathrm{sc}^- f_{\mathrm{hom}}^q)^\infty (\nabla^s u) d|D^s u| + \\ \qquad + \lambda \int_\Omega |u|^r dx + \int_\Omega \beta u dx \quad \text{if } u \in BV(\Omega) \\ +\infty \qquad\qquad\qquad\qquad\qquad\qquad\qquad\ \text{if } u \in L^r(\Omega) \setminus BV(\Omega) \end{cases}$$

for every $u \in L^r(\Omega)$.

Finally, we observe that, by virtue of (12.1.22) with $p = 1$, and Proposition 4.4.1, the functionals $u \in L^r(\Omega) \mapsto G_{1/\varepsilon}(\Omega, u) + \lambda \int_\Omega |u|^r dx + \int_\Omega \beta u dx$ are equi-coercive.

Because of this, and of (13.4.3), the assumptions of Theorem 3.5.6 are fulfilled with $U = L^r(\Omega)$, and the proof follows from Theorem 3.5.6, once we observe that obviously

$$\limsup_{\varepsilon \to 0^+} i_\varepsilon(\Omega, \lambda, \beta) \leq \mathcal{L}^n(\Omega) \int_Y f(y, z) dy + \lambda \int_\Omega |u_z|^r dx + \int_\Omega \beta u_z dx < +\infty$$

for every $z \in C$. ∎

We now treat the case of Neumann minimum problems in Sobolev spaces.

Theorem 13.4.2. Let f be as in (12.0.1), $p \in\ ]1, +\infty]$, $C \subseteq \mathbf{R}^n$ be convex, $q \in [p, +\infty]$, and let f_{hom}^q be defined in (12.1.18). Assume that (12.1.22), (13.1.1), (13.1.2) and (13.1.3) hold. For every $\varepsilon > 0$, $\Omega \in \mathcal{A}_0$ convex, $\lambda \in\ ]0, +\infty[$, $r \in\ ]1, p^*[$, $\beta \in L^{p'}(\Omega)$, and $\vartheta \in L^{p'}(\partial\Omega)$ let

$$(13.4.4) \qquad i_\varepsilon(\Omega, \lambda, \beta, \vartheta) = \inf \left\{ \int_\Omega f\left(\frac{x}{\varepsilon}, \nabla u\right) dx + \right.$$

$$\left. + \lambda \int_\Omega |u|^r dx + \int_\Omega \beta u dx + \int_{\partial\Omega} \vartheta \gamma_\Omega u d\mathcal{H}^{n-1} : u \in W^{1,q}(\Omega) \right\},$$

$$m_{\mathrm{hom}}(\Omega, \lambda, \beta, \vartheta) = \min \left\{ \int_\Omega \mathrm{sc}^- f_{\mathrm{hom}}^q(\nabla u) dx + \right.$$

$$\left. + \lambda \int_\Omega |u|^r dx + \int_\Omega \beta u dx + \int_{\partial\Omega} \vartheta \gamma_\Omega u d\mathcal{H}^{n-1} : u \in W^{1,p}(\Omega) \right\},$$

and let $\{u_\varepsilon\}_{\varepsilon>0} \subseteq W^{1,q}(\Omega)$ be such that

$$\lim_{\varepsilon \to 0^+} \left(\int_\Omega f\left(\frac{x}{\varepsilon}, \nabla u_\varepsilon\right) dx + \lambda \int_\Omega |u_\varepsilon|^r dx + \int_\Omega \beta u_\varepsilon dx + \int_{\partial\Omega} \vartheta \gamma_\Omega u_\varepsilon d\mathcal{H}^{n-1} -$$

$$-i_\varepsilon(\Omega, \lambda, \beta, \vartheta) \right) = 0.$$

Then f_{hom}^q is convex and satisfies (12.7.2), $\{i_\varepsilon(\Omega, \lambda, \beta, \vartheta)\}_{\varepsilon>0}$ converges as $\varepsilon \to 0^+$ to $m_{\text{hom}}(\Omega, \lambda, \beta, \vartheta)$, $\{u_\varepsilon\}_{\varepsilon>0}$ has cluster points in $L^p(\Omega)$ as $\varepsilon \to 0^+$, and every such point is a solution of $m_{\text{hom}}(\Omega, \lambda, \beta, \vartheta)$.

Moreover, if $q = p$ and (12.1.23) too holds, then $\text{sc}^- f_{\text{hom}}^p = f_{\text{hom}}^p$, for every $z \in \mathbf{R}^n$ the infimum in the definition of $f_{\text{hom}}^p(z)$ is attained, problems in (13.4.4) have solutions, and for every $\varepsilon > 0$ one can take as u_ε a solution of $i_\varepsilon(\Omega, \lambda, \beta, \vartheta)$.

Finally, if $\int_\Omega \beta dx + \int_{\partial\Omega} \vartheta d\mathcal{H}^{n-1} = 0$, one can also take $\lambda = 0$.

Proof. The properties of f_{hom}^q follow from (12.0.1) and Proposition 12.1.3.

Let Ω, λ, r, β, ϑ be as above, and set $s = \max\{p, r\}$.

For every $\varepsilon > 0$ let $G_{1/\varepsilon}$ be defined in (13.1.4), and $B(\Omega, \cdot)$ be given by

$$B(\Omega, \cdot) : u \in L^1(\Omega) \mapsto \begin{cases} \int_{\partial\Omega} \vartheta \gamma_\Omega u d\mathcal{H}^{n-1} & \text{if } u \in W^{1,1}(\Omega) \\ +\infty & \text{if } u \in L^1(\Omega) \setminus W^{1,1}(\Omega). \end{cases}$$

Let let us prove that

(13.4.5) $$\Gamma^-(L^s(\Omega)) \lim_{\varepsilon \to 0^+} \{G_{1/\varepsilon}(\Omega, u) + B(\Omega, u)\} =$$

$$= \begin{cases} \int_\Omega \text{sc}^- f_{\text{hom}}^q(\nabla u) dx + \int_{\partial\Omega} \vartheta \gamma_\Omega u d\mathcal{H}^{n-1} & \text{if } u \in W^{1,p}(\Omega) \\ +\infty & \text{if } u \in L^s(\Omega) \setminus W^{1,p}(\Omega) \end{cases}$$

for every $u \in L^s(\Omega)$.

To do this, let $u \in W^{1,p}(\Omega)$ with $\int_\Omega \text{sc}^- f_{\text{hom}}^q(\nabla u) dx < +\infty$, and let $\{\varepsilon_h\} \subseteq]0, +\infty[$ with $\varepsilon_h \to 0$. Then, by Theorem 13.1.7, and Proposition 3.2.3, it follows that there exists $\{u_h\} \subseteq W_{\text{loc}}^{1,q}(\mathbf{R}^n)$ such that $u_h \to u$ in $L^1(\Omega)$, and

$$\limsup_{h \to +\infty} \int_\Omega f\left(\frac{x}{\varepsilon_h}, \nabla u_h\right) dx \le \int_\Omega \text{sc}^- f_{\text{hom}}^q(\nabla u) dx.$$

We now observe that, by (12.1.22), and Rellich-Kondrachov Compactness Theorem, it follows that $u_h \to u$ in weak-$W^{1,p}(\Omega)$ (weak*-$W^{1,\infty}(\Omega)$ if $p = +\infty$) and in $L^s(\Omega)$, and that the restriction of $B(\Omega, \cdot)$ to $W^{1,p}(\Omega)$ is continuous with respect to the weak-$W^{1,p}(\Omega)$ (weak*-$W^{1,\infty}(\Omega)$ if $p = +\infty$) topology. Consequently, we have that

$$\Gamma^-(L^s(\Omega)) \limsup_{h \to +\infty} \left\{ G_{1/\varepsilon_h}(\Omega, u) + B(\Omega, u) \right\} \le$$

$$\le \int_\Omega \mathrm{sc}^- f^q_{\mathrm{hom}}(\nabla u) dx + \int_{\partial\Omega} \vartheta \gamma_\Omega u d\mathcal{H}^{n-1},$$

from which, together with Proposition 3.2.3, we conclude that

$$(13.4.6) \qquad \Gamma^-(L^s(\Omega)) \limsup_{\varepsilon \to 0^+} \left\{ G_{1/\varepsilon}(\Omega, u) + B(\Omega, u) \right\} \le$$

$$\le \begin{cases} \int_\Omega \mathrm{sc}^- f^q_{\mathrm{hom}}(\nabla u) dx + \int_{\partial\Omega} \vartheta \gamma_\Omega u d\mathcal{H}^{n-1} & \text{if } u \in W^{1,p}(\Omega) \\ +\infty & \text{if } u \in L^s(\Omega) \setminus W^{1,p}(\Omega) \end{cases}$$

$$\text{for every } u \in L^s(\Omega).$$

On the other side, if $u \in L^s(\Omega)$ is such that $\Gamma^-(L^s(\Omega)) \liminf_{\varepsilon \to 0^+}$ $\left\{ G_{1/\varepsilon}(\Omega, u) + B(\Omega, u) \right\} < +\infty$, by virtue of Proposition 3.2.6, let $\{\varepsilon_h\} \subseteq$ $]0, +\infty[$ with $\varepsilon_h \to 0$, and $\{u_h\} \subseteq W^{1,q}_{\mathrm{loc}}(\mathbf{R}^n)$ be such that $u_h \to u$ in $L^s(\Omega)$, and

$$(13.4.7) \qquad \Gamma^-(L^s(\Omega)) \liminf_{\varepsilon \to 0^+} \left\{ G_{1/\varepsilon}(\Omega, u) + B(\Omega, u) \right\} =$$

$$= \lim_{h \to +\infty} \left\{ \int_\Omega f\left(\frac{x}{\varepsilon_h}, \nabla u_h \right) dx + \int_{\partial\Omega} \vartheta \gamma_\Omega u_h d\mathcal{H}^{n-1} \right\}.$$

Then, by (13.4.7), (12.1.22), and the Trace Theorem for Sobolev Functions, we infer that (we treat only the case in which $p \in]1, +\infty[$, the one where $p = +\infty$ being trivial)

$$\limsup_{h \to +\infty} \left\{ \|\nabla u_h\|^p_{L^p(\Omega)} - C_\Omega \|\nabla u_h\|_{L^p(\Omega)} - C_\Omega \|u_h\|_{L^p(\Omega)} \right\} \le$$

$$\le \limsup_{h \to +\infty} \left\{ \|\nabla u_h\|^p_{L^p(\Omega)} - C_\Omega \|\vartheta\|_{L^{p'}(\partial\Omega)} \|\gamma_\Omega u\|_{L^p(\partial\Omega)} \right\} < +\infty$$

for some $C_\Omega > 0$, from which, since $\{u_h\}$ is bounded in $L^p(\Omega)$, we conclude that $\{u_h\}$ is bounded in $W^{1,p}(\Omega)$, and hence that $u_h \to u$ in weak-$W^{1,p}(\Omega)$.

We now observe that the functional $u \in W^{1,p}(\Omega) \mapsto \int_{\partial\Omega} \vartheta \gamma_\Omega u_h d\mathcal{H}^{n-1}$ is weak-$W^{1,p}(\Omega)$-continuous. Therefore, by using Theorem 13.1.7, we conclude that

$$\Gamma^-(L^s(\Omega)) \liminf_{\varepsilon \to 0^+} \left\{ G_{1/\varepsilon}(\Omega, u) + B(\Omega, u) \right\} \ge$$

$$\ge \begin{cases} \int_\Omega \mathrm{sc}^- f^q_{\mathrm{hom}}(\nabla u) dx & \text{if } u \in W^{1,p}(\Omega) \\ +\infty & \text{if } u \in L^s(\Omega) \setminus W^{1,p}(\Omega) \end{cases} \quad \text{for every } u \in L^s(\Omega),$$

from which, together with (13.4.6), equality (13.4.5) follows.

Moreover, by (13.4.5), the $L^s(\Omega)$-continuity of the functional $u \in L^s(\Omega) \mapsto \lambda \int_\Omega |u|^r dx + \int_\Omega \beta u dx$, (12.1.22), and Proposition 3.5.2, it is straightforward to verify that

$$(13.4.8) \qquad \Gamma^-(L^s(\Omega)) \lim_{\varepsilon \to 0^+} \left\{ G_{1/\varepsilon}(\Omega, u) + \right.$$

$$\left. + \lambda \int_\Omega |u|^r dx + \int_\Omega \beta u dx + B(\Omega, u) \right\} =$$

$$= \begin{cases} \int_\Omega \mathrm{sc}^- f_{\mathrm{hom}}^q(\nabla u) dx + \lambda \int_\Omega |u|^r dx + \\ \qquad + \int_\Omega \beta u dx + \int_{\partial\Omega} \vartheta \gamma_\Omega u d\mathcal{H}^{n-1} & \text{if } u \in W^{1,p}(\Omega) \\ +\infty & \text{if } u \in L^s(\Omega) \setminus W^{1,p}(\Omega) \end{cases}$$

for every $u \in L^s(\Omega)$.

To complete the proof, we now prove that the functionals $u \in L^s(\Omega) \mapsto G_{1/\varepsilon}(\Omega, u) + \lambda \int_\Omega |u|^r dx + \int_\Omega \beta u dx + B(\Omega, u)$ are equi-coercive.

In fact, when $p < +\infty$, by (12.1.22), Hölder's inequality, and the Trace Theorem for Sobolev Functions it is easy to see that there exists $C_\Omega \in {]0, +\infty[}$ such that (here we denote by C_Ω various constants, all depending on the same quantities)

$$G_{1/\varepsilon}(\Omega, u) + \lambda \int_\Omega |u|^r dx + \int_\Omega \beta u dx + B(\Omega, u) \geq$$

$$\geq \int_\Omega |\nabla u|^p dx + \lambda \int_\Omega |u|^r dx - \|\beta\|_{L^{p'}(\Omega)} \|u\|_{L^p(\Omega)} - \|\vartheta\|_{L^{p'}(\partial\Omega)} \|\gamma_\Omega u\|_{L^p(\partial\Omega)} \geq$$

$$\geq \|\|\nabla u\|\|_{L^p(\Omega)}^p + \lambda \|u\|_{L^r(\Omega)}^r - C_\Omega \left(\|\beta\|_{L^{p'}(\Omega)} + \|\vartheta\|_{L^{p'}(\partial\Omega)} \right) \|u\|_{W^{1,p}(\Omega)}$$

for every $\varepsilon > 0$, $u \in W^{1,p}(\Omega)$,

or, when $p = +\infty$,

$$G_{1/\varepsilon}(\Omega, u) + \lambda \int_\Omega |u|^r dx + \int_\Omega \beta u dx + B(\Omega, u) \geq$$

$$\geq \lambda \|u\|_{L^r(\Omega)}^r - \|\beta\|_{L^1(\Omega)} \|u\|_{L^\infty(\Omega)} - \|\vartheta\|_{L^1(\partial\Omega)} \|\gamma_\Omega u\|_{L^\infty(\partial\Omega)} \geq$$

$$\geq \lambda \|u\|_{L^r(\Omega)}^r - C_\Omega \left(\|\beta\|_{L^1(\Omega)} + \|\vartheta\|_{L^1(\partial\Omega)} \right) \|u\|_{W^{1,\infty}(\Omega)}$$

for every $\varepsilon > 0$, $u \in W^{1,\infty}(\Omega)$.

By the above inequalities, and by Proposition 4.4.3, once we recall that every $u \in L^s(\Omega)$ satisfying $G_{1/\varepsilon}(\Omega, u) < +\infty$ actually is in $W^{1,p}(\Omega)$, the desired equi-coerciveness follows.

Because of this, and of (13.4.8), the assumptions of Theorem 3.5.6 are fulfilled with $U = L^s(\Omega)$, and the proof follows from Theorem 3.5.6, once we observe that obviously

$$\limsup_{\varepsilon \to 0^+} i_\varepsilon(\Omega, \lambda, \beta, \vartheta) \le$$

$$\le \mathcal{L}^n(\Omega) \int_Y f(y, z) dy + \lambda \int_\Omega |u_z|^r dx + \int_\Omega \beta u_z dx + \int_{\partial\Omega} \vartheta \gamma_\Omega u_z d\mathcal{H}^{n-1} < +\infty$$

for every $z \in C$.

Finally, if $\int_\Omega \beta dx + \int_{\partial\Omega} \vartheta d\mathcal{H}^{n-1} = 0$ and $\lambda = 0$, the proof follows by arguing as above, once we observe that

$$i_\varepsilon(\Omega, 0, \beta, \vartheta) = \inf \left\{ \int_\Omega f\left(\frac{x}{\varepsilon}, \nabla u\right) dx + \int_\Omega \beta u dx + \int_{\partial\Omega} \vartheta \gamma_\Omega u d\mathcal{H}^{n-1} : \right.$$

$$\left. u \in W^{1,q}(\Omega), \int_\Omega u dx = 0 \right\} \text{ for every } \varepsilon > 0,$$

$$m_{\text{hom}}(\Omega, 0, \beta, \vartheta) = \min \left\{ \int_\Omega \text{sc}^- f_{\text{hom}}^q(\nabla u) dx + \int_\Omega \beta u dx + \int_{\partial\Omega} \vartheta \gamma_\Omega u d\mathcal{H}^{n-1} : \right.$$

$$\left. u \in W^{1,p}(\Omega), \int_\Omega u dx = 0 \right\},$$

and that by (12.1.22), Theorem 4.3.19, and Proposition 4.4.3, the above functionals are equi-coercive in $L^s(\Omega)$. ■

We now pass to the case of Dirichlet minimum problems. We start with the one in BV spaces.

Theorem 13.4.3. *Let f be as in (12.0.1), $C \subseteq \mathbf{R}^n$ be convex, $q \in [1, +\infty]$, and let f_{hom}^q be defined in (12.1.18). Assume that (12.1.22) with $p = 1$, (13.1.1), (13.1.2) and (13.1.3) hold. For every $\varepsilon > 0$, $\Omega \in \mathcal{A}_0$ convex, $\lambda \in]0, +\infty[$, $r \in]1, 1^*[$, $\beta \in L^{r'}(\Omega)$, $z_0 \in \text{int}(C)$, and $c \in \mathbf{R}$ let*

(13.4.9) $$i_\varepsilon^0(\Omega, \lambda, \beta) =$$

$$= \inf \left\{ \int_\Omega f\left(\frac{x}{\varepsilon}, \nabla u\right) dx + \lambda \int_\Omega |u|^r dx + \int_\Omega \beta u dx : u \in u_{z_0} + c + W_0^{1,q}(\Omega) \right\},$$

$$m_{\text{hom}}^0(\Omega, \lambda, \beta) = \min \left\{ \int_\Omega \text{sc}^- f_{\text{hom}}^q(\nabla u) dx + \int_\Omega (\text{sc}^- f_{\text{hom}}^q)^\infty (\nabla^s u) d|D^s u| + \right.$$

$$+ \int_{\partial\Omega} (\text{sc}^- f_{\text{hom}}^q)^\infty ((u_{z_0} + c - \gamma_\Omega u)\mathbf{n}_\Omega) d\mathcal{H}^{n-1} +$$

$$+\lambda \int_\Omega |u|^r dx + \int_\Omega \beta u dx : u \in BV(\Omega) \Big\},$$

and let $\{u_\varepsilon\}_{\varepsilon>0} \subseteq u_{z_0} + c + W_0^{1,q}(\Omega)$ satisfy

$$\lim_{\varepsilon\to 0^+} \left(\int_\Omega f(hx, \nabla u_\varepsilon) dx + \lambda \int_\Omega |u_\varepsilon|^r dx + \int_\Omega \beta u_\varepsilon dx - i_\varepsilon^0(\Omega, \lambda, \beta) \right) = 0.$$

Then f_{hom}^q is convex and satisfies (12.7.2) with $p = 1$, $\{i_\varepsilon^0(\Omega, \lambda, \beta)\}_{\varepsilon>0}$ converges as $\varepsilon \to 0^+$ to $m_{\text{hom}}^0(\Omega, \lambda, \beta)$, $\{u_\varepsilon\}_{\varepsilon>0}$ has cluster points in $L^r(\Omega)$ as $\varepsilon \to 0^+$, and every such point is a solution of $m_{\text{hom}}^0(\Omega, \lambda, \beta)$.

Proof. Similar to the one of Theorem 13.4.1 with the necessary changes. In particular, with the functionals $G_{1/\varepsilon}(\Omega, \partial\Omega, u_{z_0} + c, \cdot)$ defined in (13.1.5) in place of those given by (13.1.4), and with Theorem 13.2.3 in place of Theorem 13.1.7. ∎

The following result deals with the case of Dirichlet minimum problems in Sobolev spaces.

Theorem 13.4.4. Let f be as in (12.0.1), $p \in]1, +\infty]$, $C \subseteq \mathbf{R}^n$ be convex, $q \in [p, +\infty]$, and let f_{hom}^q be defined in (12.1.18). Assume that (12.1.22), (13.1.1), (13.1.2) and (13.1.3) hold. For every $\varepsilon > 0$, $\Omega \in \mathcal{A}_0$ convex, $\beta \in L^{p'}(\Omega)$, $z_0 \in \text{int}(C)$, and $c \in \mathbf{R}$ let $i_\varepsilon^0(\Omega, 0, \beta)$ be defined in (13.4.9),

$$m_{\text{hom}}^0(\Omega, 0, \beta) =$$

$$= \min \left\{ \int_\Omega \text{sc}^- f_{\text{hom}}^q(\nabla u) dx + \int_\Omega \beta u dx : u \in u_{z_0} + c + W_0^{1,p}(\Omega) \right\},$$

and let $\{u_\varepsilon\}_{\varepsilon>0} \subseteq u_{z_0} + c + W_0^{1,q}(\Omega)$ be such that

$$\lim_{\varepsilon\to 0^+} \left(\int_\Omega f\left(\frac{x}{\varepsilon}, \nabla u_\varepsilon\right) dx + \int_\Omega \beta u_\varepsilon dx - i_\varepsilon^0(\Omega, 0, \beta) \right) = 0.$$

Then f_{hom}^q is convex and satisfies (12.7.2), $\{i_\varepsilon^0(\Omega, 0, \beta)\}_{\varepsilon>0}$ converges as $\varepsilon \to 0^+$ to $m_{\text{hom}}^0(\Omega, 0, \beta)$, $\{u_\varepsilon\}_{\varepsilon>0}$ has cluster points in $L^p(\Omega)$ as $\varepsilon \to 0^+$, and every such point is a solution of $m_{\text{hom}}^0(\Omega, 0, \beta)$.

Moreover, if $q = p$ and (12.1.23) too holds, then $\text{sc}^- f_{\text{hom}}^p = f_{\text{hom}}^p$, for every $z \in \mathbf{R}^n$ the infimum in the definition of $f_{\text{hom}}^p(z)$ is attained, problems in (13.4.9) have solutions, and for every $\varepsilon > 0$ one can take as u_ε a solution of $i_\varepsilon^0(\Omega, 0, \beta)$.

Proof. Similar to the one of Theorem 13.4.2 with the necessary changes. In particular, with the functionals $G_{1/\varepsilon}(\Omega, \partial\Omega, u_{z_0} + c, \cdot)$ defined in (13.1.5) in place of those given by (13.1.4), and with Theorem 13.2.3 in place of Theorem 13.1.7. ∎

Finally, we treat the case of mixed minimum problems.

Theorem 13.4.5. Let f be as in (12.0.1), and f_{hom}^1 be defined in (12.1.18) with $q = 1$. Assume that (13.1.12) holds with $\phi \colon \mathbf{R}^n \to [0, +\infty]$ convex and satisfying $\text{int}(\text{dom}\phi) \neq \emptyset$, $\lim_{z \to +\infty} \frac{\phi(z)}{|z|} = +\infty$, $a \in L_{\text{loc}}^1(\mathbf{R}^n)$ Y-periodic, and $M \geq 0$. For every $\varepsilon > 0$, $\Omega \in \mathcal{A}_0$ convex, $\Gamma \in \mathcal{B}(\partial\Omega)$ with $H^{n-1}(\Gamma) > 0$, $\beta \in L^\infty(\Omega)$, $\vartheta \in L^\infty(\partial\Omega)$, $z_0 \in \text{int}(\text{dom}\phi)$, and $c \in \mathbf{R}$ let

$$(13.4.10) \qquad i_\varepsilon(\Omega, \Gamma, \beta, \vartheta) = \inf \left\{ \int_\Omega f\left(\frac{x}{\varepsilon}, \nabla u\right) dx + \right.$$

$$\left. + \int_\Omega \beta u \, dx + \int_{\partial\Omega} \vartheta \gamma_\Omega u \, dH^{n-1} : u \in u_{z_0} + c + W_{0,\Gamma}^{1,1}(\Omega) \right\},$$

$$m_{\text{hom}}(\Omega, \Gamma, \beta, \vartheta) = \min \left\{ \int_\Omega \text{sc}^- f_{\text{hom}}^1(\nabla u) dx + \right.$$

$$\left. + \int_\Omega \beta u \, dx + \int_{\partial\Omega} \vartheta \gamma_\Omega u \, dH^{n-1} : u \in u_{z_0} + c + W_{0,\Gamma}^{1,1}(\Omega) \right\},$$

and let $\{u_\varepsilon\}_{\varepsilon>0} \subseteq u_{z_0} + c + W_{0,\Gamma}^{1,1}(\Omega)$ be such that

$$\lim_{\varepsilon \to 0^+} \left(\int_\Omega f\left(\frac{x}{\varepsilon}, \nabla u_\varepsilon\right) dx + \int_\Omega \beta u_\varepsilon dx + \int_{\partial\Omega} \vartheta \gamma_\Omega u_\varepsilon dH^{n-1} - \right.$$

$$\left. - i_\varepsilon(\Omega, \Gamma, \beta, \vartheta) \right) = 0.$$

Then f_{hom}^1 is convex and satisfies

$$(13.4.11) \quad \text{sc}^- \phi(z) \leq f_{\text{hom}}^1(z) \leq \int_Y a(y) dy + M \text{sc}^- \phi(z) \text{ for every } z \in \mathbf{R}^n,$$

$\{i_\varepsilon(\Omega, \Gamma, \beta, \vartheta)\}$ converges as $\varepsilon \to 0^+$ to $m_{\text{hom}}(\Omega, \Gamma, \beta, \vartheta)$, $\{u_\varepsilon\}_{\varepsilon>0}$ has cluster points in $L^1(\Omega)$ as $\varepsilon \to 0^+$, and every such point is a solution of $m_{\text{hom}}(\Omega, \Gamma, \beta, \vartheta)$.

Moreover, if (12.1.23) too holds, then $\text{sc}^- f_{\text{hom}}^1 = f_{\text{hom}}^1$, for every $z \in \mathbf{R}^n$ the infimum in the definition of $f_{\text{hom}}^1(z)$ is attained, problems in (13.4.10) have solutions, and for every $\varepsilon > 0$ one can take as u_ε a solution of $i_\varepsilon(\Omega, \Gamma, \beta, \vartheta)$.

Proof. The proof of the estimates in (13.4.11) follows as in the one of Proposition 12.1.3, and the properties of f_{hom}^q follow from (12.0.1) and Proposition 12.1.3.

Let Ω, Γ, β, ϑ, z_0, c be as above.

For every $\varepsilon > 0$ let $G_{1/\varepsilon}(\Omega, \Gamma, u_{z_0} + c, \cdot)$ be defined in (13.1.5), and $B(\Omega, \cdot)$ be given by

$$B(\Omega, \cdot) \colon u \in L^1(\Omega) \mapsto \begin{cases} \int_{\partial\Omega} \vartheta \gamma_\Omega u \, dH^{n-1} & \text{if } u \in W^{1,1}(\Omega) \\ +\infty & \text{if } u \in L^1(\Omega) \setminus W^{1,1}(\Omega). \end{cases}$$

Then, by following the outlines of the proof of Theorem 13.4.2, and by using Theorem 13.3.1 in place of Theorem 13.1.7, one proves that

$$(13.4.12) \qquad \Gamma^-(L^1(\Omega)) \limsup_{\varepsilon \to 0^+} \{G_{1/\varepsilon}(\Omega, \Gamma, u_{z_0} + c, u) + B(\Omega, u)\} \le$$

$$\le \begin{cases} \int_\Omega \mathrm{sc}^- f^q_{\mathrm{hom}}(\nabla u)dx + \int_{\partial\Omega} \vartheta \gamma_\Omega u d\mathcal{H}^{n-1} \\ \qquad\qquad\qquad\qquad \text{if } u \in u_{z_0} + c + W^{1,1}_{0,\Gamma}(\Omega) \\ +\infty \qquad\qquad\qquad \text{if } u \in L^1(\Omega) \setminus (u_{z_0} + c + W^{1,1}_{0,\Gamma}(\Omega)) \end{cases}$$

for every $u \in L^1(\Omega)$.

On the other side, if $u \in L^1(\Omega)$ is such that $\Gamma^-(L^1(\Omega)) \liminf_{\varepsilon \to 0^+} \{G_{1/\varepsilon}(\Omega, \Gamma, u_{z_0} + c, u) + B(\Omega, u)\} < +\infty$, by virtue of Proposition 3.2.6, let $\{\varepsilon_h\} \subseteq]0, +\infty[$ with $\varepsilon_h \to 0$, and $\{u_h\} \subseteq u_{z_0} + c + W^{1,q}_{0,\Gamma}(\Omega)$ be such that $u_h \to u$ in $L^1(\Omega)$, and

$$(13.4.13) \qquad \Gamma^-(L^1(\Omega)) \liminf_{h \to +\infty} \{G_{1/\varepsilon_h}(\Omega, \Gamma, u_{z_0} + c, u) + B(\Omega, u)\} =$$

$$= \lim_{h \to +\infty} \left\{ \int_\Omega f\left(\frac{x}{\varepsilon_h}, \nabla u_h\right) dx + \int_{\partial\Omega} \vartheta \gamma_\Omega u_h d\mathcal{H}^{n-1} \right\}.$$

Then, by (13.4.13), (13.1.12), and the Trace Theorem for Sobolev Functions, we infer that (we treat only the case in which $p \in]1, +\infty[$, the one where $p = +\infty$ being trivial)

$$\limsup_{h \to +\infty} \left\{ \int_\Omega \phi(\nabla u_h)dx - C_\Omega \|\nabla u_h\|_{L^1(\Omega)} - C_\Omega \|u_h\|_{L^1(\Omega)} \right\} \le$$

$$\le \limsup_{h \to +\infty} \left\{ \int_\Omega \phi(\nabla u_h)dx - C_\Omega \|\vartheta\|_{L^\infty(\partial\Omega)} \|\gamma_\Omega u\|_{L^1(\partial\Omega)} \right\} < +\infty$$

for some $C_\Omega > 0$, from which, since $\{u_h\}$ is bounded in $L^1(\Omega)$, we conclude that $\{\int_\Omega \phi(\nabla u_h)dx\}$ is bounded. Because of this, and of the Dunford-Pettis-de la Vallée Poussin Compactness Theorem, we conclude that $u_h \to u$ in weak-$W^{1,1}(\Omega)$.

At this point, as in the proof of Theorem 13.4.2, we conclude that

$$\Gamma^-(L^1(\Omega)) \liminf_{\varepsilon \to 0^+} \{G_{1/\varepsilon}(\Omega, \Gamma, u_{z_0} + c, u) + B(\Omega, u)\} \ge$$

$$\ge \begin{cases} \int_\Omega \mathrm{sc}^- f^q_{\mathrm{hom}}(\nabla u)dx + \int_{\partial\Omega} \vartheta \gamma_\Omega u d\mathcal{H}^{n-1} \\ \qquad\qquad\qquad\qquad \text{if } u \in u_{z_0} + c + W^{1,1}_{0,\Gamma}(\Omega) \\ +\infty \qquad\qquad\qquad \text{if } u \in L^1(\Omega) \setminus (u_{z_0} + c + W^{1,1}_{0,\Gamma}(\Omega)) \end{cases}$$

for every $u \in L^1(\Omega)$,

from which, together with (13.4.12), the $L^1(\Omega)$-continuity of the functional $u \in L^1(\Omega) \mapsto \int_\Omega \beta u dx$, and Proposition 3.5.2, it is straightforward to verify that

$$\Gamma^-(L^1(\Omega)) \lim_{\varepsilon \to 0^+} \left\{ G_{1/\varepsilon}(\Omega, \Gamma, u_{z_0} + c, u) + \int_\Omega \beta u dx + B(\Omega, u) \right\} =$$

$$= \begin{cases} \int_\Omega \mathrm{sc}^- f^q_{\hom}(\nabla u) dx + \int_\Omega \beta u dx + \int_{\partial\Omega} \vartheta \gamma_\Omega u d\mathcal{H}^{n-1} \\ \qquad\qquad \text{if } u \in u_{z_0} + c + W^{1,1}_{0,\Gamma}(\Omega) \\ +\infty \qquad\qquad \text{if } u \in L^1(\Omega) \setminus (u_{z_0} + c + W^{1,1}_{0,\Gamma}(\Omega)) \end{cases}$$

for every $u \in L^1(\Omega)$.

Finally, again as in Theorem 13.4.2, by (13.1.12), the growth conditions of ϕ, and Proposition 4.4.5, the functionals $u \in L^1(\Omega) \mapsto G_{1/\varepsilon}(\Omega, \Gamma, u_{z_0} + c, u) + \int_\Omega \beta u dx + B(\Omega, u)$ turn out to be equi-coercive, and the proof completes as in the one of Theorem 13.4.2. ∎

§13.5 Homogenization with Oscillating Special Constraints

The techniques developed in these last chapters seem to be flexible enough to be applied to other homogenization problems.

In particular, they can be exploited to study some homogenization problem in electrostatics that are, in some sense, intermediate between those treated until now, and that are concerned with materials with periodically distributed conductors, possibly also "thin," namely with null Lebesgue measure. In fact, whilst in Chapter 12 the constraints were allowed to quickly oscillate, and in this one they were fixed, the problems that we are going to treat involve constraints that can oscillate, but only in some restricted ways.

In this section we report briefly on this approach by quickly describing the main steps needed to get the homogenization result, and by emphasizing the crucial points. We refer to [DAGP] for a complete exposition of the matter.

The case that we are going to treat deals with energy densities of the following type

(13.5.1) $$f \colon (x, z) \in \mathbf{R}^n \times \mathbf{R}^n \mapsto g(x, z) + I_{B_{\varphi(x)}(0)}(z),$$

where φ assumes only the values 0 and $+\infty$. More precisely, g and φ satisfy

(13.5.2) $$\begin{cases} g \colon (x, z) \in \mathbf{R}^n \times \mathbf{R}^n \mapsto g(x, z) \in [0, +\infty[\\ g \; (\mathcal{L}_n(\mathbf{R}^n \times \mathcal{B}(\mathbf{R}^n))\text{-measurable} \\ g(\cdot, z) \; Y\text{-periodic and in } L^1(Y) \text{ for every } z \in \mathbf{R}^n \\ g(x, \cdot) \text{ convex for a.e. } x \in \mathbf{R}^n \\ \varphi \colon x \in \mathbf{R}^n \mapsto \varphi(x) \in \{0, +\infty\} \\ \varphi \; Y\text{-periodic and measurable.} \end{cases}$$

It is clear that, if g, φ, and f are as above, then f is $(\mathcal{L}_n(\mathbf{R}^n \times \mathcal{B}(\mathbf{R}^n))$-measurable, and $\mathrm{dom}f(x, \cdot)$ can oscillate between only two sets as x varies in $\mathbf{R}^n$, namely

(13.5.3). $\mathrm{dom}f(x, \cdot) \in \{\{0\}, \mathbf{R}^n\}$ for a.e. $x \in \mathbf{R}^n$

Of course, the admissible configurations for an energy functional with density f satisfying (13.5.3) turn out to be subject to the extreme constraint $\nabla u(x) = 0$ for a.e. x in some given zones, and to no constraints in their complements.

In this setting, the crucial point that distinguish the present homogenization problem relies on the behaviour of $f_{\mathrm{hom}}^{+\infty}$ defined by (12.1.18) with $q = +\infty$. It is analyzed in the result below, that hold under the assumption

(13.5.4) $\mathrm{int}\left(\mathrm{dom}f_{\mathrm{hom}}^{+\infty}\right) \neq \emptyset.$

Proposition 13.5.1. *Let g, φ be as in (13.5.2), f be defined in (13.5.1), and $f_{\mathrm{hom}}^{+\infty}$ in (12.1.18) with $q = +\infty$. Assume that (13.5.4) holds. Then $\mathrm{dom}f_{\mathrm{hom}}^{+\infty} = \mathbf{R}^n$.*

Proof. First of all, we observe that $\mathrm{dom}f_{\mathrm{hom}}^{+\infty}$ is symmetric with respect to 0. In fact, if $z \in \mathrm{dom}f_{\mathrm{hom}}^{+\infty}$ there exists $v \in W_{\mathrm{per}}^{1,\infty}(Y)$ such that $|z + \nabla v(y)| \leq \varphi(y)$ for a.e. $y \in Y$, and $\int_Y g(y, z + \nabla v)dy < +\infty$. Consequently, it is easy to verify that $\int_Y g(y, -z - \nabla v)dy < +\infty$ too, and therefore that $-z \in \mathrm{dom}f_{\mathrm{hom}}^{+\infty}$.

We now prove that if $z \in \mathrm{dom}f_{\mathrm{hom}}^{+\infty}$, and $t > 0$ then $tz \in \mathrm{dom}f_{\mathrm{hom}}^{+\infty}$. In fact, if $z \in \mathrm{dom}f_{\mathrm{hom}}^{+\infty}$, there exists $v \in W_{\mathrm{per}}^{1,\infty}(Y)$ such that $|z + \nabla v(y)| \leq \varphi(y)$ for a.e. $y \in Y$, and $\int_Y g(y, z + \nabla v)dy < +\infty$. Consequently, $|tz + t\nabla v(y)| \leq \varphi(y)$ for a.e. $y \in Y$, and by using the summability properties of g, it follows that also $\int_Y g(y, tz + t\nabla v)dy < +\infty$, and the desired property follows.

Because of this, and of the symmetry of $\mathrm{dom}f_{\mathrm{hom}}^{+\infty}$, the proof follows. ∎

If on one side Proposition 13.5.1 simplifies the application of the homogenization techniques, conditions (13.5.3) and (13.5.4) have another crucial consequence that allows the setting of the homogenization process in L^1 topologies.

Let g, φ be as in (13.5.2), and f be defined in (13.5.1). For every $r \in]0, +\infty[$ and $\{r_h\} \subseteq [0, +\infty[$, let G_r be given by (13.1.4), $\widetilde{G}'$ and $\widetilde{G}''$ by (13.1.6), and G' and G'' by (13.1.8).

Then, by a direct verification and by using the same arguments exploited in the proof of Proposition 12.1.1, it can be verified that

(13.5.5) $\widetilde{G}'_-(\Omega, u + c) = \widetilde{G}'_-(\Omega, u), \ \widetilde{G}''_-(\Omega, u + c) = \widetilde{G}''(\Omega, u)$

for every $\Omega \in \mathcal{A}_0$, $u \in L^1_{\mathrm{loc}}(\mathbf{R}^n)$, $c \in \mathbf{R}$,

(13.5.6) $\widetilde{G}'_-$ and $\widetilde{G}''_-$ are translation invariant,

(13.5.7) $\widetilde{G}'_-$ and $\widetilde{G}''_-$ are increasing,

(13.5.8) $\widetilde{G}'_-$ is weakly superadditive,

(13.5.9) $\widetilde{G}''_-$ is convex.

The role played by (13.5.3) and (13.5.4) is particularly clear when gradients of convex combinations of admissible configurations are taken into account.

To see this, we first observe that (13.5.4) implies (11.1.3) with $q = +\infty$, and consequently that Lemma 11.1.1 holds. Once obtained the cut-off functions, if $\Omega \in \mathcal{A}_0$, ψ is one of these, and u, $v \in W^{1,\infty}_{\text{loc}}(\mathbf{R}^n)$ are such that $\int_\Omega f(x, \nabla u)dx < +\infty$ and $\int_\Omega f(x, \nabla v)dx < +\infty$, then the particular shape of φ allows to prove that $\nabla(\psi + (1 - \psi)v)(x) = \psi(x)\nabla u(x) + (1 - \psi(x))\nabla v(x) + (u(x) - v(x))\nabla\psi(x) \in \text{dom}f(x, \cdot)$ for a.e. $x \in \Omega$, from which a condition like $\int_\Omega f(x, \nabla(\psi + (1 - \psi)v))dx < +\infty$ follows.

Remarks of this type, coupled with the technique used in the proof of Proposition 12.2.1 and in §12.4, provide that

(13.5.10) $\widetilde{G}''_-$ is weakly subadditive,

and that

(13.5.11) $G''(Y, u_z) = G'_-(Y, u_z) = f^{+\infty}_{\text{hom}}(z)$ for every $z \in \mathbf{R}^n$.

In addition, since (13.5.4) implies also (11.2.3) with $q = +\infty$, Lemma 11.2.2 too holds. Consequently, (13.5.11) and an argument similar to the one exploited in the proof of Lemma 12.3.1 provide that

(13.5.12) $G''_-(\Omega, u) \leq \int_\Omega f^{+\infty}_{\text{hom}}(\nabla u)dx$ for every $\Omega \in \mathcal{A}_0$, $u \in PA(\mathbf{R}^n)$.

Finally, again (13.5.11) and an argument similar to the one exploited in the proof of Lemma 12.5.2 yield the following blow-up condition

(13.5.13) $\limsup\limits_{r \to 0+} \dfrac{1}{r^n} G'_-(Q_r(x_0), u) \geq G'_-(Q_1(x_0), u(x_0) + \nabla u(x_0)\cdot(\cdot - x_0))$

for every $u \in C^1(\mathbf{R}^n)$, and $x_0 \in \mathbf{R}^n$.

At this point, conditions (13.5.5)÷(13.5.13) and an argument similar to the one used in the proof of Proposition 12.6.1, allow us to apply Theorem 9.9.4, and to deduce that

(13.5.14) $G'_-(\Omega, u) = G''_-(\Omega, u) =$

$$= \int_\Omega f^{+\infty}_{\text{hom}}(\nabla u)dx + \int_\Omega (f^{+\infty}_{\text{hom}})^\infty(\nabla^s u)d|D^s u|$$

for every $\Omega \in \mathcal{A}_0$, $u \in BV_{\text{loc}}(\mathbf{R}^n)$.

In conclusion, the following results can be proved from (13.5.14).

Theorem 13.5.2. Let g, φ be as in (13.5.2), and f as in (13.5.1). Let $f_{\text{hom}}^{+\infty}$ be defined in (12.1.18) with $q = +\infty$, and G_r in (13.1.4) for every $r \in]0, +\infty[$. Assume that (13.5.4) holds. Then $f_{\text{hom}}^{+\infty}$ is convex and finite on $\mathbf{R}^n$, and

$$\Gamma^-(L^1(\Omega)) \liminf_{\varepsilon \to 0^+} G_{1/\varepsilon}(\Omega, u) = \Gamma^-(L^1(\Omega)) \limsup_{\varepsilon \to 0^+} G_{1/\varepsilon}(\Omega, u) =$$

$$= \int_\Omega f_{\text{hom}}^{+\infty}(\nabla u)dx + \int_\Omega (f_{\text{hom}}^{+\infty})^\infty(\nabla^s u)d|D^s u|$$

for every $\Omega \in \mathcal{A}_0$ with Lipschitz boundary, $u \in BV(\Omega)$.

Theorem 13.5.3. Let g, φ be as in (13.5.2), f as in (13.5.1), and let $f_{\text{hom}}^{+\infty}$ be defined in (12.1.18) with $q = +\infty$. Let $\Omega \in \mathcal{A}_0$ have Lipschitz boundary, and $G_r(\Omega, \partial\Omega, 0, \cdot)$ in (13.1.5) for every $r \in]0, +\infty[$. Assume that (13.5.4) holds. Then $f_{\text{hom}}^{+\infty}$ is convex and finite on $\mathbf{R}^n$, and

$$\Gamma^-(L^1(\Omega)) \liminf_{\varepsilon \to 0^+} G_{1/\varepsilon}(\Omega, \partial\Omega, 0, u) =$$

$$= \Gamma^-(L^1(\Omega)) \limsup_{\varepsilon \to 0^+} G_{1/\varepsilon}(\Omega, \partial\Omega, 0, u) =$$

$$= \int_\Omega f_{\text{hom}}^{+\infty}(\nabla u)dx + \int_\Omega (f_{\text{hom}}^{+\infty})^\infty(\nabla^s u)d|D^s u| + \int_{\partial\Omega} (f_{\text{hom}}^{+\infty})^\infty(-\gamma_\Omega u \mathbf{n}_\Omega)d\mathcal{H}^{n-1}$$

for every $u \in BV(\Omega)$.

For what concerns condition (13.5.4), we remark that, for example, it is fulfilled provided

$$\varphi(y) = +\infty \text{ for a.e. } y \text{ in a neighborhood of } \partial Y.$$

By Theorems 13.5.2 and 13.5.3, and by means of arguments already exploited in the above chapters, the following results on the convergence of minima and of minimizers in BV spaces can be proved under the coerciveness assumption below

(13.5.15) $|z|^p \le g(x, z)$ for a.e. $x \in \mathbf{R}^n$, and every $z \in \mathbf{R}^n$,

for some $p \in [1, +\infty[$.

Theorem 13.5.4. Let g, φ be as in (13.5.2), f as in (13.5.1), and let $f_{\mathrm{hom}}^{+\infty}$ be defined in (12.1.18) with $q = +\infty$. Assume that (13.5.4) and (13.5.15) with $p = 1$ hold. For every $\varepsilon > 0$, every $\Omega \in \mathcal{A}_0$ with Lipschitz boundary, $\lambda \in \,]0, +\infty[$, $r \in \,]1, 1^*[$, and $\beta \in L^\infty(\Omega)$ let

$$i_\varepsilon(\Omega, \lambda, \beta) =$$

$$= \inf \left\{ \int_\Omega f\left(\frac{x}{\varepsilon}, \nabla u\right) dx + \lambda \int_\Omega |u|^r dx + \int_\Omega \beta u dx : u \in W^{1,\infty}(\Omega) \right\},$$

$$m_{\mathrm{hom}}(\Omega, \lambda, \beta) =$$

$$= \min \left\{ \int_\Omega f_{\mathrm{hom}}^{+\infty}(\nabla u) dx + \int_\Omega (f_{\mathrm{hom}}^{+\infty})^\infty(\nabla^s u) d|D^s u| + \right.$$

$$\left. + \lambda \int_\Omega |u|^r dx + \int_\Omega \beta u dx : u \in BV(\Omega) \right\},$$

and let $\{u_\varepsilon\}_{\varepsilon > 0} \subseteq W^{1,\infty}(\Omega)$ be such that

$$\lim_{\varepsilon \to 0^+} \left(\int_\Omega f\left(\frac{x}{\varepsilon}, \nabla u_\varepsilon\right) dx + \lambda \int_\Omega |u_\varepsilon|^r dx + \int_\Omega \beta u_\varepsilon dx - i_\varepsilon(\Omega, \lambda, \beta) \right) = 0.$$

Then $f_{\mathrm{hom}}^{+\infty}$ is convex, finite on $\mathbf{R}^n$, and satisfies (12.7.2) with $p = 1$, $\{i_\varepsilon(\Omega, \lambda, \beta)\}_{\varepsilon > 0}$ converges as $\varepsilon \to 0^+$ to $m_{\mathrm{hom}}(\Omega, \lambda, \beta)$, $\{u_\varepsilon\}_{\varepsilon > 0}$ has cluster points in $L^1(\Omega)$ as $\varepsilon \to 0^+$, and every such point is a solution of $m_{\mathrm{hom}}(\Omega, \lambda, \beta)$.

Theorem 13.5.5. Let g, φ be as in (13.5.2), f as in (13.5.1), and let $f_{\mathrm{hom}}^{+\infty}$ be defined in (12.1.18) with $q = +\infty$. Assume that (13.5.4) and (13.5.15) with $p \in \,]1, +\infty[$ hold. For every $\varepsilon > 0$, every $\Omega \in \mathcal{A}_0$ with Lipschitz boundary, $\lambda \in \,]0, +\infty[$, and $\beta \in L^{p'}(\Omega)$ let

$$i_\varepsilon(\Omega, \lambda, \beta) =$$

$$= \inf \left\{ \int_\Omega f\left(\frac{x}{\varepsilon}, \nabla u\right) dx + \lambda \int_\Omega |u|^p dx + \int_\Omega \beta u dx : u \in W^{1,\infty}(\Omega) \right\},$$

$$m_{\mathrm{hom}}(\Omega, \lambda, \beta) =$$

$$= \min \left\{ \int_\Omega f_{\mathrm{hom}}^{+\infty}(\nabla u) dx + \lambda \int_\Omega |u|^p dx + \int_\Omega \beta u dx : u \in W^{1,p}(\Omega) \right\},$$

and let $\{u_\varepsilon\}_{\varepsilon > 0} \subseteq W^{1,\infty}(\Omega)$ be such that

$$\lim_{\varepsilon \to 0^+} \left(\int_\Omega f\left(\frac{x}{\varepsilon}, \nabla u_\varepsilon\right) dx + \lambda \int_\Omega |u_\varepsilon|^p dx + \int_\Omega \beta u_\varepsilon dx - i_\varepsilon(\Omega, \lambda, \beta) \right) = 0.$$

Then $f_{\text{hom}}^{+\infty}$ is convex, finite on $\mathbf{R}^n$, and satisfies (12.7.2), $\{i_\varepsilon(\Omega, \lambda, \beta)\}_{\varepsilon>0}$ converges as $\varepsilon \to 0^+$ to $m_{\text{hom}}(\Omega, \lambda, \beta)$, $\{u_\varepsilon\}_{\varepsilon>0}$ has cluster points in $L^p(\Omega)$ as $\varepsilon \to 0^+$, and every such point is a solution of $m_{\text{hom}}(\Omega, \lambda, \beta)$.

Theorem 13.5.6. *Let g, φ be as in (13.5.2), f as in (13.5.1), and let $f_{\text{hom}}^{+\infty}$ be defined in (12.1.18) with $q = +\infty$. Assume that (13.5.4) and (13.5.15) with $p = 1$ hold. For every $\varepsilon > 0$, every $\Omega \in \mathcal{A}_0$ with Lipschitz boundary, $\lambda \in]0, +\infty[$, $r \in]1, 1^*[$, $\beta \in L^\infty(\Omega)$ let*

$$i_\varepsilon^0(\Omega, \lambda, \beta) =$$

$$= \inf\left\{ \int_\Omega f\left(\frac{x}{\varepsilon}, \nabla u\right) dx + \lambda \int_\Omega |u|^r dx + \int_\Omega \beta u\, dx : u \in W_0^{1,\infty}(\Omega) \right\},$$

$$m_{\text{hom}}^0(\Omega, \lambda, \beta) =$$

$$= \min\left\{ \int_\Omega f_{\text{hom}}^{+\infty}(\nabla u) dx + \int_\Omega (f_{\text{hom}}^{+\infty})^\infty(\nabla^s u) d|D^s u| + \right.$$

$$\left. + \int_{\partial\Omega} (f_{\text{hom}}^{+\infty})^\infty(-\gamma_\Omega u \mathbf{n}_\Omega) d\mathcal{H}^{n-1} + \lambda \int_\Omega |u|^r dx + \int_\Omega \beta u\, dx : u \in BV(\Omega) \right\},$$

and let $\{u_\varepsilon\}_{\varepsilon>0} \subseteq W_0^{1,\infty}(\Omega)$ satisfy

$$\lim_{\varepsilon \to 0^+} \left(\int_\Omega f(hx, \nabla u_\varepsilon) dx + \lambda \int_\Omega |u_\varepsilon|^r dx + \int_\Omega \beta u_\varepsilon dx - i_\varepsilon^0(\Omega, \lambda, \beta) \right) = 0.$$

Then $f_{\text{hom}}^{+\infty}$ is convex, finite on $\mathbf{R}^n$, and satisfies (12.7.2) with $p = 1$, $\{i_\varepsilon^0(\Omega, \lambda, \beta)\}_{\varepsilon>0}$ converges as $\varepsilon \to 0^+$ to $m_{\text{hom}}^0(\Omega, \lambda, \beta)$, $\{u_\varepsilon\}_{\varepsilon>0}$ has cluster points in $L^1(\Omega)$ as $\varepsilon \to 0^+$, and every such point is a solution of $m_{\text{hom}}^0(\Omega, \lambda, \beta)$.

Theorem 13.5.7. *Let g, φ be as in (13.5.2), f as in (13.5.1), and let $f_{\text{hom}}^{+\infty}$ be defined in (12.1.18) with $q = +\infty$. Assume that (13.5.4) and (13.5.15) with $p \in]1, +\infty[$ hold. For every $\varepsilon > 0$, every $\Omega \in \mathcal{A}_0$ with Lipschitz boundary, $\beta \in L^\infty(\Omega)$ let*

$$i_\varepsilon^0(\Omega, \beta) = \inf\left\{ \int_\Omega f\left(\frac{x}{\varepsilon}, \nabla u\right) dx + \int_\Omega \beta u\, dx : u \in W_0^{1,\infty}(\Omega) \right\},$$

$$m_{\text{hom}}^0(\Omega, \beta) = \min\left\{ \int_\Omega f_{\text{hom}}^{+\infty}(\nabla u) dx + \int_\Omega \beta u\, dx : u \in W_0^{1,p}(\Omega) \right\},$$

and let $\{u_\varepsilon\}_{\varepsilon>0} \subseteq W_0^{1,\infty}(\Omega)$ satisfy

$$\lim_{\varepsilon \to 0^+} \left(\int_\Omega f(hx, \nabla u_\varepsilon) dx + \int_\Omega \beta u_\varepsilon dx - i_\varepsilon^0(\Omega, \beta) \right) = 0.$$

Then $f_{\text{hom}}^{+\infty}$ is convex, finite on $\mathbf{R}^n$, and satisfies (12.7.2), $\{i_\varepsilon^0(\Omega, \beta)\}_{\varepsilon>0}$ converges as $\varepsilon \to 0^+$ to $m_{\text{hom}}^0(\Omega, \beta)$, $\{u_\varepsilon\}_{\varepsilon>0}$ has cluster points in $L^1(\Omega)$ as $\varepsilon \to 0^+$, and every such point is a solution of $m_{\text{hom}}^0(\Omega, \beta)$.

§13.6 Final Remarks

In this section we make some comments to some of the results described in the book.

First of all, we recall that, in the elastic-plastic torsion context, in [Ca2] the interaction between gradient constraints and obstacle conditions has been treated, and some stability criteria and counterexamples have been discussed.

In [CS1] some asymptotic behaviour results of the type of those in Chapter 12 have been obtained in the general not necessarily periodic case and when the gradient constraints are constant. Of course, in this case the limit density is no more constant with respect to the space variables.

The gradient constrained homogenization for Dirichlet problems with nonhomogeneous boundary data has been treated in [CS4], and, in the case of electrostatic type problems, in [DA4].

We also point out that an another approach to the homogenization of the electrostatic screening problem, that covers also the case of conductors with zero Lebesgue measure has been developed in [CDADM] and [DA4].

To describe it, let $\mathcal{C}_Y$ be a collection of subsets of Y, and define the set $\mathcal{C}$ of the periodically distributed conductors as

$$(13.6.1) \qquad \mathcal{C} = \{(i_1, \ldots, i_n) + C : (i_1, \ldots, i_n) \in \mathbf{Z}^n, \, C \in \mathcal{C}_Y\}.$$

Let g be as in (13.5.2). Then, for a given $\Omega \in \mathcal{A}(\mathbf{R}^n)$, the approach relies on the study of the asymptotic behaviour as $\varepsilon \to 0^+$ of minimum problems for energies of the type

$$u \in W^{1,\infty}(\Omega) \mapsto \int_\Omega g\left(\frac{x}{\varepsilon}, \nabla u\right) dx$$

under the constraint

$$u \text{ is constant in } \Omega \cap \varepsilon S \text{ for every } S \in \mathcal{C}.$$

Note that, in this case, only the constancy zones of the admissible configurations are determined a priori, not the constant values that remain undetermined.

For the treatment of this problem, the techniques proposed in Chapter 13 can be suitably adapted to produce homogenization results and formulas, in the same order of ideas of those already obtained in the book, in both the frameworks of Sobolev and BV spaces.

Finally, we point out that the homogenization of energies slightly less general than those considered in the present book, have been approached in [At1] and [CCEDA] by using techniques based on perturbations of the densities, and on a representation result for piecewise affine functions stating that every $u = \sum_{j=1}^{m}(u_{z_j} + s_j)\chi_{P_j} \in PA(\mathbf{R}^n)$ can be represented on a convex open set Ω as the maximum among a finite number of minima of its components $u_{z_1} + s_1, \ldots, u_{z_m} + s_m$.

Such approach does not require conditions like (12.7.1), but seem to be limited to the treatment of homogenization problems only in Sobolev spaces, and not in BV ones. Moreover it seems to work only when the admissible configurations are in the same space in which the energies are coercive, to fix ideas in the case in which $q = p \in]1, +\infty]$ according to the notations used here.

Nevertheless, such approach seems to be quite general, and we address as an open problem its development for the treatment in general settings of integral representation, relaxation and homogenization problems.

Chapter 14

Some Explicit Computations of Homogenized Energies in Mathematical Models Originating Unbounded Functionals

In this chapter we intend to give the physical flavour of the results obtained in Chapters 12 and 13. We discuss the homogenization of some simple energies of the type of those appearing in the elastic-plastic torsion problem, in the modelling of rubber-like nonlinear elastomers, and in the electrostatic screening problem described in §6.5. To this aim, we derive explicit calculations of the homogenization formula, together with some convergence results for minima and minimizers.

Our examples also show that the homogenization formula can exhibit some surprising features, even when the constraints on the admissible deformations are fixed.

§14.1 Homogenization in Elastic-Plastic Torsion

In this section we discuss the some examples of homogenization of energies of the type of those appearing in the elastic-plastic torsion problem, among which the one of the elastic-plastic torsion problem in one space dimension, and with a fixed constraint.

The proposed examples show that, even in simple cases, the features of the energy densities are not inherited by homogenized ones.

Example 14.1.1. Let $n = 1$, α, $\beta \in \mathbf{R}$ with $0 < \alpha < \beta$, $m > 0$, and let $\mu = 2 \left(\frac{1}{\alpha} + \frac{1}{\beta} \right)^{-1}$ be the harmonic mean of α and β. Let a be $]0, 1[$-periodic and satisfying

$$a(y) = \begin{cases} \alpha & \text{if } 0 < y < 1/2 \\ \beta & \text{if } 1/2 \le y < 1 \end{cases} \text{ for every } y \in \,]0,1[,$$

and set

$$f \colon (x, z) \in \mathbf{R} \times \mathbf{R} \mapsto a(x)z^2 + I_{[0,m]}(|z|).$$

Then, it is clear that f fulfils (12.0.1), (13.1.1) with $C = [-m, m]$, (13.1.2), (13.1.3), and (12.1.23). We also recall that in this case

$$f_{\text{hom}}^{+\infty} \colon z \in \mathbf{R} \mapsto \min \left\{ \int_0^1 a(y)(z + v')^2 dy : \right.$$

$$\left. v \in W_{\text{per}}^{1,\infty}(]0,1[), \ |z + v'(y)| \le m \text{ for a.e. } y \in \,]0,1[\right\}.$$

Let us prove that

$$(14.1.1) \qquad f_{\text{hom}}^{+\infty}(z) = \begin{cases} \mu z^2 & \text{if } |z| \le \frac{\alpha m}{\mu} \\ \frac{\alpha}{2}m^2 + \frac{\beta}{2}(2z - m)^2 & \text{if } \frac{\alpha m}{\mu} < |z| \le m \\ +\infty & \text{if } |z| > m \end{cases}$$

$$\text{for every } z \in \mathbf{R}.$$

To do this, we first observe that it is straightforward to verify that $f_{\text{hom}}^{+\infty}$ is symmetric with respect to 0.

Let now $z \in \mathbf{R}$ be such that $f_{\text{hom}}^{+\infty}(z) < +\infty$, then there exists $v \in W_{\text{per}}^{1,\infty}(]0,1[)$ with $|z + v'(y)| \le m$ for a.e. $y \in \,]0,1[$. Consequently, it follows that

$$|z| = \left| \int_0^1 (z + v')dy \right| \le \int_0^1 |z + v'| dy \le m,$$

from which we conclude that

$$(14.1.2) \qquad f_{\text{hom}}^{+\infty}(z) = +\infty \text{ for every } z \in \mathbf{R} \text{ with } |z| > m.$$

Let now $z \in [0, m]$, then it is clear that

$$(14.1.3) \qquad f_{\text{hom}}^{+\infty}(z) = \min \left\{ \int_0^1 a(y)(v')^2 dy : \right.$$

$$\left. v \in u_z + W_0^{1,\infty}(]0,1[), \ |v'(y)| \le m \text{ for every } y \in \,]0,1[\right\}.$$

Let v_0 be a solution of the right-hand side of (14.1.3). Then, by (14.1.3), it is follows that

$$(14.1.4) \qquad f_{\text{hom}}^{+\infty}(z) = \min\left\{\alpha \int_0^{1/2} (v')^2 dy : v \in W^{1,\infty}(]0,1/2[),\right.$$

$$\left. v(0) = 0, \ v(1/2) = v_0(1/2), \ |v'(y)| \le m \text{ for every } y \in]0,1/2[\right\} +$$

$$+ \min\left\{\beta \int_{1/2}^1 (v')^2 dy : v \in W^{1,\infty}(]1/2,1[), \ v(1/2) = v_0(1/2), \ v(1) = z,\right.$$

$$\left. |v'(y)| \le m \text{ for every } y \in]1/2,1[\right\}.$$

We now observe that, due to the presence in the right-hand side of (14.1.4) of the constraint condition and of the boundary data, it turns out that $|v_0(1/2)| \le m/2$ and that $|z - v_0(1/2)| \le m/2$. Therefore, once we recall that $z \in [0,m]$, by (14.1.4) we deduce that

$$(14.1.5) \qquad\qquad\qquad\qquad f_{\text{hom}}^{+\infty}(z) =$$

$$= \min_{t \in [z-\frac{m}{2},\frac{m}{2}]} \left\{\min\left\{\alpha \int_0^{1/2} (v')^2 dy : v \in W^{1,\infty}(]0,1/2[),\right.\right.$$

$$\left. v(0) = 0, \ v(1/2) = t, \ |v'(y)| \le m \text{ for every } y \in]0,1/2[\right\} +$$

$$+ \min\left\{\beta \int_{1/2}^1 (v')^2 dy : v \in W^{1,\infty}(]1/2,1[),\right.$$

$$\left.\left. v(1/2) = t, \ v(1) = z, \ |v'(y)| \le m \text{ for every } y \in]1/2,1[\right\}\right\}.$$

Now, for every $t \in [z - \frac{m}{2}, \frac{m}{2}]$, the functions $y \in]0,1/2[\mapsto 2ty$ and $y \in]1/2,1[\mapsto t+2(z-t)(y-1/2)$ are the solutions of the problems in the right-hand side of (14.1.5). In fact they satisfy the gradient constraint conditions, and solve the corresponding problems without gradient constraints. This implies that

$$(14.1.6) \qquad f_{\text{hom}}^{+\infty}(z) = \min_{t \in [z-\frac{m}{2},\frac{m}{2}]} \{2(\alpha + \beta)t^2 - 4\beta zt + 2\beta z^2\}.$$

Because of this the expression of $f_{\text{hom}}^{+\infty}(z)$ can be easily determined. In fact, if $\frac{\beta z}{\alpha+\beta} > \frac{m}{2}$, i.e. if $z > \frac{m\alpha}{\mu}$, then $t = \frac{m}{2}$ is a solution of the problem in (14.1.6), and $f_{\text{hom}}^{+\infty}(z) = \frac{\alpha}{2}m^2 + \frac{\beta}{2}(2z-m)^2$. If $z - \frac{m}{2} \le \frac{\beta z}{\alpha+\beta} \le \frac{m}{2}$, then $t = \frac{\beta z}{\alpha+\beta}$ is a solution of the problem in (14.1.6), and $f_{\text{hom}}^{+\infty}(z) = \mu z^2$.

Finally, it cannot be $\frac{\beta z}{\alpha+\beta} < z - \frac{m}{2}$, otherwise it would result $z > \frac{m\beta}{\mu} > m$, contrary to the choice $z \in [0, m]$.

Because of this, and of (14.1.2), formula (14.1.1) follows.

We point out that, due to the presence of the constraint term $I_{[0,m]}$, the above $f_{\text{hom}}^{+\infty}$ is no more a quadratic form in its effective domain, contrary to what happens when $m = +\infty$ (cf. [S]). Nevertheless it agrees, at least for z small, with the homogenized density deduced from f when $m = +\infty$.

As corollary, from Theorem 12.7.1 and (14.1.1) the following homogenization result can be deduced.

Theorem 14.1.2. *Let a, m and $f_{\text{hom}}^{+\infty}$ be as in Example 14.1.1. For every $\varepsilon > 0$, every bounded open interval I of $\mathbf{R}$, and $\beta \in L^1(I)$ let*

$$m_\varepsilon^0(I, \beta) = \min\left\{ \int_I a\left(\frac{x}{\varepsilon}\right)(u')^2 dx + \int_I \beta u dx : \right.$$

$$\left. u \in W_0^{1,\infty}(I),\ |u'(x)| \le m \text{ for a.e. } x \in I \right\},$$

$$m_{\text{hom}}^0(I, \beta) = \min\left\{ \int_I f_{\text{hom}}^{+\infty}(u') dx + \int_I \beta u dx : u \in W_0^{1,\infty}(I) \right\},$$

and let for every $\varepsilon > 0$, u_ε be the unique solution of $m_\varepsilon^0(I, \beta)$. Then $\{m_\varepsilon^0(I, \beta)\}_{\varepsilon > 0}$ converges as $\varepsilon \to 0^+$ to $m_{\text{hom}}^0(I, \beta)$, and $\{u_\varepsilon\}_{\varepsilon > 0}$ converges as $\varepsilon \to 0^+$ in $L^\infty(\mathcal{I})$ to the unique solution of $m_{\text{hom}}^0(I, \beta)$.

We now discuss an example showing that the loss of properties pointed out in Example 14.1.1 can be even more shrinking.

Example 14.1.3. Let $n = 1$, and

$$f \colon (x, z) \in \mathbf{R} \times \mathbf{R} \mapsto z^2 + I_{[0,\varphi(x)]}(|z|),$$

where φ is $]0, 1[$-periodic, and satisfies $\varphi(y) = \frac{1}{y^2}$ for every $y \in]0, 1[$. Then it is clear that f fulfils (12.0.1) and (12.1.23). Moreover, it turns out that $\tilde{C}^2(0) = \mathbf{R}$, and that

(14.1.7) $$\tilde{f}_{\text{hom}}^2(z) = f_{\text{hom}}^2(z) = \frac{1}{6}(z+1)^3 - \frac{1}{3} \text{ for every } z \ge 1,$$

$\tilde{f}_{\text{hom}}^2(z)$ and f_{hom}^2 being given by (12.1.1) and (12.1.18) relatively to the f above.

To see this, we first observe that the first equality in (14.1.7) is trivial since $n = 1$.

Let now $z \geq 1$, then the solution $v \in W^{1,2}_{\text{per}}(Y)$ of the minimum problem defining $f^2_{\text{hom}}(z)$ exists and satisfies

(14.1.8) $\qquad v'(y) = \begin{cases} \left(\frac{z+1}{2}\right)^2 - z & \text{if } y \in]0, \frac{2}{z+1}[\\ \frac{1}{y^2} - z & \text{if } y \in]\frac{2}{z+1}, 1[. \end{cases}$

In fact, if $w \in W^{1,2}_{\text{per}}(Y)$ satisfies $|z + v'(y) + w'(y)| \leq \varphi(y)$ for a.e. $y \in Y$, it results that $\int_0^1 w' dy = 0$, that $w'(y) \leq 0$ for a.e. $y \in]\frac{2}{z+1}, 1[$, and that

$$\int_0^1 (z + v' + w')^2 dy - \int_0^1 (z + v')^2 dy =$$

$$= \int_0^1 (z + v')^2 dy + 2 \int_0^{2/(z+1)} \left(\frac{z+1}{2}\right)^2 w' dy + 2 \int_{2/(z+1)}^1 \frac{1}{y^2} w' dy =$$

$$= \int_0^1 (z + v')^2 dy + \int_{2/(z+1)}^1 \left(\frac{1}{y^2} - \left(\frac{z+1}{2}\right)^2\right) w' dy \geq 0.$$

Because of (14.1.8), the second equality in (14.1.7) follows.

The example below describes a surprising feature of the homogenization of unbounded functionals. It proves that, if for every $x \in \mathbf{R}^n$ the elastic-plastic constraint is described by a ball with centre in 0, but with radius depending on x, then the global homogenized elastic-plastic constraint can be no more a ball (cf. also [CS1]).

Example 14.1.4. Let $n = 2$, and

$$f: (x, z) \in \mathbf{R}^2 \times \mathbf{R}^2 \mapsto I_{[0,\varphi(x)]}(|z|),$$

where φ is Y-periodic, and satisfies

$$\varphi(y_1, y_2) = \begin{cases} 1 & \text{if } 0 < y_1 < 1/2 \\ 2 & \text{if } 1/2 \leq y_1 < 1 \end{cases} \quad \text{for every } (y_1, y_2) \in Y.$$

Moreover, let

$$\psi: t \in [0,1] \mapsto \frac{1}{2}\left(\sqrt{1-t^2} + \sqrt{4-t^2}\right),$$

and

$$K = \left\{(z_1, z_2) \in \mathbf{R}^2 : |z_2| \leq 1, \ |z_1| \leq \psi(z_2)\right\}.$$

Then f trivially fulfils (12.0.1) and (12.1.23). Moreover, it results that $\tilde{C}^{+\infty}(0) = K$, and that

(14.1.9) $\qquad \tilde{f}^{+\infty}_{\text{hom}}(z) = f^{+\infty}_{\text{hom}}(z) = I_K(z) \quad \text{for every } z \in \mathbf{R}^2.$

To see this, we first observe that the first equality in (14.1.9) is trivial, and that the identity between $\widetilde{C}^{+\infty}(0)$ and K follows from (14.1.9). Therefore, we need to prove only the last equality in (14.1.9).

To do this, since $f_{\text{hom}}^{+\infty} = I_{\text{dom}f_{\text{hom}}^{+\infty}}$, we just have to verify that $\text{dom}f_{\text{hom}}^{+\infty} = K$.

Let us prove that

$$(14.1.10) \qquad\qquad \text{dom}f_{\text{hom}}^{+\infty} \subseteq K.$$

Let $z = (z_1, z_2) \in \text{dom}f_{\text{hom}}^{+\infty}$, and let $v \in W_{\text{per}}^{1,\infty}(Y)$ be such that $|z + \nabla v(y)| \leq \varphi(y)$ for a.e. $y \in Y$. Let us fix $\overline{y}_1 \in]0, \frac{1}{2}[$, and recall that $\varphi(\overline{y}_1, y_2) = 1$ for every $y_2 \in]0, 1[$. Then, by the Y-periodicity of v we deduce that

$$|z_2| = |u_z(\overline{y}_1, 1) + v(\overline{y}_1, 1) - u_z(\overline{y}_1, 0) - v(\overline{y}_1, 0)| =$$

$$= \left| \int_0^1 (z_2 + \nabla_2 v(\overline{y}_1, y_2)) dy_2 \right| \leq \int_0^1 \varphi(\overline{y}_1, y_2) dy_2 = 1.$$

In order to prove the second one, once we recall that $|z_1 + \nabla_1 v(y_1, y_2)|^2 + |z_2 + \nabla_2 v(y_1, y_2)|^2 \leq \varphi(y_1, y_2)$ for a.e. $(y_1, y_2) \in Y$, we get that

$$(14.1.11) \qquad |z_1| = \left| \int_0^1 \int_0^1 (z_1 + \nabla_1 v(y_1, y_2)) dy_2 dy_1 \right| \leq$$

$$\leq \int_0^{1/2} \int_0^1 |z_1 + \nabla_1 v(y_1, y_2)| dy_2 dy_1 + \int_{1/2}^1 \int_0^1 |z_1 + \nabla_1 v(y_1, y_2)| dy_2 dy_1 \leq$$

$$\leq \int_0^{1/2} \int_0^1 \sqrt{1 - (z_2 + \nabla_2 v(y_1, y_2))^2} dy_2 dy_1 +$$

$$+ \int_{1/2}^1 \int_0^1 \sqrt{4 - (z_2 + \nabla_2 v(y_1, y_2))^2} dy_2 dy_1.$$

Now, since the functions $t \in [-1, 1] \mapsto -\sqrt{1 - t^2}$ and $t \in [-2, 2] \mapsto -\sqrt{4 - t^2}$ are convex, by (14.1.11), Jensen's inequality, and the Y-periodicity of v, we deduce that

$$|z_1| \leq \frac{1}{2} \sqrt{2 \int_0^{1/2} \int_0^1 (1 - (z_2 + \nabla_2 v(y_1, y_2))^2) \, dy_2 dy_1} +$$

$$+ \frac{1}{2} \sqrt{2 \int_{1/2}^1 \int_0^1 (4 - (z_2 + \nabla_2 v(y_1, y_2))^2) \, dy_2 dy_1} =$$

$$= \frac{1}{2} \left(\sqrt{1 - z_2^2} + \sqrt{4 - z_2^2} \right) = \psi(z_2),$$

from which, together with (14.1.11), we conclude that $z \in K$, and therefore that (14.1.10) holds.

To prove that

$$(14.1.12) \qquad\qquad K \subseteq \mathrm{dom} f_{\mathrm{hom}}^{+\infty},$$

it suffices to remark that, for every $z \in K$, the function

$$v \colon (y_1, y_2) \in Y \mapsto \begin{cases} \sqrt{1 - z_2^2} y_1 - \sqrt{1 - z_2^2} & \text{if } (y_1, y_2) \in {]}0, \tfrac{1}{2}[\times {]}0, 1[\\ \sqrt{1 - z_2^2} y_1 - \psi(z_2) & \text{if } (y_1, y_2) \in {]}\tfrac{1}{2}, 1[\times {]}0, 1[\end{cases}$$

is in $W_{\mathrm{per}}^{1,\infty}(Y)$, and that $|z + \nabla v(y)| \leq \varphi(y)$ for a.e. $y \in Y$. In fact this implies that is $z \in \mathrm{dom} f_{\mathrm{hom}}^{+\infty}$, and thus (14.1.12) follows.

§14.2 Homogenization in the Modelling of Nonlinear Elastomers

We now analyze the homogenized integrands relative to the energy densities derived by Treloar in the modelling of rubber-like nonlinear elastomers in the one dimensional case (cf. also [CCDAG2]).

Example 14.2.1. We take $n = 1$, α, $\beta \in \mathbf{R}$ with $0 < \alpha < \beta$, $G \colon \mathbf{R} \to [0, +\infty[$ measurable and $]0, 1[$-periodic with $\alpha \leq G(x) \leq \beta$ for a.e. x in $[0, 1]$, and $q = 2$.

First of all we treat the case of the so called simple shear, in which

$$(14.2.1) \qquad f \colon (x, z) \in \mathbf{R} \times \mathbf{R} \mapsto \begin{cases} \tfrac{1}{2} G(x) \left(z - \tfrac{1}{z} \right)^2 & \text{if } z > 0 \\ +\infty & \text{if } z \leq 0. \end{cases}$$

It is clear that f in (14.2.1) satisfies (12.0.1), (13.1.1) with $C = {]}0, +\infty[$, (13.1.2), (13.1.3), (12.1.23), and

$$(14.2.2) \qquad \frac{1}{2} \alpha z^2 - \alpha \leq f(x, z) \text{ for a.e. } x \in \mathbf{R} \text{ and every } z \in \mathbf{R}.$$

Then by (14.2.1) and (12.1.18), f_{hom} (for simplicity we write f_{hom} in place of f_{hom}^2) is given by

$$(14.2.3) \qquad f_{\mathrm{hom}} \colon z \in \mathbf{R} \mapsto \min \left\{ \frac{1}{2} \int_0^1 G(x) \left(z + u' - \frac{1}{z + u'} \right)^2 dx : \right.$$

$$\left. u \in W_{\mathrm{per}}^{1,2}({]}0, 1[), \ z + u'(x) > 0 \text{ a.e. in } {]}0, 1[\right\},$$

where, for every $z \in \mathbf{R}$, the minimum exists because of (12.0.1), (12.1.23), and (14.2.2).

Our aim consists in trying to describe f_{hom}, or at least its properties.

A first remark in this direction is that

(14.2.4) $f_{\text{hom}}(z) < +\infty$ if and only if $z > 0$,

and, consequently, that problems in (14.2.3) have solutions for every $z > 0$.

Let $z > 0$, ϕ be the inverse function of $\zeta \in]0, +\infty[\mapsto \zeta - \frac{1}{\zeta^3}$, and $c(z) \in \mathbf{R}$ be the only solution of the equation

(14.2.5) $$\int_0^1 \phi\left(\frac{c}{G(t)}\right) dt = z.$$

We point out that ϕ is explicitly computable, although with a complicated expression, and that $c(z)$ exists since $c \in \mathbf{R} \mapsto \int_0^1 \phi(\frac{c}{G(t)}) dt$ is strictly increasing and $\min\{\phi(\frac{c}{\alpha}), \phi(\frac{c}{\beta})\} \leq \int_0^1 \phi(\frac{c}{G(t)}) dt \leq \max\{\phi(\frac{c}{\alpha}), \phi(\frac{c}{\beta})\}$ for every $c \in \mathbf{R}$.

Let u be given by

(14.2.6) $$u: x \in]0, 1[\mapsto \int_0^x \phi\left(\frac{c(z)}{G(t)}\right) dt - zx,$$

then $u \in W^{1,\infty}_{\text{per}}(]0, 1[)$, and by (14.2.5), it results that $z + u'(x) \geq \phi(\frac{c(z)}{G(x)}) \geq \min\{\phi(\frac{c(z)}{\beta}), \phi(\frac{c(z)}{\alpha})\} > 0$ for a.e. $x \in]0, 1[$. Consequently, u turns out to be a weak solution of the Euler equation

$$\left(G(x)\left(z + u' - \frac{1}{(z+u')^3}\right)\right)' = 0,$$

from which we conclude that u is actually a solution of the problem defining $f_{\text{hom}}(z)$.

In conclusion, by (14.2.4) and (14.2.6), we infer that

(14.2.7) $$f_{\text{hom}}(z) = \begin{cases} \frac{1}{2}\int_0^1 G(x)\left(\phi\left(\frac{c(z)}{G(x)}\right) - \frac{1}{\phi\left(\frac{c(z)}{G(x)}\right)}\right)^2 dx & \text{if } z > 0 \\ +\infty & \text{if } z \leq 0 \end{cases}$$

for every $z \in \mathbf{R}$.

Since $\phi(0) = 1$, we have obviously that $c(1) = 0$, and consequently, by (14.2.7), that

$$f_{\text{hom}}(1) = 0.$$

In addition, since $\lim_{z \to +\infty} c(z) = +\infty$ and $\lim_{y \to +\infty} \frac{\phi(y)}{y} = 1$, by (14.2.5) and the estimates on G we infer that

$$\lim_{z \to +\infty} \frac{c(z)}{z} = \lim_{z \to +\infty} \frac{c(z)}{\int_0^1 \phi\left(\frac{c(z)}{G(t)}\right) dt} = \lim_{z \to +\infty} \frac{c(z)}{\int_0^1 \frac{c(z)}{G(t)} dt} = \left(\int_0^1 \frac{1}{G(t)} dt\right)^{-1}.$$

Moreover, again by the asymptotic behaviour of $\phi(y)$ as y increases, we conclude that

$$\lim_{z \to +\infty} \frac{f_{\hom}(z)}{z^2} = \lim_{z \to +\infty} \frac{1}{2z^2} \int_0^1 G(x) \left(\frac{c(z)}{G(x)} - \frac{G(x)}{c(z)} \right)^2 dx =$$

$$= \frac{1}{2} \left(\int_0^1 \frac{1}{G(x)} dx \right)^{-1}.$$

On the other side, since $\lim_{z \to 0+} c(z) = -\infty$, and $\lim_{y \to -\infty} y^{1/3} \phi(y) = -1$, by (14.2.5) and the estimates on G, we infer that $\lim_{z \to 0+} z^3 c(z) = -(\int_0^1 G(t)^{1/3} dt)^3$. Moreover, again by the asymptotic behaviour of $\phi(y)$ as y decreases to $-\infty$, we get that

$$\lim_{z \to 0+} z^2 f_{\hom}(z) =$$

$$= \lim_{z \to 0+} z^2 \frac{1}{2} \int_0^1 G(x) \left(-\left(\frac{G(x)}{c(z)} \right)^{1/3} + \left(\frac{c(z)}{G(x)} \right)^{1/3} \right)^2 dx =$$

$$= \frac{1}{2} \left(\int_0^1 G(x)^{1/3} dx \right)^3.$$

Finally, since $\lim_{z \to 1} c(z) = 0$ and $\lim_{y \to 0} \frac{\phi(y)-1}{y} = \frac{1}{4}$, by (14.2.5) and the estimates on G it follows that $\lim_{z \to 1} \frac{c(z)}{z-1} = 4(\int_0^1 \frac{1}{G(t)} dt)^{-1}$, and, again by the asymptotic behaviour of $\phi(y)$ as y approaches 0, that

$$\lim_{z \to 1} \frac{f_{\hom}(z)}{(z-1)^2} = \lim_{z \to 1} \frac{1}{(z-1)^2} \frac{1}{2} \int_0^1 G(x) \left(1 + \frac{1}{4} \frac{c(z)}{G(x)} - \frac{1}{1 + \frac{1}{4} \frac{c(z)}{G(x)}} \right)^2 dx =$$

$$= 2 \left(\int_0^1 \frac{1}{G(x)} dx \right)^{-1}.$$

We now treat the case of the so called simple extension, in which

$$f \colon (x, z) \in \mathbf{R} \times \mathbf{R} \mapsto \begin{cases} \frac{1}{2} G(x) \left(z^2 + \frac{2}{z} - 3 \right) & \text{if } z > 0 \\ +\infty & \text{if } z \leq 0. \end{cases}$$

It is clear that the above f satisfies (12.0.1), (13.1.1) with $C = \,]0, +\infty[$, (13.1.2), (13.1.3), (12.1.23), and

$$\frac{1}{2} \alpha z^2 - \frac{3}{2} \alpha \leq f(x, z) \quad \text{for a.e. } x \in \mathbf{R} \text{ and every } z \in \mathbf{R}.$$

This time let ϕ be the inverse function of $\zeta \in \,]0, +\infty[\, \mapsto \zeta - \frac{1}{\zeta^2}$, and $c(z) \in \mathbf{R}$ be the only solution of the equation in (14.2.5). Then arguments similar to the ones already used in the previous case yield that ϕ is explicitly computable, and

$$f_{\text{hom}}(z) = \begin{cases} \frac{1}{2}\int_0^1 G(x)\left(\phi\left(\frac{c(z)}{G(x)}\right)^2 + \frac{2}{\phi\left(\frac{c(z)}{G(x)}\right)} - 3\right)dx & \text{if } z > 0 \\ +\infty & \text{if } z \leq 0 \end{cases}$$

for every $z \in \mathbf{R}$.

Analogously, also in this case it follows that

$$f_{\text{hom}}(1) = 0,$$

and

$$\lim_{z\to+\infty} \frac{f_{\text{hom}}(z)}{z^2} = \frac{1}{2}\left(\int_0^1 \frac{1}{G(x)}dx\right)^{-1},$$

$$\lim_{z\to 0^+} zf_{\text{hom}}(z) = \left(\int_0^1 G(x)^{1/2}dx\right)^2,$$

$$\lim_{z\to 1} \frac{f_{\text{hom}}(z)}{(z-1)^2} = \frac{3}{2}\left(\int_0^1 \frac{1}{G(x)}dx\right)^{-1}.$$

Finally we consider $C_1, C_2 \colon \mathbf{R} \to [0, +\infty[$ measurable and $]0, 1[$-periodic with $\alpha \leq C_1(x) \leq \beta$, $\alpha \leq C_2(x) \leq \beta$ for a.e. x in $[0, 1]$, and

$$f \colon (x, z) \in \mathbf{R} \times \mathbf{R} \mapsto$$

$$\begin{cases} \frac{1}{2}C_1(x)\left(z^2 + \frac{2}{z} - 3\right) + \frac{1}{2}C_2(x)\left(\frac{1}{z^2} + 2z - 3\right) & \text{if } z > 0 \\ +\infty & \text{if } z \leq 0. \end{cases}$$

It is clear that f satisfies (12.0.1), (13.1.1) with $C =]0, +\infty[$, (13.1.2), (13.1.3), (12.1.23), and

$$\frac{1}{2}\alpha z^2 - \frac{3}{2}\alpha \leq f(x, z) \text{ for a.e. } x \in \mathbf{R} \text{ and every } z \in \mathbf{R}.$$

Due to the presence of the two coefficients C_1 and C_2, in this case we define, for a.e. $x \in]0, 1[$, $\phi(x, \cdot)$ as the inverse function of $\zeta \in]0, +\infty[\mapsto C_1(x)(\zeta - \frac{1}{\zeta^2}) + C_2(x)(1 - \frac{1}{\zeta^3})$, and $c(z) \in \mathbf{R}$ as the only solution of

$$\int_0^1 \phi(t, c)dt = z.$$

Then arguments similar to the ones already used before imply that for a.e. $x \in]0, 1[$, $\phi(x, \cdot)$ is explicitly computable, and

$$f_{\text{hom}}(z) = \begin{cases} \frac{1}{2}\int_0^1 \left\{C_1(x)\left(\phi(x, c(z))^2 + \frac{2}{\phi(x, c(z))} - 3\right) + \\ \quad + C_2(x)\left(\frac{1}{\phi(x, c(z))^2} + 2\phi(x, c(z)) - 3\right)\right\}dx & \text{if } z > 0 \\ +\infty & \text{if } z \leq 0 \end{cases}$$

for every $z \in \mathbf{R}$.

Analogously, also in this case it follows that

$$f_{\hom}(1) = 0,$$

and

$$\lim_{z \to +\infty} \frac{f_{\hom}(z)}{z^2} = \frac{1}{2}\left(\int_0^1 \frac{1}{C_1(x)} dx\right)^{-1},$$

$$\lim_{z \to 0^+} z^2 f_{\hom}(z) = \frac{1}{2}\left(\int_0^1 C_2(x)^{1/3} dx\right)^3,$$

$$\lim_{z \to 1} \frac{f_{\hom}(z)}{(z-1)^2} = \frac{3}{2}\left(\int_0^1 \frac{1}{C_1(x) + C_2(x)} dx\right)^{-1}.$$

In conclusion, as it can be easily deduced looking at the asymptotic behaviours of the above functions $f_{\hom}$, in all the examples considered the shape of the integrands f is not preserved in the homogenization process.

Nevertheless, at macroscopic level, the mesoscale behaviour remains the same. In fact, in all the three cases, the behaviours of the homogenized functions $f_{\hom}$ close to $z = +\infty$ and $z = 0$ are the same of those of the homogenized functions of the leading parts of the corresponding integrands, namely of $\frac{1}{2}G(x)z^2$ or $\frac{1}{2}C_1(x)z^2$ (cf. [S]), and of $\frac{1}{2}G(x)\frac{1}{z^2}$ or $G(x)\frac{1}{z}$ or $\frac{1}{2}C_2(x)\frac{1}{z^2}$ (for which simple calculations of the kind of the above ones can be carried out).

Analogously, since $\frac{1}{2}G(x)(z - \frac{1}{z})^2 = \frac{1}{2}G(x)(\frac{z+1}{z})^2(z-1)^2$, $\frac{1}{2}G(x)(z^2 + \frac{2}{z} - 3) = \frac{1}{2}G(x)\frac{z+2}{z}(z-1)^2$, $\frac{1}{2}C_1(x)(z^2 + \frac{2}{z} - 3) + \frac{1}{2}C_2(x)(\frac{1}{z^2} + 2z - 3) = \frac{1}{2}(C_1(x)\frac{z+2}{z} + C_2(x)\frac{1+2z}{z^2})(z-1)^2$ for a.e. $x \in \mathbf{R}$ and every $z \in \mathbf{R}$, the relative homogenized integrands behave, close to $z = 1$, like the homogenized functions of the leading parts of the corresponding integrands, namely of $2G(x)(z-1)^2$ or $\frac{3}{2}G(x)(z-1)^2$ or $\frac{3}{2}(C_1(x) + C_2(x))(z-1)^2$.

In this weak sense, the discussed models are stable with respect to the homogenization process.

From Theorem 12.7.4 and (14.2.7) the following homogenization result can be deduced.

Theorem 14.2.2. *Let G be as in Example 14.2.1, and let $f_{\hom}$ be given by (14.2.7). For every $\varepsilon > 0$, $a, b \in \mathbf{R}$, $\beta \in L^1(\Omega)$, and $c \in \mathbf{R}$ let*

$$m_\varepsilon(a, b, \beta, c) = \min\left\{\int_a^b G\left(\frac{x}{\varepsilon}\right)\left(u' - \frac{1}{u'}\right)^2 dx + \int_a^b \beta u\, dx + c u(b) : \right.$$

$$\left. u \in W^{1,2}(]a, b[), \ u(a) = 0, \ u'(x) > 0 \text{ for a.e. } x \in]a, b[\right\},$$

$$m_{\text{hom}}(a, b, \beta, c) = \min \left\{ \int_a^b f_{\text{hom}}(u')dx + \int_a^b \beta u dx + cu(b) : \right.$$

$$\left. u \in W^{1,2}(]a, b[), \ u(a) = 0 \right\},$$

and let for every $\varepsilon > 0$, u_ε be the unique solution of $m_\varepsilon(a, b, \beta, c)$. Then $\{m_\varepsilon(a, b, \beta, c)\}_{\varepsilon>0}$ converges as $\varepsilon \to 0^+$ to $m_{\text{hom}}(a, b, \beta, c)$, and $\{u_\varepsilon\}_{\varepsilon>0}$ converges as $\varepsilon \to 0^+$ in $L^\infty(]a, b[)$ to the unique solution of $m_{\text{hom}}(a, b, \beta, c)$.

§14.3 Homogenization in Electrostatic Screening

Finally, we examine the densities relative to the electrostatic screening problem (cf. also [CS3], [DAGP]).

Example 14.3.1. In this example we first study the case of a generic quadratic energy density, from which we then deduce the results for the problem under consideration.

Let $\{a_{ij}\}_{i,j\in\{1,...,n\}}$ be a $n \times n$ symmetric matrix of measurable Y-periodic functions on $\mathbf{R}^n$ satisfying for some $0 < \lambda \leq \Lambda < +\infty$

$$(14.3.1) \qquad \lambda|z|^2 \leq \sum_{i,j=1}^n a_{ij}(x)z_i z_j \leq \Lambda|z|^2$$

for a.e. $x \in \mathbf{R}^n$, and every $z \in \mathbf{R}^n$,

let φ be a measurable Y-periodic function on $\mathbf{R}^n$ taking only the values 0 and $+\infty$, and set

$$f: (x, z) \in \mathbf{R}^n \times \mathbf{R}^n \mapsto \sum_{i,j=1}^n a_{ij}(x)z_i z_j + I_{B_{\varphi(x)}(0)}(z),$$

and

$$(14.3.2) \qquad K_\varphi = \{z \in \mathbf{R}^n : \text{there exists } v \in W_{\text{per}}^{1,\infty}(Y), \text{ such that}$$

$$|z + \nabla v(y)| \leq \varphi(y) \text{ for a.e. } y \in Y\}.$$

We prove that, if

$$(14.3.3) \qquad \text{int}(K_\varphi) \neq \emptyset,$$

then there exists a constant $n \times n$ symmetric matrix $\{a_{ij}^{\text{hom}}\}_{i,j\in\{1,...,n\}}$ such that

$$(14.3.4) \qquad \lambda|z|^2 \leq \sum_{i,j=1}^n a_{ij}^{\text{hom}} z_i z_j \text{ for every } z \in \mathbf{R}^n,$$

and

(14.3.5) $$f_{\mathrm{hom}}^{+\infty}(z) = \sum_{i,j=1}^{n} a_{ij}^{\mathrm{hom}} z_i z_j \text{ for every } z \in \mathbf{R}^n,$$

$f_{\mathrm{hom}}^{+\infty}$ being given by (12.1.18) relatively to the f above.

First of all, we observe that Jensen's inequality and (14.3.1) yield

(14.3.6) $$\lambda |z|^2 = \min \left\{ \int_Y \lambda |z + \nabla v|^2 dy : v \in W_{\mathrm{per}}^{1,\infty}(Y) \right\} \leq f_{\mathrm{hom}}^{+\infty}(z)$$

for every $z \in \mathbf{R}^n$.

Consequently, (14.3.4) follows from (14.3.6), once we prove (14.3.5).

To do this, let us observe that

$$K_\varphi = \mathrm{dom} f_{\mathrm{hom}}^{+\infty},$$

consequently, (14.3.3) ensures that

(14.3.7) $$\mathrm{int}(\mathrm{dom} f_{\mathrm{hom}}^{+\infty}) \neq \emptyset.$$

We now observe that (14.3.7) and Proposition 13.5.1 imply that $f_{\mathrm{hom}}^{+\infty}$ is finite on $\mathbf{R}^n$. Therefore, in order to prove (14.3.5) we can use standard characterizations of quadratic forms. So, we just need to verify that

(14.3.8) $$f_{\mathrm{hom}}^{+\infty}(\lambda z) = \lambda^2 f_{\mathrm{hom}}^{+\infty}(z) \text{ for every } z \in \mathbf{R}^n \text{ and } \lambda \in \mathbf{R},$$

and that

(14.3.9) $$f_{\mathrm{hom}}^{+\infty}(z_1 + z_2) + f_{\mathrm{hom}}^{+\infty}(z_1 - z_2) = 2 f_{\mathrm{hom}}^{+\infty}(z_1) + 2 f_{\mathrm{hom}}^{+\infty}(z_2)$$

for every $z_1, z_2 \in \mathbf{R}^n$.

To prove (14.3.8), we take $z \in \mathbf{R}^n$, $v \in W_{\mathrm{per}}^{1,\infty}(Y)$ such that $|z + \nabla v(y)| \leq \varphi(y)$ for a.e. $y \in Y$, and $\lambda \in \mathbf{R}$. Then $\lambda v \in W_{\mathrm{per}}^{1,\infty}(Y)$, $|\lambda z + \lambda \nabla v(y)| = \lambda |z + \nabla v(y)| \leq \lambda \varphi(y) = \varphi(y)$ for a.e. $y \in Y$, and

$$f_{\mathrm{hom}}^{+\infty}(\lambda z) \leq \int_Y f(y, \lambda z + \lambda \nabla v) dy = \lambda^2 \int_Y f(y, z + \nabla v) dy,$$

from which it follows that

$$f_{\mathrm{hom}}^{+\infty}(\lambda z) \leq \lambda^2 f_{\mathrm{hom}}^{+\infty}(z).$$

By replacing λ with $1/\lambda$, and z with λz in the above inequality, condition (14.3.8) follows.

To prove (14.3.9), we take z_1, $z_2 \in \mathbf{R}^n$, and v_1, $v_2 \in W^{1,\infty}_{\text{per}}(Y)$ such that $|z_1 + \nabla v_1(y)| \leq \varphi(y)$ and $|z_2 + \nabla v_2(y)| \leq \varphi(y)$ for a.e. $y \in Y$. Then $v_1 \pm v_2 \in W^{1,\infty}_{\text{per}}(Y)$, $|z_1 \pm z_2 + \nabla(v_1 \pm v_2)(y)| \leq |z_1 + \nabla v_1(y)| + |z_2 + \nabla v_2(y)| \leq 2\varphi(y) = \varphi(y)$ for a.e. $y \in Y$, and

$$f^{+\infty}_{\text{hom}}(z_1 + z_2) + f^{+\infty}_{\text{hom}}(z_1 - z_2) \leq$$

$$\leq \int_Y f(y, z_1 + z_2 + \nabla(v_1 + v_2))dy + \int_Y f(y, z_1 - z_2 + \nabla(v_1 - v_2))dy \leq$$

$$\leq 2 \int_Y f(y, z_1 + \nabla v_1)dy + 2 \int_Y f(y, z_2 + \nabla v_2)dy,$$

from which it follows that

$$f^{+\infty}_{\text{hom}}(z_1 + z_2) + f^{+\infty}_{\text{hom}}(z_1 - z_2) \leq 2f^{+\infty}_{\text{hom}}(z_1) + 2f^{+\infty}_{\text{hom}}(z_2).$$

Let now w_1, $w_2 \in W^{1,\infty}_{\text{per}}(Y)$ be such that $|(z_1 + z_2) + \nabla w_1(y)| \leq \varphi(y)$ and $|(z_1 - z_2) + \nabla w_2(y)| \leq \varphi(y)$ for a.e. $y \in Y$. Then $w_1 \pm w_2 \in W^{1,\infty}_{\text{per}}(Y)$, $|2z_1 + \nabla(w_1 + w_2)(y)| \leq |z_1 + z_2 + \nabla w_1(y)| + |z_1 - z_2 + \nabla w_2(y)| \leq 2\varphi(y) = \varphi(y)$, $|2z_2 + \nabla(w_1 - w_2)(y)| \leq |z_1 + z_2 + \nabla w_1(y)| + |-(z_1 - z_2) - \nabla w_2(y)| \leq 2\varphi(y) = \varphi(y)$ for a.e. $y \in Y$, and

$$f^{+\infty}_{\text{hom}}(2z_1) + f^{+\infty}_{\text{hom}}(2z_2) \leq$$

$$\leq \int_Y f(y, 2z_1 + \nabla(w_1 + w_2))dy + \int_Y f(y, 2z_2 + \nabla(w_1 - w_2))dy \leq$$

$$\leq 2 \int_Y f(y, z_1 z_2 + \nabla w_1)dy + 2 \int_Y f(y, z_1 - z_2 + \nabla w_2)dy,$$

from which it follows that

$$f^{+\infty}_{\text{hom}}(2z_1) + f^{+\infty}_{\text{hom}}(2z_2) \leq 2f^{+\infty}_{\text{hom}}(z_1 + z_2) + 2f^{+\infty}_{\text{hom}}(z_1 - z_2).$$

This, together with (14.3.8), completes the proof of (14.3.9).

By virtue of (14.3.8) and (14.3.9) the existence of a constant $n \times n$ symmetric matrix $\{a^{\text{hom}}_{ij}\}_{i,j \in \{1,\dots,n\}}$ satisfying (14.3.5) follows.

We now examine the case in which $a_{ij}(x) = \begin{cases} 1 & \text{if } i = j \\ 0 & \text{if } i \neq j \end{cases}$ for a.e. $x \in \mathbf{R}^n$, under various sets of assumptions on the constraint φ.

Let φ be a measurable Y-periodic function on $\mathbf{R}^n$ taking only the values 0 and $+\infty$, and set

(14.3.10) $$d\colon (x, z) \in \mathbf{R}^n \times \mathbf{R}^n \mapsto |z|^2 + I_{B_{\varphi(x)}(0)}(z).$$

Let $d_{\mathrm{hom}}^{+\infty}$ be the homogenized density of d defined by means of (12.1.18). We first prove that, provided (14.3.3) holds and φ satisfies suitable invariance conditions with respect to reflections, then there exist $d_1^{\mathrm{hom}}, \ldots, d_n^{\mathrm{hom}} \in [1, +\infty[$ such that

$$(14.3.11) \qquad d_{\mathrm{hom}}^{+\infty}(z) = \sum_{j=1}^{n} d_{ij}^{\mathrm{hom}} |z|^2 \text{ for every } z \in \mathbf{R}^n.$$

In order to describe precisely the above mentioned invariance conditions, we denote for every $i \in \{1, \ldots, n\}$, by R_i be the $n \times n$ matrix associated to the reflection with respect to the hyperplane orthogonal to the i-th coordinate axis, i.e. the matrix such that

$$R_i(z_1, \ldots, z_i, \ldots, z_n) = (z_1, \ldots, -z_i, \ldots, z_n)$$

$$\text{for every } (z_1, \ldots, z_i, \ldots, z_n) \in \mathbf{R}^n.$$

Then, we assume that

$$(14.3.12) \quad \varphi(R_i y) = \varphi(y) \text{ for a.e. } y \in \mathbf{R}^n, \text{ and every } i \in \{1, \ldots, n-1\}.$$

Since

$$K_\varphi - \mathrm{dom} d_{\mathrm{hom}}^{+\infty},$$

by (14.3.3), we conclude that

$$\mathrm{int}(\mathrm{dom} d_{\mathrm{hom}}^{+\infty}) \neq \emptyset.$$

Consequently, because of the results in the general case, there exists a constant $n \times n$ symmetric matrix $\{d_{ij}^{\mathrm{hom}}\}_{i,j \in \{1, \ldots, n\}}$ such that

$$(14.3.13) \qquad d_{\mathrm{hom}}^{+\infty}(z) = \sum_{i,j=1}^{n} d_{ij}^{\mathrm{hom}} z_i z_j \text{ for every } z \in \mathbf{R}^n.$$

Let us prove that

$$(14.3.14) \quad d_{\mathrm{hom}}^{+\infty}(R_i z) = d_{\mathrm{hom}}^{+\infty}(z) \text{ for every } z \in \mathbf{R}^n, \text{ and } i \in \{1, \ldots, n-1\}.$$

Let $z \in \mathbf{R}^n$, and $v \in W_{\mathrm{per}}^{1,\infty}(Y)$ be such that $|z + \nabla v(y)| \leq \varphi(y)$ for a.e. $y \in Y$, and let $i \in \{1, \ldots, n-1\}$. Then, once we observe that $R_i \mathbf{Z}^n = \mathbf{Z}^n$, we get that $v \circ R_i \in W_{\mathrm{per}}^{1,\infty}(Y)$, and, from (14.3.12), that $|R_i z + \nabla(v \circ R_i)(y)| = |R_i z + R_i \nabla v(R_i y)| = |z + \nabla v(R_i y)| \leq \varphi(R_i y) = \varphi(y)$ for a.e. $y \in Y$. Consequently, we infer that

$$d_{\mathrm{hom}}^{+\infty}(R_i z) \leq \int_Y |R_i z + \nabla(v \circ R_i)|^2 dy = \int_Y |R_i(z + \nabla v(R_i y))|^2 dy =$$

$$= \int_{R_i Y} |z + \nabla v(y)|^2 dy = \int_Y |z + \nabla v|^2 dy,$$

from which it follows that

$$d_{\text{hom}}^{+\infty}(R_i z) \le d_{\text{hom}}^{+\infty}(z) \text{ for every } z \in \mathbf{R}^n, \text{ and } i \in \{1, \dots, n-1\}.$$

Let now $i \in \{1, \dots, n-1\}$. Then, once we recall that R_i^2 agrees with the identity matrix, an iterated use of the above inequality, yields

$$d_{\text{hom}}^{+\infty}(z) = d_{\text{hom}}^{+\infty}(R_i^2 z) \le d_{\text{hom}}^{+\infty}(R_i z) \le d_{\text{hom}}^{+\infty}(z)$$

$$\text{for every } z \in \mathbf{R}^n, \text{ and } i \in \{1, \dots, n-1\},$$

from which (14.3.14) follows.

Consequently, because of (14.3.14), of (14.3.13), and of elementary linear algebra arguments, we infer that (14.3.11) holds.

Finally, because of (14.3.4), it soon follows that $d_1^{\text{hom}}, \dots, d_n^{\text{hom}} \in [1, +\infty[$.

In particular, from (14.3.11) we deduce that

(14.3.15)
$$d_j^{\text{hom}} = \inf \left\{ \int_Y |\mathbf{e}_j + \nabla v|^2 dy : v \in W_{\text{per}}^{1,\infty}(Y), \right.$$

$$\left. |\mathbf{e}_j + \nabla v(y)| \le \varphi(y) \text{ for a.e. } y \in Y \right\} \text{ for every } j \in \{1, \dots, n\}.$$

We now prove that, if (14.3.3) holds and φ satisfies suitable invariance conditions with respect to rotations, then there exists $d^{\text{hom}} \in [1, +\infty[$ such that

(14.3.16)
$$d_{\text{hom}}^{+\infty}(z) = d^{\text{hom}} |z|^2 \text{ for every } z \in \mathbf{R}^n.$$

This time the invariance conditions are described as follows. For every $i, j \in \{1, \dots, n\}$ with $i < j$, let R_{ij} be the $n \times n$ matrix associated to the clockwise $\frac{\pi}{2}$-rotation in the (i, j) plane, i.e. the matrix such that

$$R_{ij}(z_1, \dots, z_i, \dots, z_j, \dots, z_n) = (z_1, \dots, z_j, \dots, -z_i, \dots, z_n)$$

$$\text{for every } (z_1, \dots, z_i, \dots, z_j, \dots, z_n) \in \mathbf{R}^n.$$

We recall that for every $i, j \in \{1, \dots, n\}$ with $i < j$, R_{ij}^4 agrees with the identity matrix.

Then, we assume that

(14.3.17)
$$\varphi(R_{ij} y) = \varphi(y)$$

$$\text{for a.e. } y \in \mathbf{R}^n, \text{ and every } i, j \in \{1, \dots, n\} \text{ with } i < j.$$

As before, it turns out that (14.3.13) holds.

Let us prove that

(14.3.18) $$d_{\text{hom}}^{+\infty}(R_{ij}z) = d_{\text{hom}}^{+\infty}(z)$$

for every $z \in \mathbf{R}^n$, and $i, j \in \{1, \ldots, n\}$ with $i < j$.

Let $z \in \mathbf{R}^n$, and $v \in W_{\text{per}}^{1,\infty}(Y)$ be such that $|z + \nabla v(y)| \leq \varphi(y)$ for a.e. $y \in Y$, and let $i, j \in \{1, \ldots, n\}$ with $i < j$. Then, once we observe that $R_{ij}\mathbf{Z}^n = \mathbf{Z}^n$, we get that $v \circ R_{ij} \in W_{\text{per}}^{1,\infty}(Y)$, and, from (14.3.17), that $|R_{ij}z + \nabla(v \circ R_{ij})(y)| = |R_{ij}z + R_{ij}\nabla v(R_{ij}y)| = |z + \nabla v(R_{ij}y)| \leq \varphi(R_{ij}y) = \varphi(y)$ for a.e. $y \in Y$. Consequently, we infer that

$$d_{\text{hom}}^{+\infty}(R_{ij}z) \leq \int_Y |R_{ij}z + \nabla(v \circ R_{ij})|^2 dy = \int_Y |R_{ij}(z + \nabla v(R_{ij}y))|^2 dy =$$

$$= \int_{R_{ij}Y} |z + \nabla v(y)|^2 dy = \int_Y |z + \nabla v|^2 dy,$$

from which it follows that

$$d_{\text{hom}}^{+\infty}(R_{ij}z) \leq d_{\text{hom}}^{+\infty}(z) \text{ for every } z \in \mathbf{R}^n, \text{ and } i, j \in \{1, \ldots, n\} \text{ with } i < j.$$

Let now $i, j \in \{1, \ldots, n\}$ with $i < j$. By an iterated use of the above inequality, we obtain that

$$d_{\text{hom}}^{+\infty}(z) = d_{\text{hom}}^{+\infty}(R_{ij}^4 z) \leq d_{\text{hom}}^{+\infty}(R_{ij}^3 z) \leq d_{\text{hom}}^{+\infty}(R_{ij}^2 z) \leq d_{\text{hom}}^{+\infty}(R_{ij}z) \leq$$

$$\leq d_{\text{hom}}^{+\infty}(z) \text{ for every } z \in \mathbf{R}^n, \text{ and } i, j \in \{1, \ldots, n\} \text{ with } i < j,$$

from which (14.3.18) follows.

Consequently, because of (14.3.18), of (14.3.13), and of elementary linear algebra arguments, we infer that (14.3.16) holds.

Finally, because of (14.3.4), it soon follows that $d^{\text{hom}} \in [1, +\infty[$.

In particular, from (14.3.16) we deduce that

(14.3.19) $$d^{\text{hom}} = \inf \left\{ \int_Y |\mathbf{e}_1 + \nabla v|^2 dy : v \in W_{\text{per}}^{1,\infty}(Y), \right.$$

$$\left. |\mathbf{e}_1 + \nabla v(y)| \leq \varphi(y) \text{ for a.e. } y \in Y \right\}.$$

As final remark, we point out that if φ satisfies (14.3.12) and some of the equalities in (14.3.17), then it is easy to see that some of coefficients d_j^{hom} in (14.3.11) coincide.

We now want to deduce some estimates on the constants $d_1^{\text{hom}}, \ldots, d_n^{\text{hom}}$ and d^{hom} appearing in (14.3.11) and (14.3.16), for some special choices of the constraint function φ.

To do this, we take $l_1, \ldots, l_n \in \,]0,1[$ with $l_1 \leq l_2 \leq \ldots \leq l_n$, set $P = \prod_{i=1}^{n}]\frac{1-l_i}{2}, \frac{1+l_i}{2}[$, and define φ as the measurable Y-periodic function on $\mathbf{R}^n$ taking only the values 0 and $+\infty$ such that

$$(14.3.20) \qquad \varphi(y) = \begin{cases} 0 & \text{if } y \in P \\ +\infty & \text{if } y \in Y \setminus P \end{cases} \quad \text{for every } y \in Y.$$

Then, φ fulfils (14.3.3).

If φ is given by (14.3.20), we prove that

$$(14.3.21) \qquad 1 + \frac{\prod_{i=1}^{n} l_i}{1 - l_j} \leq \inf \left\{ \int_Y |\mathbf{e}_j + \nabla v|^2 dy : v \in W_{\text{per}}^{1,\infty}(Y), \right.$$

$$\left. |\mathbf{e}_1 + \nabla v(y)| \leq \varphi(y) \text{ for a.e. } y \in Y \right\} \leq \frac{1}{1 - l_j} \quad \text{for every } j \in \{1, \ldots, n\}.$$

Let us fix $j \in \{1, \ldots, n\}$. For the sake of simplicity, we consider only the case in which $n \geq 3$ and $j \in \{2, \ldots, n-1\}$. The remaining cases can be treated analogously with few formal changes.

Let us prove the left-hand side of (14.3.21).

Let us set $S_j^- = \prod_{i=1}^{j-1}]\frac{1-l_i}{2}, \frac{1+l_i}{2}[$, $S_j^+ = \prod_{i=j+1}^{n}]\frac{1-l_i}{2}, \frac{1+l_i}{2}[$, and $P_j = S_j^- \times \,]0,1[\times S_j^+$. Moreover, for every $y \in \mathbf{R}^n$ let us set $\tilde{y}_j^- = (y_1, \ldots, y_{j-1})$ and $\tilde{y}_j^+ = (y_{j+1}, \ldots, y_n)$

Let $v \in W_{\text{per}}^{1,\infty}(Y)$ be such that $|\mathbf{e}_j + \nabla v(y)| \leq \varphi(y)$ for a.e. $y \in Y$, then

$$(14.3.22) \qquad \int_Y |\mathbf{e}_j + \nabla v|^2 dy \geq \int_{Y \setminus P_j} (1 + \nabla_j v)^2 dy + \int_{P_j} (1 + \nabla_j v)^2 dy \geq$$

$$\geq \inf \left\{ \int_{Y \setminus P_j} (\nabla_j u)^2 dy : u \in W^{1,\infty}(Y \setminus P_j), \ u(\tilde{y}_j^-, 1, \tilde{y}_j^+) = u(\tilde{y}_j^-, 0, \tilde{y}_j^+) + 1 \right.$$

$$\text{for every } (\tilde{y}_j^-, \tilde{y}_j^+) \in (]0,1[^{n-1} \setminus S_j^-) \times (]0,1[^{n-1} \setminus S_j^+) \right\} +$$

$$+ \inf \left\{ \int_{P_j} (\nabla_j u)^2 dy : u \in W^{1,\infty}(P_j), \ u \text{ constant in } P, \right.$$

$$u(\tilde{y}_j^-, 1, \tilde{y}_j^+) = u(\tilde{y}_j^-, 0, \tilde{y}_j^+) + 1 \text{ for every } (\tilde{y}_j^-, \tilde{y}_j^+) \in S_j^- \times S_j^+ \right\}.$$

Now, by using Jensen's inequality, it is easy to verify that

$$\inf \left\{ \int_{Y \setminus P_j} (\nabla_j u)^2 dy : u \in W^{1,\infty}(Y \setminus P_j), \ u(\tilde{y}_j^-, 1, \tilde{y}_j^+) = u(\tilde{y}_j^-, 0, \tilde{y}_j^+) + 1 \right.$$

$$\text{for every } (\tilde{y}_j^-, \tilde{y}_j^+) \in (]0,1[^{n-1} \setminus S_j^-) \times (]0,1[^{n-1} \setminus S_j^+) \Big\} =$$

$$= \int_{Y \setminus P_j} (\nabla_j u_{\mathbf{e}_j})^2 dy = 1 - \prod_{\substack{i=1 \\ i \neq j}}^{n} l_i.$$

Moreover, again by Jensen's inequality, and since $\frac{1}{1-l_j} u_{\mathbf{e}_j}$ is constant in $S_j^- \times \{\frac{1-l_j}{2}\} \times S_j^+$, it is easy to prove that

$$\inf \Big\{ \int_{P_j} (\nabla_j u)^2 dy : u \in W^{1,\infty}(P_j), \ u \text{ constant in } P,$$

$$u(\tilde{y}_j^-, 1, \tilde{y}_j^+) = u(\tilde{y}_j^-, 0, \tilde{y}_j^+) + 1 \text{ for every } (\tilde{y}_j^-, \tilde{y}_j^+) \in S_j^- \times S_j^+ \Big\} =$$

$$= \inf \Big\{ \int_{S_j^- \times]0,1-l_j[\times S_j^+} (\nabla_j u)^2 dy : u \in W^{1,\infty}(S_j^- \times]0, 1 - l_j[\times S_j^+),$$

$$u \text{ constant in } S_j^- \times \Big\{ \frac{1-l_j}{2} \Big\} \times S_j^+,$$

$$u(\tilde{y}_j^-, 1 - l_j, \tilde{y}_j^+) = u(\tilde{y}_j^-, 0, \tilde{y}_j^+) + 1 \text{ for every } (\tilde{y}_j^-, \tilde{y}_j^+) \in S_j^- \times S_j^+ \Big\} \geq$$

$$\geq \inf \Big\{ \int_{S_j^- \times]0,1-l_j[\times S_j^+} (\nabla_j u)^2 dy : u \in W^{1,\infty}(S_j^- \times]0, 1 - l_j[\times S_j^+),$$

$$u(\tilde{y}_j^-, 1 - l_j, \tilde{y}_j^+) = u(\tilde{y}_j^-, 0, \tilde{y}_j^+) + 1 \text{ for every } (\tilde{y}_j^-, \tilde{y}_j^+) \in S_j^- \times S_j^+ \Big\} =$$

$$= \int_{S_j^- \times]0,1-l_j[\times S_j^+} \Big(\nabla_j \frac{1}{1-l_j} u_{\mathbf{e}_j} \Big)^2 dy \geq$$

$$= \inf \Big\{ \int_{S_j^- \times]0,1-l_j[\times S_j^+} (\nabla_j u)^2 dy : u \in W^{1,\infty}(S_j^- \times]0, 1 - l_j[\times S_j^+),$$

$$u \text{ constant in } S_j^- \times \Big\{ \frac{1-l_j}{2} \Big\} \times S_j^+,$$

$$u(\tilde{y}_j^-, 1 - l_j, \tilde{y}_j^+) = u(\tilde{y}_j^-, 0, \tilde{y}_j^+) + 1 \text{ for every } (\tilde{y}_j^-, \tilde{y}_j^+) \in S_j^- \times S_j^+ \Big\}.$$

Consequently, by (14.3.22), we deduce that

$$\int_Y |\mathbf{e}_j + \nabla v|^2 dy \geq 1 - \prod_{\substack{i=1 \\ i \neq j}}^{n} l_i + \frac{1}{1-l_j} \prod_{\substack{i=1 \\ i \neq j}}^{n} l_i,$$

from which the left-hand side of (14.3.21) follows.

In order to prove the right-hand side of (14.3.21), let $v_j \in W_{\mathrm{per}}^{1,\infty}(Y)$ be such that

$$
v_j(y) = \begin{cases} \frac{l_j}{1-l_j} y_j & \text{if } y_j \in \,]0, \frac{1-l_j}{2}[\\ -y_j + \frac{1}{2} & \text{if } y_j \in [\frac{1-l_j}{2}, \frac{1+l_j}{2}] \\ \frac{l_j}{1-l_j} y_j - \frac{l_j}{1-l_j} & \text{if } y_j \in \,]\frac{1+l_j}{2}, 1] \end{cases} \quad \text{for every } y \in Y.
$$

Then $|e_j + \nabla v_j(y)| \leq \varphi(y)$ for a.e. $y \in Y$, and

$$
\inf \left\{ \int_Y |e_j + \nabla v|^2 dy : v \in W_{\mathrm{per}}^{1,\infty}(Y), \right.
$$

$$
\left. |e_1 + \nabla v(y)| \leq \varphi(y) \text{ for a.e. } y \in Y \right\} \leq \int_Y |e_j + \nabla v_j|^2 dy = \frac{1}{1 - l_j},
$$

from which also the right-hand side of (14.3.21) follows.

In particular, we can consider d in (14.3.10) written with φ given by (14.3.20). In this case, we observe that φ fulfils also (14.3.12) but not (14.3.17), unless $l_1 = l_2 = \ldots, l_n$. Then, by (14.3.15) and (14.3.21) we obtain that

$$
(14.3.23) \qquad 1 + \frac{\prod_{i=1}^n l_i}{1 - l_j} \leq d_j^{\mathrm{hom}} \leq \frac{1}{1 - l_j} \quad \text{for every } j \in \{1, \ldots, n\},
$$

from which we conclude that the values $d_1^{\mathrm{hom}}, \ldots, d_n^{\mathrm{hom}}$ can also be different, even being the coefficients of d all equal to 1.

This actually happens. In fact if $n = 2$ and $l_2 > \frac{1}{2 - l_1}$, by (14.3.23) it follows that

$$
d_1^{\mathrm{hom}} \leq \frac{1}{1 - l_1} < 1 + \frac{l_1 l_2}{1 - l_2} \leq d_2^{\mathrm{hom}}.
$$

Finally, if $l_1 = l_2 = \ldots, l_n = l$, then φ in (14.3.20) fulfils also (14.3.17). Consequently, (14.3.19) and (14.3.21) yield

$$
(14.3.24) \qquad 1 + \frac{l^n}{1 - l} \leq d^{\mathrm{hom}} \leq \frac{1}{1 - l},
$$

from which we conclude that in this case d^{hom} is strictly larger than 1.

In particular, if $n = 1$, then (14.3.24) implies that

$$
d^{\mathrm{hom}} = \frac{1}{1 - l}.
$$

As corollary, from Theorem 13.5.7, (14.3.11), and (14.3.16) the following homogenization result can be deduced.

Theorem 14.3.2. *Let* $A \in \mathcal{A}_0$ *with* $A \subset\subset Y$, φ *be a measurable Y-periodic function on* $\mathbf{R}^n$ *taking only the values 0 and $+\infty$ and such that* $\varphi(y) = +\infty$ *for a.e.* $y \in Y \setminus A$. *For every* $\varepsilon > 0$, *every* $\Omega \in \mathcal{A}_0$, *and* $\beta \in L^2(\Omega)$ *let*

$$m_{\varepsilon}^0(\Omega, \beta) = \min \left\{ \int_{\Omega} |\nabla u|^2 dx + \int_{\Omega} \beta u dx : \right.$$

$$\left. u \in W_0^{1,2}(\Omega), \ |\nabla u(x)| \le \varphi\left(\frac{x}{\varepsilon}\right) \text{ for a.e. } x \in \Omega \right\},$$

and let for every $\varepsilon > 0$, u_ε *be the unique solution of* $m_{\varepsilon}^0(\Omega, \beta)$. *Then there exists a constant matrix* $\{d_{ij}^{\mathrm{hom}}\}$ *satisfying*

$$|z|^2 \le \sum_{i,j=1}^{n} d_{ij}^{\mathrm{hom}} z_i z_j \text{ for every } z \in \mathbf{R}^n$$

such that $\{m_{\varepsilon}^0(\Omega, \beta)\}_{\varepsilon>0}$ *converges as* $\varepsilon \to 0^+$ *to*

$$m_{\mathrm{hom}}^0(\Omega, \beta) = \min \left\{ \int_{\Omega} \sum_{i,j=1}^{n} d_{ij}^{\mathrm{hom}} \nabla_i u \nabla_j u 2 dx + \int_{\Omega} \beta u dx : u \in W_0^{1,2}(\Omega) \right\},$$

and $\{u_\varepsilon\}_{\varepsilon>0}$ *converges as* $\varepsilon \to 0^+$ *in* $L^2(\Omega)$ *to the unique solution of* $m_{\mathrm{hom}}^0(\Omega, \beta)$.

In addition, if φ *satisfies (14.3.12) and* $d_1^{\mathrm{hom}}, \dots, d_n^{\mathrm{hom}}$ *are given by (14.3.11), then*

$$d_{ij}^{\mathrm{hom}} = \begin{cases} d_i^{\mathrm{hom}} & \text{if } i = j \\ 0 & \text{if } i \ne j \end{cases} \text{ for every } i, j \in \{1, \dots, n\},$$

whilst, if φ *satisfies (14.3.17) and* d^{hom} *is given by (14.3.19), then*

$$d_{ij}^{\mathrm{hom}} = \begin{cases} d^{\mathrm{hom}} & \text{if } i = j \\ 0 & \text{if } i \ne j \end{cases} \text{ for every } i, j \in \{1, \dots, n\}.$$

These results can be qualitatively interpreted in the following way. The presence of conductors in the void, at mesoscopic level, has a distortion effect. Generally, anisotropy can be generated, and the space at macroscopic level can appear as filled with crystals.

In particular, let us consider a regular distribution of conductors that enjoy reflection invariance properties. Then, at macroscopic level, the dielectric susceptibility tensor, which connects polarization field to the electric one, is diagonal, and "optical axes" coincide with the coordinate ones.

On the contrary, if the distribution of the conductors has rotational invariance properties, then the effect of distortion is just a change in the dielectric constant, and the space, at macroscopic level, appears as filled with a homogeneous medium.

In conclusion, there is some suggestion that, from the point of view of electrostatics, every dielectric could be described by a certain mixture of conductors and void. So conductors and void could be thought as the only components of matter.

Bibliography

[ADM] ACERBI E., DAL MASO G.: *New Lower Semicontinuity Results for Polyconvex Integrals*; Calc. Var. Partial Differential Equations **2**, (1994), 329-371.

[AF] ACERBI E., FUSCO N.: *Semicontinuity Problems in the Calculus of Variations*; Arch. Rational Mech. Anal. **86**, (1984), 125-145.

[A] ADAMS R.A.: "Sobolev Spaces;" Pure Appl. Math. **65**, Academic Press, New York (1975).

[AM] ALBERTI G., MAJER P.: *Gap Phenomenon for Some Autonomous Functionals*; J. Convex Anal. **1**, (1994), 31-45.

[ASC] ALBERTI G., SERRA CASSANO F.: *Non-Occurrence of Gap for One-Dimensional Autonomous Functionals*; in "Calculus of Variations, Homogenization, and Continuum Mechanics," CIRM, Marseille-Luminy, 21-25 June 1993, Ser. Adv. Math. Appl. Sci. **18**, World Scientific Publishing, Singapore (1994), 1-17.

[Am] AMBROSIO L.: *New Lower Semicontinuity Results for Integral Functionals*; Rend. Accad. Naz. Sci. XL Mem. Mat. **11**, (1987), 1-42.

[AAB] AMBROSIO L., ASCENZI O., BUTTAZZO G.: *Lipschitz Regularity for Minimizers of Integral Functionals with Highly Discontinuous Integrands*; J. Math. Anal. Appl. **142**, (1989), 301-316.

[ADM1] AMBROSIO L., DAL MASO G.: *A General Chain Rule for Distributional Derivatives*; Proc. Amer. Math. Soc. **108**, (1990), 691 -702.

[ADM2] AMBROSIO L., DAL MASO G.: *On the Relaxation in $BV(\Omega; \mathbf{R}^m)$ of Quasi-Convex Integrals*; J. Funct. Anal. **109**, (1992), 76-97.

[AFP] AMBROSIO L., FUSCO N., PALLARA D.: "Functions of Bounded Variation and Free Discontinuity Problems;" Oxford Math. Monogr., Claredon Press, Oxford (2000).

[An] ANGELL T.S.: *A Note on the Approximation of Optimal Solutions of Free Problems of the Calculus of Variations*; Rend. Circ. Mat. Palermo (2), **28**, (1979), 258-272.

[Ai] ANNIN B.D.: *Existence and Uniqueness of the Solution of the Elastic-Plastic Torsion Problem for a Cylindrical Bar of Oval Cross-Section*; J. Appl. Math. Mech. **29**, (1965), 1038-1047.

[AG] ANZELLOTTI G., GIAQUINTA M.: *Funzioni BV e tracce*; Rend. Sem. Mat. Univ. Padova **60**, (1978), 1-22.

[At1] ATTOUCH H.: *Introduction a l'homogénéisation d'inéquations variationnelles*; Rend. Sem. Mat. Univ. Politecn. Torino **40**, (1982), 1-23.

[At2] ATTOUCH H.: "Variational Convergence for Functions and Operators;" Pitman, London (1984).

[BBGT] BAIOCCHI C., BUTTAZZO G., GASTALDI F., TOMARELLI F.: *General Existence Theorems for Unilateral Problems in Continuum Mechanics*; Arch. Rational Mech. Anal. **100**, (1988), 149-189.

[BM1] BALL J.M., MIZEL V.J.: *Singular Minimizers for Regular One-Dimensional Problems in the Calculus of Variations*; Bull. Amer. Math. Soc. **11**, (1984), 143-146.

[BM2] BALL J.M., MIZEL V.J.: *One Dimensional Variational Problems whose Minimizers Do not Satisfy the Euler-Lagrange Equation*; Arch. Rational Mech. Anal. **90**, (1985), 325-388.

[BP] BAKHVALOV N.S., PANASENKO G.P.: "Homogenization: Averaging Processes in Periodic Media;" Math. Appl. (Soviet Ser.) **36**, Kluwer Academic Publishers, Dordrecht (1989).

[BLMY] BANKS A.H.T., LYBECK N.J., MUNOZ B., YANYO L.: *Nonlinear Elastomers: Modelling and Estimation*; Proceedings of the "Third IEEE Mediterranean Symposium on New Directions in Control and Automation," vol. 1, Limassol, Cyprus, (1995), 1-7.

[BLP] BENSOUSSAN A., LIONS J.L., PAPANICOLAOU G.: "Asymptotic Analysis for Periodic Structures;" Stud. Math. Appl. **5**, North-Holland, Amsterdam (1978).

[BB] BETHUEL F., BREZIS H.: *Minimisation de $\int |\nabla(u-(\frac{x}{|x|}))|^2$ et divers phénomènes de gap*; C. R. Acad. Sci. Paris Sér. I Math. **310**, (1990), 859-864.

[BK] BLATZ P.J., KO W.L.: *Application of Finite Elastic Theory to Deformation of Rubbery Materials*; Trans. Soc. Rheology **6**, (1962), 223-251.

[BM] BOCCARDO L., MARCELLINI P.: *Sulla convergenza delle soluzioni di disequazioni variazionali*; Ann. Mat. Pura Appl. (4), **110**, (1976), 137-159.

[B] BOUCHITTÉ G., *Convergence et relaxation de fonctionnelles du calcul des variations à croissance linéaire. Applications à l'homogénéisation en plasticité*; Ann. Fac. Sci. Univ. Toulouse Math. **8**, (1986/87), 7-36.

[BoB] BOUCHITTÉ G., BUTTAZZO G.: *New Lower Semicontinuity Results for Nonconvex Functionals Defined on Measures*; Nonlinear Anal. **15**, (1990), 679-692.

[BDM] BOUCHITTÉ G., DAL MASO G.: *Integral Representation and Relaxation of Convex Local Functionals on $BV(\Omega)$*; Ann. Scuola Norm. Sup. Pisa Cl. Sci. (4), **20**, (1993), 483 -533.

[BFM] BOUCHITTÉ G., FONSECA I., MASCARENHAS L.: *A Global Method for Relaxation*; Arch. Rational Mech. Anal. **145**, (1998), 51-98.

[BD] BRAIDES A., DEFRANCESCHI A.: "Homogenization of Multiple Integrals;" Oxford Lecture Ser. Math. Appl. **12**, Oxford University Press, Oxford (1998).

[BG] BRAIDES A., GARRONI A.: *Homogenization of Periodic Nonlinear Media with Stiff and Soft Inclusions*; Math. Models Methods Appl. Sci. **5**, (1995), 543-564.

[Br1] BREZIS H.: *Multiplicateur de Lagrange en Torsion Elasto-Plastique*; Arch. Rational Mech. Anal. **49**, (1973), 32-40.

[Br2] BREZIS H.: "Analyse Fonctionnelle - Théorie et Applications;" Masson, Paris (1983).

[BS] BREZIS H., SIBONY M.: *Équivalence de deux inéquations variationnelles et applications*; Arch. Rational Mech. Anal. **41**, (1971), 254-265.

[Bi] BRIANE M.: *Homogenization in Some Weakly Connected Domains*; Ricerche Mat. **47**, (1998), 51-94.

[BDD] BRIANE M., DAMLAMIAN A., DONATO P.: *H-Convergence for Perforated Domains*; in "Nonlinear Partial Differential Equations and Their Applications. Collège de France Seminar. Volume XIII, (Paris 1994/1996)," H. Brezis and J.L. Lions editors, Pitman Res. Notes Math. Ser. **391**, Longman Scientific & Technical, Harlow (1999), 62-100.

[Bu1] BUTTAZZO G.: *Su una definizione generale dei Γ-limiti*; Boll. Un. Mat. Ital. (5), **14**-B, (1977), 722-744.

[Bu2] BUTTAZZO G.: "Semicontinuity, Relaxation and Integral Representation in the Calculus of Variations;" Pitman Res. Notes Math. Ser. **207**, Longman Scientific & Technical, Harlow (1989).

[BuB] BUTTAZZO G., BELLONI M.: *A Survey on Old and Recent Results about the Gap Phenomenon in the Calculus of Variations*; in "Recent Developments in Well-Posed Variational Problems," Math. Appl. **331**, Kluwer Academic Publishers, Dordrecht (1995), 1-27.

[BDM1] BUTTAZZO G., DAL MASO G.: *Γ-Limits of Integral Functionals*; J. Analyse Math. **37**, (1980), 145-185.

[BDM2] BUTTAZZO G., DAL MASO G.: *Integral Representation and Relaxation of Local Functionals*; Nonlinear Anal. **9**, (1985), 515-532.

[BDM3] BUTTAZZO G., DAL MASO G.: *A Characterization of Nonlinear Functionals on Sobolev Spaces which Admit an Integral Representation with a Carathéodory Integrand*; J. Math. Pures Appl. (9), **64**, (1985), 337-361.

[BuM1] BUTTAZZO G., MIZEL V.J.: *Interpretation of the Lavrentiev Phenomenon by Relaxation*; J. Funct. Anal. **110**, (1992), 434-460.

[BuM2] BUTTAZZO G., MIZEL V.J.: *On a Gap Phenomenon for Isoperimetrically Constrained Variational problems*; J. Convex Anal. **2**, (1995), 87-101.

[C1] CACCIOPPOLI R.: *Trasformazioni piane, superficie quadrabili, integrali di superficie*; Rend. Circ. Mat. Palermo (1), **54**, (1930), 217-262. Also in Renato Caccioppoli: "Opere;" Vol. I, Edizioni Cremonese, Roma (1963), 191-244.

[C2] CACCIOPPOLI R.: *Misura e integrazione sugli insiemi dimensionalmente orientati*. Notes I, II; Rend. Acc. Naz. Lincei (8), **12**, (1952), 3-11, 137-146. Also in Renato Caccioppoli: "Opere;" Vol. I, Edizioni Cremonese, Roma (1963), 358-380.

[C3] CACCIOPPOLI R.: "Opere;" Vol. I, Edizioni Cremonese, Roma (1963).

[CR] CAFFARELLI L.A., RIVIERE N.M.: *On the Lipschitz Character of the Stress Tensor when Twisting an Elastic-Plastic Bar*; Arch. Rational Mech. Anal. **69**, (1979), 31-36.

[Ca1] CARBONE L.: *Sur la convergence des intégrales du type de l'énergie sur des fonctions a gradient borné*; J. Math. Pures Appl. (9) **56**, (1977), 79-84.

[Ca2] CARBONE L.: *Γ-convergence d'intégrales sur des functions avec des contraintes sur le gradient*; Comm. Partial Differential Equations **2**, (1977), 627-651.

[Ca3] CARBONE L.: *Sull'omogeneizzazione di un problema variazionale con vincoli sul gradiente*; Atti Accad. Naz. Lincei Cl. Sci. Fis. Mat. Natur. Rend. Lincei (8) Mat. Appl. **63**, (1977), 10-14.

[Ca4] CARBONE L.: *Sur un problème d'homogénéisation avec des contraintes sur le gradient*; J. Math. Pures Appl. (9) **58**, (1979), 275-297.

[CCDAG1] CARBONE L., CIORANESCU D., DE ARCANGELIS R., GAUDIELLO A.: *An Approach to the Homogenization of Nonlinear Elastomers via the Theory of Unbounded Functionals*; C. R. Acad. Sci. Paris Sér. I Math. **332**, (2001), 283-288.

[CCDAG2] CARBONE L., CIORANESCU D., DE ARCANGELIS R., GAUDIELLO A.: *Homogenization of Unbounded Functionals and Non-*

linear Elastomers. The General Case; Asymptotic Anal. (in course of publication).

[CCDAG3] CARBONE L., CIORANESCU D., DE ARCANGELIS R., GA-UDIELLO A.: *An Approach to the Homogenization of Nonlinear Elastomers in the Case of the Fixed Constraints Set*; Rend. Accad. Sci. Fis. Mat. Napoli (4), **67**, (2000), 235-244.

[CCDAG4] CARBONE L., CIORANESCU D., DE ARCANGELIS R., GA-UDIELLO A.: *Homogenization of Unbounded Functionals and Nonlinear Elastomers. The Case of the Fixed Constraints Set*; (in course of publication).

[CCEDA] CARBONE L., CORBO ESPOSITO A., DE ARCANGELIS R.: *Homogenization of Neumann Problems for Unbounded Integral Functionals*; Boll. Un. Mat. Ital. (8), **2**-B, (1999), 463-491.

[CDA1] CARBONE L., DE ARCANGELIS R.: *On the Relaxation of Some Classes of Unbounded Integral Functionals*; Matematiche **51**, (1996), 221-256; Special Issue in Honour of Francesco Guglielmino.

[CDA2] CARBONE L., DE ARCANGELIS R.: *On Integral Representation, Relaxation and Homogenization for Unbounded Functionals*; Atti Accad. Naz. Lincei Cl. Sci. Fis. Mat. Natur. Rend. Lincei (9) Mat. Appl. **8**, (1997), 129-135.

[CDA3] CARBONE L., DE ARCANGELIS R.: *Integral Representation for Some Classes of Unbounded Functionals*; Atti Sem. Mat. Fis. Univ. Modena **46**-Suppl., (1998), 533-567; Special Issue Dedicated to Prof. Calogero Vinti.

[CDA4] CARBONE L., DE ARCANGELIS R.: *Unbounded Functionals: Applications to the Homogenization of Gradient Constrained Problems*; Ricerche Mat. **48**, (1999), 139-182.

[CDA5] CARBONE L., DE ARCANGELIS R.: *On the Relaxation of Dirichlet Minimum Problems for Some Classes of Unbounded Integral Functionals*; Ricerche Mat. **48**-Suppl., (1999), 347-372; Special Issue in Memory of Ennio De Giorgi.

[CDA6] CARBONE L., DE ARCANGELIS R.: *On a Non-Standard Convex Regularization and the Relaxation of Unbounded Functionals of the Calculus of Variations*; J. Convex Anal. **6**, (1999), 141-162.

[CDA7] CARBONE L., DE ARCANGELIS R.: *On the Unique Extension Problem for Functionals of the Calculus of Variations*; Atti Accad. Naz. Lincei Cl. Sci. Fis. Mat. Natur. Rend. Lincei (9) Mat. Appl. **12**, (2001).

[CDADM] CARBONE L., DE ARCANGELIS R., DE MAIO U.: *Homogenization of Media with Periodically Distributed Conductors*; Asymptotic Anal. **23**, (2000), 157-194.

[CS1] CARBONE L., SALERNO S.: *On a Problem of Homogenization with Quickly Oscillating Constraints on the Gradient*; J. Math. Anal. Appl. **90**, (1982), 219-250.

[CS2] CARBONE L., SALERNO S.: *Further Results on a Problem of Homogenization with Constraints on the Gradient*; J. Analyse Math. **44**, (1984/85), 1-20.

[CS3] CARBONE L., SALERNO S.: *Homogenization with Unbounded Constraints on the Gradient*; Nonlinear Anal. **9**, (1985), 431-444.

[CS4] CARBONE L., SALERNO S.: *Some Remarks on a Problem of Homogenization with Fixed Traces*; Appl. Anal. **22**, (1986), 71-86.

[CSb] CARBONE L., SBORDONE C.: *Some Properties of Γ-Limits of Integral Functionals*; Ann. Mat. Pura Appl. (4), **122**, (1979), 1-60.

[CDMLP] CARRIERO M., DAL MASO G., LEACI A., PASCALI E.: *Relaxation of the Nonparametric Plateau Problem with an Obstacle*; J. Math. Pures Appl. (9), **67**, (1988), 359-396.

[CLP] CARRIERO M., LEACI A., PASCALI E.: *On the Semicontinuity and Relaxation for Integrals with Respect to the Lebesgue Measure Added to Integrals with Respect to a Radon Measure*; Ann. Mat. Pura Appl. (4), **149**, (1987), 1-21.

[Ce] CECCONI J.: *Sull'area di Peano e sulla definizione assiomatica dell'area di una superficie*; Rend. Sem. Mat. Univ. Padova **20**, (1951), 307-314.

[Cs1] CESARI L.: "Surface Area;" Ann. of Math. Stud. **35**, Princeton University Press, Princeton (1956).

[Cs2] CESARI L.: "Optimization-Theory and Applications;" Springer, Berlin (1983).

[CA] CESARI L., ANGELL T.S.: *On the Lavrentiev Phenomenon*; Calcolo **22**, (1985), 17-29.

[CPSC] CHIADÒ PIAT V., SERRA CASSANO F.: *Some Remarks about the Density of Smooth Functions in Weighted Sobolev Spaces*; J. Convex Anal. **1**, (1994), 135-142.

[CD] CIORANESCU D., DONATO P.: "An Introduction to Homogenization;" Oxford Lecture Ser. Math. Appl. **17**, Oxford University Press, Oxford (1999).

[CSJP1] CIORANESCU D., SAINT JEAN PAULIN J.: *Homogenization in Open Sets with Holes*; J. Math. Anal. Appl. **71**, (1979), 590-607.

[CSJP2] CIORANESCU D., SAINT JEAN PAULIN J.: "Homogenization of Reticulated Structures;" Appl. Math. Sci. **136**, Springer, Berlin (1999).

[CV] CLARKE F.H., VINTER R.B.: *Regularity Properties of Solutions to the Basic Problem in the Calculus of Variations*; Trans. Amer. Mat. Soc. **291**, (1985), 73-98.

[Co] COHN D.L.: "Measure Theory;" Birkhäuser, Boston (1980).

[CEDA1] CORBO ESPOSITO A., DE ARCANGELIS R.: *The Lavrenti-eff Phenomenon and Different Processes of Homogenization*; Comm. Partial Differential Equations **17**, (1992), 1503-1538.

[CEDA2] CORBO ESPOSITO A., DE ARCANGELIS R.: *Comparison Results for Some Types of Relaxation of Variational Integral Functionals*; Ann. Mat. Pura Appl. (4), **164**, (1993), 155-193.

[CEDA3] CORBO ESPOSITO A., DE ARCANGELIS R.: *A Characterization of Families of Function Sets Described by Constraints on the Gradient*; Ann. Inst. H. Poincaré Anal. Non Linéaire **11**, (1994), 553-609.

[CEDA4] CORBO ESPOSITO A., DE ARCANGELIS R.: *Homogenization of Dirichlet Problems with Nonnegative Bounded Constraints on the Gradient*; J. Analyse Math. **64**, (1994), 53-96.

[CEDA5] CORBO ESPOSITO A., DE ARCANGELIS R.: *Further Results on Comparison for Some Types of Relaxation of Variational Integrals Functionals*; in "Nonlinear Variational Problems and Partial Differential Equations," Pitman Res. Notes Math. Ser. **320**, Longman Scientific & Technical, Harlow (1995), 157-181.

[CEDA6] CORBO ESPOSITO A., DE ARCANGELIS R.: *Some Notes on a Characterization of Function Sets Described by Constraints on the Gradient*; in "Nonlinear Variational Problems and Partial Differential Equations," A. Marino and M.K.V. Murthy editors, Pitman Res. Notes Math. Series **320**, Longman Scientific & Technical, Harlow (1995), 182-196.

[CEDA7] CORBO ESPOSITO A., DE ARCANGELIS R.: *A Characterization of Sets of Functions and Distributions on $\mathbf{R}^n$ Described by Constraints on the Gradient*; J. Convex Anal. **3**, (1996), 167-194.

[CESC] A. CORBO ESPOSITO, F. SERRA CASSANO: *A Lavrentieff Phenomenon for Problems of Homogenization with Constraints on the Gradient*; Ricerche Mat. **46**, (1997), 127-159.

[D] DACOROGNA B.: "Direct Methods in the Calculus of Variations;" Appl. Math. Sci. **78**, Springer, Berlin (1989).

[DM1] DAL MASO G.: *Integral Representation on $BV(\Omega)$ of Γ-Limits of Variational Integrals*; Manuscripta Math. **30**, (1980), 387-413.

[DM2] DAL MASO G.: "An Introduction to Γ-Convergence;" Progr. Nonlinear Differential Equations Appl. **8**, Birkhäuser, Boston (1993).

[DMM] DAL MASO G., MODICA L.: *A General Theory of Variational Functionals*; in "Topics in Functional Analysis 1980/81," Quaderni Scuola Norm. Sup. Pisa, Pisa (1982), 149-221.

[DDG] D'APICE C., DURANTE T., GAUDIELLO A.: *Some New Results on a Lavrentieff Phenomenon for Problems of Homogenization with Constraints on the Gradient*; Matematiche **54**, (1999), 3-47.

[DA1] DE ARCANGELIS R.: *Some Remarks on the Identity between a Variational Integral and its Relaxed Functional*; Ann. Univ. Ferrara Sez. VII (N.S.) **35**, (1989), 135-145.

[DA2] DE ARCANGELIS R.: *A General Homogenization Result for Almost Periodic Functionals*; J. Math. Anal. Appl. **156**, (1991), 358-380.

[DA3] DE ARCANGELIS R.: *The Lavrentieff Phenomenon for Quadratic Functionals*; Proc. Roy. Soc. Edinburgh Sect. A **125A**, (1995), 329-339.

[DA4] DE ARCANGELIS R.: *Homogenization of Dirichlet Minimum Problems with Conductor Type Periodically Distributed Constraints*; in "Nonlinear Partial Differential Equations and Their Applications. Collège de France Seminar. Volume XIV, (Paris 1996/2000)," H. Brezis and J.L. Lions editors, Elsevier Science (in course of publication).

[DAG] DE ARCANGELIS R., GARGIULO G.: *Homogenization of Integral Functionals with Linear Growth Defined on Vector-valued Functions*; NoDEA Nonlinear Differential Equations Appl. **2**, (1995), 371-416.

[DAGP] DE ARCANGELIS R., GAUDIELLO A., PADERNI G.: *Some Cases of Homogenization of Linearly Coercive Gradient Constrained Variational Problems*; Math. Models Methods Appl. Sci **6**, (1996), 901-940.

[DAT1] DE ARCANGELIS R., TROMBETTI C.: *On the Relaxation of Some Classes of Dirichlet Minimum Problems*; Comm. Partial Differential Equations **24**, (1999), 975-1006.

[DAT2] DE ARCANGELIS R., TROMBETTI C.: *On the Lavrentieff Phenomenon for Some Classes of Dirichlet Minimum Problems*; J. Convex Anal. **7**, (2000), 271-297.

[DAV] DE ARCANGELIS R., VITOLO A.: *Some Cases of Homogenization with Unbounded Oscillating Constraints on the Gradient*; Asymptotic Anal. **5**, (1992), 397-428.

[DG1] DE GIORGI E.: *Definizione ed espressione analitica del perimetro di un insieme*; Atti Accad. Naz. Lincei Rend. Cl. Sci. Fis. Mat. Natur. (8), **14**, (1953), 390-393.

[DG2] DE GIORGI E.: *Su una teoria generale della misura $(r-1)$-dimensionale in uno spazio ad r dimensioni*; Ann. Mat. Pura Appl. (4) **36**, (1954), 191-213.

[DG3] DE GIORGI E.: *Sulla convergenza di alcune successioni d'integrali del tipo dell'area*; Rend. Mat. (6), **8**, (1975), 277-294.

[DG4] DE GIORGI E.: *Generalized Limits in Calculus of Variations*; in "Topics in Functional Analysis 1980/81," Quaderni Scuola Norm. Sup. Pisa, Pisa (1982), 117-148.

[DG5] DE GIORGI E.: *Γ-convergenza e G-convergenza*; Boll. Un. Mat. Ital. (4), **14**-A, (1977), 213-220.

[DG6] DE GIORGI E.: *G-operators and Γ-convergence*; Proceedings of the International Congress of Mathematicians, Warszawa, August 16-24, 1983, PWN, Warszawa (1984), 1175-1191.

[DG7] DE GIORGI E.: *Some Semi-Continuity and Relaxation Problems*; in "Ennio De Giorgi Colloquium," Pitman Res. Notes Math. Ser. **125**, Pitman, London (1985), 1-11.

[DGCP] DE GIORGI E., COLOMBINI F., PICCININI L.C.: "Frontiere orientate di misura minima e questioni collegate;" Quaderni Scuola Norm. Sup. Pisa, Pisa (1972).

[DGF1] DE GIORGI E., FRANZONI T.: *Su un tipo di convergenza variazionale*; Atti Accad. Naz. Lincei Rend. Cl. Sci. Fis. Mat. Natur. (8), **58**, (1975), 842-850.

[DGF2] DE GIORGI E., FRANZONI T.: *Su un tipo di convergenza variazionale*; Rend. Sem. Mat. Brescia **3**, (1979), 63-101.

[DGL] DE GIORGI E., LETTA G.: *Une notion générale de convergence faible pour des fonctions croissantes d'ensemble*; Ann. Scuola Norm. Sup. Pisa Cl. Sci. (4), **4**, (1977), 61-99.

[DGS] DE GIORGI E., SPAGNOLO S.: *Sulla convergenza degli integrali dell'energia per operatori ellittici del secondo ordine*; Boll. Un. Mat. Ital. (4), **8**, (1973), 391-411.

[DMD] DE MAIO U., DURANTE T.: *Homogenization of Dirichlet Problems for Some Types of Integral Functionals*; Ricerche Mat. **46**, (1997), 177-202.

[DS] DEMENGEL F., SUQUET P.: *On Locking Materials*; Acta Appl. Math. **6**, (1986), 185-211.

[DuS] DUNFORD N., SCHWARTZ J.: "Linear Operators. Part I: General Theory;" Wiley Classics Lib., Wiley Interscience, New York (1988).

[DL] DUVAUT G., LANCHON H.: *Sur la solution du problème de torsion élastoplastique d'une barre cylindrique de section quelconque*; C. R. Acad. Sci. Paris Sér. I Math. **264**, (1967), 520-523.

[DLi] DUVAUT G., LIONS J.L.: "Inequalities in Mechanics and Physics;" Grundlehren Math. Wiss. **219**, Springer, Berlin (1976).

[ET] EKELAND I., TEMAM R.: "Convex Analysis and Variational Problems;" Stud. Math. Appl. **1**, North-Holland, Amsterdam (1976).

[EG] EVANS L.C., GARIEPY R.F.: "Measure Theory and Fine Properties of Functions;" Stud. Adv. Math., CRC Press, Boca Raton (1992).

[F] FEDERER H.: "Geometric Measure Theory;" Grundlehren Math. Wiss. **153**, Springer, Berlin (1969).

[Fe] FERRY J.D.: "Viscoelastic Properties of Polymers;" J. Wiley & Sons, New York (1980).

[Fr] FRÉCHET M.: *Sur le prolongement des fonctionnelles semi-continues et sur l'aire des surfaces courbes*; Fund. Math. **7**, (1925), 210-224.

[Fu] FUSCO N.: Γ-*convergenza unidimensionale*; Boll. Un. Mat. Ital. (5) **16**-B, (1979), 74-86.

[G] GAGLIARDO E.: *Caratterizzazioni delle tracce sulla frontiera relative ad alcune classi di funzioni in n variabili*; Rend. Sem. Mat. Univ. Padova **27**, (1957), 284-305.

[Gi] GIAQUINTA M.: *Growth Conditions and Regularity, a Counterexample*; Manuscripta Math. **59**, (1987), 245-248.

[GMS1] GIAQUINTA M., MODICA G., SOUČEK J.: *Functionals with Linear Growth in the Calculus of Variations*; Comment. Math. Univ. Carolinae **20**, (1979), 143-156.

[GMS2] GIAQUINTA M., MODICA G., SOUČEK J.: "Cartesian Currents in the Calculus of Variations I;" Ergeb. Math. Grenzgeb. (3), **37**, Springer, Berlin (1998).

[GMS3] GIAQUINTA M., MODICA G., SOUČEK J.: "Cartesian Currents in the Calculus of Variations II;" Ergeb. Math. Grenzgeb. (3), **38**, Springer, Berlin (1998).

[Gu] GIUSTI E.: "Minimal Surfaces and Functions of Bounded Variation;" Monogr. Math. **80**, Birkhäuser, Boston (1984).

[GL] GLOWINSKI R., LANCHON H.: *Torsion élastoplastique d'une barre cylindrique de section multiconnexe*; J. Mécanique **12**, (1973), 151-171.

[Go] GOFFMAN C.: *Lower Semi-Continuity and Area Functionals, I. The Non-Parametric Case*; Rend. Circ. Mat. Palermo (2) **2**, (1954), 203-235.

[GLi] GOFFMAN C., LIU F.-C.: *Discontinuous Mappings and Surface Area*; Proc. London Math. Soc. (3) **20**, (1970), 237-248.

[GS] GOFFMAN C., SERRIN J.: *Sublinear Functions of Measures and Variational Integrals*; Duke Math. J. **31**, (1964), 159-178.

[GK] GONCHARENKO M., KHRUSLOV E. Ya.: *Homogenization of Electrostatic Problems in Domains with Nets*; in "Homogenization and Applications to Material Sciences," Nice, June 6-10, 1995, D. Cioranescu, A. Damlamian and P. Donato editors, GAKUTO Internat. Ser. Math. Sci. Appl. **9**, Gakkotosho, Tokyo (1995), 215-223.

[HM] HEINRICHER A.C., MIZEL V.J.: *The Lavrentiev Phenomenon for Invariant Variational Problems*; Arch. Rational Mech. Anal. **102**, (1988), 57-93.

[JKO] JIKOV V.V., KOZLOV S.M., OLEINIK O.A.: "Homogenization of Differential Operators and Integral Functionals;" Springer, Berlin (1994).

[J] JORDAN C.: "Cours d'Analyse;" Gauthier-Villars, Paris (1909-1915).

[K] KOLMOGOROFF A.: *Beitraege zur Masstheorie*; Math. Ann. **107**, (1933), 351-366.

[L1] LANCHON H.: *Solution du problème de torsion élastoplastique d'une barre cylindrique de section quelconque*; C. R. Acad. Sci. Paris Sér. I Math. **269**, (1969), 791-794.

[L2] LANCHON H.: *Sur la solution du problème de torsion élastoplastique d'une barre cylindrique de section multiconnexe*; C. R. Acad. Sci. Paris Sér. I Math. **271**, (1970), 1137-1140.

[L3] LANCHON H.: *Torsion élastoplastique d'une barre cylindrique de section simplement ou multiplement connexe*; J. Mécanique **13**, (1974), 267-320.

[La] LAVRENTIEFF M.: *Sur quelques problèmes du calcul des variations*; Ann. di Mat. (at present Ann. Mat. Pura Appl.) (4), **4**, (1926), 7-28.

[Le] LEBESGUE H.: *Intégrale, Longueur, Aire*; Ann. di Mat. (at present Ann. Mat. Pura Appl.) (3), **7**, (1902), 231-359.

[M1] MANIA' B.: *Sull'approssimazione delle curve e degli integrali*; Boll. Un. Mat. Ital. **13**, (1934), 36-41.

[M2] MANIA' B.: *Sopra un esempio di Lavrentieff*; Boll. Un. Mat. Ital. **13**, (1934), 147-153.

[Ma1] MARCELLINI P.: *Periodic Solutions and Homogenization of Non Linear Variational Problems*; Ann. Mat. Pura Appl. (4), **117**, (1978), 139-152.

[Ma2] MARCELLINI P.: *On the Definition and the Lower Semicontinuity of Certain Quasi-Convex Integrals*; Ann. Inst. H. Poincaré Anal. Non Linéaire **3**, (1986), 391-409.

[Ma3] MARCELLINI P.: *Regularity and Existence of Solutions of Elliptic Equations with p, q-Growth Conditions*; J. Differential Equations **90**, (1991), 1-30.

[MS1] MARCELLINI P., SBORDONE C.: *Homogenization of Non-Uniformly Elliptic Operators*, Appl. Anal. **8**, (1978), 101-113.

[MS2] MARCELLINI P., SBORDONE C.: *Semicontinuity Problems in the Calculus of Variations*; Nonlinear Anal. **4**, (1980), 241-257.

[MM] MARCUS M., MIZEL J.: *Representation Theorems for Nonlinear Disjointly Additive Functionals and Operators on Sobolev Spaces*; Trans. Amer. Mat. Soc. **228**, (1977), 1-45.

[MaS] MARINO A., SPAGNOLO S.: *Un tipo di approssimazione dell'operatore* $\sum_{ij} D_i(a_{ij}(x)D_j)$ *con operatori* $\sum_j D_j(\beta(x)D_j)$; Ann. Scuola Norm. Sup. Pisa (3), **23**, (1969), 657-673.

[MaM] MASSARI U., MIRANDA M.: "Minimal Surfaces of Codimension One;" North-Holland Math. Stud. **91**, North-Holland, Amsterdam (1984).

[Me] MEYERS N.G.: *Integral Inequalities of Poincaré and Wirtinger Type*; Arch. Rational Mech. Anal. **68**, (1978), 113-120.

[MS] MEYERS N.G., SERRIN J.: *H=W*; Proc. Nat. Acad. Sci. U.S.A. **51**, (1964), 1055-1056.

[Mi1] MIRANDA M.: *Distribuzioni aventi derivate misure ed insiemi di perimetro localmente finito*; Ann. Scuola Norm. Sup. Pisa (3), **18**, (1964), 27-56.

[Mi2] MIRANDA M.: *Superfici cartesiane generalizzate ed insiemi di perimetro localmente finito sui prodotti cartesiani*; Ann. Scuola Norm. Sup. Pisa (3), **18**, (1964), 515-542.

[Mi3] MIRANDA M.: *Renato Caccioppoli e la teoria geometrica della misura*; Ricerche Mat. **40**, (1991), 111-119.

[M] MORREY C.B.: "Multiple Integrals in the Calculus of Variations;" Grundlehren Math. Wiss. **130**, Springer, Berlin (1966).

[MT] MURAT F., TARTAR L.: *H-Convergence*; in "Topics in the Mathematical Modelling of Composite Materials," A. Cherkaev and R. Kohn editors, Progr. Nonlinear Differential Equations Appl. **31**, Birkhäuser, Boston (1997), 21-44.

[P] PANKOV A.: "*G*-Convergence and Homogenization of Nonlinear Partial Differential Operators;" Math. Appl. **422**, Kluwer Academic Publishers, Dordrecht (1997).

[Ph] PHILLIPS R.S.: *Integration in a Convex Linear Topological Space*; Trans. Amer. Mat. Soc. **47**, (1940), 114-145.

[RT] RAUCH J., TAYLOR M.: *Electrostatic Screening*; J. Math. Phys. **16**, (1975), 284-288.

[ReT] REDDY B.D., TOMARELLI F.: *The Obstacle Problem for an Elastoplastic Body*; Appl. Math. Optim. **21**, (1990), 89-110.

[RHN] RENARDY M., HRUSA W.J., NOHEL J.A.: "Mathematical Problems in Viscoelasticity;" Pitman Monogr. Surveys Pure Appl. Math. **35**, Longman Scientific & Technical, Harlow, J. Whiley & Sons Inc., New York (1987).

[R] ROCKAFELLAR R.T.: "Convex Analysis;" Princeton Math. Ser. **28**, Princeton University Press, Princeton (1972).

[RW] ROCKAFELLAR R.T., WETS R.J-B.: "Variational Analysis;" Grundlehren Math. Wiss. **317**, Springer, Berlin (1998).

[Ro] ROYDEN H.L.: "Real Analysis;" Second Edition, Collier Macmillan International Editions, New York (1968).

[Ru] RUDIN W.: "Real and Complex Analysis;" McGraw-Hill Ser. Higher Math., New York (1966).

[SP] SANCHEZ-PALENCIA E.: "Nonhomogeneous Media and Vibration Theory;" Lecture Notes in Phys. **127**, Springer, Berlin (1980).

[S] SBORDONE C.: *Su alcune applicazioni di un tipo di convergenza variazionale*; Ann. Scuola Norm. Sup. Pisa Cl. Sci. (4), **2**, (1975), 617-638.

[S1] SERRIN J.: *A New Definition of the Integral for Non-Parametric Problems in the Calculus of Variations*; Acta Math. **102**, (1959), 23-32.

[S2] SERRIN J.: *On the Definition and Properties of Certain Variational Integrals*; Trans. Amer. Math. Soc. **101**, (1961), 139-167.

[S] SPAGNOLO S.: *Sulla convergenza di soluzioni di equazioni paraboliche ed ellittiche*; Ann. Scuola Norm. Sup. Pisa Cl. Sci. (3), **22**, (1968), 577-597.

[T] TARTAR L.: *Remarks on Homogenization*, in "Homogenization and Effective Moduli of Materials and Media," J.L. Ericksen, D. Kinderlehrer, R.V. Kohn and J.L. Lions editors, IMA Volumes in Mathematics and its Applications 1, Springer, Berlin (1986), 228-246.

[T1] TING T.W.: *Elastic-Plastic Torsion Problem III*; Arch. Rational Mech. Anal. **34**, (1969), 228-244.

[T2] TING T.W.: *Elastic-Plastic Torsion of Convex Cylindrical Bars*; J. Math. Mech. **19**, (1969), 531-551.

[T3] TING T.W.: *Elastic-Plastic Torsion of Simply Connected Cylindrical Bars*; Indiana Univ. Math. J. **20**, (1971), 1047-1076.

[To] TONELLI L.: "Fondamenti di Calcolo delle Variazioni;" Zanichelli, Bologna (1921-23).

[Tr] TRELOAR L.R.G.: "The Physics of Rubber Elasticity;" Clarendon Press, Oxford (1975).

[Tv] TREVES F.: "Topological Vector Spaces, Distributions and Kernels;" Pure Appl. Math. **25**, Academic Press, New York (1967).

[Tm] TROMBETTI C.: *On the Lower Semicontinuity and Relaxation Properties of Certain Classes of Variational Integrals*; Rend. Accad. Naz. Sci. XL Mem. Mat. **115**, (1997), 25-51.

[V] VOL'PERT A.I.: *The Spaces BV and Quasi-Linear Equations*; Math. USSR Sbornik **2**, (1967), 225-267.

[W] WARD I.M.: "Mechanical Properties of Solid Polymers;" J. Wiley & Sons, New York (1983).

[Z1] ZHIKOV V.V.: *Averaging of Functionals of the Calculus of Variations and Elasticity Theory*; Math. USSR Izv. **29**, (1987), 33-66.

[Z2] ZHIKOV V.V.: *On Lavrentiev's Phenomenon*; Russian J. Math. Phys. **3**, (1994), 249-269.

[ZKON] ZHIKOV V.V., KOZLOV S.M., OLEINIK O.A., KHA T'EN NGO-AN: *Averaging and G-convergence of Differential Operators*; Russian Math. Surveys **34**, (1979), 69-147.

[Z] ZIEMER W.P.: "Weakly Differentiable Functions;" Grad. Texts in Math. **120**, Springer, Berlin (1989).

List of Symbols

Index

Index

393

of convex functions..........21
Locally bounded from below
    function 36
Locally compact space...........3
Locally convex space............10
Lower semicontinuity 3
Lower semicontinuity at a point..3
Lower semicontinuous envelope..28
Lusin's theorem.................64

M

Measurable function 51
Measure.......................46
Measure space.................46
Minimizer 1
Minimizing sequence 1
Mollifier......................107
Monotone convergence theorem . 53
Monotonicity of the integral.....51
Multiindex......................1

N

Negative part of a measure......46
Non-trivial supporting
    hyperplane..................13
Norm..........................5
Normed space...................5

O

Operator norm 128
Outer regular measure 63

P

Partition of unity..............270
Perfect family of sets 72
Periodic function 58
Piecewise affine function 8
Poincaré inequality 130
Poincaré-Wirtinger inequality..130
Polyhedral set 2
Positive measure...............46
Positive part of a measure 46
Positively 1-homogeneous
    function 22

Precompact set 3
Product σ-algebra 47
Product measure space 47
Projection....................128

R

Radon measure................48
Radon positive measure.........47
Radon real measure............48
Radon vector measure 48
Radon-Nikodym derivative......61
Radon-Nikodym theorem 61
Real measure..................46
Recession function.............22
Regular measure...............64
Regularization in locally convex
    subspaces of $L^1_{loc}(\mathbf{R}^n)$......167
Regularization of a function....108
Regularization of a measure....108
Relative boundary..............12
Relative interior 12
Relatively compact set 3
Relatively countably compact...84
Relatively sequentially
    compact set..................3
Relaxed functional............102
Rellich-Kondrachov
    compactness theorem 126
Rescaled homothety
    of a function 80
Riesz representation theorem....64

S

σ-algebra......................45
σ-algebra of Borel sets..........46
σ-compact space...............48
σ-finite measure 47
Seminorm 5
Separation theorem............10
Sequential lower value 105
Sequential lower semicontinuity .. 4
Sequentially closed set 3
Sequentially coercive function...85
Sequentially compact set.........3